College Algebra with Applications

José Barros-Neto

Rutgers University

West Publishing Company

St. Paul   *New York*   *Los Angeles*   *San Francisco*

Alverno College
Library Media Center
Milwaukee, Wisconsin

Cover Art: #9 by Teiji Ono
Production Coordination: Editing, Design & Production, Inc.

COPYRIGHT © 1985   By WEST PUBLISHING CO.
                            50 West Kellogg Boulevard
                            P.O. Box 43526
                            St. Paul, Minnesota 55164

Library of Congress Cataloging in Publication Data

Barros-Neto, José. 1927–
  Algebra for college students.

  Includes index.
  1. Algebra.   I. Title.
QA154.2.B385   1985   512.9   84-19508
ISBN 0-314-85217-4

EQUATIONS OF CONICS IN STANDARD POSITION

Circle: $x^2 + y^2 = r^2$
Parabola: $y^2 = 4px$, $x^2 = 4py$
Ellipse: $x^2/a^2 + y^2/b^2 = 1$
Hyperbola: $x^2/a^2 - y^2/b^2 = 1$

FUNCTIONS

Function notation: $y = f(x)$
Composite function: $(f \circ g)(x) = f(g(x))$.
Inverse function notation: $y = f^{-1}(x)$
The basic property: $(f \circ f^{-1})(x) = (f^{-1} \circ f)(x) = x$.

EXPONENTIAL FUNCTIONS

$f(x) = a^x$, $a > 0$, $a \neq 1$

Domain: the set of all real numbers
Range: the set of all positive real numbers

PROPERTIES

$a^x a^y = a^{x+y}$
$(a^x)^y = a^{xy}$
$a^{-x} = 1/a^x$

If $a > 1$ and $x < y$, then $a^x < a^y$ (increasing)
If $0 < a < 1$ and $x < y$, then $a^x > a^y$ (decreasing)

THE NATURAL EXPONENTIAL FUNCTION

$f(x) = e^x$, $e \simeq 2.7128 \ldots$

LOGARITHMIC FUNCTIONS

$y = \log_a x$ if and only if $x = a^y$, $a > 0$, $a \neq 1$.

Domain: the set of all positive real numbers
Range: the set of all real numbers.

FACTORIAL AND BINOMIAL FORMULAS

$n! = 1 \cdot 2 \cdot \ldots \cdot (n-1)(n)$
$1! = 1$, $0! = 1$
$\binom{n}{k} = n!/k!(n-k)!$
$(a + b)^n = \sum_{k=0}^{n} \binom{n}{k} a^{n-k} b^k$

PROPERTIES OF LOGARITHMS

$\log_a(xy) = \log_a x + \log_a y$
$\log_a(x/y) = \log_a x - \log_a y$
$\log_a x^m = m \log_a x$

Change of base formula:
$\log_a x = \log_b x/\log_b a$
Common logarithms:
$y = \log x$ (base 10)
Natural logarithms:
$y = \ln x = \log_e x$ (base e)

PERMUTATIONS OF k OBJECTS FROM A SET OF n OBJECTS

$P(n, k) = n!/(n - k)!$

COMBINATIONS OF k OBJECTS FROM A SET OF n OBJECTS

$C(n, k) = n!/(n - k)k!$

PARTITION OF n OBJECTS IN k CELLS

$\Gamma(n, n_1, \ldots, n_k) = n!/n_1! n_2! \ldots n_k!$

ARITHMETIC SEQUENCE

$a_1, a_2, \ldots, a_n, \ldots$
Common difference: $a_n - a_{n-1} = d$
Formula for the nth term: $a_n = a_1 + (n - 1)d$
Sum of the first n terms: $s_n = (a_1 + a_n)/2$

GEOMETRIC SEQUENCE

$a_1, a_2, \ldots, a_n, \ldots$
Common ratio: $r = a_{n+1}/a_n$
Formula for the nth term: $a_n = a_1 r^{n-1}$
Sum of the first n terms: $s_n = a_1(1 - r^n)/(1 - r)$
Sum of a geometric series:
$s = a_1 + a_2 + \ldots + a_n + \ldots = a_1/(1 - r)$, $|r| < 1$

College Algebra with Applications

Table of Contents

1

Fundamentals of Algebra   1

2

Linear and Quadratic Equations and Inequalities   66

Preface

This book is designed to prepare students for the study of mathematically oriented courses in fields such as business and economics, and in the physical and biologic sciences.

The first two chapters contain a detailed review of algebra that includes the real number systems, algebra of polynomials, and the study of linear and quadratic equations and inequalities. They are followed by a chapter on coordinate geometry, containing the equations of a straight line, parallel and perpendicular lines, linear relations, graphs of equations, symmetry, and an optional section on conic sections. This chapter prepares students for the chapters on functions and their properties.

Functions are introduced by using the notion of related variables, a concept derived from many examples encountered in the applied sciences and mathematics. Properties of functions and graph sketching techniques are then discussed. The study of variation, a notion associated to that of related variables, completes the first of the two chapters on functions. In the second, polynomial and rational functions are discussed, and graphing techniques for these functions are described. The difficult notions of composite and inverse functions make up the last section of this chapter. Exponential and logarithmic functions form another chapter. These important functions are presented with a large number of examples and applications to economics, biology, geology, and chemistry.

The chapter on systems of equations contains the Gaussian elimination method, matrices, determinants, Cramer's rule, nonlinear systems of equations, systems of inequalities, and linear programming. Properties of polynomials, such as the division algorithm, the Remainder and Factor theorems and their consequences are discussed in the chapter on zeros of polynomials. Finally, the last chapter covers the principle of mathematical induction, arithmetic and geometric sequences and series, permutations and combinations, and elementary probability.

BOOK ORGANIZATION

The book is written in a concise and informal style. Lists of axioms or formal statements of theorems are avoided as much as possible. Occasionally, elementary proofs are presented, but only in situations where the learning of such proofs may improve the students' understanding of the subject matter. Explanations always proceed from the particular to the general case. Selected examples, often taken from the applied sciences, are used to motivate and prepare students for new concepts and definitions. The various topics discussed throughout the book are never presented as isolated abstract mathematical entities. Whenever possible, we show how concrete ideas studied and developed in the natural sciences evolve into mathematical concepts. Also, when a link between two or more mathematical concepts exist, such a link is discussed and explained.

A typical chapter or section of the book starts with one or more examples taken from the applied sciences, or with a relevant historical note. This is done as a motivation and to show the interplay between mathematics and the applied sciences.

Examples, practice exercises, and exercises

Every section of the book contains a good number of examples whose aim is to illustrate the ideas and techniques being discussed. Each example is followed by a practice exercise whose level of difficulty is the same as that of the example. By solving the practice exercise, students will have the opportunity to check their knowledge immediately before advancing to the next subject.

The exercises at the ends of the sections vary from simple exercises to more challenging ones. Exercises that present a degree of difficulty are indicated with the symbol ●.

Chapter summaries

Detailed chapter summaries, placed at the ends of chapters, review the main definitions and results, and include pertinent comments relating the various ideas discussed throughout the chapter.

Review exercises

Following each chapter summary, students will find a comprehensive list of review exercises with varying levels of difficulty.

Calculators

Calculators are important computational tools, and their use should be encouraged. They expedite calculations and improve accuracy. Because some instructors may object to the use of calculators at this level of instruction, we include a large number of examples, practice exercises, and exercises that do not require the use of a calculator and may require the use of tables. These always precede corresponding examples and exercises where calculators are necessary. All examples and exercises requiring the use of calculators are indicated by the symbol **c**.

ACKNOWLEDGMENTS

A book at this level personifies the teaching experience, taste, and pedagogical ideas of the author. However, it would have been impossible to have written this book without the multitude of comments, criticisms, and suggestions of several reviewers.

I am indebted to the following individuals who reviewed all or parts of the manuscript at various stages of preparation: Dennis Allison, Austin Community College; Mary Jane Causey, University of Mississippi; Mary Marsha Cupitt, North Carolina State University; Frank Gilfeather, University of Nebraska, Lincoln; Ray Guzman, Pasadena City College; Doug Hall, Michigan State University; Stanley Luckwecki, Clemson University; Betty Miller, West Virginia University; Eric Nussbaum, SUNY-Albany; Dan Weeks, Rutgers University. Special appreciation goes to John Tobey, Jr., from Northern Essex Community College, for the care he took in reading the complete manuscript and for his thoughtful comments.

A preliminary version of this book was used during the academic year 1983–84, at Rutgers University. I wish to express my gratitude to colleagues Richard Bumby, Steve Greenfield, Michael O'Nan, and Bert Walsh for many useful observations and criticisms. In particular, my deepest appreciation to Bert Walsh, who carefully read the complete manuscript and contributed in many ways to the improvement of its quality, level, and style.

I am also very thankful to the staff of West Publishing Company and to my editor, Pat Fitzgerald, for the care, support, and assistance they provided in the preparation of this book.

Last but not least, my gratitude goes to my wife, Iva, for her help, patience, and encouragement.

Jose Barros-Neto

College Algebra with Applications

1

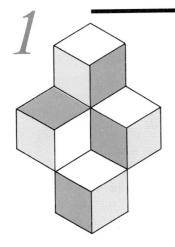

Fundamentals of Algebra

Throughout this book, we shall consider mostly real numbers. The chapter begins with an informal discussion about *natural numbers, integers, rational numbers*, and *irrational numbers* that form the real number system. You should pay special attention to the distinction between a rational and an irrational number. Real numbers can be represented by the points of a line. After the real number system is introduced, we discuss the basic concepts of algebra. Polynomials are then defined, and some of the algebraic operations on polynomials are briefly reviewed. We also discuss special products and the factoring of polynomials. Next we introduce rational expressions, which are formal quotients of polynomials. Their relationship to polynomials is analogous to that of fractions a/b to integers a and b. The algebraic operations on rational expressions are reviewed with some detail. Next, radicals and fractional powers of a real number are discussed. The chapter ends with a section on complex numbers and their properties.

1.1 NATURAL NUMBERS, INTEGERS, RATIONAL AND IRRATIONAL NUMBERS

Natural Numbers

The most common number system consists of the *natural numbers*

$$1, 2, 3, \ldots, n, \ldots$$

also called *counting numbers*.

Two fundamental operations are defined on natural numbers: *addition* (or *sum*) and *multiplication* (or *product*). The symbol $+$ is used to indicate addition, while the symbols $\times$ and $\cdot$ are used to indicate multiplication. The product of two numbers may also be written in different ways. For example, 2(5) and (2)(5) represent the product 2×5 of 2 by 5. When letters such as a, b, c, m, n, p, q, x, y, and z are used to represent unspecified numbers, it is not necessary

1

to use multiplication symbols to indicate a product. For example, $2n$ represents the product of 2 and n, while apq represents the product of a, p, and q.

The natural numbers are *closed* under the addition of multiplication operations. This means that every time we add or multiply two natural numbers, the result is a natural number.

Prime Numbers

The natural number 60 can be written as the product of 4 and 15, that is, $60 = 4 \cdot 15$. The numbers 4 and 15 are called *factors* of 60, and 60 is said to be a *multiple* of 4 and 15.

If the only factors of a natural number greater than 1 are the number itself and 1, then the natural number is said to be a *prime number*. Some examples of prime numbers are 2, 3, 5, 7, 11, 13, 17, and 19. It can be shown that there are infinitely many prime numbers. Going back to the number 60, since $4 = 2 \cdot 2$ and $15 = 3 \cdot 5$, we can write $60 = 2 \cdot 2 \cdot 3 \cdot 5$; the numbers 2, 3, and 5 are the *prime factors* of 60.

Every natural number greater than 1 that is not a prime number is called a *composite number*. Some composite numbers are 4, 6, 8, 9, 10, 15, 30, and 60. The following theorem about composite numbers is important in many applications.

Fundamental Theorem of Arithmetic

Every composite number can be written in a unique way, except for the order of the factors, as a product of prime factors.

When a number is written as a product of all its prime factors, we obtain the *prime factorization* of the number.

◆ **Example 1.**   Find the prime factorizations of 180 and 210.

Solution.   First find as many factors 2 as possible, then factors 3, factors 5, and so on.

$$180 = 2 \cdot 90 = 2 \cdot 2 \cdot 45 = 2 \cdot 2 \cdot 3 \cdot 15 = 2 \cdot 2 \cdot 3 \cdot 3 \cdot 5$$
$$210 = 2 \cdot 105 = 2 \cdot 3 \cdot 35 = 2 \cdot 3 \cdot 5 \cdot 7.$$

◇ **Practice Exercise 1.**   Write 360 as a product of prime numbers.

Answer. $360 = 2 \cdot 2 \cdot 2 \cdot 3 \cdot 3 \cdot 5.$

Integers

Subtraction (denoted by the symbol $-$) is the inverse of the addition operation. For example, $5 - 3 = 2$, because $5 = 3 + 2$; the number 2 is the *difference* of 5 and 3.

The natural numbers are not closed under subtraction. For instance, since there is no natural number that can be added to 6 to give 4, the difference $4 - 6$ cannot be defined as a natural number. Similarly, the difference $1 - 1$ cannot be a natural number.

In order to have a number system in which the difference of any two numbers is always defined, the *integers* are introduced; they consist of the number 0 and the *signed numbers* $\pm n$, where n is a natural number. An integer $+n$ (with a plus sign) is called *positive*, and an integer $-n$ (with a minus sign) is called *negative*. We also have $0 = +0 = -0$. Every natural number n is identified with a positive integer $+n$; thus natural numbers are *positive integers*. The addition and multiplication operations extend to the integers, and the integers are closed under the operations of addition, subtraction, and multiplication.

Sets

Throughout this book we shall use some set notation. A *set* is a collection of objects, which are called the *elements* of the set. Sets are usually denoted by uppercase letters, such as A, B, S, X, and Y, while the elements of a set are denoted by lowercase letters, such as a, b, s, x, and y. The *empty* or *null* set $\varnothing$ is the set that contains no element.

A pair of braces $\{\ \}$ used with words or symbols can describe a set. For example, the set of natural numbers is denoted by

$$\mathbb{N} = \{1, 2, 3, \ldots, n, \ldots\},$$

and the set of integers will be denoted by

$$\mathbb{Z} = \{\ldots, -n, \ldots, -2, -1, 0, 1, 2, \ldots, n, \ldots\}.$$

If a is an element of set A, we write $a \in A$; the notation $a \notin A$ means that a is not an element of set A.

Given two sets A and B, if every element of A is an element of B, then A is said to be a *subset* of B, and we write $A \subset B$. Set B is said to contain set A, or set A is said to be contained in set B. As an example, the set $\mathbb{N}$ of natural numbers is a subset of the set $\mathbb{Z}$ of integers, and we write $\mathbb{N} \subset \mathbb{Z}$.

Rational Numbers

Division (indicated with the symbol $\div$) is the inverse of multiplication. For example, $12 \div 3 = 4$ because $12 = 3 \cdot 4$. The number 12 is called the *dividend*, 3 is called the *divisor*, and 4 is called the *quotient*. Instead of writing $12 \div 3$ to denote the quotient of 12 by 3 we may also write $12/3$ or $\frac{12}{3}$.

The integers are not closed under the operation of division. For example, since there is no integer which when multiplied by 3 gives 2, the quotient $2 \div 3$ cannot be defined as an integer. Similarly, $-\frac{4}{5}$ and $\frac{8}{21}$ are not integers. In order to have them defined, the set of integers is again enlarged by introducing *rational numbers*.

If m and n denote integers with $n \neq 0$, then m/n is said to represent a *rational number*. The number m/n is also called a *fraction*, where m is the numerator and n the *denominator*. It is necessary to assume $n \neq 0$, because division by zero is undefined.

We denote the set of all rational numbers by

$$\mathbb{Q} = \left\{ \frac{m}{n} : m, n \in \mathbb{Z} \text{ with } n \neq 0 \right\}.$$

(The colon is read ''such that.'')

Two rational numbers m/n and p/q are said to be *equal* exactly when

$$(1.1) \quad mq = np.$$

For example, $2/3 = 6/9$ because $2 \cdot 9 = 18 = 3 \cdot 6$; and $-5/7 = 5/-7$ because $(-5)(-7) = 35 = 7 \cdot 5$. However, $3/5 \neq 7/8$.

From this definition it follows that when the numerator and denominator of a fraction have a common factor, we can *simplify* the fraction by cancelling the common factor. For example,

$$\frac{18}{30} = \frac{3 \cdot 6}{5 \cdot 6} = \frac{3}{5}.$$

If a rational number m/n, with $n > 0$, is such that m and n have no common factor other than 1, then m/n is said to be in *reduced form*. For example, $3/5$ is the reduced form of $18/30$, and $2/3$ is the reduced form of $30/45$. Notice that the reduced form of $12/-24$ is $-1/2$ and not $1/-2$.

By identifying every integer m with the rational number $m/1$, we make the set $\mathbb{Z}$ of integers a subset of $\mathbb{Q}$, the set of rational numbers. Thus, we have the following inclusions among the three different systems of numbers: $\mathbb{N} \subset \mathbb{Z} \subset \mathbb{Q}$.

Any two rational numbers can be written in equivalent form with the same denominator. For example, $3/4$ and $5/6$ can be written as $9/12$ and $10/12$, two fractions with the same denominator. This is used to add or subtract fractions.

The four arithmetic operations extend to the rational number as follows. If m/n and p/q are rational numbers, then their *sum* is defined by

$$(1.2) \quad \frac{m}{n} + \frac{p}{q} = \frac{mq}{nq} + \frac{pn}{nq} = \frac{mq + pn}{nq}.$$

The *difference* is defined by

$$(1.3) \quad \frac{m}{n} - \frac{p}{q} = \frac{mq}{nq} - \frac{pn}{nq} = \frac{mq - pn}{nq}.$$

The *product* is defined by

$$(1.4) \quad \frac{m}{n} \times \frac{p}{q} = \frac{mp}{nq}.$$

Finally, the *quotient* is defined by

$$(1.5) \quad \frac{m}{n} \div \frac{p}{q} = \frac{m}{n} \times \frac{q}{p} = \frac{mq}{np}.$$

The set of rational numbers is closed under the operations of addition, subtraction, multiplication, and division. The basic properties of these operations will be reviewed in detail after we introduce real numbers.

◆ **Example 2.**   Perform the indicated operations and simplify:

$$\textbf{a)}\ \frac{2}{3}+\frac{3}{4}\quad \textbf{b)}\ \frac{3}{4}-\frac{2}{3}\quad \textbf{c)}\ \frac{2}{3}\times\frac{3}{4}\quad \textbf{d)}\ \frac{2}{3}\div\frac{3}{4}.$$

Solution.   We have

$$\textbf{a)}\ \frac{2}{3}+\frac{3}{4}=\frac{8}{12}+\frac{9}{12}=\frac{8+9}{12}=\frac{17}{12}\qquad \textbf{b)}\ \frac{3}{4}-\frac{2}{3}=\frac{9}{12}-\frac{8}{12}=\frac{9-8}{12}=\frac{1}{12}$$

$$\textbf{c)}\ \frac{2}{3}\times\frac{3}{4}=\frac{2\cdot 3}{3\cdot 4}=\frac{6}{12}=\frac{1}{2}\qquad\qquad \textbf{d)}\ \frac{2}{3}\div\frac{3}{4}=\frac{2}{3}\times\frac{4}{3}=\frac{8}{9}$$

◇ **Practice Exercise 2.**   Perform the following operations:

$$\textbf{a)}\ \frac{5}{6}+\frac{3}{8}\quad \textbf{b)}\ \frac{5}{6}-\frac{3}{8}\quad \textbf{c)}\ \frac{5}{6}\times\frac{3}{8}\quad \textbf{d)}\ \frac{5}{6}\div\frac{3}{8}.$$

Answer. $\textbf{a)}\ \dfrac{29}{24}\quad \textbf{b)}\ \dfrac{11}{24}\quad \textbf{c)}\ \dfrac{5}{16}\quad \textbf{d)}\ \dfrac{20}{9}.$

◆ **Example 3.**   Calculate the sum $\dfrac{7}{12}+\dfrac{5}{20}$.

Solution.   According to formula (1.2), we have

$$\frac{7}{12}+\frac{5}{20}=\frac{140}{240}+\frac{60}{240}=\frac{140+60}{240}=\frac{200}{240}=\frac{5}{6}.$$

Notice that when adding or subtracting fractions we can always use the product of the denominators as the common denominator. But the best choice of common denominator is the *least common multiple* (LCM) of the denominators. Recall that the least common multiple of several numbers is the *smallest natural number* which is a multiple of all of them. To find the LCM of two (or even more) integers, write their prime factorizations and then form a new integer whose prime factorization contains each of the primes from the given integers the *largest* number of times that it occurs in any of the given integers. For example,

$$12 = 2\cdot 2\cdot 3\quad\text{and}\quad 20 = 2\cdot 2\cdot 5$$

so their LCM is $2\cdot 2\cdot 3\cdot 5 = 60$. We now have

$$\frac{7}{12}+\frac{5}{20}=\frac{35}{60}+\frac{15}{60}=\frac{50}{60}=\frac{5}{6}.$$

◇ **Practice Exercise 3.**   Calculate the difference $\dfrac{17}{12}-\dfrac{3}{4}$. Give your answer in simplified form.

Answer. $\dfrac{2}{3}.$

Properties of the Equality

The equality between rational numbers (also integers and rational numbers) satisfy two basic properties which we now describe.

If x and y are rational numbers such that

$$x = y,$$

then

$$x + z = y + z \quad \text{and} \quad xz = yz,$$

for every rational number z.

In other words, if we add or multiply both sides of an equality by the same number, the equality remains unchanged.

Decimal Numbers

A rational number m/n denotes the division of m by n. By performing a long division we obtain the *decimal representation* of the rational number. For example,

$$\frac{1}{2} = 0.5 \quad \text{and} \quad \frac{3}{4} = 0.75.$$

Also,

$$\frac{1}{3} = 0.333\ldots, \quad \frac{2}{11} = 0.1818\ldots, \quad \frac{1}{6} = 0.1666\ldots,$$

where a digit or a group of digits repeats without end. These examples show that decimal representations of rational numbers are of two types:

1) *terminating* or *finite decimals*, such as $1/2 = 0.5$, $3/4 = 0.75$, and $1/8 = 0.125$.

2) *repeating* or *periodic decimals*, such as $1/3 = 0.333\ldots$, $2/11 = 0.1818\ldots$, and $1/6 = 0.1666\ldots$

Repeating decimals are called *nonterminating* or *infinite* decimals. The digit or group of digits that repeats indefinitely is called the *period* of the repeating decimal. Repeating decimals can be written as follows: $1/3 = 0.\overline{3}$, $2/11 = 0.\overline{18}$, and $1/6 = 0.1\overline{6}$, where the bar indicates the period.

Every rational number can be represented by a terminating or a repeating decimal number. If the only factors of the denominator of a rational number are 2s and 5s, then its decimal representation is a terminating decimal. Indeed, we can always multiply the numerator and the denominator by 2s and 5s so that the denominator becomes a power of 10. For example,

$$\frac{1}{2} = \frac{1 \cdot 5}{2 \cdot 5} = \frac{5}{10} = 0.5,$$

$$\frac{21}{20} = \frac{21}{2 \cdot 2 \cdot 5} = \frac{21 \cdot 5}{2 \cdot 2 \cdot 5 \cdot 5} = \frac{105}{100} = 1.05,$$

$$\frac{1}{8} = \frac{1}{2 \cdot 2 \cdot 2} = \frac{5 \cdot 5 \cdot 5}{2 \cdot 2 \cdot 2 \cdot 5 \cdot 5 \cdot 5} = \frac{125}{1000} = 0.125.$$

However, if the denominator of a rational number reduced to its simplest form contains at least one prime factor other than 2 or 5, then the decimal representation is a repeating decimal. For example, $4/9 = 0.\overline{4}$, $7/6 = 1.1\overline{6}$, and $50/11 = 4.\overline{54}$.

Conversely, *every terminating or repeating decimal represents a rational number*, as shown in the examples that follow.

◆ **Example 4.**   Write 0.375 as a rational number.

Solution.   We have

$$0.375 = \frac{3}{10} + \frac{7}{100} + \frac{5}{1000}$$
$$= \frac{300}{1000} + \frac{70}{1000} + \frac{5}{1000} = \frac{375}{1000} = \frac{3}{8}.$$

◇ **Practice Exercise 4.**   Write 0.675 as a fraction.

Answer. $\frac{27}{40}$.

◆ **Example 5.**   Transform $0.\overline{3}$ into a fraction.

Solution.   Set $r = 0.333\ldots$. If we multiply both sides of this equality by 10, the equality remains unchanged (second property of the equality as stated above), and we obtain

$$10r = 3.333\ldots .$$

The reason we have multiplied by 10 is because the period of the given decimal number contains only one digit. As a consequence, the decimal parts of r and $10r$ are the same. Now, if we add $-r = -0.333\ldots$ to both sides of the last equality, it remains unchanged (first property of the equality as stated above) and we get

$$\begin{array}{rl} 10r = & 3.333\ldots \\ -r = & -0.333\ldots \\ \hline 9r = & 3. \end{array}$$

Finally, multiplying both sides by 1/9, we obtain

$$r = \frac{3}{9} = \frac{1}{3}.$$

◇ **Practice Exercise 5.**   What fraction has decimal representation 0.666. . .?

Answer. $\frac{2}{3}$.

◆ **Example 6.**   Find a fraction whose decimal representation is $0.1\overline{36}$.

Solution.   Set $r = 0.13636\ldots$ and multiply both sides by 10 and 1000, respectively, to get

$$10r = 1.36\ldots \quad \text{and} \quad 1000r = 136.36\ldots.$$

The reason why one multiplies both sides of the original equality by 10 and 1000 is to obtain two expressions whose right-hand sides have the same decimal part. Subtract $10r$ from $1000r$ to cancel out the decimal part and get

$$\begin{array}{r} 1000r = 136.36\ldots \\ -10r = -1.36\ldots \\ \hline 990r = 135 \end{array}$$

$$r = \frac{135}{990} = \frac{3}{22}.$$

◇ **Practice Exercise 6.**   Write $0.31818\ldots$ as a fraction.

Answer. $\dfrac{7}{22}.$

Irrational Numbers

We have seen that every rational number has a terminating or a repeating decimal representation and vice-versa. Thus, if we can produce a decimal number that is *nonterminating* and *nonrepeating*, such a number cannot represent a rational number. The infinite decimal number

$$0.101001000100001\ldots$$

is nonterminating and nonrepeating. Notice that it contains only the digits 0 and 1 distributed according to the following pattern: one zero follows the first digit 1, two zeros follow the second digit 1, three zeros follow the third digit 1, and so on. By using different digits or creating new patterns of distribution of digits, it is possible to write many examples of nonterminating and nonrepeating decimals, such as

$$0.010110111011110\ldots,$$
$$0.120120012000120000012\ldots,$$
$$0.10110111011110\ldots.$$

None of these decimal numbers can be a rational number, because they are all nonterminating and nonrepeating decimals. Such decimals are called *irrational numbers*.

In order to give examples of irrational numbers, it is not necessary to rely on the decimal notation as we did above. Irrational numbers have been known for more than 2,000 years, while the decimal notation, as it is used today, was introduced in the 16th century by the Dutch mathematician Simon Stevin (1548–1620).

To describe other examples of irrational numbers, we recall the Pythagorean theorem. Among the many mathematical discoveries of the early Greek mathematicians, the Pythagorean theorem stands out as one of the simplest and most beautiful results in geometry. The theorem, leading to the discovery of irrational numbers, was destined to shatter the Greek concept of numbers based entirely on natural and rational numbers.

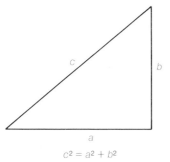

$c^2 = a^2 + b^2$

Figure 1.1

HISTORICAL NOTE

About the Number π
The Babylonians used the fraction 25/8 as an approximation for π.

The Pythagorean Theorem

In a right triangle, the square of the length c of the hypotenuse is equal to the sum of the squares of the lengths a and b of the other two sides ($c^2 = a^2 + b^2$).

If $a = b = 1$, then by the Pythagorean theorem we have

$$c^2 = 1^2 + 1^2$$

or

$$c^2 = 2.$$

That is, the length c is a number that multiplied by itself is equal to 2. The Greek mathematicians, whose number system contained only natural and rational numbers, believed that c was a rational number and tried, without success, to find such a number. Later, they realized that *there is no rational number whose square can be equal to 2*. This led to the invention of $\sqrt{2}$, an irrational number.

Other examples of irrational numbers are $\sqrt{3}, \sqrt{5}, \sqrt{7}, \sqrt{10}, \sqrt{11}, \sqrt{12}$, the number π—which gives the ratio between the length of a circle and its diameter—and the number e, which is used in the definition of *natural logarithms*.

To decide whether a given number is irrational may not be an easy task. Sometimes, quite involved proofs are required, as in the cases of the numbers π and e. Such proofs are beyond the scope of this book and will not be discussed here.

Every irrational number can be represented as a decimal number. For example, $\sqrt{2} = 1.41421\ 35624\ldots$, $\sqrt{3} = 1.73205\ 08076\ldots$, $\pi = 3.14159\ 26535\ldots$, and $e = 2.71828\ 28284.\ldots$ The decimal form of an irrational number is always nonterminating and nonrepeating.

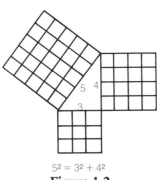

$5^2 = 3^2 + 4^2$

Figure 1.2

🔳 **EXERCISES 1.1**

Find the prime factorization of each of the following.

1. 240   **2.** 480   **3.** 432
4. 567   **5.** 2310   **6.** 4725

Calculate each of the following and simplify your answer.

7. $\dfrac{25}{120} + \dfrac{42}{80}$ **8.** $\dfrac{18}{45} - \dfrac{7}{120}$ **9.** $\dfrac{21}{60} \times \dfrac{24}{84}$ **10.** $\dfrac{7}{15} \div \dfrac{35}{45}$

Find the decimal representation of the given fractions.

11. $\dfrac{3}{7}$ **12.** $\dfrac{5}{8}$ **13.** $\dfrac{21}{40}$ **14.** $\dfrac{4}{15}$

15. $\dfrac{3}{11}$ **16.** $\dfrac{12}{25}$ **17.** $\dfrac{15}{120}$ **18.** $\dfrac{18}{240}$

Write the following decimal numbers as fractions.

19. 0.15 **20.** 1.25 **21.** $0.\overline{123}$ **22.** $0.\overline{321}$

23. $0.4\overline{18}$ **24.** $0.05\overline{3}$ **25.** $1.1\overline{06}$ **26.** $2.0\overline{5}$

Write the following sums as repeating decimals.

27. $0.\overline{4} + 0.\overline{15}$ **28.** $0.\overline{3} + 0.\overline{21}$ **29.** $0.\overline{12} + 0.\overline{203}$ **30.** $0.\overline{25} + 0.\overline{123}$

1.2   REAL NUMBERS

A *real number* is either a rational or an irrational number. According to Section 1.1, every real number has a decimal representation. If the real number is rational, then its decimal representation is terminating or repeating; if the real number is irrational, then its decimal representation is nonterminating and nonrepeating.

The set of all real numbers, denoted by $\mathbb{R}$, contains as subsets the natural numbers, the integers, the rational and irrational numbers. We have the following inclusions:

$$\mathbb{N} \subset \mathbb{Z} \subset \mathbb{Q} \subset \mathbb{R}.$$

The real number system and its properties form the basis of calculus. Throughout this book, unless otherwise stated, the word *number* will always signify *real number*.

The operations of addition and multiplication defined for rational numbers extend to real numbers. Also, the equality between real numbers satisfies the same properties which hold for the equality between rational numbers (see page 6). The following list of six properties of $\mathbb{R}$ provides, in principle, everything we need to know about the arithmetic in $\mathbb{R}$.

Properties of Real Numbers

Let $\mathbb{R}$ denote the set of all real numbers and let x, y, and z be arbitrary elements of $\mathbb{R}$.

I. *Closure.*

The set $\mathbb{R}$ is closed under the addition and multiplication operations. That is, $x + y$ and xy are both real numbers.

II. *Commutative properties.*

$$x + y = y + x \quad \text{and} \quad xy = yx.$$

III. *Associative properties.*

$$(x + y) + z = x + (y + z) \quad \text{and} \quad x(yz) = (xy)z.$$

IV. *Existence of identities.*

The number 0, called the *additive identity*, is the unique number such that $x + 0 = x$, for every $x \in \mathbb{R}$.

The number 1, called the *multiplicative identity*, is the unique number such that $x \cdot 1 = x$, for every $x \in \mathbb{R}$.

V. *Existence of additive and multiplicative inverses.*
  For every $x \in \mathbb{R}$, there is a unique $y \in \mathbb{R}$ such that $x + y = 0$.
  For every $x \in \mathbb{R}$, with $x \neq 0$, there is a unique $y \in \mathbb{R}$
such that $xy = 1$.
VI. *Distributive property of the product over the sum.*

$$x(y + z) = xy + xz.$$

From properties I through VI we can derive the rules for the arithmetic of real numbers, as well as the rules used to simplify algebraic expressions whose terms represent numbers.

Difference

The unique number y, in property V, such that $x + y = 0$ is called the *additive inverse* or *opposite* of x and is denoted by $y = -x$. The existence of the additive inverse allows us to define the *difference* of any two real numbers. If x and y are numbers, their difference is defined by

$$(1.6) \quad x - y = x + (-y).$$

Quotient

When $x \neq 0$, the unique number y, in property V, such that $xy = 1$ is called the *multiplicative inverse* or *reciprocal* of x. The multiplicative inverse of x is denoted by $\frac{1}{x}$, or $1/x$, or x^{-1}, and satisfies the relations

$$xx^{-1} = x^{-1}x = 1.$$

The division operation is now defined as multiplication by a reciprocal. If x and y are numbers with $y \neq 0$, then the *quotient* $\frac{x}{y}$ (also denoted by x/y or $x \div y$) is defined by

$$(1.7) \quad \frac{x}{y} = x\left(\frac{1}{y}\right) = x \cdot y^{-1}.$$

The quotient x/y is also called a *fraction*. By using properties 1 through VI above, you can check that no matter what the real numbers x, y, z, and w are (as long as y and w are nonzero), the fractions formed from them will obey the following rules.

Operations on Fractions

$$\frac{x}{y} + \frac{z}{w} = \frac{xw}{yw} + \frac{yz}{yw} = \frac{xw + yz}{yw}$$

$$\frac{x}{y} - \frac{z}{w} = \frac{xw}{yw} - \frac{yz}{yw} = \frac{xw - yz}{yw}$$

$$\frac{x}{y} \cdot \frac{z}{w} = \frac{xz}{yw}$$

$$\frac{x}{y} \div \frac{z}{w} = \frac{x}{y} \cdot \frac{w}{z} = \frac{xw}{yz} \quad \text{provided that } z \neq 0.$$

These rules are easy to remember; they are analogous to the rules of arithmetic of the rational numbers given on page 4.

The number zero (additive identity) has two very important properties, called *multiplicative properties of zero*, that can be derived from properties I through VI above.

Multiplicative Properties of Zero

$$x \cdot 0 = 0 \cdot x = 0$$

If $xy = 0$, then either $x = 0$ or $y = 0$.

◆ **Example 1.**   What properties are being illustrated in each of the following examples?

a) $\dfrac{3}{5} + \left(15 + \dfrac{4}{9}\right) = \left(\dfrac{3}{5} + 15\right) + \dfrac{4}{9}$

b) $a(7 + 8) = a \cdot 7 + a \cdot 8 = 7a + 8a$

c) $a(3 - \pi) = a(3 + (-\pi)) = a \cdot 3 + a \cdot (-\pi)$.

Solution.   **a)** Associativity of the sum.
b) Distributivity of the product over the sum and commutativity of the product.
c) The definition of difference and the distributive property.

◇ **Practice Exercise 1.**   What properties are being illustrated in each of the following?
a) $(a - 1.36) + 5.12 = 5.12 + (a - 1.36)$
b) $(\sqrt{2} + \sqrt{3}) + 0 = 0 + (\sqrt{2} + \sqrt{3}) = \sqrt{2} + \sqrt{3}$
c) $1(r + 2) = r + 2$.

Answer.   **a)** Commutativity of the sum.   **b)** Commutativity of the sum; the fact that 0 is the additive identity.   **c)** The fact that 1 is the multiplicative identity.

◆ **Example 2.**   Compute the following expression

$$\dfrac{\dfrac{2}{3} - \dfrac{3}{4} + \dfrac{5}{12}}{\dfrac{5}{9} - \dfrac{1}{4}}.$$

Solution.   We have

$$\dfrac{\dfrac{2}{3} - \dfrac{3}{4} + \dfrac{5}{12}}{\dfrac{5}{9} - \dfrac{1}{4}} = \dfrac{\dfrac{8}{12} - \dfrac{9}{12} + \dfrac{5}{12}}{\dfrac{20}{36} - \dfrac{9}{36}} = \dfrac{\dfrac{4}{12}}{\dfrac{11}{36}} = \dfrac{4}{12} \times \dfrac{36}{11} = \dfrac{12}{11}.$$

◇ **Practice Exercise 2.**   Compute and give your answer in simplified form

$$\frac{\dfrac{4}{5} + \dfrac{7}{10} - \dfrac{3}{4}}{\dfrac{2}{5} + \dfrac{1}{2}}.$$

Answer. $\dfrac{5}{6}$.

The Coordinate Line

Real numbers can be put into a *one-to-one correspondence* with the points of a line. That is, each real number corresponds to a unique point on a line, and each point on a line corresponds to a unique real number.

On a line (Figure 1.2), fix an arbitrary point O, the *origin*, and a *unit segment OU*. The point O corresponds to the number 0, and the point U corresponds to the number 1. By repeating the unit segment, moving from left to right, obtain the points corresponding to the positive integers 1, 2, 3, ... Starting at the origin and repeating the unit segment, moving now from right to left, obtain the points corresponding to the negative integers -1, -2, -3, ...

To find the point on the line corresponding to a fraction m/n, with n a natural number, divide the unit segment OU in n equal parts. Starting at the origin, repeat the first subdivision m times to the right if m is a positive integer, or m times to the left of the origin if m is a negative integer. Figure 1.3 shows the location of 5/4, 5/2, and $-3/2$.

Figure 1.3

Thus, each rational number is the *coordinate* of a unique point on the line. However, not every point on the line has a rational coordinate. There are points that do not correspond to rational numbers. Such points correspond to irrational numbers.

Points on the line associated with irrational numbers can, in certain cases, be located by geometric construction with a ruler and a compass. In Figure 1.4, the triangle OUA is a right isosceles triangle whose sides OU and UA both measure one unit of length. According to the Pythagorean theorem, the length of the hypotenuse OA is equal to $\sqrt{2}$ units. If we draw an arc of a circle with center at the origin and radius OA, the arc intersects the line at a point P that has coordinate $\sqrt{2}$.

A similar procedure could be used to locate the points corresponding to irrational numbers such as $\sqrt{3}$, $\sqrt{5}$, and $\sqrt{7}$. However, the points associated with π and e cannot be located with a ruler and a compass.

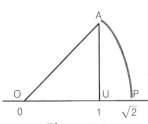

Figure 1.4

In establishing the one-to-one correspondence between real numbers and points of a line, we can say that every point P to the right of the origin corresponds to a unique *positive real number r*. Points to the left of the origin correspond to *negative real numbers*. The number r is called the *coordinate* or *abscissa* of the point P. A *real axis* or *coordinate axis* (or simply *axis*) is a line on which a coordinate system has been defined.

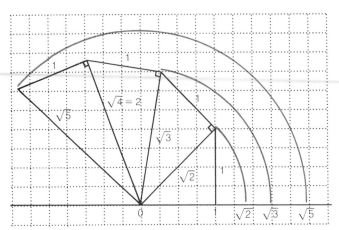

Figure 1.5   Locating $\sqrt{2}$, $\sqrt{3}$, and $\sqrt{5}$ with the help of a ruler and a compass.

 EXERCISES 1.2

Indicate what properties of the sum and product are being illustrated in each of Exercises 1–8.

1. $\left(a + \dfrac{3}{8}\right) + b = a + \left(\dfrac{3}{8} + b\right)$

2. $r + \dfrac{3}{11} = \dfrac{3}{11} + r$

3. $\dfrac{1}{5} \cdot (-x) = (-x) \cdot \dfrac{1}{5}$

4. $\left(\dfrac{7}{8} - a\right) \cdot \dfrac{9}{15} = \dfrac{9}{15} \cdot \left(\dfrac{7}{8} - a\right)$

5. $\dfrac{3}{4} \cdot \left(\dfrac{1}{2} + \dfrac{10}{11}\right) = \dfrac{3}{4} \cdot \dfrac{1}{2} + \dfrac{3}{4} \cdot \dfrac{10}{11}$

6. $\dfrac{8}{12} \cdot \left(\dfrac{3}{5} + 0\right) = \dfrac{8}{12} \cdot \dfrac{3}{5}$

7. $(a + b) \cdot \dfrac{3}{10} = \dfrac{3}{10} a + \dfrac{3}{10} b$

8. $x \cdot \left(\dfrac{2}{3} - y\right) = \dfrac{2}{3} x - xy$

In Exercises 9–14, compute the given expressions in two ways: a) using the distributive property; and b) first performing the operations inside the parentheses. Give your answer in simplified form.

9. $\dfrac{3}{10}\left(\dfrac{2}{3} + \dfrac{1}{4}\right)$

10. $\dfrac{4}{5}\left(\dfrac{3}{7} + \dfrac{2}{9}\right)$

11. $\dfrac{5}{12}\left(\dfrac{-4}{5} + \dfrac{8}{12}\right)$

12. $\dfrac{6}{7}\left(\dfrac{1}{2} - \dfrac{8}{11}\right)$

13. $\left(-\dfrac{4}{9}\right)\left(\dfrac{1}{3} - \dfrac{2}{5}\right)$

14. $(-5)\left(\dfrac{4}{7} + \dfrac{5}{9}\right)$

In Exercises 15–28, compute and give your answers in simplified form.

15. $\dfrac{2 + \dfrac{1}{2}}{2 - \dfrac{1}{2}}$

16. $\dfrac{5 - \dfrac{1}{3}}{2 + \dfrac{4}{5}}$

17. $\dfrac{\dfrac{2}{3} + \dfrac{1}{2}}{\dfrac{7}{9} - \dfrac{2}{3}}$

18. $\dfrac{\dfrac{4}{5} - \dfrac{2}{3}}{\dfrac{3}{4} + \dfrac{2}{5}}$

19. $\dfrac{\dfrac{1}{3} - \dfrac{3}{4}}{\dfrac{1}{2} + \dfrac{2}{3}}$

20. $\dfrac{\dfrac{3}{5} - \dfrac{1}{4}}{\dfrac{3}{8} + \dfrac{2}{3}}$

21. $\dfrac{\dfrac{2}{3} + \dfrac{3}{6} - \dfrac{3}{12}}{\dfrac{7}{8} - \dfrac{4}{3}}$

22. $\dfrac{\dfrac{3}{4} - \dfrac{4}{10} + \dfrac{2}{5}}{\dfrac{1}{3} - \dfrac{4}{5}}$

23. $\dfrac{\dfrac{4}{9} - \dfrac{1}{6} + \dfrac{5}{18}}{\dfrac{2}{3} + \dfrac{5}{6}}$

24. $\dfrac{\dfrac{3}{8} + \dfrac{5}{6} - \dfrac{1}{4}}{\dfrac{3}{4} + \dfrac{5}{9}}$

25. $\dfrac{\dfrac{3}{5} \times \left(\dfrac{2}{3} - \dfrac{3}{4}\right)}{\dfrac{4}{6} + \dfrac{2}{3} - \dfrac{2}{5}}$

26. $\dfrac{\dfrac{3}{8} \times \left(\dfrac{3}{5} - \dfrac{2}{3}\right)}{\dfrac{5}{6} - \dfrac{1}{2} + \dfrac{3}{4}}$

27. $\dfrac{\dfrac{3}{5} - \dfrac{3}{4} + \dfrac{6}{8}}{\dfrac{1}{3} \times \dfrac{3}{7} + \dfrac{5}{4}}$

28. $\dfrac{\left(\dfrac{5}{6} - \dfrac{1}{4}\right)\left(\dfrac{2}{3} + \dfrac{1}{8}\right)}{\dfrac{3}{4} \times \dfrac{2}{5} - \dfrac{5}{6}}$

● 29. The sum of two rational numbers is necessarily a rational number. Is the sum of two irrational numbers necessarily an irrational number?

● 30. Is the product of two irrational numbers necessarily an irrational number?

● 31. Show that $\sqrt{2} + \dfrac{5}{6}$ is irrational. (*Hint*: If $x = \sqrt{2} + \dfrac{5}{6}$, then $x - \dfrac{5}{6} = \sqrt{2}$. Can x be rational?)

● 32. Show that $\dfrac{4}{5}\sqrt{2}$ is irrational.

● 33. Show that if a is a rational number and b is an irrational number, then $a + b$ is irrational.

● 34. Show that if $a \neq 0$ is a rational number and b is an irrational number, then ab is irrational.

35. Which of the following are rational numbers? Irrational numbers?
 a) $(2\sqrt{3})\sqrt{3}$ 　　　　 b) $4 + \sqrt{2}$
 c) $(2 + 3\sqrt{2}) + (1 - 3\sqrt{2})$ 　 d) $0.23023002300023\ldots$

36. Which of the following are rational numbers? Irrational numbers?
 a) $\sqrt{12/27}$ 　 b) $\dfrac{3}{4}\sqrt{2} + 1$
 c) $(0.5)\sqrt{5}$ 　 d) $0.\overline{23}$

c 37. The fraction 22/7 was used by Archimedes as an approximation of π. Is 22/7 greater than or less than π?

c 38. The fractions 41/29 and 99/70 are approximations of $\sqrt{2}$. Are they greater than or less than $\sqrt{2}$?

● 39. Show that if $\dfrac{a}{b} = \dfrac{c}{d}$, then $\dfrac{a + b}{b} = \dfrac{c + d}{d}$.

● 40. Show that if $\dfrac{a}{b} = \dfrac{c}{d}$, then $\dfrac{a + b}{a - b} = \dfrac{c + d}{c - d}$.

1.3   INTEGRAL EXPONENTS

The multiplication operation leads to the following definition of *powers of a real number.*

Positive Exponents

If a is a real number and n is a natural number, then we define

(1.8) $a^n = \underbrace{a \cdot a \cdot \ldots \cdot a.}_{n \text{ factors}}$

HISTORICAL NOTE

Powers of a Number

The notation x^n, denoting an integral power of a variable, originated with René Descartes (1596–1650), a French mathematician. Prior to its introduction, notations used for a variable and powers of a variable were somewhat clumsy. For example, Francois Viète (1540–1603), a French lawyer and mathematician, introduced letters to denote constants and variables. He would have written the powers of a number A as follows: Aq (square of A), Ac (cube of A), Aqq (fourth power of A), and so on.

For example

$$2^5 = 2 \cdot 2 \cdot 2 \cdot 2 \cdot 2 = 32$$
$$\left(-\frac{1}{5}\right)^2 = \left(-\frac{1}{5}\right)\left(-\frac{1}{5}\right) = \frac{1}{25}$$
$$\left(\frac{3}{4}\right)^3 = \frac{3}{4} \cdot \frac{3}{4} \cdot \frac{3}{4} = \frac{27}{64}$$
$$(-1)^5 = (-1)(-1)(-1)(-1)(-1) = -1.$$

In the expression a^n, called a *power* or an *exponential*, a is the *base* and n is the *exponent*. We read it "nth power of a," "a to the nth power," or simply, "a to the n."

Powers of a real number have a few simple properties. First, consider multiplication of exponentials with the same base. The product of 3^2 by 3^4 is

$$3^2 \cdot 3^4 = \underbrace{(3 \cdot 3)}_{2 \text{ factors}}\underbrace{(3 \cdot 3 \cdot 3 \cdot 3)}_{4 \text{ factors}}$$
$$= \underbrace{3 \cdot 3 \cdot 3 \cdot 3 \cdot 3 \cdot 3}_{6 \text{ factors}} = 3^6.$$

This suggests that to multiply powers of 3, we should keep the same base and *add* the exponents. The same is true for any other base such as 4/5, 8, π, or e. In general, the product of a^n by a^m contains a total of $m + n$ factors equal to a; thus

$$a^n \cdot a^m = a^{n+m}.$$

For example, $x^6 \cdot x^{10} = x^{16}$. The last property can be generalized further:

$$a^n \cdot a^m \cdot a^p = a^{n+m+p}.$$

For example, $y^3 \cdot y^8 \cdot y^7 = y^{18}$.

Sometimes we have to take a power of an exponential. For example, suppose that we want to raise 2^3 to the fourth power. We have

$$(2^3)^4 = 2^3 \cdot 2^3 \cdot 2^3 \cdot 2^3$$
$$= 2^{3+3+3+3}$$
$$= 2^{3 \cdot 4} = 2^{12}.$$

This indicates that to raise a power to a power we should keep the same base and *multiply* the exponents. Symbolically:

$$(a^n)^m = a^{nm}.$$

For example, $(x^5)^3 = x^{15}$. This property can also be generalized:

$$((a^n)^m)^p = a^{nmp}.$$

For example, $((y^2)^4)^3 = y^{24}$.

Next, consider division of powers. The quotients $3^4/3^4$, $5^8/5^6$, and $7^2/7^5$ can be simplified as follows:

$$\frac{3^4}{3^4} = 1,$$

$$\frac{5^8}{5^6} = \frac{5^6 \cdot 5^2}{5^6} = 5^2,$$

and

$$\frac{7^2}{7^5} = \frac{7^2}{7^2 \cdot 7^3} = \frac{1}{7^3}.$$

These examples illustrate the following property of exponents.

If $a \neq 0$, then

$$\frac{a^n}{a^m} = \begin{cases} a^{n-m} & \text{if} \quad n > m \\ 1 & \text{if} \quad n = m \\ \dfrac{1}{a^{m-n}} & \text{if} \quad n < m \end{cases}$$

In many instances, it is necessary to take powers of products or quotients such as

$$\begin{aligned}(2a)^4 &= (2a)(2a)(2a)(2a) \\ &= (2 \cdot 2 \cdot 2 \cdot 2)(a \cdot a \cdot a \cdot a) \\ &= 16a^4,\end{aligned}$$

and

$$\left(\frac{x}{4}\right)^3 = \frac{x}{4} \cdot \frac{x}{4} \cdot \frac{x}{4} = \frac{x^3}{64}.$$

These two examples illustrate the following properties of exponents:

$$(ab)^n = a^n b^n$$

and

$$\left(\frac{a}{b}\right)^n = \frac{a^n}{b^n} \quad \text{if} \quad b \neq 0.$$

Up to now we have considered only exponents that are natural numbers (i.e., positive integers). It is necessary to extend the definition to exponents that are negative integers or 0.

Zero or Negative Exponents

If a is a real number and $a \neq 0$ then by definition

$$(1.9) \quad a^0 = 1$$

and

$$(1.10) \quad a^{-n} = \frac{1}{a^n}, \quad \text{for all natural numbers } n.$$

Note that the restriction $a \neq 0$ is essential. Otherwise, the right-hand side of (1.10) is not defined. Also, the symbol 0^0 is not defined.

Definitions (1.9) and (1.10) can be justified as follows. If n and m are natural numbers with $m > n$, we have

$$(1.11) \quad \frac{a^m}{a^n} = a^{m-n}.$$

If we set $n = m$, then the left-hand side of (1.11) becomes 1 while the right-hand side equals a^0. Thus, $a^0 = 1$. On the other hand, if we set $m = 0$ in (1.11) we get

$$\frac{1}{a^n} = \frac{a^0}{a^n} = a^{0-n} = a^{-n}$$

which is formula (1.10).

Some numerical examples are:

$$7^0 = 1, \quad \left(-\frac{2}{3}\right)^0 = 1, \quad (12.5)^0 = 1,$$

$$3^4 = \frac{1}{3^4} = \frac{1}{81},$$

$$\left(\frac{3}{5}\right)^{-2} = \frac{1}{\left(\frac{3}{5}\right)^2} = \frac{1}{\frac{9}{25}} = \frac{25}{9},$$

and

$$\left(-\frac{1}{2}\right)^{-3} = \frac{1}{\left(-\frac{1}{2}\right)^3} = \frac{1}{-\frac{1}{8}} = -8.$$

It can be shown that the properties we have described for powers in which the exponents are natural numbers are also true when the exponents are integers. For future reference we list the properties of integral exponents.

Properties of Exponents

If a and b are numbers and if n and m are integers, then

1. $a^n \cdot a^m = a^{n+m}$
2. $(a^n)^m = a^{nm}$
3. $\dfrac{a^n}{a^m} = a^{n-m}$, $a \neq 0$
4. $(ab)^n = a^n b^n$
5. $\left(\dfrac{a}{b}\right)^n = \dfrac{a^n}{b^n}$, $b \neq 0$
6. $a^0 = 1$, $a \neq 0$
7. $a^{-n} = \dfrac{1}{a^n}$, $a \neq 0$.

◆ **Example 1.** Use the properties of exponents to simplify each expression:

a) $(2ax^2)(5ay^4)$ **b)** $\left(\dfrac{3a^2}{5bc^3}\right)^2 \left(\dfrac{5ab^2}{c}\right)$.

Solution. We have
a) $(2ax^2)(5ay^4) = 2 \cdot 5 \cdot a \cdot a \cdot x^2 \cdot y^4 = 10a^2x^2y^4$

b) $\left(\dfrac{3a^2}{5bc^3}\right)^2 \cdot \left(\dfrac{5ab^2}{c}\right) = \dfrac{9a^4}{25b^2c^6} \cdot \dfrac{5ab^2}{c} = \dfrac{45a^5b^2}{25b^2c^7} = \dfrac{9a^5}{5c^7}$.

◇ **Practice Exercise 1.** Simplify each of the following expressions:

a) $(3ax^2y^3)^2$ **b)** $\left(\dfrac{3x^2}{yz}\right)^4$.

Answer. **a)** $9a^2x^4y^6$,   **b)** $\dfrac{81x^8}{y^4z^4}$.

◆ **Example 2.** Write each of the following expressions with positive exponents only and simplify:

a) $(4a^{-2}x^3)^{-2}$ **b)** $\dfrac{6a^2b^{-3}}{2a^{-3}b}$

Solution. **a)** According to the definition (1.10) and properties of exponents, we get

$$(4a^{-2}x^3)^{-2} = \frac{1}{(4a^{-2}x^3)^2} = \frac{1}{4^2(a^{-2})^2(x^3)^2}$$

$$= \frac{1}{16a^{-4}x^6} = \frac{1}{16x^6} \cdot \frac{1}{a^{-4}} = \frac{a^4}{16x^6}.$$

b) Also

$$\frac{6a^2 b^{-3}}{2a^{-3}b} = \frac{6}{2} \cdot \frac{a^2}{a^{-3}} \cdot \frac{b^{-3}}{b} = 3 \cdot a^2 \cdot a^3 \cdot \frac{1}{b \cdot b^3}$$

$$= \frac{3a^5}{b^4}.$$

◇ **Practice Exercise 2.** Write with positive exponents only and simplify:

a) $(5b^2 y^{-4})^3$ **b)** $\dfrac{12u^3 v^4}{4u^{-1}v^2}$.

Answer. **a)** $\dfrac{125b^6}{y^{12}}$ **b)** $3u^4 v^2$.

Scientific Notation

In the applied sciences we often work with extremely large or small numbers. For example, one light-year (the distance traveled by a ray of light in one year) is equal to

9,460,000,000,000 kilometers;

the mass of a proton is equal to

0. 000 000 000 000 000 000 000 000 001 673 kilograms.

In order to simplify notation and calculations, such numbers are usually expressed in scientific notation.

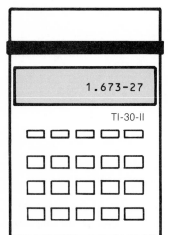

1.673-27

TI-30-II

Figure 1-6

> A number is said to be written in *scientific notation* when it is written in the form
>
> $$a \times 10^n$$
>
> where a is a decimal number such that $1 \le a < 10$ and n is an integer.

For example, in scientific notation one light-year is written as 9.46×10^{12} kilometers and the mass of a proton as 1.673×10^{-27} kilograms.

◆ **Example 3.** Write the following numbers in scientific notation:
a) 25×10^{-7} **b)** 0.112×10^6.

Solution. **a)** $25 \times 10^{-7} = (2.5 \times 10) \times 10^{-7} = 2.5 \times (10 \times 10^{-7}) = 2.5 \times 10^{-6}$
b) $0.112 \times 10^6 = (1.12 \times 10^{-1}) \times 10^6 = 1.12 \times (10^{-1} \times 10^6) = 1.12 \times 10^5$

◇ **Practice Exercise 3.** Rewrite in scientific notation:
a) 0.034×10^8 and **b)** 416.3×10^{-10}.

Answer.   **a)** 3.4×10^6 **b)** 4.163×10^{-8}.

As the following example shows, some calculations are greatly simplified by using scientific notation and the properties of exponents.

◆ **Example 4.**   Evaluate the expression

$$\frac{(1.21 \times 10^5)(8.1 \times 10^{-8})}{(3.3 \times 10^7)^2}$$

and give your answer in scientific notation.

Solution.   Using the associativity of the product and the properties of exponents, we write

$$\frac{(1.21 \times 10^5)(8.1 \times 10^{-8})}{(3.3 \times 10^7)^2} = \frac{(1.21)(8.1)(10^5 \times 10^{-8})}{(3.3)^2(10^7)^2}$$

$$= \frac{(1.21)(8.1)(10^5 \times 10^{-8})}{(10.89)(10^{14})}$$

$$= \frac{(1.21)(8.1)}{10.89} \times 10^{5-8-14}$$

$$= \frac{9.801}{10.89} \times 10^{-17} = 0.9 \times 10^{-17}$$

$$= 9.0 \times 10^{-18}.$$

◇ **Practice Exercise 4.**   Perform the indicated operations and give your answer in scientific notation.

$$\frac{(9.23 \times 10^{-7})(2.3 \times 10^8)}{(1.15 \times 10^{-12})(2.4 \times 10^4)}$$

Answer. 7.6×10^9.

Significant Digits

The result of a measurement is a number. Since measurements are never perfectly accurate, numbers obtained by measurement are only approximations to some exact value which we really cannot determine.

There is a standard language that is used to describe the accuracy of measurements. Let us give an example using the meter as unit of length. If we say that 12 meters is the measurement of a pole *to the nearest meter*, this really means that the pole has length between 11.5 meters and 12.5 meters. That is, the *exact value* of the length is no less than 11.5 meters and no more than 12.5 meters. If the measurement is done more accurately and we find that 12.4 meters is the measurement of the pole *to the nearest tenth of a meter*, then the length of the pole is really between 12.35 meters and 12.45 meters. Also, if we say that the pole measures 12.00 meters, then we mean that the pole measures 12 meters to the nearest hundredth of a meter and, in this case, the measure of the pole is between 11.995 meters and 12.005 meters.

The first measurement of 12 meters is said to have two significant digits of accuracy; the second of 12.4 meters has three significant digits of accuracy; and the third of 12.00 meters has four significant digits of accuracy.

The number of *significant digits* of accuracy in a measurement is the number of digits from the leftmost nonzero digit to the rightmost digit.

The table that follows shows some numbers, how many significant digits there are in each number, and the measurement range of each number.

Number	Number of significant digits	Range of measurement
0.2016	4	0.20155 to 0.20165
1.36	3	1.355 to 1.365
0.04	1	0.035 to 0.045
0.0012	2	0.00115 to 0.00125

Final zeros of a number may or may not be significant digits. For example, suppose that a road measures 12,000 meters, to the nearest meter; then all the zeros are significant digits. If the road measures 12,000 meters to the nearest ten meters, then the first two zeros are significant digits, but the last one is not. Finally, if the road measures 12,000 meters to the nearest thousand meters, then the three zeros are not significant digits. To avoid ambiguity, we write the three measurements in scientific notation, as follows: 1.2000×10^4 meters represents a measurement to the nearest meter (five significant digits); 1.200×10^4 meters is the measurement to the nearest ten meters (four significant digits); and 1.2×10^4 meters is the measurement to the nearest thousand meters (two significant digits).

In future calculations we shall use the following rules concerning significant digits:

1) When adding or subtracting numbers obtained by measurements, round off the answer so that it will show no more decimal places than the measurement with fewest decimal places.

2) When multiplying or dividing numbers representing measurements, round off the answer so that it has the same number of significant digits as the number with fewest significant digits.

3) When computing powers or roots, round off the answer so that it has the same number of significant digits as the number whose power or root we are computing.

◆ Example 5.   In physics, the formula

$$F = G \frac{mm'}{r^2}$$

gives the magnitude of the force of attraction between two masses m and m' placed at a distance r from each other, where G is a physical constant. Find F, knowing that $m = 3.9 \times 10^{20}$ units of mass, $m' = 1.5 \times 10^{20}$ units of mass, and $r = 1.0 \times 10^{6}$ units of distance, and assuming that $G = 1$. Give your answer in scientific notation.

Solution. We have

$$
\begin{aligned}
F &= \frac{(3.9 \times 10^{20})(1.5 \times 10^{20})}{(1.0 \times 10^{6})^2} \\
&= \frac{(3.9)(1.5)(10^{20} \times 10^{20})}{10^{12}} \\
&= (3.9)(1.5) \times 10^{28} \\
&= 5.85 \times 10^{28} \\
&\simeq 5.9 \times 10^{28}.
\end{aligned}
$$

◇ **Practice Exercise 5.** Two spherical lead balls of masses 4.50 kg and 5.00 kg are placed so that their centers are 0.500 m apart. Find the magnitude of the force of attraction F between these masses if the constant G is equal to 6.67×10^{-11} N·m^2/kg^2. [Here N is the abbreviation of Newton, a unit of force.]

Answer. $F = 6.00 \times 10^{-9}$ N.

Using Your Calculator

Scientific calculators allow the entry of a number in scientific notation. Calculations can then be quickly performed, and the answer is displayed in scientific notation. There are two principal standard ways (called *logic modes*) to enter data into calculators: the *algebraic* mode and the *reverse Polish notation* (RPN) mode. Algebraic logic uses parentheses and follows the rules of ordinary algebra. The RPN logic avoids parentheses and, perhaps, is a little more difficult to learn. All the sample computations shown in this book were performed on the Texas Instrument TI-30-II calculator, which operates in the algebraic mode. When we list a sequence of keystrokes, we will be using the TI-30-II. The same sequence of keystrokes will probably work (perhaps after a few minor changes) on most calculators using the algebraic mode. However, if you use a different calculator, you should consult the owner's manual for details on how to operate the calculator.

C Here are the keystrokes you should perform in order to enter the number 1.673×10^{-27}:

ON/C   1.673   EE   27   +/−

The calculator also has a key $\boxed{y^x}$ to compute powers of a number. For example, $20^8 = 2.56 \times 10^{10}$. The keystrokes 20 $\boxed{y^x}$ 8 $\boxed{=}$ will give you the answer in scientific notation.

SOME FUNDAMENTAL
CONSTANTS

Mass of an electron	9.107×10^{-31} kg
Charge of an electron	4.802×10^{-10} statcoulomb
Mass of a proton	1.673×10^{-27} kg
Mass of the Sun	1.999×10^{30} kg
Avogadro's number	6.022×10^{23}
Earth-Moon mean distance	3.843×10^{8} m
Earth-Sun mean distance	1.496×10^{11} m

 EXERCISES 1.3

In Exercises 1–16, simplify each expression and write all answers without exponents.

1. 11^4
2. 25^3
3. 5^{-3}
4. 3^{-4}

5. $\left(\dfrac{3}{4}\right)^{-2}$
6. $\left(\dfrac{3}{4}\right)^{-3}$
7. $(-8)^3$
8. $(-5)^4$

9. $(-7)^{-3}$
10. $(-9)^{-3}$
11. $3(-4)^{-3}$
12. $2\left(\dfrac{1}{2}\right)^{-4}$

13. $3^{-2} + 4^{-1}$
14. $2^{-4} + 5^{-2}$
15. $\left(\dfrac{2}{3}\right)^{-1} + 3^{-1}$
16. $\left(\dfrac{3}{4}\right)^{-2} + 9^{-1}$

In Exercises 17–34, write each expression with positive exponents only.

17. $(2xy^3)^{-2}$
18. $\dfrac{u^{15}}{u^8}$
19. $x^{-2} + \dfrac{1}{x^{-2}}$

20. $\dfrac{(x+y)^{10}}{(x+y)^{13}}$
21. $\dfrac{3x^2y^{-4}}{z^{-5}}$
22. $\dfrac{3}{a^{-2}bc^{-1}}$

23. $(4x^{-8}y^{-3})x^2y^2(3y)$
24. $(3x^2y^{-1})^2$
25. $\left(\dfrac{4ab^2}{c^2}\right)^{-4}$

26. $\left(\dfrac{5a^2x^3}{b}\right)^{-3}$
27. $\left(\dfrac{x^{-2}y^{-3}}{2}\right)^2$
28. $x(y-x)^{-2}$

29. $\dfrac{3}{(ab)^{-4}(ab)^{-5}}$
30. $\dfrac{a^2b^{-5}}{(a^4b^{-2})^{-3}}$
31. $\left(\dfrac{a^{-2}b^{-3}c^{-4}}{3ab}\right)$

32. $\dfrac{xyz^{-2}}{x^2yz}$
33. $\left[\left(\dfrac{a^{-2}x}{y^2}\right)^2\right]^2$
34. $\left(\dfrac{x^{-1}y^4}{z^3}\right)^{-2}$

● **35.** Show that if a and b are nonzero, then $\dfrac{a^{-n}}{b^{-n}} = \dfrac{b^n}{a^n}$.

● **36.** Show that if a and b are nonzero, then $\left(\dfrac{a}{b}\right)^{-1} = \dfrac{b}{a}$.

Write each of the following numbers in scientific notation.

37. 5,150,000,000   **38.** 12,015,000,000   **39.** 0.000 000 001 8

40. 0.000 000 000 025   **41.** 0.000 000 000 514 2   **42.** 0.000 000 010 342

In Exercises 43–48, the given numbers represent approximate measurements. Find the range of each measurement.

43. 5.6 ft   **44.** 7.8 m   **45.** 15.76 km

46. 105.7 mi   **47.** 21.315 kg   **48.** 13.27 lb

State the number of significant digits in each of the following numbers.

49. 31.6   **50.** 6.05   **51.** 56.012   **52.** 0.5014

53. 0.01506   **54.** 0.00015   **55.** 3.2×10^{-8} **56.** 1.83×10^{21}

c *In Exercises 57–60, perform the indicated operations and give your answers in scientific notation.*

57. $\dfrac{3.1 \times 10^{-12}}{9.0 \times 10^{6}}$ **58.** $\dfrac{1.25 \times 10^{11}}{0.60 \times 10^{5}}$

59. $\dfrac{(4.0 \times 10^{12})(3.5 \times 10^{-9})}{(2.0 \times 10^{-4})^{2}}$ **60.** $\dfrac{(8.15 \times 10^{-6})(3.2 \times 10^{-3})}{(1.2 \times 10^{5})(2.71 \times 10^{8})}$

c *In Exercises 61–64, convert each of the numbers to scientific notation, simplify, and give your answer in scientific notation.*

61. $\dfrac{(0.000\ 000\ 012)(41000)}{0.000\ 000\ 001\ 5}$ **62.** $\dfrac{(560\ 000\ 000)(0.000\ 002)}{0.000\ 000\ 021}$

63. $\dfrac{(3210)(0.000\ 000\ 000\ 18)}{(321\ 000\ 000)(0.000\ 516)}$ **64.** $\dfrac{(1\ 200\ 000\ 000)(0.000\ 25)}{(0.000\ 000\ 000\ 004)}$

c 65. The mass of the earth is 5.983×10^{24} kilograms (kg) and a metric ton is 10^{6} grams (g). Express the mass of the earth in metric tons.

c 66. The mass of the moon is 1.228×10^{-2} times the mass of the earth. Find the mass of the moon in kg. (See Exercise 65.)

c 67. Find the force of attraction between the earth and the moon, knowing that the mass of the earth is 5.983×10^{24} kg, the mass of the moon is 7.337×10^{22} kg, the mean distance between the earth and the moon is 3.843×10^{8} m, and the gravitational constant G is 6.673×10^{-11} N·m²/kg².

c 68. The speed of light is 186,000 miles per second. Find
a) the distance traveled by a ray of light in one year;
b) the distance it travels in four years.

c 69. The speed of light (in a vacuum) is 2.998×10^{8} meters per second. If 1 meter is equal to 6.214×10^{-4} miles, find the speed of light in miles per second.

c 70. At standard temperature and pressure, 1 g mol (or 32 g) of oxygen, O_2, occupies a volume of 22.4 liters $= 22.4 \times 10^{-3}$ m³. The number of molecules of oxygen in 1 g mol is constant and equal to 6.02×10^{23} (Avogadro's number). Find the volume of each oxygen molecule in liters and in m³.

1.4 POLYNOMIALS

When a letter, such as x, is used to denote a real number, without any particular number being assigned to it, we say that x is a *real variable*. Other letters, such as t, u, v, w, y, z, and so on, can also be used to denote real variables.

Let x be a variable in $\mathbb{R}$ and consider the powers

$$x^0 = 1, \quad x^1 = x, \quad x^2, \ldots, \quad x^n, \ldots$$

where the exponents are *nonnegative integers*. These powers of x are the basic elements used to define polynomials.

Polynomial in One Variable

A *polynomial* in a single variable x over $\mathbb{R}$ is an expression of the form

$$(1.12) \quad a_n x^n + a_{n-1} x^{n-1} + \cdots + a_1 x + a_0,$$

where $a_0, a_1, \ldots, a_n$ are real numbers, called *coefficients* of the polynomial, and n is a natural number.

If $a_n \neq 0$, then the *degree* of the polynomial is n. If $a_n = a_{n-1} = \cdots = a_1 = 0$, but $a_0 \neq 0$, then the polynomial has degree 0. Every nonzero constant defines a polynomial of degree 0. When *all* coefficients of a polynomial are zero, the polynomial is said to be the *zero polynomial*. By convention, no degree is assigned to the zero polynomial.

The term $a_k x^k$ in a given polynomial is called the *term of degree k* or *kth-degree term*; for example, the term

$$-x^2 \quad \text{in the polynomial} \quad 4x^3 - x^2 + 5x + 7$$

is called its *second-degree term*. If $a_n \neq 0$, then $a_n x^n$, the term of degree n is called the *leading term* and a_n the *leading coefficient* of the polynomial (1.12). The term a_0 is the *constant term*. For example,

$$4x^3 - x^2 + 5x + 7$$

is a polynomial of degree 3 with leading coefficient 4 and constant term 7. In a polynomial, terms with zero coefficients are frequently left unwritten, e.g.,

$$x^3 - 3x + 1 \quad \text{is short for} \quad x^3 + 0x^2 - 3x + 1.$$

A polynomial containing exactly one nonzero term, such as $5x^2$, is called a *monomial*. A *binomial* is a polynomial containing exactly two terms: $3x^6 - 8$ is a binomial of degree 6. A polynomial containing exactly three terms is a *trinomial*. For example, $5x^2 - 2x + 3$ is a trinomial of degree 2 with leading coefficient 5 and constant term 3.

A *linear polynomial* is a polynomial of degree 1:

$$a_1 x + a_0, \quad \text{with } a_1 \neq 0.$$

A polynomial of degree 2 such as

$$a_2x^2 + a_1x + a_0, \quad \text{with } a_2 \neq 0,$$

is called a *quadratic polynomial*. Polynomials of degree 3 are called *cubic polynomials*.

Polynomials can be added, subtracted, and multiplied by using the associative and commutative properties for the sum and product of numbers, the distributive property of the product over the sum, and the properties of exponents.

◆ **Example 1.** Add the polynomials $x^3 + 4x + 2$ and $3x^2 - 2x + 1$.

Solution. We have

$$(x^3 + 4x + 2) + (3x^2 - 2x + 1)$$
$$= x^3 + 3x^2 + (4x - 2x) + (2 + 1) \qquad \text{[Associative and commutative properties]}$$
$$= x^3 + 3x^2 + (4 - 2)x + 3 \qquad \text{[Distributive property]}$$
$$= x^3 + 3x^2 + 2x + 3.$$

Sometimes it is convenient to use the following scheme to add polynomials.

$$\begin{array}{r} x^3 \qquad\quad + 4x + 2 \\ 3x^2 - 2x + 1 \\ \hline x^3 + 3x^2 + 2x + 3 \end{array}$$

◇ **Practice Exercise 1.** Add the polynomials $x^4 - 2x^3 + 5x^2 - x + 4$ and $2x^3 - x^2 - 5$.

Answer. $x^4 + 4x^2 - x - 1$.

◆ **Example 2.** Find the difference $(x^4 - 5x^3 + x^2 - 3x + 1) - (2x^3 - 5x^2 + 2x - 4)$.

Solution. The polynomial $-(2x^3 - 5x^2 + 2x - 4) = -2x^3 + 5x^2 - 2x + 4$ is called the additive inverse of the polynomial $2x^3 - 5x^2 + 2x - 4$. To subtract the two polynomials you may add the first polynomial to the additive inverse of the second one:

$$(x^4 - 5x^3 + x^2 - 3x + 1) - (2x^3 - 5x^2 + 2x - 4)$$
$$= (x^4 - 5x^3 + x^2 - 3x + 1) + (-2x^3 + 5x^2 - 2x + 4)$$
$$= x^4 + (-5 - 2)x^3 + (1 + 5)x^2 + (-3 - 2)x + (1 + 4)$$
$$= x^4 - 7x^3 + 6x^2 - 5x + 5.$$

You should also try arranging the computations to subtract the given polynomials in a way similar to the scheme to add polynomials used in Example 1.

◇ **Practice Exercise 2.**   Subtract $x^4 - x^3 + 5x - 1$ from $4x^3 - 5x^2 + 6x - 2$.

Answer. $-x^4 + 5x^3 - 5x^2 + x - 1$.

Note that to *add* or *subtract* polynomials we *add* or *subtract coefficients of terms of the same degree.*

◆ **Example 3.**   Multiply the polynomials $4x + 2$ and $2x^2 + 3x - 5$.

Solution.   We have

$$(4x + 2)(2x^2 + 3x - 5)$$
$$= 4x(2x^2 + 3x - 5) + 2(2x^2 + 3x - 5) \quad \text{[Distributive property]}$$
$$= (4x)(2x^2) + (4x)(3x) + (4x)(-5)$$
$$\quad + 2(2x^2) + 2(3x) + 2(-5) \quad \text{[Distributive property]}$$
$$= 8x^3 + 12x^2 - 20x + 4x^2 + 6x - 10 \quad \text{[Computing the products]}$$
$$= 8x^3 + 16x^2 - 14x - 10.$$

When multiplying polynomials, the following scheme may be used:

$$
\begin{array}{l}
2x^2 + \ 3x \ - \ 5 \\
4x \ + \ 2 \\
\hline
8x^3 + 12x^2 - 20x \qquad [= 4x(2x^2 + 3x - 5)] \\
\qquad\quad 4x^2 + \ 6x - 10 \quad [= 2(2x^2 + 3x - 5)] \\
\hline
8x^3 + 16x^2 - 14x - 10
\end{array}
$$

Notice that the degree of the product of polynomials is equal to the sum of the degrees of the polynomials; the leading term of the product is the *product* of the leading terms, and the constant term of the product is the *product* of the constant terms.

◇ **Practice Exercise 3.**   Compute the product

$$(4x^3 - 3x^2 + 5)(x^2 - 2)$$

Answer. $4x^5 - 3x^4 - 8x^3 + 11x^2 - 10$.

Polynomials in More Than One Variable

We may also consider polynomials in two or more variables. A *polynomial in two variables x and y over $\mathbb{R}$ is a finite sum of terms of the form $ax^n y^m$,* where n and m are nonnegative integers and a is a real number. For example, $x + y - 1$, $x^2 - y^2$, and $3x^2 + y^2 + 4xy - 3x + 5y - 6$ are polynomials in x and y. Similarly, there are polynomials in three or more variables.

The sum of polynomials in two variables is found by adding the coefficients of terms of the same degree in x and in y.

◆ **Example 4.**   Add the polynomials $x^2 + xy + y^2 + 5x - y + 3$ and $3x^2 + 6y^2 - 7x + 9y + 11$.

Solution.   We may arrange our work as follows:

$$x^2 + xy + \ y^2 + 5x - \ y + \ 3$$
$$3x^2 \qquad + 6y^2 - 7x + 9y + 11$$
$$\overline{4x^2 + xy + 7y^2 - 2x + 8y + 14}$$

◇ **Practice Exercise 4.**   Find the sum of $3u^2 - 2uv + v^2 - 2u + 3v - 1$ and $u^2 - 3v^2 - 4v + 3$.

Answer. $4u^2 - 2uv - 2v^2 - 2u - v + 2$.

Polynomials in two or more variables are multiplied in the same way as are polynomials of one variable.

◆ **Example 5.**   Multiply $x + y$ and $x^2 - xy + y^2$.

Solution.   We have

$$(x + y)(x^2 - xy + y^2) = x(x^2 - xy + y^2) + y(x^2 - xy + y^2)$$
$$= x^3 - x^2y + xy^2 + x^2y - xy^2 + y^3$$
$$= x^3 + y^3.$$

The following scheme may also be used:

$$x^2 - xy + y^2$$
$$x \ + \ y$$
$$\overline{x^3 - x^2y + xy^2}$$
$$x^2y - xy^2 + y^3$$
$$\overline{x^3 \qquad\qquad\quad + y^3}$$

◇ **Practice Exercise 5.**   Find the product of $x^3 + x^2y + xy^2 + y^3$ and $x - y$.

Answer. $x^4 - y^4$.

Now, we briefly consider the division of a polynomial by a monomial. If the degree of the monomial is less than or equal to the degrees of each term of the polynomial, then the division is a simple operation: just divide each term of the polynomial by the monomial.

◆ **Example 6.**   Divide $6x^4y^2 - 8x^3y^3 + 2x^2y^4$ by $2x^2y$.

Solution.   Write

$$\frac{6x^4y^2 - 8x^3y^3 + 2x^2y^4}{2x^2y} = \frac{6x^4y^2}{2x^2y} - \frac{8x^3y^3}{2x^2y} + \frac{2x^2y^4}{2x^2y}$$
$$= 3x^2y - 4xy^2 + y^3.$$

◇ **Practice Exercise 6.**   Perform the division $\dfrac{6x^8 + 12x^6 + 15x^4}{3x^2}$.

Answer. $2x^6 + 4x^4 + 5x^2$.

Division of one polynomial by another will be reviewed in detail in Chapter 8.

⬡ EXERCISES 1.4

In Exercises 1–24, perform the indicated operations.

1. $(2x^2 - 3x + 1) + (-x^2 + x - 4)$
2. $(x^3 - 5x + 2) + (x^2 + 7x + 8)$
3. $\left(\dfrac{1}{2}x^4 + 2x^2 - \dfrac{1}{5}\right) - \left(\dfrac{1}{3}x^4 - 3x^2 + \dfrac{1}{2}\right)$
4. $(3x^2 + 5x + 4) - (x^4 + 3x^2 - 4)$
5. $(x^7 + 5x^5 + 3x^3 + x + 1) + (4x^4 + 2x^2 + 1)$
6. $(2t^3 + 3t^2 + 4t + 1) + (-5t^2 - 6t - 4)$
7. $(t^5 - t^4 + t^3 - t^2 - 1) + (2 + 3t + 4t^2 + 5t^3)$
8. $(5x^2 - 6x + 1) + (3x - 8) - (x^3 - 2x^2 + 4x - 1)$
9. $(x^4 - 1) + (x^4 - 3x^2) - (-x^4 - 2x^3 + x^2 + x - 5)$
10. $(3x^2 - 2xy + 4y^2 - 6y + 3) - (-3y^2 + 6xy + 3x - 4)$
11. $(5x^3 - 4x^2y + 3xy^2 - 4y^3 + xy + 1) - (3x^2y - 6xy^2 + 4x^2 - 6xy + x - 4)$
12. $(7x^3 - 5x^2y + 10xy^2 - 6y^3 + 3x^2 - 5xy - 3y^2 + 3x - 4y + 7)$
    $+ (8x^2y - 6xy^2 + x^2 - 6xy + 7y^2 - 5x + 8y - 3)$
13. $(2x^2 - 3x + 1)(4x - 2)$
14. $(5t^3 - 3t)(t^2 - t + 1)$
15. $(3x^2 - 2x + 1)(x^2 - 2x + 3)$
16. $(x^2 + x + 1)(x^2 + x - 1)$
17. $\left(\dfrac{1}{3}x^2 - \dfrac{2}{3}x + 1\right)\left(\dfrac{1}{2}x^2 + \dfrac{3}{4}x - 2\right)$
18. $(x^4 - 3x^2 + 1)(x^3 + x)$
19. $(x^5 - x^3 + 1)(x^2 + x + 1)$
20. $(x - y)(x^2 + xy + y^2)$
21. $(x + y + 1)(x + y - 1)$
22. $(2x - 3y + 4)(2x + 3y - 4)$
23. $(x^2 + y^2 + xy)(x^2 - y^2 - xy)$
24. $(2x^2 - y^2 + 4x)(2x^2 - 3y^2 - 6x)$

In Exercises 25–30, you are being asked to multiply three or more polynomials. Multiply two of them, then multiply that result by the third polynomial, and so on.

25. $(x - 1)(x - 2)(x - 3)$
26. $(2x + 1)(2x + 2)(2x + 3)$
27. $(2x + 1)(2x + 1)(2x + 1)$
28. $(x - 1)(x - 1)(x - 1)$
29. $(3x + 2)(3x - 2)(x + 1)(x - 1)$
30. $(4x - 1)(4x + 1)(x - 3)(x + 3)$

In Exercises 31–40, perform the indicated divisions.

31. $(6x^4 + 8x^2 - 10x) \div 2x$
32. $(12x^3 - 4x^2 - 8x) \div 4x$
33. $(9x^7 - 12x^5 - 3x^4 + 6x^3) \div 3x^2$
34. $(10x^8 - 15x^6 + 20x^4) \div 5x^2$

35. $\dfrac{16x^9 - 20x^7 + 12x^5}{4x^3}$

36. $\dfrac{15x^8 - 10x^7 + 5x^5 + 25x^3}{5x^3}$

37. $(10x^3y^2 - 6x^2y^3) \div 2xy^2$

38. $(9u^3v^3 - 6u^2v^4 + 12u^4v^2) \div 3u^2v$

39. $\dfrac{3x^2yz - 6x^2yz^3 + 9x^3y^2z^3}{3xyz}$

40. $\dfrac{4ax^2y^3 + 6a^2xy^4 - 10ax^3y^2}{2axy}$

HISTORICAL NOTE

Algebra

The word *algebra* was part of the title of an Arabic manuscript from about 800 AD describing certain rules to solve equations. A literal translation of the word *al-jabr* (algebra) is *the restoration*, meaning the operations that are necessary in order to "restore" or "balance" an equation. Today, the word is used to describe the branch of mathematics dealing with statements of relations utilizing letters and other symbols to represent numbers or values.

1.5 SPECIAL PRODUCTS; FACTORING

In our dealings with polynomials, we frequently encounter certain products called *special products*. They deserve special attention, and you should readily recognize them, because of their numerous applications. Here is a list of some of the most common ones.

SPECIAL PRODUCTS

   I. *Square of a binomial.*
$$(x + y)^2 = x^2 + 2xy + y^2.$$

  II. *Product of a sum by a difference.*
$$(x + y)(x - y) = x^2 - y^2.$$

 III. *Product of two binomials.*
$$(x + a)(x + b) = x^2 + (a + b)x + ab.$$

 IV. *Sum of cubes.*
$$x^3 + y^3 = (x + y)(x^2 - xy + y^2).$$

  V. *Difference of cubes.*
$$x^3 - y^3 = (x - y)(x^2 + xy + y^2).$$

 VI. *Product of two binomials.*
$$(mx + n)(px + q) = mpx^2 + (mq + np)x + nq.$$

You should check the validity of formulas I through VI by computing the indicated products. Formulas I, II, IV, and V definitely need to be memorized in view of the many applications. Formulas III and VI are similar to patterns which follow and don't require memorization. Also, notice that formula III is a particular case of formula VI, when $m = p = 1$, $n = a$, and $q = b$.

◆ Example 1.   Find each of the following products:
a) $(2a + 7)^2$ **b)** $(3u^2 + 4)(3u^2 - 4)$
c) $(3a + 2)(9a^2 - 6a + 4)$ **d)** $(3x + 1)(4x + 3)$.

Solution.   **a)** By special product I,

$$(2a + 7)^2 = (2a)^2 + 2(2a) \cdot 7 + 7^2$$
$$= 4a^2 + 28a + 49$$

b) This is a product of a sum by a difference,

$$(3u^2 + 4)(3u^2 - 4) = (3u^2)^2 - 4^2$$
$$= 9u^4 - 16.$$

c) You may use special product IV,

$$(3a + 2)(9a^2 - 6a + 4) = (3a + 2)[(3a)^2 - (3a) \cdot 2 + 2^2]$$
$$= (3a)^3 + 2^3$$
$$= 27a^3 + 8.$$

d) This is a product of two binomials,

$$(3x + 1)(4x + 3) = 3 \cdot 4 \cdot x^2 + (3 \cdot 3 + 1 \cdot 4)x + 1 \cdot 3$$
$$= 12x^2 + 13x + 3.$$

◇ Practice Exercise 1.   Use special products to perform each multiplication:
a) $(2b - 7)^2$ **b)** $(2v + 5)(2v - 5)$
c) $(4u - 3)(16u^2 + 12u + 9)$ **d)** $(3x + 2)(x + 6)$.

Answer.   **a)** $4b^2 - 28b + 49$ **b)** $4v^2 - 25$ **c)** $64u^3 - 27$ **d)** $3x^2 + 20x + 12$.

Factoring Polynomials

In Section 1.1, we mentioned that each natural number can be written as a product of prime factors. For example, $60 = 2 \cdot 2 \cdot 3 \cdot 5$, where 2, 3, and 5 are the prime factors of 60.

The situation is analogous for polynomials. For example, direct multiplication shows that the quadratic polynomial $2x^2 + 10x + 12$ can be written as the following product:

$$2x^2 + 7x + 3 = (2x + 1)(x + 3)$$

In the product $(2x + 1)(x + 3)$, the polynomials $2x + 1$ and $x + 3$ are called *factors* of the product. The process of finding the factors of a polynomial is called *factoring*.

A polynomial is said to be *prime* or *irreducible* if it cannot be written as a product of two polynomials of positive degrees. For example, $2x + 1$, and $x + 3$ are prime polynomials.

A polynomial is said to be in *completely factored* form when it is written as a product of prime polynomials.

When factoring polynomials, it is necessary to specify *over* what set of numbers we want to carry out the factorization. For example, the polynomial $x^2 - 4$ with *integer* coefficients can be written as a product of polynomials with integer coefficients,

$$x^2 - 4 = (x - 2)(x + 2).$$

We say that $x^2 - 4$ has been factored *over the integers*. On the other hand, it is not possible to write $x^2 - 2$ as a product of two polynomials of degree 1 with integer coefficients. Thus, $x^2 - 2$ is said to be *irreducible over the integers*. However, since

$$x^2 - 2 = (x - \sqrt{2})(x + \sqrt{2}),$$

the polynomial $x^2 - 2$ can be factored *over the real numbers*.

There are polynomials that cannot be factored over the real numbers, such as

$$x^2 + 1, \quad x^2 + 4, \quad 3x^2 + 1, \quad \text{and} \quad x^2 - 2x + 5.$$

These polynomials are said to be *irreducible over the real* numbers. Later, after introducing the concept of *complex numbers,* we shall see that these polynomials can be factored *over* the complex numbers.

In this chapter, polynomials with integer coefficients will be factored so that the factors contain only integer coefficients.

Now we discuss several examples. If all terms of a polynomial contain a *common factor,* then we can factor the polynomial by using the distributive property.

◆ **Example 2.**   Factor each polynomial:
a) $6ax^3 + 4ax^2$ **b)** $2a(x + y) - 5b(x + y)$.

Solution.   **a)** We look for the "largest" common factor of the two monomials. Since $6ax^3 = (2ax^2)3x$ and $4ax^2 = (2ax^2)2$, it follows that $2ax^2$ is a common factor. Thus

$$6ax^3 + 4ax^2 = (2ax^2)3x + (2ax^2)2$$
$$= 2ax^2(3x + 2).$$

b) In this case, $x + y$ is the largest common factor of $2a(x + y)$ and $5b(x + y)$. Thus

$$2a(x + y) - 5b(x + y) = (x + y)(2a - 5b).$$

◇ **Practice Exercise 2.**   Factor each polynomial:
a) $4u^2x^2 - 8u2x$ **b)** $(u - v)5x + 3(u - v)$.

Answer.   **a)** $4u^2x(x - 2)$ **b)** $(u - v)(5x + 3)$.

Most polynomials that can be factored at all are factored by recognizing them as special products.

◆ **Example 3.**   Factor the polynomial $4y^2 - 25$.

Solution.   Since

$$4y^2 - 25 = (2y)^2 - 5^2,$$

we see that the given polynomial is a difference of two squares. According to special product II, we write

$$4y^2 - 25 = (2y)^2 - 5^2,$$
$$= (2y + 5)(2y - 5).$$

◇ **Practice Exercise 3.**   Factor the expression $64x^2 - 49$.

Answer. $(8x + 7)(8x - 7)$.

◆ **Example 4.**   Factor the polynomial $2x^3 + 16$.

Solution.   First, notice that 2 is a common factor of both coefficients. Using the distributive property, write

$$2x^3 + 16 = 2(x^3 + 8).$$

Since $8 = 2^3$, apply special product IV to obtain

$$2x^3 + 16 = 2(x^3 + 8)$$
$$= 2(x + 2)(x^2 - 2x + 4).$$

◇ **Practice Exercise 4.**   Factor $3x^3 - 24$.

Answer. $3(x - 2)(x^2 + 2x + 4)$.

◆ **Example 5.**   Factor the quadratic polynomial $2x^2 + 10x + 12$.

Solution.   Since 2 is a common factor, write

$$2x^2 + 10x + 12 = 2(x^2 + 5x + 6).$$

Next, look for two numbers a and b such that $a + b = 5$ and $ab = 6$. Since 2 and 3 satisfy these requirements, write

$$2x^2 + 10x + 12 = 2(x^2 + 5x + 6)$$
$$= 2(x + 2)(x + 3).$$

◇ **Practice Exercise 5.**   Factor $3x^2 + 24x + 45$.

Answer. $3(x + 3)(x + 5)$.

◆ **Example 6.**   Factor the polynomial $2x^2 - 5x - 3$.

Solution.   Using special product VI, we first look for two integers m and p so that $mp = 2$. Since the only factors of 2 are 1 and 2, the factorization of the given polynomial is of the form

$$2x^2 - 5x - 3 = (2x + \quad)(x + \quad)$$

where the blank spaces will be occupied by two integers n and q such that $nq = -3$ and $mq + np = -5$. Since the integral factors of -3 are -1 and 3 or 1 and -3, trial and error lead us to the factorization

$$2x^2 - 5x - 3 = (2x + 1)(x - 3).$$

Inspection of the other products

$$(2x - 1)(x + 3)$$
$$(2x + 3)(x - 1)$$
$$(2x - 3)(x + 1)$$

shows that none of them equals $2x^2 - 5x - 3$. As an exercise, you should check that in each of the three products above, the term of highest degree is $2x^2$ and the constant term is -3. However, the term of degree one fails to be $-5x$.

◇ **Practice Exercise 6.**   Write $3x^2 + 11x - 4$ as a product of two linear polynomials.

Answer. $(3x - 1)(x + 4)$.

◆ **Example 7.**   Factor completely the expression $81a^4 - 1$.

Solution.   The given expression is a difference of two squares,

$$81a^4 - 1 = (9a^2)^2 - 1^2.$$

Thus, using special product II, write

$$81a^4 - 1 = (9a^2)^2 - 1.$$
$$= (9a^2 + 1)(9a^2 - 1).$$

But $9a^2 - 1 = (3a)^2 - 1^2$ is also the difference of two squares. Repeating the process, obtain the complete factorization of the given expression:

$$81a^4 - 1 = (9a^2 + 1)(9a^2 - 1)$$
$$= (9a^2 + 1)[(3a)^2 - 1^2]$$
$$= (9a^2 + 1)(3a + 1)(3a - 1).$$

◇ **Practice Exercise 7.**   Factor completely $16u^4 - 1$.

Answer. $(4u^2 + 1)(2u + 1)(2u - 1)$.

Factoring by Grouping

To factor certain expressions, it is sometimes necessary to group some of their terms, factor out any common factors that may be present, and use the distributive property. This technique, illustrated in the next example, is called *factoring by grouping*.

◆ **Example 8.**   Factor the following expressions:
a) $6xy + 2ay - 3bx - ab$,   **b)** $2ax^2 - 2ay + x^2b - by$.

Solution.   **a)** First, arrange the terms in two groups,

$$6xy + 2ay - 3bx - ab = (6xy - 3bx) + (2ay - ab).$$

Next, factor each group

$$6xy + 2ay - 3bx - ab = 3x(2y - b) + a(2y - b).$$

Since $2y - b$ is a common factor of the last two terms, obtain

$$6xy + 2ay - 3bx - ab = (2y - b)(3x + a).$$

b) Analogously,

$$2ax^2 - 2ay^2 + x^2b - by = (2ax^2 - 2ay) + (x^2b - by)$$
$$= 2a(x^2 - y) + b(x^2 - y)$$
$$= (x^2 - y)(2a + b).$$

◇ **Practice Exercise 8.**   Factor each of the expressions:
a) $2au^2 + bu^2 - 4av - 2bv$ **b)** $3mn - 6mq - pn + 2pq$.

Answer.   **a)** $(u^2 - 2v)(2a + b)$ **b)** $(3m - p)(n - 2q)$.

We end up this section with the following hints that may help you to factor a polynomial.

FACTORING HINTS

1. Look for a common factor.
2. See if you can use one of the special products.
3. Try factoring by grouping.
4. Look to see if further factoring can be performed.
5. Remember that a *completely factored* polynomial must be written as a product of *prime* factors.

EXERCISES 1.5

In Exercises 1–16, use special products to expand each of the following.

1. $(2x + 1)(2x + 3)$ **2.** $(2x - 4)(2x + 3)$ **3.** $(2x + 3a)(2x - a)$

4. $(3y + b)(3y + 2b)$ **5.** $(2ax - 5b)^2$ **6.** $(3r + as)^2$

7. $(4y - 3b)^2$ **8.** $(3au - 4b)^2$ **9.** $(x^2 + 5)(x^2 - 5)$

10. $(5a^2 + 3)(5a^2 - 3)$ **11.** $(x + \sqrt{2})(x - \sqrt{2})$ **12.** $(t - \sqrt{5})(t + \sqrt{5})$

13. $(\sqrt{a} + \sqrt{b})(\sqrt{a} - \sqrt{b})$ **14.** $(\sqrt{2}x + 1)(\sqrt{2}x - 1)$ ● **15.** $(x - y + a)^2$

● **16.** $(u - v + 2b)^2$

In Exercises 17–58, factor completely each of the following.

17. $x^2 + 6x + 8$ **18.** $y^2 - 6y + 5$ **19.** $2x^2 + 16x + 30$

20. $3x^2 - 6x - 45$ **21.** $5x^2 - 20$ **22.** $4x^2 - 36$

23. $3ax^2 + 6ax - 9a$ **24.** $2mu^2 - 10mu + 8m$ **25.** $64a^2 - 9b^2$

26. $b^2 - 81c^2$ **27.** $16x^4 - 81$ **28.** $81y^4 - 625$

29. $x^3 - 27$ **30.** $r^3 - 64$ **31.** $8x^3 + 64$

32. $64x^3 + 27$ **33.** $16ax^3 + 2a$ **34.** $3x^3 - 24$

35. $2x^2 + 5x - 3$ **36.** $2x^2 - 7x + 6$ **37.** $4x^2 - 4x + 1$

38. $4x^2 - 20x + 25$ **39.** $6x^2 - 5x + 1$ **40.** $6x^2 + 7x + 2$

41. $3x^2 + 11x + 6$ **42.** $5x^2 + 9x - 2$ **43.** $2ac - 6ad + bc - 3bd$

44. $3mu - 6nu - 5mv + 10nv$ **45.** $6ax - 9ay - 2bx + 3by$ **46.** $8rt - 2pt + 12rq - 3pq$

47. $x^2y + 2ay - 3bx^2 - 6ab$ **48.** $2uy^2 + 6bu - 5y^2 - 15b$ **49.** $(a + b)^3 - 8$

50. $(x + y)^4 - 16$ **51.** $x^6 - 64$ **52.** $y^6 - 729$

● **53.** $(x + 2y)^2 - 4z^2$ ● **54.** $(2a - b)^2 - 16c^2$ ● **55.** $(2x - 1)^3 + 8$

● **56.** $(2x + 1)^3 - 8$ ● **57.** $(a + b)^2 - (a - b)^2$ ● **58.** $(a - b)^2 - (a + b)^2$

In Exercises 59–64, the given polynomials have rational coefficients. Factor each polynomial over the reals. [Hint: Write $\dfrac{u^2}{2} - 8 = \dfrac{1}{2}(u^2 - 16)$ and complete the factorization.]

59. $\dfrac{u^2}{2} - 8$ **60.** $\dfrac{v^2}{5} - 5$ **61.** $x^2 - \dfrac{1}{6}x - \dfrac{1}{6}$

62. $x^2 + \dfrac{7x}{12} + \dfrac{1}{12}$ **63.** $\dfrac{x^3}{2} + 4$ **64.** $\dfrac{x^3}{3} - 9$

65. Without using a calculator, compute the difference $(321.4)^2 - (320.4)^2$. (*Hint*: Use special product II.)

66. Without a calculator, find the difference $(162.25)^2 - (160.25)^2$.

● **67.** Show that any odd number (that is, a number of the form $2n + 1$ with n a natural number) can be written as the difference of squares of two consecutive natural numbers.

● **68.** Show that if you square an integer and subtract 1, the result equals the product of the integer preceding and the integer following that integer.

● **69.** Show that one-fourth of the difference between the squares of the integer following and the integer preceding a given integer is the integer itself.

● **70.** Show that if you take a number, multiply it by the number four units larger, add four units to the product, take the square root and subtract 2 from the result, you get the number you started with.

1.6 RATIONAL EXPRESSIONS

A *rational expression* is the quotient of two polynomials. For example,

$$\frac{3x+1}{5}, \quad \frac{x+5}{2x^2-3x-1}, \quad \text{and} \quad \frac{x^4+x+2}{x^2+5x+1}$$

are rational expressions. More generally, if $a_n x^n + a_{n-1} x^{n-1} + \cdots + a_1 x + a_0$ is a polynomial of degree n and $b_m x^m + b_{m-1} x^{m-1} + \cdots + b_1 x + b_0$ is a polynomial of degree m, we may form the rational expression

$$\frac{a_n x^n + a_{n-1} x^{n-1} + \cdots + a_1 x + a_0}{b_m x^m + b_{m-1} x^{m-1} + \cdots + b_1 x + b_0}.$$

Since in this expression x denotes an arbitrary real number while the coefficients $a_0, a_1, \ldots, a_n$ and $b_0, b_1, \ldots, b_n$ denote fixed real numbers, the rational expression represents the quotient of two real numbers, *except for those values of x for which the denominator is equal to zero*. Thus, when we deal with rational expressions it will always be understood that *we are excluding the values of x for which the denominator is equal to zero*. For example, $\dfrac{2x+1}{x-3}$ has a meaning for all real numbers x such that $x \neq 3$.

Rational expressions can be added, subtracted, multiplied, divided, and simplified in the same manner as fractions [Section 1.2].

Simplifying Rational Expressions

To simplify a rational expression, we divide both numerator and denominator by any factor they have in common. In analogy with rational numbers (or fractions), rational expressions whose numerator and denominator have no common factors are said to be in *lowest terms*.

◆ **Example 1.**   Simplify $\dfrac{x^2-4}{2x^2-3x-2}$.

Solution.   Factor both numerator and denominator and cancel the common factor:

$$\frac{x^2-4}{2x^2-3x-2} = \frac{(x+2)(x-2)}{(2x+1)(x-2)} = \frac{x+2}{2x+1}.$$

Note that the equality

$$\frac{x^2-4}{2x^2-3x-2} = \frac{x+2}{2x+1}$$

holds *only for $x \neq 2$*. This is because if $x = 2$, the denominator of the first rational expression is equal to zero.

◇ **Practice Exercise 1.**   Reduce $\dfrac{2x^2+5x-3}{x^2+x-6}$ to lowest terms.

Answer. $\dfrac{2x-1}{x-2}$.

Addition and Subtraction

To add or subtract two rational expressions with the same denominator, we add or subtract the numerators and keep the same denominator.

◆ **Example 2.** Perform the indicated sum and simplify: $\dfrac{3x^2 + 2x - 6}{x^2 - 1} + \dfrac{2 - x}{x^2 - 1}$.

Solution. We have

$$\frac{3x^2 + 2x - 6}{x^2 - 1} + \frac{2 - x}{x^2 - 1} = \frac{3x^2 + 2x - 6 + 2 - x}{x^2 - 1}$$

$$= \frac{3x^2 + x - 4}{x^2 - 1}$$

$$= \frac{(3x + 4)(x - 1)}{(x + 1)(x - 1)}$$

$$= \frac{3x + 4}{x + 1}.$$

◇ **Practice Exercise 2.** Subtract as indicated and simplify: $\dfrac{2x^2 + 2x - 5}{x^2 - 4} - \dfrac{3x + 5}{x^2 - 4}$

Answer. $\dfrac{2x - 5}{x - 2}$.

In order to add or subtract rational expressions with different denominators, we have to replace them, as we do with ordinary fractions, by equivalent rational expressions with the same denominator. We can always choose as common denominator the product of the denominators. However, computation is simpler if we use the *least common denominator* (LCD) of the rational expressions.

Least Common Denominator

To find the LCD of several rational expressions, proceed as follows:
1. Factor completely each of the denominators.
2. Multiply all prime factors, each one raised to the highest power of that factor occurring in any one factorization.

◆ **Example 3.** Subtract $\dfrac{3x - 1}{x - 1} - \dfrac{2x}{2x + 3}$.

Solution. Take the product $(x - 1)(2x + 3)$ of the denominators as the common denominator and write

$$\frac{3x-1}{x-1} - \frac{2x}{2x+3} = \frac{(3x-1)(2x+3)}{(x-1)(2x+3)} - \frac{2x(x-1)}{(x-1)(2x+3)}$$

$$= \frac{6x^2+7x-3}{(x-1)(2x+3)} - \frac{2x^2-2x}{(x-1)(2x+3)}$$

$$= \frac{6x^2+7x-3-2x^2+2x}{(x-1)(2x+3)}$$

$$= \frac{4x^2+9x-3}{(x-1)(2x+3)}.$$

◇ **Practice Exercise 3.**   Add $\dfrac{4x}{3x+1} + \dfrac{x+1}{x-2}$.

Answer. $\dfrac{7x^2-4x+1}{(3x+1)(x-2)}$.

◆ **Example 4.**   Perform the indicated operations and simplify:

$$\frac{2}{x^2+2x} + \frac{6}{2x+4} - \frac{3x+1}{x^2}.$$

Solution.   First, we find the LCD of the denominators according to the following scheme:

$$x^2+2x = x(x+2)$$
$$2x+4 = 2(x+2)$$
$$\frac{x^2 = x^2}{\text{LCD} = 2(x+2)x^2}$$

Observe that there are three prime factors: 2, x, and $x+2$. The highest exponent of the factors 2 and $x+2$ is 1, and the highest exponent of the factor x is 2. Thus, the LCD is the product of 2, $x+2$, and x^2. Next, write

$$\frac{2}{x^2+2x} + \frac{6}{2x+4} - \frac{3x+1}{x^2}$$

$$= \frac{2}{x(x+2)} + \frac{6}{2(x+2)} - \frac{3x+1}{x^2}$$

$$= \frac{2(2x)}{2(x+2)x^2} + \frac{6(x^2)}{2(x+2)x^2} - \frac{2(x+2)(3x+1)}{2(x+2)x^2}$$

$$= \frac{4x}{2(x+2)x^2} + \frac{6x^2}{2(x+2)x^2} - \frac{2(3x^2+7x+2)}{2(x+2)x^2}$$

$$= \frac{4x+6x^2-6x^2-14x-4}{2(x+2)x^2}$$

$$= \frac{-10x-4}{2(x+2)x^2}$$

$$= -\frac{\cancel{2}(5x+2)}{\cancel{2}(x+2)x^2}$$

$$= -\frac{5x+2}{(x+2)x^2}.$$

◇ **Practice Exercise 4.**   Add and simplify: $\dfrac{2x-1}{x^2+2x+1}+\dfrac{3x-4}{x^2-1}$.

Answer. $\dfrac{5x^2-4x-3}{(x+1)^2(x-1)}$.

Multiplication and Division

The product of rational expressions is obtained by multiplying their numerators and multiplying their denominators. In order to simplify computations, it is advisable to factor the polynomials and to cancel out common factors that appear in the numerator and denominator.

◆ **Example 5.**   Multiply $\dfrac{x^2-6x+9}{2x+10}\cdot\dfrac{2x+6}{x^2-9}$.

Solution.   First, factor completely the polynomials appearing in both expressions and cancel out common factors.

$$\frac{x^2-6x+9}{2x+10}\cdot\frac{2x+6}{x^2-9}=\frac{(x-3)^2}{2(x+5)}\cdot\frac{2(x+3)}{(x+3)(x-3)}$$
$$=\frac{2(x-3)^2(x+3)}{2(x+5)(x+3)(x-3)}$$
$$=\frac{x-3}{x+5}.$$

◇ **Practice Exercise 5.**   Multiply $\dfrac{3x-6}{2x^2-5x+2}\cdot\dfrac{x^2+6x+8}{3x+6}$.

Answer. $\dfrac{x+4}{2x-1}$.

The pattern for dividing one rational expression by another is the same as it is for fractions: invert the divisor and multiply. Just as in any other multiplication, cancel out any common factors that may appear in the (new) numerators and denominators before actually carrying out the multiplication.

◆ **Example 6.**   Divide $\dfrac{2x^2+5x-3}{x^2+10x+25}\div\dfrac{2x-1}{x^2+5x}$. Give your answer in simplified form.

Solution.   We have

$$\frac{2x^2 + 5x - 3}{x^2 + 10x + 25} \div \frac{2x - 1}{x^2 + 5x}$$

$$= \frac{2x^2 + 5x - 3}{x^2 + 10x + 25} \cdot \frac{x^2 + 5x}{2x - 1} \qquad \text{[Invert the divisor]}$$

$$= \frac{(2x - 1)(x + 3)x(x + 5)}{(x + 5)^2(2x - 1)} \qquad \begin{array}{l}\text{[Factor numerators}\\ \text{and denominators]}\end{array}$$

$$= \frac{x(x + 3)}{x + 5}. \qquad \text{[Cancel out common factors]}$$

◇ **Practice Exercise 6.**   Divide and simplify: $\dfrac{2x^2 - 5x + 3}{x^2 + 2x} \div \dfrac{x^2 - 5x + 4}{x + 2}$.

Answer. $\dfrac{2x - 3}{x(x - 4)}$.

Compound Expressions

On many occasions, we have to deal with quotients in which the numerator and denominator are neither polynomials nor rational expressions but can be reduced to rational expressions.

◆ **Example 7.**   Simplify the expression $\dfrac{1 - \dfrac{7}{x^2 - 9}}{1 - \dfrac{1}{x - 3}}$.

Solution.   First write the numerator and denominator as rational expressions, and then divide.

$$\frac{1 - \dfrac{7}{x^2 - 9}}{1 - \dfrac{1}{x - 3}} = \frac{\dfrac{x^2 - 9 - 7}{x^2 - 9}}{\dfrac{x - 3 - 1}{x - 3}}$$

$$= \frac{\dfrac{x^2 - 16}{x^2 - 9}}{\dfrac{x - 4}{x - 3}} = \frac{x^2 - 16}{x^2 - 9} \cdot \frac{x - 3}{x - 4}$$

$$= \frac{(x + 4)(x - 4)(x - 3)}{(x + 3)(x - 3)(x - 4)} = \frac{x + 4}{x + 3}.$$

◇ **Practice Exercise 7.**   Simplify the compound expression $\dfrac{\dfrac{2}{x+3}+1}{1-\dfrac{2}{x+7}}$.

Answer. $\dfrac{x+7}{x+3}$.

 EXERCISES 1.6

Simplify each of the following.

1. $\dfrac{15a^2x^4}{3ax^2}$

2. $\dfrac{18m^3n^6}{9m^2n^2}$

3. $\dfrac{10x^6+15x^4-5x^2}{5x}$

4. $\dfrac{18a^4+27a^5}{9a^3}$

5. $\dfrac{6x-18}{6x-24}$

6. $\dfrac{12y^2-8y}{4y^2-16y}$

7. $\dfrac{x+1}{x^2+9x+8}$

8. $\dfrac{x^2-1}{x^2+5x+4}$

9. $\dfrac{u^2-9}{u^2+6u+9}$

10. $\dfrac{z^2-16}{2z^2-5z-12}$

11. $\dfrac{2x^2+10x}{2x^2+9x-5}$

12. $\dfrac{2x^2+x-3}{2x^2+5x+3}$

13. $\dfrac{4x^2+7x+4}{4x^2+11x+6}$

14. $\dfrac{10x^2+26x-12}{5x^2+18x-8}$

Perform the operations and simplify.

15. $3+\dfrac{x-1}{x+2}$

16. $5-\dfrac{x-3}{x-4}$

17. $x-2-\dfrac{x-2}{x+3}$

18. $x-5+\dfrac{x-1}{x+4}$

19. $\dfrac{x+4}{x-4}-\dfrac{x-1}{x+4}$

20. $\dfrac{5}{3x+1}-\dfrac{x}{x+2}$

21. $\dfrac{5x+10}{x^2+6x+8}+\dfrac{3}{x+4}$

22. $\dfrac{3x+6}{x^2-4}+\dfrac{5x}{x-2}$

23. $\dfrac{2x-1}{x^2+6x+9}+\dfrac{x}{x+3}$

24. $\dfrac{x-1}{x^2+5x+6}+\dfrac{x-2}{x^2+4x+4}$

25. $\dfrac{x-2}{2x^2-x-1}+\dfrac{x+2}{2x^2+3x+1}$

26. $\dfrac{2x}{3x^2+x-2}-\dfrac{2x-8}{3x^2-8x+4}$

27. $\dfrac{1}{x-1}-\dfrac{1}{3x-1}+\dfrac{1}{4-4x}$

28. $\dfrac{x^2}{(x+1)^2}+\dfrac{1}{x-1}-\dfrac{1}{2}$

29. $\dfrac{3}{x^2+3x+2}+\dfrac{2}{x^2-x-6}-\dfrac{1}{x^2-2x-3}$

30. $\dfrac{2}{x-5}-\dfrac{2}{x+5}-\dfrac{15}{x^2-25}$

Perform the multiplications and simplify.

31. $\dfrac{x+1}{x^2+x}\cdot\dfrac{x}{x-1}$

32. $\dfrac{x^4-4x^2}{x+2}\cdot\dfrac{x-2}{x^2-2x}$

33. $\dfrac{4y-8}{4(y+2)}\cdot\dfrac{y}{2y-4}$

34. $\dfrac{x^2-16}{x^2-9}\cdot\dfrac{3-x}{2x+8}$

35. $\dfrac{5}{8b^2} \cdot \dfrac{4b + 6}{10b + 15}$

36. $\dfrac{2x - 5}{x^2 - 4} \cdot \dfrac{3x - 6}{4x - 10}$

37. $\dfrac{2x^2 + 5x + 2}{x^2 + 5x - 6} \cdot \dfrac{x^2 + 7x + 6}{4x^2 + 4x + 1}$

38. $\dfrac{2x^2 - 3x - 2}{x^2 + x - 2} \cdot \dfrac{x^2 + 3x + 2}{4x^2 - 1}$

39. $\dfrac{x^3 - x^2 - 2x}{x^2 - 1} \cdot \dfrac{2x + 1}{x^2 - 2x}$

40. $\dfrac{x^3 - 9x}{x^2 - x - 2} \cdot \dfrac{x^2 - 1}{x^2 + 2x - 3}$

Perform the divisions and simplify.

41. $\dfrac{\dfrac{2x}{x^3}}{\dfrac{4x}{x^5}}$

42. $\dfrac{\dfrac{x + 5}{8}}{\dfrac{x - 5}{16}}$

43. $\dfrac{\dfrac{x^2 - 1}{x}}{\dfrac{x + 1}{x^2}}$

44. $\dfrac{\dfrac{x - 6}{2x - 12}}{\dfrac{3x}{}}$

45. $\dfrac{\dfrac{5x^2}{4x^2 - 1}}{\dfrac{10x}{2x + 1}}$

46. $\dfrac{\dfrac{9a^2 - 1}{4 - a}}{\dfrac{3a + 1}{a^2 - 16}}$

47. $\dfrac{\dfrac{2x^2 - 7x + 3}{2x + 1}}{\dfrac{x^2 - 9}{10x + 5}}$

48. $\dfrac{\dfrac{x - 5}{x^2 + 8x + 15}}{\dfrac{x - 2}{x^2 + 6x + 9}}$

49. $\dfrac{\dfrac{3x^2 - 8x - 3}{4x^2 - 1}}{\dfrac{x^2 + x - 12}{2x^2 + 3x + 1}}$

50. $\dfrac{\dfrac{2}{3}(x^2 + 2x - 15)}{\dfrac{5}{6}(x^2 - 9)}$

51. $\dfrac{\dfrac{1}{x + 1} + \dfrac{1}{x - 1}}{\dfrac{2}{x}}$

52. $\dfrac{1 + \dfrac{1}{2x}}{\dfrac{2}{3x} - 1}$

53. $\dfrac{\dfrac{3}{x - 2} - 1}{3 - \dfrac{1}{x - 2}}$

54. $\dfrac{\dfrac{x}{x + 1} - 2}{3 - \dfrac{x}{x + 1}}$

55. $\dfrac{\dfrac{2x}{x - 1} + \dfrac{3}{x + 1}}{\dfrac{4x}{x + 1} - \dfrac{2}{x - 1}}$

56. $\dfrac{\dfrac{x}{x + 2} - \dfrac{2}{x - 3}}{\dfrac{3}{x + 2} - \dfrac{x}{x - 3}}$

57. $\dfrac{1 - \dfrac{5}{x^2 - 4}}{\dfrac{x - 3}{4 - x}}$

58. $\dfrac{\dfrac{5}{x^2 - 9} + 1}{\dfrac{5}{x - 3} + 1}$

59. $\dfrac{\dfrac{7}{x^2 - 9} - 1}{\dfrac{1}{x - 3} - 1}$

60. $1 - \dfrac{x - \dfrac{1}{x}}{1 - \dfrac{1}{x}}$

1.7 RADICALS

Let a be a real number and n be a natural number. Can we find a real number x such that the statement

$$(1.13) \quad x^n = a$$

is true?

   Before answering this question, let us discuss a few examples. If $a = 4$ and $n = 2$, then both $x = 2$ and $x = -2$ satisfy equation (1.13), because $2^2 = 4$ and $(-2)^2 = 4$.

   If $a = -27$ and $n = 3$, then $x = -3$ is the unique number such that $(-3)^3 = -27$.

   Finally, if $a = -1$ and $n = 2$, there is *no* real number x such that $x^2 = -1$.

These examples indicate that the following general results are true.

n even	$a > 0$	There are *two* real numbers with opposite signs satisfying the statement (1.13)
n odd	$a > 0$ or $a < 0$	There is a *unique* real number satisfying (1.13)
n even	$a < 0$	*No* real number satisfies (1.13)

We now make the following definition:

---

nth Root of a Real Number

A real number x satisfying equation (1.13) is called an nth root of a.

For example, 2 and -2 are *second* (or *square*) *roots* of 4; -3 is the *third* (or *cube*) *root* of -27; and 5 and -5 are *fourth roots* of 625.

The Principal nth Root of a Real Number

The principal nth root of a real number a is the *positive* nth root of a if $a > 0$ and n is even, or it is the *unique* nth root of a if n is odd.

For example, 2 is the *principal square root* of 4; 3 is the *principal cube root* of 27; and 5 is the *principal fourth root* of 625.

Notation

The principal nth root of a is denoted $\sqrt[n]{a}$.

HISTORICAL NOTE

The Radical Sign
It seems that the radical sign $\sqrt{}$ originated from the letter r used as an abbreviation of the word *radix*.

In this notation, the symbol $\sqrt{}$ is called a *radical*, the natural number n is the *index* of the radical, and a is the *radicand*.

When $n = 2$, it is customary to omit the index from the radical. We write $\sqrt{a}$, and read the *principal square root* of a or simply the *square root* of a. When $n = 3$, $\sqrt[3]{a}$ is called the *principal third root* of a or simply the *cube root* of a.

From the definition of principal nth root, it follows that:

$$\sqrt[n]{a} = b \quad \text{if and only if} \quad a = b^n.$$

◆ **Example 1.**   Compute the radicals.

a) $\sqrt{64}$ b) $\sqrt[5]{-\dfrac{32}{243}}$.

Solution.   **a)** The integers 8 and -8 are such that $8^2 = (-8)^2 = 64$. Since $\sqrt{64}$ denotes the principal square root of 64, it follows that $\sqrt{64} = 8$.

b) $\sqrt[5]{-\dfrac{32}{243}} = -\dfrac{2}{3}$, because $\left(-\dfrac{2}{3}\right)^5 = -\dfrac{32}{248}$.

◇ **Practice Exercise 1.**   Compute the radicals.

a) $\sqrt[3]{-64}$ b) $\sqrt[4]{625}$.

Answer.   **a)** -4 **b)** 5.

Listed below are all the properties of radicals that you must know to work with them. We assume that all of the numbers are such that the expressions are defined.

Properties of Radicals

1. $(\sqrt[n]{a})^n = a$
2. $\sqrt[n]{ab} = \sqrt[n]{a}\,\sqrt[n]{b}$
3. $\sqrt[n]{\dfrac{a}{b}} = \dfrac{\sqrt[n]{a}}{\sqrt[n]{b}}$ $(b \neq 0)$
4. $\sqrt[m]{\sqrt[n]{a}} = \sqrt[mn]{a}$.

Property 1 is a restatement of the definition of the principal nth root of a. We prove property 2, leaving the others as exercises. Let $x = \sqrt[n]{a}$ and $y = \sqrt[n]{b}$. By the definition of nth root,

$$x^n = a \quad \text{and} \quad y^n = b.$$

Multiplying these two expressions, we have

$$x^n \cdot y^n = (xy)^n = ab.$$

Thus, xy is the nth root of ab, that is, $xy = \sqrt[n]{ab}$. Substituting $\sqrt[n]{a}$ for x and $\sqrt[n]{b}$ for y, we obtain

$$\sqrt[n]{a} \cdot \sqrt[n]{b} = \sqrt[n]{ab}.$$

Simplifying Radicals

A radical is said to be in *simplified form* when
1. All possible factors have been eliminated from under the radical.
2. The index of the radical is the smallest possible.

◆ **Example 2.**   Use the properties of radicals to simplify each expression. Assume that all letters represent positive numbers.

a) $\sqrt[3]{216}$ **b)** $\sqrt{\dfrac{15}{144}}$ **c)** $\sqrt{\dfrac{8a^4}{b^6}}$ **d)** $\sqrt[3]{\sqrt{64x^8}}$.

Solution.   **a)** Since $216 = 8 \cdot 27$, write

$$\sqrt[3]{216} = \sqrt[3]{8 \cdot 27} = \sqrt[3]{8} \cdot \sqrt[3]{27} = 2 \cdot 3 = 6.$$

b) Using property 3, obtain

$$\sqrt{\frac{15}{144}} = \frac{\sqrt{15}}{\sqrt{144}} = \frac{\sqrt{15}}{12}.$$

c) Using properties 2 and 3, write

$$\sqrt{\frac{8a^4}{b^6}} = \frac{\sqrt{8a^4}}{\sqrt{b^6}} = \frac{\sqrt{2 \cdot 4 \cdot a^4}}{\sqrt{b^6}} = \frac{2a^2\sqrt{2}}{b^3}.$$

d) Here, use properties 4 and 2:

$$\sqrt[3]{\sqrt{64x^8}} = \sqrt[6]{64x^8} = \sqrt[6]{2^6 x^6 x^2} = 2x\sqrt[6]{x^2}.$$

◇ **Practice Exercise 2.**   Use the properties above to simplify each radical:

a) $\sqrt{\dfrac{32}{81}}$ **b)** $\sqrt[3]{-512}$ **c)** $\sqrt{\dfrac{16x^3}{y^4}}$ **d)** $\sqrt{\sqrt{81a^4b^5}}$

Answer.   **a)** $4\sqrt{2}/9$ **b)** -8 **c)** $\dfrac{4x\sqrt{x}}{y^2}$ **d)** $3ab\sqrt[4]{b}$.

Sum and Difference of Radicals

The distributive property of the product over the sum allows us to combine terms of a sum or difference of radicals with the *same index* and *same radicand*.

◆ **Example 3.**   Perform the following operations and simplify.

a) $3\sqrt{2} + 4\sqrt{18} - \sqrt{32}$ **b)** $5\sqrt[3]{x^4} - \sqrt[3]{x^7} + \sqrt[3]{8x^4}$.

Solution.   **a)** We have

$$3\sqrt{2} + 4\sqrt{18} - \sqrt{32} = 3\sqrt{2} + 4\sqrt{2 \cdot 9} - \sqrt{2 \cdot 16}$$
$$= 3\sqrt{2} + 4 \cdot 3\sqrt{2} - 4\sqrt{2}$$
$$= 3\sqrt{2} + 12\sqrt{2} - 4\sqrt{2}$$
$$= (3 + 12 - 4)\sqrt{2}   \quad \text{[Distributive property]}$$
$$= 11\sqrt{2}.$$

b) First, remove all third powers from under each radical; then use the distributive property.

$$5\sqrt[3]{x^4} - 3\sqrt[3]{x^7} + \sqrt[3]{8x^4} = 5\sqrt[3]{x^4 x} - 3\sqrt[3]{x^6 x} + \sqrt[3]{2^3 x^3 x}$$
$$= 5x\sqrt[3]{x} - 3x^2\sqrt[3]{x} + 2x\sqrt[3]{x}$$
$$= (5x - 3x^2 + 2x)\sqrt[3]{x}$$
$$= (7x - 3x^2)\sqrt[3]{x}.$$

◇ **Practice Exercise 3.**   Perform the indicated operations and simplify:

a) $2\sqrt[3]{16} - 5\sqrt[3]{2} + 2\sqrt[3]{128}$ **b)** $\sqrt{8x^5} + \sqrt{18x^3} - \sqrt{2x^5}$.

Answer.   **a)** $7\sqrt[3]{2}$ **b)** $(x^2 + 3x)\sqrt{2x}$.

Rationalizing Denominators

Property 3 of radicals tells us that the radical of a quotient is the quotient of radicals. For instance,

$$\sqrt{\frac{15}{2}} = \frac{\sqrt{15}}{\sqrt{2}}$$

and we have a fraction whose denominator contains a radical. In most computations, it is preferable to replace such a fraction by an equivalent one whose denominator does not contain a radical. In this example, this can be achieved by multiplying both numerator and denominator by $\sqrt{2}$. Thus,

$$\sqrt{\frac{15}{2}} = \frac{\sqrt{15}}{\sqrt{2}} = \frac{\sqrt{15} \cdot \sqrt{2}}{\sqrt{2} \cdot \sqrt{2}} = \frac{\sqrt{30}}{\sqrt{4}} = \frac{\sqrt{30}}{2}.$$

This process is known as *rationalizing the denominator.*

◆ **Example 4.**   Rationalize the denominator in each of the following expressions:

a) $\dfrac{\sqrt{2} + \sqrt{3}}{\sqrt{5}}$ **b)** $\dfrac{\sqrt{x} - \sqrt{y}}{\sqrt{x}}$ **c)** $\dfrac{\sqrt[3]{3}}{\sqrt[3]{2}}$.

Solution.   **a)** Multiply both numerator and denominator by $\sqrt{5}$, obtaining

$$\frac{\sqrt{2} + \sqrt{3}}{\sqrt{5}} = \frac{(\sqrt{2} + \sqrt{3})\sqrt{5}}{\sqrt{5} \cdot \sqrt{5}} = \frac{\sqrt{10} + \sqrt{15}}{5}.$$

b) Multiplying the numerator and denominator by $\sqrt{x}$ gives us

$$\frac{\sqrt{x} - \sqrt{y}}{\sqrt{x}} = \frac{(\sqrt{x} - \sqrt{y})\sqrt{x}}{\sqrt{x}\sqrt{x}} = \frac{x - \sqrt{xy}}{x}.$$

c) Since $\sqrt[3]{2} \cdot \sqrt[3]{2^2} = \sqrt[3]{2^3} = 2$, we multiply both numerator and denominator by $\sqrt[3]{2^2}$ to obtain

$$\frac{\sqrt[3]{3}}{\sqrt[3]{2}} = \frac{\sqrt[3]{3} \cdot \sqrt[3]{2^2}}{\sqrt[3]{2} \cdot \sqrt[3]{2^2}} = \frac{\sqrt[3]{3 \cdot 2^2}}{\sqrt[3]{2^3}} = \frac{\sqrt[3]{12}}{2}.$$

◇ **Practice Exercise 4.**   Rationalize the denominators:

a) $\dfrac{\sqrt{2} - \sqrt{5}}{\sqrt{3}}$ **b)** $\dfrac{\sqrt{a} + \sqrt{b}}{\sqrt{c}}$ **c)** $\dfrac{\sqrt[4]{2}}{\sqrt[4]{3}}.$

Answer.   **a)** $\dfrac{\sqrt{6} - \sqrt{15}}{3}$ **b)** $\dfrac{\sqrt{ac} + \sqrt{bc}}{c}$ **c)** $\dfrac{\sqrt[4]{54}}{3}.$

◆ **Example 5.**   Rationalize the denominators of

a) $\dfrac{5 - \sqrt{2}}{3 + \sqrt{2}}$ **b)** $\dfrac{1}{\sqrt{x} + \sqrt{y}}.$

Solution.   **a)** Recalling the special product $(x + y)(x - y) = x^2 - y^2$, we see that $(3 + \sqrt{2})(3 - \sqrt{2}) = 9 - 2 = 7$. Thus, multiplying both numerator and denominator by $3 - \sqrt{2}$, we obtain

$$\begin{aligned}
\frac{5 - \sqrt{2}}{3 + \sqrt{2}} &= \frac{(5 - \sqrt{2})(3 - \sqrt{2})}{(3 + \sqrt{2})(3 - \sqrt{2})} \\
&= \frac{5 \cdot 3 - 5\sqrt{2} - 3\sqrt{2} + \sqrt{2} \cdot \sqrt{2}}{3^2 - (\sqrt{2})^2} \\
&= \frac{15 - 5\sqrt{2} - 3\sqrt{2} + 2}{9 - 2} \\
&= \frac{17 - 8\sqrt{2}}{7}.
\end{aligned}$$

The expression $3 - \sqrt{2}$ is often called the *conjugate* of $3 + \sqrt{2}$.

b) In this case $\sqrt{x} - \sqrt{y}$ is the conjugate of $\sqrt{x} + \sqrt{y}$. Multiplying the numerator and denominator by $\sqrt{x} - \sqrt{y}$ gives us

$$\frac{1}{\sqrt{x} + \sqrt{y}} = \frac{\sqrt{x} - \sqrt{y}}{(\sqrt{x} + \sqrt{y})(\sqrt{x} - \sqrt{y})} = \frac{\sqrt{x} - \sqrt{y}}{x - y}.$$

◇ **Practice Exercise 5.**   Rationalize the denominators of the following expressions:

a) $\dfrac{1}{\sqrt{5} - \sqrt{2}}$ **b)** $\dfrac{\sqrt{a} + \sqrt{b}}{\sqrt{a} - \sqrt{b}}.$

Answer.   **a)** $\dfrac{\sqrt{5} + \sqrt{2}}{3}$ **b)** $\dfrac{a + b + 2\sqrt{ab}}{a - b}$.

Simplifying Expressions Containing Radicals

An expression containing radicals is in *simplified form* if
1. All radicals have been simplified.
2. All radicals have been removed from denominators.
3. All possible operations have been performed.

EXERCISES 1.7

In Exercises 1–40, simplify each expression. Assume that all letters represent nonnegative real numbers and that all denominators are different from zero.

1. $\sqrt{81}$
2. $\sqrt[3]{125}$
3. $\sqrt[5]{-32}$
4. $\sqrt[7]{128}$

5. $\sqrt{\dfrac{16}{25}}$
6. $\sqrt[4]{\dfrac{81}{625}}$
7. $(\sqrt{3} + 2)(\sqrt{3} - 2)$
8. $(\sqrt{5} + 3)(\sqrt{5} - \sqrt{3})$

9. $(3\sqrt{5} + 4)(3\sqrt{5} - 4)$
10. $(6 + 2\sqrt{3})(6 - 2\sqrt{3})$
11. $(\sqrt{3} - \sqrt{7})^2$
12. $(\sqrt{8} + \sqrt{2})^2$

13. $\sqrt[3]{8a^6}$
14. $\sqrt[3]{216x^3y^6}$
15. $\sqrt[3]{-27a^5}$
16. $\sqrt[4]{32p^5q^6}$

17. $\sqrt{2x}\sqrt{8x^3}$
18. $\sqrt{3y^2}\sqrt{27y^4}$
19. $\sqrt{3ab^3}\sqrt{6a^3b}$
20. $\sqrt{4u^3v^5}\sqrt{5v^5u^3}$

21. $\sqrt{2ax^2}\sqrt{8a^3x^4}$
22. $\sqrt{5u^2v^3}\sqrt{10u^4v^7}$
23. $\sqrt[3]{2xy^2}\sqrt[3]{4x^5y^7}$
24. $\sqrt[5]{9a^3b}\sqrt[5]{27a^2b^9}$

25. $\sqrt[6]{4a^2bc^3}\sqrt[6]{16a^4b^5c^9}$
26. $\sqrt[3]{3r^2s}\sqrt[3]{-9r}$
27. $\sqrt{\dfrac{16a^2b^4}{c^6}}$
28. $\sqrt[3]{\dfrac{27ax^4}{8b^6}}$

29. $\sqrt[5]{\dfrac{32a^6b^7}{c^5}}$
30. $\sqrt{\dfrac{50m^2n^4}{p^3}}$
31. $\sqrt{\dfrac{1}{2a^3b}}$
32. $\sqrt[3]{\dfrac{3b^3x}{2a}}$

33. $\sqrt[3]{\dfrac{4am^2}{9n}}$
34. $\sqrt{\dfrac{3x^2}{yz}}$
35. $\sqrt{\sqrt[3]{128a^7b^8x^8}}$
36. $\sqrt{\sqrt{16m^2p^4q^6}}$

37. $\sqrt{(m + n)^2}$
38. $(\sqrt[4]{2ab^2x^3})^4$
39. $\sqrt{25a\sqrt{25a^3}}$
40. $\sqrt[3]{27m^2\sqrt{9n^5}}$

In Exercises 41–52, perform the indicated operations and simplify.

41. $3\sqrt{2} + 5\sqrt{8}$
42. $5\sqrt{3} - 4\sqrt{12}$
43. $2\sqrt{3} - 4\sqrt{27}$

44. $6\sqrt{5} - 3\sqrt{20}$
45. $\sqrt{2} + \sqrt{5} - 6\sqrt{8} + 3\sqrt{45}$
46. $5\sqrt{2} - 4\sqrt{27} + 6\sqrt{3} + 3\sqrt{8}$

47. $\sqrt[3]{16} + 2\sqrt[3]{54}$
48. $\sqrt[3]{40} - 2\sqrt[3]{135}$
49. $\sqrt{4x} - \sqrt{16x} + \sqrt{25x}$

50. $\sqrt{8a} + \sqrt{18a} + \sqrt{50a}$
51. $\sqrt{8a^3} + \sqrt{18a^3} + \sqrt{50a^3}$
52. $\sqrt[3]{8x^4} - \sqrt[3]{27x^7} + \sqrt[3]{64x^4}$

In Exercises 53–76 rationalize the denominators.

53. $\dfrac{1 + \sqrt{3}}{\sqrt{2}}$
54. $\dfrac{2 - \sqrt{2}}{\sqrt{3}}$
55. $\dfrac{a + \sqrt{b}}{\sqrt{c}}$
56. $\dfrac{m - \sqrt{n}}{\sqrt{p}}$

57. $\dfrac{3}{1 - \sqrt{2}}$
58. $\dfrac{4}{2 + \sqrt{2}}$
59. $\dfrac{\sqrt{5} + \sqrt{2}}{\sqrt{3}}$
60. $\dfrac{\sqrt{7} - \sqrt{3}}{\sqrt{6}}$

61. $\dfrac{\sqrt{a} - \sqrt{b}}{\sqrt{c}}$ **62.** $\dfrac{\sqrt{2x} - \sqrt{y}}{\sqrt{xy}}$ **63.** $\dfrac{\sqrt{x^3} + \sqrt{x^5}}{\sqrt{x^3}}$ **64.** $\dfrac{2\sqrt{a^3} + \sqrt{a^5}}{\sqrt{a^3}}$

65. $\dfrac{\sqrt{6} - 4}{2 + \sqrt{6}}$ **66.** $\dfrac{6}{2 + 3\sqrt{2}}$ **67.** $\dfrac{a}{\sqrt{a} + 2}$ **68.** $\dfrac{3}{2 - \sqrt{x}}$

69. $\dfrac{\sqrt{x}}{\sqrt{x} + \sqrt{y}}$ **70.** $\dfrac{2a}{\sqrt{a} + \sqrt{b}}$ **71.** $\dfrac{\sqrt{x} - 1}{1 - \sqrt{x - 1}}$ **72.** $\dfrac{2}{\sqrt{x + 1} - \sqrt{x}}$

73. $\dfrac{1}{a + \sqrt{b}}$ **74.** $\dfrac{1}{\sqrt{a + b + c}}$

● **75.** $\dfrac{1}{\sqrt{x^2 + 1} + x}$ ● **76.** $\dfrac{h}{\sqrt{2(x + h) + 1} - \sqrt{2x + 1}}$

● **77.** Show that $\sqrt{3} + \sqrt{5} = \sqrt{8 + 2\sqrt{15}}$.   ● **78.** Show that $\sqrt{3} - \sqrt{2} = \sqrt{5 - 2\sqrt{6}}$.

● **79.** Prove that $\sqrt[n]{\dfrac{a}{b}} = \dfrac{\sqrt[n]{a}}{\sqrt[n]{b}}$.   ● **80.** Prove that $\sqrt[m]{\sqrt[n]{a}} = \sqrt[mn]{a}$.

1.8   RATIONAL EXPONENTS

In this section, we extend the definition of powers to *rational* values of the exponents.

Rational Exponents

If a is a *nonnegative real number* and m and n are *natural numbers*, define

$$(1.14) \quad a^{m/n} = (\sqrt[n]{a})^m = \sqrt[n]{a^m}$$

The definition tells us that a rational power $a^{m/n}$ can be computed in two different ways: we may *raise the nth root of a to the mth power*, or we can *take the nth root of the mth power of a.*

◆ **Example 1.**   Find the value of each expression:
a) $16^{3/4}$ **b)** $(8a^6)^{2/3}$.

Solution.   We have
a) $16^{3/4} = (\sqrt[4]{16})^3 = 2^3 = 8$
or
$16^{3/4} = \sqrt[4]{16^3} = \sqrt[4]{4096} = 8.$
b) $(8a^6)^{2/3} = (\sqrt[3]{8a^6})^2 = (2a^2)^2 = 4a^4$
or
$(8a^6)^{2/3} = \sqrt[3]{(8a^6)^2} = \sqrt[3]{64a^{12}} = 4a^4.$

Notice that both expressions $(\sqrt[4]{16})^8$ and $(\sqrt[3]{8a^6})^2$ are easier to evaluate than their counterparts.

◇ **Practice Exercise 1.**   Compute each of the following expressions:
a) $8^{2/3}$ **b)** $(81a^8)^{3/4}$.

Answer.   **a)** 4   **b)** $27a^6$.

Setting $m = 1$ in the definition of rational exponents gives us the formula

$$(1.15) \quad a^{1/n} = \sqrt[n]{a}.$$

That is, the principal nth root of a number can be viewed as a rational exponent. Combining formulas (1.14) and (1.15), we may write

$$a^{m/n} = (a^{1/n})^m = (a^m)^{1/n}.$$

Negative Exponents

The definition of rational exponents can be extended to the case of *negative* exponents by defining

$$a^{-m/n} = \frac{1}{a^{m/n}}.$$

◆ **Example 2.**   **a)** Evaluate $8^{-2/3}$.   **b)** Simplify $(25a^6)^{-1/2}$.

Solution.   Using the definition of negative rational exponents, we have
a) $8^{-2/3} = \dfrac{1}{8^{2/3}} = \dfrac{1}{(\sqrt[3]{8})^2} = \dfrac{1}{4}$.

b) $(25a^6)^{-1/2} = \dfrac{1}{(25a^6)^{1/2}} = \dfrac{1}{\sqrt{25a^6}} = \dfrac{1}{5a^3}$.

◇ **Practice Exercise 2.**   **a)** Evaluate $81^{-3/4}$.   **b)** Simplify $(8a^6)^{-1/3}$.

Answer.   **a)** 1/27   **b)** $1/(2a^2)$.

Properties 1 through 5 of integer exponents (Section 1.3) are valid for rational exponents, too. For future reference, we list them here.

Properties of Rational Exponents

Let a and b be real numbers and let r and s be rational numbers. The following properties hold whenever the indicated powers are defined.

1. $a^r \cdot a^s = a^{r+s}$
2. $(a^r)^s = a^{rs}$
3. $\dfrac{a^r}{a^s} = a^{r-s}$, $a \neq 0$
4. $(ab)^r = a^r \cdot b^r$
5. $\left(\dfrac{a}{b}\right)^r = \dfrac{a^r}{b^r}$, $b \neq 0$
6. $a^0 = 1$, $a \neq 0$
7. $a^{-r} = \dfrac{1}{a^r}$, $a \neq 0$.

◆ **Example 3.** Simplify and write your answers using only positive exponents. Assume that all letters represent positive numbers.

a) $(2a^{1/3})(4a^{1/2})$ **b)** $\left(\dfrac{4x^{-2}}{y^{-4}}\right)^{1/2}$

Solution. We have

a) $(2a^{1/3})(4a^{1/2}) = 8a^{1/3 + 1/2} = 8a^{5/6}$.

b) $\left(\dfrac{4x^{-2}}{y^{-4}}\right)^{1/2} = \left(\dfrac{4y^4}{x^2}\right)^{1/2} = \dfrac{\sqrt{4y^4}}{\sqrt{x^2}} = \dfrac{2y^2}{x}$.

◇ **Practice Exercise 3.** Assuming that all variables represent positive real numbers, simplify each expression and write your answers with positive exponents only.

a) $(3a^{1/4}b^{-3/4})^4$ **b)** $\left(\dfrac{2u^{1/2}}{u^{1/3}}\right)^3$.

Answer. **a)** $\dfrac{81a}{b^3}$ **b)** $8u^{1/2}$.

Many expressions involving radicals can best be simplified by changing the radicals to rational exponents, performing the operations, and changing back to radicals.

◆ **Example 4.** Express as one radical in simplest form:

a) $\sqrt{2} \cdot \sqrt[3]{4}$ **b)** $\dfrac{\sqrt[3]{4a^2}}{\sqrt{2a}}$.

Solution.   We have

a) $\sqrt{2} \cdot \sqrt[3]{4} = \sqrt{2} \cdot \sqrt[3]{2^2} = 2^{1/2} \cdot 2^{2/3} = 2^{1/2 + 2/3} = 2^{7/6} = 2 \cdot 2^{1/6} = 2\sqrt[6]{2}$

b) $\dfrac{\sqrt[3]{4a^2}}{\sqrt{2a}} = \dfrac{\sqrt[3]{(2a)^2}}{\sqrt{2a}} = \dfrac{(2a)^{2/3}}{(2a)^{1/2}} = (2a)^{2/3 - 1/2} = (2a)^{1/6} = \sqrt[6]{2a}.$

◇ **Practice Exercise 4.**   Write as one radical in simplest form.

a) $\dfrac{\sqrt[4]{8}}{\sqrt{2}}$ **b)** $\sqrt{2x^2} \cdot \sqrt[4]{8x}.$

Answer.   **a)** $\sqrt[4]{2}$ **b)** $2x\sqrt[4]{2x}.$

 EXERCISES 1.8

Write each of the following using radicals instead of rational exponents. Assume that all variables represent positive real numbers.

1. $8^{3/2}$ **2.** $27^{4/3}$ **3.** $64^{-1/4}$ **4.** $81^{-3/2}$
5. $(a^2 b^3)^{3/5}$ **6.** $(4xy^2)^{2/3}$ **7.** $(x^2 + y^2)^{1/2}$ **8.** $(x^4 + xy)^{1/4}$

Write each of the following using rational exponents instead of radicals.

9. $\sqrt[3]{5^2}$ **10.** $\sqrt[6]{7^5}$ **11.** $\sqrt[4]{x^3}$ **12.** $\sqrt[5]{y^2}$
13. $a^2 \sqrt{a}$ **14.** $b\sqrt{b^3}$ **15.** $a^3 \sqrt[4]{a^5}$ **16.** $x^2 \sqrt[3]{x^5}$

Simplify each of the following expressions. All variables represent positive real numbers.

17. $(-8)^{2/3}$ **18.** $(-27)^{4/3}$ **19.** $81^{-1/2}$ **20.** $9^{-3/2}$
21. $(-243)^{3/5}$ **22.** $128^{-5/7}$ **23.** $(5u^{1/2})(3u^{3/2})$ **24.** $2ax^{1/6}(3ax^{1/3})$
25. $(-125x^6)^{2/3}$ **26.** $(16a^2 x^4)^{3/2}$ **27.** $(m^{-2}n^{-3})^{-1/6}$ **28.** $(8a^{-3}b^{-6})^{-2/3}$
29. $\left(\dfrac{5a^{1/4}}{b^{1/2}}\right)^2$ **30.** $\left(\dfrac{a^{1/3}}{b^{3/2}}\right)^6$ **31.** $\left(\dfrac{81x^{-8}}{y^4}\right)^{3/4}$ **32.** $\left(\dfrac{-125a^3 b^6}{c^{-9}}\right)^{2/3}$

In Exercises 33–46, simplify the given expressions and write the answers with positive exponents only. All variables represent positive numbers.

33. $(36a^8 b^2)^{-1/2}$ **34.** $(125a^3)^{-2/3}$ **35.** $(64a^2 b^3)^{-2/3}$ **36.** $(32x^8 y^4)^{1/4}$
37. $\left(\dfrac{36x^2 y^4}{z^3}\right)^{-1/2}$ **38.** $\left(\dfrac{a^8 b^{12}}{c^3}\right)^{-3/4}$ **39.** $(a^{-2/3}b^{-1/2})^{-6}$ **40.** $(u^{-3}v^2)^{-1/6}$
41. $\left(\dfrac{16a^4 x^{-1}}{25a^{-2}x^3}\right)^{1/2}$ **42.** $\left(\dfrac{8a^{-1}b^2}{27a^2 b^{-4}}\right)^{1/3}$ **43.** $\left(\dfrac{4}{9}a^{-1/3}x^4\right)^{-3/2}$ **44.** $\left(\dfrac{8}{27}b^6 y^{-3/2}\right)^{2/3}$
45. $\left(\dfrac{a^2}{b^{1/4}}\right)\left(\dfrac{b^{-1/2}}{a^{3/2}}\right)$ **46.** $\dfrac{(a^{-6}x)^{-1/3}}{(a^2 x^4)^{-1/2}}$

In Exercises 47–60, simplify and write each expression as a radical with least positive index. Assume that all variables represent positive real numbers.

47. $\sqrt{2} \cdot \sqrt[3]{3}$

48. $\sqrt{3} \cdot \sqrt[4]{9}$

49. $\sqrt{a} \cdot \sqrt[4]{2a}$

50. $\sqrt[3]{2x} \cdot \sqrt[6]{2x}$

51. $\dfrac{\sqrt[4]{25}}{\sqrt{5}}$

52. $\dfrac{\sqrt[3]{4}}{\sqrt{2}}$

53. $\dfrac{\sqrt{2a^2}}{\sqrt[4]{2a}}$

54. $\dfrac{\sqrt{2a}}{\sqrt[3]{a}}$

55. $\dfrac{\sqrt[3]{4x^2y^2}}{\sqrt[6]{2xy}}$

56. $\dfrac{\sqrt{4a^3b^3}}{\sqrt[3]{2ab}}$

57. $\dfrac{\sqrt[3]{m^2v}}{\sqrt[6]{m^3v}}$

58. $\dfrac{\sqrt{2ax^3}}{\sqrt[3]{4a^2x}}$

59. $\sqrt[3]{2a\sqrt{2a}}$

60. $\sqrt{2a\sqrt[3]{2a}}$

1.9 COMPLEX NUMBERS

Although the real number system is more than adequate for most of the applications discussed in this book, the concept of complex numbers is needed in a few instances such as solving quadratic equations. Moreover, complex numbers are indispensable in certain branches of mathematics and are often used in physics and engineering.

In order to understand why we need complex numbers, notice that the square x^2 of an arbitrary real number x is *always nonnegative*. Thus, there is *no* real number x such that $x^2 = -1$. In other words, it is impossible to find a real number satisfying the statement $x^2 + 1 = 0$. In view of this impossibility complex numbers were invented.

The Imaginary Unit

The *imaginary unit i* is defined by requiring that

$$i^2 = -1.$$

We also write i as $\sqrt{-1}$. If the imaginary unit is combined with two real numbers by addition and multiplication, a complex number is obtained.

Definition

A *complex number* is a number of the form

$$a + bi$$

where a and b are real numbers. The set of all complex numbers is denoted by $\mathbb{C}$.

If $a = 0$, then the complex number $0 + bi = bi$ is called a *pure imaginary* number. If $b = 0$, then the complex number is equal to the real number a. Conversely, every real number can be viewed as a complex number by writing

HISTORICAL NOTE

Complex Numbers

The concept of complex numbers was first proposed by Hieronimo Cardano (1501–1576), an Italian mathematician, in his "Ars Magna," a treatise on the solution of cubic and quartic equations. Cardano's ideas were ignored until 1799, when Carl Friedrich Gauss (1777–1855), a German mathematician, revived them and used complex numbers in several of his works.

$a = a + 0i$. In this way, the set $\mathbb{R}$ of real numbers becomes a subset of the set $\mathbb{C}$ of complex numbers.

We have described several systems of numbers: *natural numbers, integers, rational, real*, and *complex numbers*. The set $\mathbb{N}$ of natural numbers is contained in the set $\mathbb{Z}$ of integers. These form a subset of $\mathbb{Q}$, the set of rational numbers. Real numbers consist of rational and irrational numbers; thus the set $\mathbb{R}$ contains the set $\mathbb{Q}$. Finally, $\mathbb{C}$, the largest set, contains $\mathbb{R}$. In set notation, the inclusion among the different systems of numbers can be written as

$$\mathbb{N} \subset \mathbb{Z} \subset \mathbb{Q} \subset \mathbb{R} \subset \mathbb{C}.$$

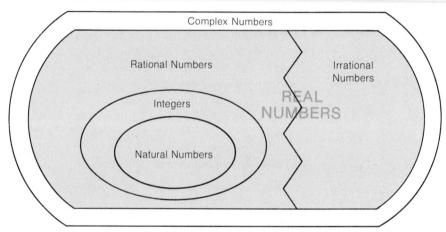

Figure 1.7

Before extending the operations of addition and multiplication to complex numbers, we introduce some definitions and terminology.

Equality

Two complex numbers $a + bi$ and $c + di$ are said to be equal and we write

$$a + bi = c + di$$

if and only if

$$a = c \quad \text{and} \quad b = d.$$

Real and Imaginary Parts of a Complex Number

Given any complex number $z = a + bi$, we call a the *real part* of z and b the *imaginary part* of z. Thus, two complex numbers are equal if and only if their real and imaginary parts are equal.

Addition and Multiplication

If $a + bi$ and $c + di$ are complex numbers, we define

$$(a + bi) + (c + di) = (a + c) + (b + d)i$$

and

$$(a + bi)(c + di) = (ac - bd) + (bc + ad)i.$$

The right-hand sides of the sum and product of two complex numbers are obtained by using the associative, commutative, and distributive properties together with the relation $i^2 = -1$. This is because we want the same arithmetic rules used for real numbers to hold for complex numbers.

◆ **Example 1.** Write each of the following numbers in the form $a + bi$:
a) $(2 + 3i) + (5 - 4i)$ **b)** $(3 + 5i)(2 + 4i)$.

Solution. **a)** We have

$$(2 + 3i) + (5 - 4i) = 2 + 3i + 5 - 4i$$
$$= (2 + 5) + (3 - 4)i$$
$$= 7 - i.$$

b) Using the distributive property, we obtain

$$(3 + 5i)(2 + 4i) = 3(2 + 4i) + 5i(2 + 4i)$$
$$= 3 \cdot 2 + 3(4i) + (5i)2 + (5i)(4i)$$
$$= 6 + 12i + 10i + 20i^2.$$

Since $i^2 = -1$, we get

$$(3 + 5i)(2 + 4i) = 6 + 22i + 20(-1)$$
$$= 6 + 22i - 20$$
$$= -14 + 22i.$$

◇ **Practice Exercise 1.** Find the sum and product of the complex numbers $2 - 3i$ and $5 + 4i$.

Answer. $7 + i, 22 - 7i$.

Complex Conjugate

If $z = a + bi$ is a complex number, then the number $\bar{z} = a - bi$ is called the *complex conjugate* of z.

Complex conjugate numbers have the same real part, but their imaginary parts have the opposite sign. Notice that the complex conjugate of $\bar{z}$ is z.

◆ **Example 2.**   Find the product $(2 + 3i)(2 - 3i)$.

Solution.   Using the distributive property, we have

$$(2 + 3i)(2 - 3i) = 2 \cdot 2 + 2(-3i) + (3i)2 + (3i)(-3i)$$
$$= 4 - \cancel{6i} + \cancel{6i} - 9i^2$$
$$= 13.$$

Thus, the product of the complex number $2 + 3i$ by its own complex conjugate is the real number 13.

◇ **Practice Exercise 2.**   Find the product $(3 + 5i)(3 - 5i)$.

Answer.   34.

In general, *the product of the complex number by its own complex conjugate is a real number.* Indeed, if $z = a + bi$, then

$$z\bar{z} = (a + bi)(a - bi)$$
$$= a \cdot a + a(-bi) + (bi)a + (bi)(-bi) \qquad \text{[Distributive property]}$$
$$= a \cdot a - a \cdot bi + a \cdot bi - (b \cdot b)i^2 \qquad \qquad [i^2 = -1]$$
$$= a^2 + b^2.$$

Thus, the product $z\bar{z}$ is a *nonnegative real number.* Furthermore, if $z = a + bi \neq 0$ (that is, if at least one of a and b is not zero), then $z\bar{z} > 0$.

The set $\mathbb{C}$ of complex numbers is closed under the operations of addition and multiplication, and these operations satisfy the commutative, associative, and distributive properties. Furthermore, 0 is the additive identity, and 1 is the multiplicative identity.

Additive Inverses

If $z = a + bi$ is a complex number, then the *additive inverse* of z is the number $-z = (-a) + (-b)i$.

Clearly, we have

$$z + (-z) = (a + bi) + ((-a) + (-b)i)$$
$$= (a + (-a)) + (b + (-b))i$$
$$= 0 + 0i$$
$$= 0.$$

◆ **Example 3.**   Find the difference $(2 + 3i) - (4 + 5i)$.

Solution.   To find the difference, we add to $2 + 3i$ the additive inverse of $4 + 5i$:

$$(2 + 3i) - (4 + 5i) = (2 + 3i) + ((-4) + (-5)i)$$
$$= (2 + (-4)) + (3 + (-5))i$$
$$= (-2) + (-2)i$$
$$= -2 - 2i.$$

◇ **Practice Exercise 3.**   Write $(7 + 2i) - (5 + 8i)$ in the form $a + bi$.

Answer. $2 - 6i$.

Multiplicative Inverses

Every complex number $z \neq 0$ has a multiplicative inverse, that is, a complex number w such that $zw = 1$. To see this, let $z = a + bi$ and suppose that w is a complex number such that

$$zw = 1.$$

Multiplying both sides of this inequality by $\bar{z}$, we obtain

$$\bar{z}(zw) = \bar{z}$$

or

$$(\bar{z}z)w = \bar{z}$$

or

$$(a^2 + b^2)w = \bar{z};$$

hence

$$w = \frac{\bar{z}}{a^2 + b^2}.$$

Recalling that $\bar{z} = a - bi$, we have

$$w = \frac{a - bi}{a^2 + b^2}$$

or

$$w = \frac{a}{a^2 + b^2} - \frac{b}{a^2 + b^2} i$$

which is the multiplicative inverse of $z = a + bi \neq 0$.

The multiplicative inverse (also called *reciprocal*) of a complex number $z = a + bi$ is denoted by z^{-1}, $1/z$, $(a + bi)^{-1}$, or $1/(a + bi)$.

Quotient

If z and w are complex numbers, with $w \neq 0$, then the *quotient* of z by w is defined by

$$\frac{z}{w} = zw^{-1}.$$

◆ **Example 4.**   Write each of the following numbers in the form $a + bi$:

   a) $\dfrac{1}{3 + 4i}$ **b)** $\dfrac{3 + 5i}{2 - 4i}$.

Solution.   **a)** According to the formula for the reciprocal of a complex number, we can write

$$\frac{1}{3 + 4i} = \frac{3}{25} - \frac{4}{25}i.$$

However, there is no need to memorize the formula. To find the reciprocal of z, just multiply the numerator and denominator of the expression $1/z$ by $\bar{z}$, as follows:

$$\frac{1}{3 + 4i} = \frac{1 \cdot (3 - 4i)}{(3 + 4i)(3 - 4i)} = \frac{3 - 4i}{9 + 16} = \frac{3}{25} - \frac{4}{25}i.$$

b) Multiplying the numerator and denominator of the given complex fraction by $2 + 4i$, which is the complex conjugate of $2 - 4i$, we obtain

$$\frac{3 + 5i}{2 - 4i} = \frac{(3 + 5i)(2 + 4i)}{(2 - 4i)(2 + 4i)} = \frac{-14 + 22i}{4 + 16}$$

$$= -\frac{14}{20} + \frac{22}{20}i = -\frac{7}{10} + \frac{11}{10}i.$$

◇ **Practice Exercise 4.**   Write each of the following numbers in the form $a + bi$:

   a) $\dfrac{1}{2 + 3i}$ **b)** $\dfrac{3 - 2i}{3 + 4i}$.

Answer.   **a)** $\dfrac{2}{13} - \dfrac{3}{12}i$ **b)** $\dfrac{1}{25} - \dfrac{18}{25}i$.

⬢ EXERCISES 1.9

In each of Exercises 1–36, write the given complex numbers in the form $a + bi$.

1. $(3 + 2i) + (-4 + 3i)$ **2.** $(-5 + 4i) + (-3 - 2i)$ **3.** $(5 - 6i) + (3 - 5i)$

4. $(12 + 5i) - (6 - 8i)$ **5.** $-(3 + 4i) + (5 - 9i)$ **6.** $(8 - 5i) - (-3 - 4i)$

7. $10 - (8 + 5i)$

8. $-9 + (-5 + 12i)$

9. $(7 - 6i) - 3i$

10. $2i - (4 - 6i)$

11. $(4 - 3i)(3 + 4i)$

12. $(5 - 5i)(3 - 8i)$

13. $(-8 - i)(4 + i)$

14. $(7 + i)(8 + 5i)$

15. $\left(\dfrac{1}{2} + \dfrac{3}{2}i\right)(3 - 4i)$

16. $\left(\dfrac{3}{4} - 5i\right)\left(1 - \dfrac{1}{4}i\right)$

17. $(3 + 5i)(3 - 5i)$

18. $(-2 + 7i)(-2 - 7i)$

19. $\left(-\dfrac{2}{3} + \dfrac{3}{4}i\right)\left(-\dfrac{2}{3} - \dfrac{3}{4}i\right)$

20. $\left(\dfrac{1}{5} + \dfrac{3}{8}i\right)\left(\dfrac{1}{5} - \dfrac{3}{8}i\right)$

21. $(2 + 5i)(2 + 5i)$

22. $(3 - 4i)(3 - 4i)$

23. $i(3 - 2i)(3 + 5i)$

24. $(1 - i)(3 - 4i)(5i)$

25. $3i(1 - 2i)(3 - 4i)$

26. $2i\left(\dfrac{1}{2} + \dfrac{3}{4}i\right)\left(\dfrac{1}{2} - \dfrac{3}{4}i\right)$

27. $\dfrac{1}{1 + i}$

28. $\dfrac{1}{5 - 3i}$

29. $\dfrac{3 - 3i}{4 + 4i}$

30. $\dfrac{1 + i}{3 + i}$

31. $\dfrac{2 + 5i}{2 - 5i}$

32. $\dfrac{5 - 3i}{5 + 3i}$

33. $\dfrac{3 - i}{2 + 4i}$

34. $\dfrac{4 - 5i}{3 - 2i}$

35. $\dfrac{2 - 3i}{3 + 6i}$

36. $\dfrac{3 + 7i}{4 - 3i}$

In Exercises 37–40, let $z = a + bi$ and $w = c + di$. Prove each of the following relations.

● **37.** $\overline{z + w} = \bar{z} + \bar{w}$

● **38.** $\overline{zw} = \bar{z}\bar{w}$

● **39.** $\overline{z^{-1}} = (\bar{z})^{-1}$

● **40.** $\overline{(-z)} = -\bar{z}$.

CHAPTER SUMMARY

Real numbers form the basis of a college algebra and trigonometry course. They consist of the *natural numbers*, the *integers*, the *rational* and *irrational numbers*. Rational numbers, also called *fractions*, are ratios of integers.

Real numbers can be represented by *decimals*. If the decimal is *terminating* or *repeating*, the number is a rational one. *Nonterminating* and *nonrepeating* decimals correspond to irrational numbers.

The set of all real numbers is closed under the addition and multiplication operations. These operations satisfy the *commutative*, *associative*, and *distributive properties*. Moreover, 0 is the *additive identity* for the sum, and 1 is the *multiplicative identity* for the product.

Real numbers can be represented by points on a *coordinate line* and points on a coordinate line are labeled by real numbers. *Positive numbers* correspond to points located to the right of the origin, while *negative numbers* are located to the left of the origin.

The multiplication operation leads to the definition of *integer powers* of a real number. After defining *radicals*, we can define *rational powers* of a real number.

The powers

$$x^0 = 1, \quad x^1 = x, \quad x^2, \ldots, \quad x^n, \ldots,$$

where x is a real number and the exponents are *nonnegative integers*, are the basic elements necessary to define *polynomials*. A *rational expression* is the quotient of two polynomials. *Factoring* is the process of determining the factors of a polynomial. To factor polynomials, it is important to know certain *special products* that occur very often in applications.

Most of the time, we shall deal with real numbers. Nonetheless the concept of *complex numbers* is needed when solving quadratic equations and, more generally, discussing properties of zeros of polynomials. Complex numbers are introduced by defining the *imaginary unit i*, which satisfies the relation $i^2 = -1$. The operations of addition and multiplication are extended to complex numbers, and the set of all complex numbers is closed under these operations. Moreover, they satisfy the commutative, associative, and distributive properties. The numbers 0 and 1 are, respectively, the additive and multiplicative identities. Every complex number has a *complex conjugate*, and the product of a complex number and its own complex conjugate is a real number.

REVIEW EXERCISES

In Exercises 1–6, write each fraction as a decimal number.

1. $\dfrac{7}{16}$ 　　　　　　　　 **2.** $\dfrac{32}{80}$ 　　　　　　　　 **3.** $\dfrac{9}{11}$

4. $\dfrac{8}{15}$ 　　　　　　　　 **5.** $\dfrac{15}{64}$ 　　　　　　　　 **6.** $\dfrac{14}{125}$

In Exercises 7–12, write each decimal number as a fraction in lowest terms.

7. 5.18 　　　　　　　　　　 **8.** 3.45 　　　　　　　　　　 **9.** $0.\overline{141}$

10. $0.\overline{45}$ 　　　　　　　　 **11.** $1.3\overline{18}$ 　　　　　　　　 **12.** $2.21\overline{6}$

In Exercises 13–16, write each sum as a repeating decimal.

13. $0.\overline{6} + 0.\overline{7}$ 　　　　 **14.** $0.\overline{3} + 0.\overline{8}$ 　　　　 **15.** $0.\overline{5} + 0.\overline{13}$ 　　　　 **16.** $0.\overline{4} + 0.\overline{18}$

In Exercises 17–22, compute each expression and give your answer in simplified form.

17. $\dfrac{\dfrac{1}{2} + \dfrac{3}{4}}{\dfrac{5}{6} - \dfrac{2}{3}}$ 　　　　　　 **18.** $\dfrac{\dfrac{4}{5} - \dfrac{2}{3}}{\dfrac{1}{2} + \dfrac{3}{7}}$ 　　　　　　 **19.** $\dfrac{\dfrac{2}{5}\left(\dfrac{3}{8} - \dfrac{1}{2}\right)}{\dfrac{4}{10} - \dfrac{3}{5}}$

20. $\dfrac{\left(\dfrac{3}{7} + \dfrac{2}{3}\right)\left(\dfrac{5}{12} - \dfrac{3}{4}\right)}{\dfrac{5}{6} - \dfrac{1}{8}}$ 　　　　 **21.** $\dfrac{\dfrac{5}{8} \div \left(\dfrac{7}{12} + 1\right)}{\dfrac{3}{14} - \dfrac{3}{7}}$ 　　　　 **22.** $\dfrac{\dfrac{3}{5} + \dfrac{5}{6}}{\dfrac{3}{4} \times \left(\dfrac{4}{5} - \dfrac{3}{8}\right)}$

23. The difference of two rational numbers is a rational number. Is the difference of two irrational numbers necessarily an irrational number?

24. Is the quotient of two irrational numbers always an irrational number?

25. Which of the following numbers are rational? Which are irrational numbers?

　　a) $(3 + 5\sqrt{2})(3 - 5\sqrt{2})$ 　　　　 b) $\sqrt{3} - 5\sqrt{3}$

　　c) $0.15015001500015000015\ldots$ 　　 d) $\sqrt{8}/\sqrt{2}$

26. Which of the following numbers are rational? Which are irrational?

a) $0.\overline{45}$ b) $(2 - 3\sqrt{2})^2$

c) $\dfrac{5 + 2\sqrt{3}}{5 - 2\sqrt{3}}$ d) $\sqrt{54}\sqrt{24}$

Simplify and write the answers with positive exponents only.

27. $(3a^{-5}x^2)(4ay^{-3})$

28. $(6x^2y^5)(4xy^{-3})^{-2}$

29. $\dfrac{6a^2b^{-5}}{4c^{-4}}$

30. $\dfrac{5au}{a^{-2}b^{-1}u^{-4}}$

31. $\left(\dfrac{3my^{-4}}{2y}\right)$

32. $\left(\dfrac{3a^2u^5}{b^{-1}}\right)^2$

33. $\left(\dfrac{m^3p^2q^5}{2mp}\right)^3$

34. $\left(\dfrac{5a^3t^6}{2s^2}\right)^2$

35. $\left(\dfrac{a^2x^{-3}}{b}\right)^{-2}$

36. $\left(\dfrac{3a^4x^2}{y^{-3}}\right)^{-3}$

37. $(25a^6x^4)^{-1/2}$

38. $(64b^3x^6y^{-9})^{-2/3}$

39. $(x^{-1/4}y^{-2/4})^{-2}$

40. $(a^{-3}m^2n^6)^{-1/6}$

41. $\left(\dfrac{8a^3x^{-1}}{27a^{-4}x^3}\right)^{1/2}$

42. $\left(\dfrac{8}{27}x^5y^{-2/3}\right)^{2/3}$

43. $\dfrac{(b^{-8}y)^{-1/4}}{(b^3y^6)^{-1/6}}$

44. $\left(\dfrac{x^4}{b^{1/2}}\right)\left(\dfrac{b^{-1/2}}{x^{3/2}}\right)^4$

c *Perform the indicated operations and give the answers in scientific notation.*

45. $\dfrac{(1.313 \times 10^{-6})(2.11 \times 10^9)}{(3.12 \times 10^3)^2(1.4 \times 10^{-8})}$

46. $\dfrac{(2.123 \times 10^{-6})(2.15 \times 10^{-5})}{(4.17 \times 10^6)^2(5.123 \times 10^8)}$

47. $\dfrac{(2.134 \times 10^{-8})(3.15 \times 10^{-3})}{(5.07 \times 10^4)(2.312 \times 10^6)}$

48. $\dfrac{(5.11 \times 10^{16})(6.12 \times 10^{-4})}{(3.25 \times 10^{-6})^3}$

c **49.** Let $V = xyz$. Find V if $x = 1.516 \times 10^3$, $y = 1.217 \times 10^4$, and $z = 1.605 \times 10^5$.

c **50.** The speed of light is approximately 186,000 miles per second. How many miles does a light signal travel in 10^{-7} seconds?

c **51.** The speed of light is approximately 300,000 kilometers per second and there are approximately 31,560,000 seconds in a mean solar year. How many kilometers does a light-year represent?

c **52.** Light emanating from the sun takes 8 minutes to reach the earth. Find the distance from the earth to the sun: a) in miles; b) in kilometers.

c **53.** One kilogram is approximately 2.203 lb. Express the weight of the earth in pounds. (Earth's mass = 5.983×10^{24} kg.)

● **54.** If a and b are positive real numbers and p and q are natural numbers, prove that

$$a^{1/p} \cdot a^{1/q} = a^{(p+q)/pq}.$$

Use special products to find each of the following products.

55. $(u - 5)(u + 2)$

56. $(v - 1)(v + 10)$

57. $(x^2 - 2)(x^2 + 3)$

58. $(a^2 + 3)(a^2 + 5)$

59. $(3a^2 + b^2)(3a^2 - b^2)$

60. $(6y^2 - z^2)(6y^2 + z^2)$

61. $(x + \sqrt{a})(x - \sqrt{a})$

62. $(2x + \sqrt{b})(2x - \sqrt{b})$

63. $(\sqrt{x} + 2)(\sqrt{x} - 2)$

64. $(\sqrt{3a} + b)(\sqrt{3a} - b)$

Factor as completely as possible each of the following.

65. $v^2 - 10v + 25$

66. $a^2 - 8a + 16$

67. $3x^2 + 10x - 8$

68. $3x^2 - 2x - 8$

69. $2ax + 5bx - 8ay - 20by$

70. $2a^3 + 10a^2 + 3a + 15$

71. $2bu - 3bv + 2au - 3av$

72. $15ax + 3bx - 20ay - 4by$

73. $8x^3 - 18x(a^2 + 2a + 1)$

74. $9(4x^2 - 4xy + y^2) - 16a^2$

Perform the indicated operations and simplify.

75. $\dfrac{5}{x-6} + \dfrac{2}{x-3}$

76. $\dfrac{8}{x-5} - \dfrac{4}{x+2}$

77. $\dfrac{5x}{x^2-4} + \dfrac{1}{x-2}$

78. $\dfrac{5}{x-3} - \dfrac{x}{x^2-6x+9}$

79. $\dfrac{4}{x^2-2x+1} - \dfrac{3}{x^2-1}$

80. $\dfrac{1}{x^2-25} + \dfrac{1}{x^2+10x+25}$

81. $\dfrac{xy}{x^3-y^3} - \dfrac{y}{x^2+xy+y^2}$

82. $\dfrac{1}{x+a} - \dfrac{a}{x^2-ax+a^2}$

83. $\dfrac{1}{3x-3} + \dfrac{1}{2x+2} - \dfrac{1}{1-x^2}$

84. $\dfrac{x}{x+y} + \dfrac{y}{x-y} - \dfrac{2xy}{x^2-y^2}$

85. $\dfrac{x - \dfrac{x^2}{x-5}}{1 + \dfrac{25}{x^2-25}}$

86. $\dfrac{3 - \dfrac{9}{3-x}}{1 + \dfrac{x^2}{9-x^2}}$

87. $\dfrac{x + \dfrac{1}{1-(1/x)}}{x - \dfrac{1}{1+(1/x)}}$

88. $\dfrac{x + \dfrac{-1}{x-(1/x)}}{x - \dfrac{1}{x+(1/x)}}$

Simplify each of the following radicals. Assume that all radicals are defined and denominators are different from zero.

89. $\sqrt{48x^3y^6}$

90. $\sqrt{81ab^5c^7}$

91. $\sqrt[4]{625a^5b^7}$

92. $\sqrt[3]{343a^6x^5y^4}$

93. $\sqrt{3ab^3}\sqrt{2a^2b^5}$

94. $\sqrt[3]{4r^4s^2}\sqrt[3]{6r^3s^5}$

95. $\sqrt[5]{\dfrac{32ab^3}{c^6}}$

96. $\sqrt{\dfrac{18a^3x^4}{75b^5}}$

97. $\sqrt{\sqrt[3]{64a^6b^8c^9}}$

98. $\sqrt{\sqrt{25m^2x^4y^6}}$

99. $(\sqrt{3ax^2y})^4$

100. $(\sqrt[3]{2a^2bx^4})^2$

In Exercises 101–112, rationalize the denominators.

101. $\dfrac{4+\sqrt{5}}{\sqrt{3}}$

102. $\dfrac{\sqrt{5}-2}{\sqrt{7}}$

103. $\dfrac{\sqrt{a}+\sqrt{b}}{\sqrt{a}}$

104. $\dfrac{\sqrt{m}-\sqrt{n}}{\sqrt{n}}$

105. $\dfrac{\sqrt{2}+\sqrt{5}}{\sqrt{3}-\sqrt{2}}$

106. $\dfrac{\sqrt{6}-\sqrt{8}}{\sqrt{5}-1}$

107. $\dfrac{2\sqrt{x}+\sqrt{y}}{\sqrt{x}-\sqrt{y}}$

108. $\dfrac{\sqrt{x}+3\sqrt{y}}{3\sqrt{x}+\sqrt{y}}$

109. $\dfrac{\sqrt{x+1}+\sqrt{x}}{\sqrt{x-1}-\sqrt{x}}$

110. $\dfrac{\sqrt{a+b}-\sqrt{b}}{\sqrt{a+b}+\sqrt{b}}$

111. $\dfrac{\sqrt{x^2+1}-\sqrt{x^2-1}}{\sqrt{x^2+1}-\sqrt{x^2-1}}$

112. $\dfrac{\sqrt{1+a^2}+\sqrt{1-a^2}}{\sqrt{1+a^2}-\sqrt{1-a^2}}$

Write each of the following complex numbers in the form a + bi.

113. $-(-3 + 4i) + (5 - 9i)$

114. $(8 - 5i) - (-3 - 4i)$

115. $(-8 - i)(4 + i)$

116. $(7 + i)(8 + 5i)$

117. $(-3 - 2i)(-3 + 2i)$

118. $(10 - 4i)(10 + 4i)$

119. $i(5 - 3i)(5 + 3i)$

120. $(5i)(3 - 6i)(-5i)$

121. $3i(-2 - 6i)(2 + 6i)$

122. $(4i)(-3 - 8i)(3 + 8i)$

123. $\dfrac{1}{4 - 5i}$

124. $\dfrac{1}{-3 - 2i}$

125. $\dfrac{i}{5 + 8i}$

126. $\dfrac{3 + 6i}{i}$

In Exercises 127–130, z = a + bi and w = c + di. Prove each of the following relations.

● **127.** $\bar{\bar{z}} = z$

● **128.** $\overline{z - w} = \bar{z} - \bar{w}$

● **129.** $\overline{z/w} = \bar{z}/\bar{w}$

● **130.** $a = \dfrac{z + \bar{z}}{2}$, $b = \dfrac{z - \bar{z}}{2i}$

2

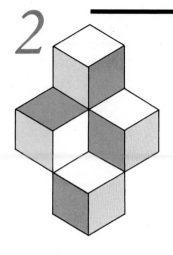

Linear and Quadratic Equations and Inequalities

The basic idea in solving equations and inequalities is to carry out a sequence of admissible operations that transform them into simpler, but equivalent, equations and inequalities that can be easily solved. In this chapter, we explain how to solve linear and quadratic equations and inequalities, and discuss several applications. One of the most important mathematical skills is the ability to translate problems that occur in real life or work situations into mathematical equations. We devote a good part of this chapter to the study of word problems that lead to linear or quadratic equations or inequalities.

2.1 SOLVING LINEAR EQUATIONS

Expressions such as

$$x + 5 = 7, \quad x^2 + 2x - 3 = 0, \quad \text{or} \quad \frac{1}{x+1} = \frac{3}{x-2}$$

are called *equations* in x. The letter x is called a *variable*. Given an equation, every number that, when substituted for x, makes the equation a true statement is called a *solution* or a *root* of the equation. For example, 2 is the only number that can be substituted for x in the equation $x + 5 = 7$ to make it a true statement. Thus 2 is the unique solution of this equation. On the other hand, since

$$1^2 + 2 \cdot 1 - 3 = 0 \quad \text{and} \quad (-3)^2 + 2(-3) - 3 = 0,$$

the numbers 1 and -3 are both solutions of the equation $x^2 + 2x - 3 = 0$. If a number is a solution of an equation, we say that the number *satisfies* the equation.

Two equations are said to be *equivalent* if they have exactly the same solutions. For example, $x + 3 = 4$ and $2x + 1 = 3$ are equivalent, because each has the number 1 as its unique solution.

If an equation is satisfied for all possible values assigned to the variable x, then the equation is called an *identity*. For example, $x^2 - 4 = (x - 2)(x + 2)$ is an identity, since every number is a solution of this equation.

Equations that are satisfied by some values of x but not by all values of x are sometimes called *conditional equations*. For example, $x + 5 = 7$ and $x^2 + 2x - 3 = 0$ are conditional equations. Throughout this book the term equation will mean, most of the time, conditional equation.

Linear Equations

A *linear equation* in one variable is an expression of the form

$$(2.1) \quad ax + b = 0,$$

where a and b are real numbers with $a \neq 0$. This equation has a unique solution

$$(2.2) \quad x = -\frac{b}{a}.$$

To find the solution, apply the properties of the real numbers listed in Section 1.2 to replace equation (2.1) by a chain of equivalent equations, as follows:

$$ax + b = 0$$
$$(ax + b) + (-b) = 0 + (-b) \qquad \text{[Add } -b \text{ to both sides]}$$
$$ax + (b + (-b)) = -b \qquad \text{[Addition is associative and 0 is the identity element for addition]}$$
$$ax = -b \qquad \text{[}-b \text{ is the additive inverse of } b\text{]}$$
$$\frac{1}{a}(ax) = \frac{1}{a}(-b) \qquad \text{[Multiply both sides by } 1/a\text{]}$$
$$\left(\frac{1}{a}a\right)x = -\frac{b}{a} \qquad \text{[Associativity of multiplication]}$$
$$x = -\frac{b}{a} \qquad \text{[}1/a \text{ is the multiplicative inverse of } a\text{]}$$

You may now verify that the number $-b/a$ is the solution of (2.1). Substituting $-b/a$ by x, obtain

$$a\left(-\frac{b}{a}\right) + b = (-b) + b = 0.$$

◆ **Example 1.** Solve the equation $(3/5)x + 2 = 0$.

Solution.   We have

$$\frac{3}{5}x + 2 = 0$$

$$\frac{3}{5}x = -2$$

$$x = \left(\frac{5}{3}\right)(-2)$$

$$x = -\frac{10}{3}.$$

◇ **Practice Exercise 1.**   Solve the equation $\frac{5}{8}x - 15 = 0$.

Answer. $x = 24$.

In some cases it is necessary to use algebraic operations to reduce the given equation to the form (2.1) before solving it.

◆ **Example 2.**   Solve the equation $\frac{(x-2)}{2} = \frac{x}{5} - \frac{1}{4}$.

Solution.   Multiplying both sides by 20 (the LCM of 2, 4, and 5) to eliminate the denominators, obtain

$$\frac{20(x-2)}{2} = 20\left(\frac{x}{5}\right) - 20\left(\frac{1}{4}\right),$$

or

$$10(x-2) = 4x - 5,$$

or

$$10x - 20 - 4x + 5 = 0.$$

Hence

$$6x - 15 = 0,$$

which is a linear equation of the form (2.1). The solution is $x = 15/6 = 5/2$. To check it, substitute 5/2 for x in the given equation and obtain

$$\frac{(5/2) - 2}{2} = \frac{5/2}{5} - \frac{1}{4}$$

$$\frac{1/2}{2} = \frac{1}{2} - \frac{1}{4}$$

$$\frac{1}{4} = \frac{1}{4},$$

a true statement. Therefore, $x = 5/2$ is the solution of the given equation.

◇ **Practice Exercise 2.**   Find the solution of the equation $\dfrac{x}{3} = \dfrac{x-1}{2} + \dfrac{4}{9}$.

Answer. $x = \dfrac{1}{3}$.

◆ **Example 3.**   Solve the equation $(x+1)(x+2) = (x-1)(x+3) + 4$.

Solution.   Computing the products and simplifying, we obtain

$$x^2 - x - 2 = x^2 + 2x - 3 + 4$$
$$-x - 2 = 2x + 1$$
$$-3x = 3$$
$$x = -1.$$

You can check this result by substituting -1 for x in the original equation.

◇ **Practice Exercise 3.**   Solve $(2x+1)(x-3) = (x-4)(2x-3) + 3$.

Answer. $x = 3$.

Some equations involving the variable in one or more denominators can be reduced to linear equations. Values of the variable for which denominators are equal to zero must be excluded. For such values the equation has no meaning.

◆ **Example 4.**   Solve $\dfrac{1}{x+1} = \dfrac{3}{x-2}$.

Solution.   We assume that $x \neq -1$ and $x \neq 2$, so that both denominators $x+1$ and $x-2$ are different from zero. In order to eliminate the denominators, we multiply both sides of the given equation by the product $(x+1)(x-2)$ and obtain

$$\frac{(x+1)(x-2)}{x+1} = \frac{3(x+1)(x-2)}{x-2}$$
$$x - 2 = 3(x+1)$$
$$x - 2 = 3x + 3$$
$$-2x = 5$$
$$x = -\frac{5}{2}.$$

We leave it to you to check that $-5/2$ is the solution of the given equation.

◇ **Practice Exercise 4.**   Solve $\dfrac{2}{x-3} = \dfrac{4}{x-1}$.

Answer. $x = 5$.

The next example shows why we must avoid values of the variable that make a denominator equal to zero.

◆ **Example 5.**   Solve $\dfrac{4x}{x-1} = 1 + \dfrac{4}{x-1}$.

Solution.   Multiplying both sides by $x - 1$ and simplifying, we obtain

$$\left[\frac{4x}{x-1}\right](x-1) = (x-1) + \left[\frac{4}{x-1}\right](x-1)$$
$$4x = x + 3$$
$$3x = 3$$
$$x = 1.$$

But 1 cannot be a solution because the denominator $x - 1$ equals 0 when $x = 1$. Therefore, the given equation has no solution.

◇ **Practice Exercise 5.**   Find the solution of $\dfrac{2-x}{1-2x} = 1 + \dfrac{3x}{1-2x}$.

Answer.   No solution. (Can you explain why?)

Literal Equations

In the applications, we often encounter equations involving several variables, and we are asked to solve a given equation for one of the variables in terms of the remaining variables. Such an equation, whose solution is an expression involving variables with unspecified values, is called a *literal equation*.

◆ **Example 6.**   In optics, the simple lens equation is $1/f = 1/d_1 + 1/d_2$, where f is the focal length, d_1 is the object distance, and d_2 is the image distance. Solve the equation for f.

Solution.   Multiplying both sides of the equation by $fd_1 d_2$, the product of all denominators, we obtain

$$d_1 d_2 = fd_2 + fd_1$$
$$d_1 d_2 = f(d_2 + d_1)$$

or

$$f(d_1 + d_2) = d_1 d_2.$$

Hence

$$f = \frac{d_1 d_2}{d_1 + d_2}.$$

◇ **Practice Exercise 6.** Solve the equation $\dfrac{1}{t} - \dfrac{1}{s} = \dfrac{1}{a}$ for s.

Answer. $s = \dfrac{at}{a - t}$.

◆ **Example 7.** The thermal expansion formula

$$l = l_0(1 + at)$$

gives the length l of a rod at a temperature of t degrees, where l_0 is the length of the rod at $0°$ and a is the linear coefficient of thermal expansion. Solve it for t.

Solution. The letter t is the unknown variable, while the other letters represent known constants. By carrying out the multiplication on the right-hand side, we have

$$l = l_0 + l_0 at.$$

Adding $-l_0$ to both sides of the last equation in order to isolate the term containing t gives us

$$l - l_0 = l_0 at,$$

or

$$l_0 at = l - l_0.$$

Multiplying both sides by $1/l_0 a$, we obtain the solution $t = (l - l_0)/l_0 a$.

◇ **Practice Exercise 7.** The area A of a trapezoid of bases b_1 and b_2 and height h is given by

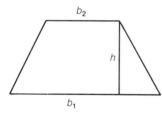

$$A = \frac{1}{2} h(b_1 + b_2).$$

Solve this formula for b_1.

Answer. $b_1 = \dfrac{2A}{h} - b_2$.

⬡ EXERCISES 2.1

Solve each of the following equations:

1. $\dfrac{3}{2} x - 5 = 0$ **2.** $\dfrac{4}{5} x + 6 = 0$ **3.** $\dfrac{2}{3} x + \dfrac{3}{4} = 0$ **4.** $\dfrac{3}{4} x - \dfrac{5}{6} = 0$

5. $\sqrt{2} x - 3 = 0$ **6.** $2x - \sqrt{3} = 0$ **7.** $5(x + 2) = 2(x - 1)$ **8.** $4(x - 3) = 5x + 4$

9. $3(x - 4) + 6 = 2(x + 1) - 7$ **10.** $3x + 5 = 2(x - 4) + 7$ **11.** $0.1(t - 4) + 0.2t = 0.5$

12. $0.2m - 0.3(m - 1) = m$ **13.** $\sqrt{3} x - 4 = 5 - \sqrt{3} x$ **14.** $\sqrt{5} x + 2 = x - 4$

15. $\sqrt{3}y - 2 = \sqrt{2}y + 6$

16. $\sqrt{2}y - 6 = \sqrt{5} - 3$

17. $\dfrac{1}{x+1} = \dfrac{2}{x+3}$

18. $\dfrac{2}{x-3} = \dfrac{5}{x+4}$

19. $\dfrac{2}{x-3} = \dfrac{3}{x-4}$

20. $\dfrac{3}{x-2} = \dfrac{5}{x+2}$

21. $\dfrac{1}{t} - \dfrac{1}{4} = \dfrac{2}{3} - \dfrac{1}{2t}$

22. $\dfrac{2}{3t} = \dfrac{3}{2t} - \dfrac{1}{5}$

23. $\dfrac{x}{5} - \dfrac{x+1}{4} = \dfrac{3x+1}{2} - 3$

24. $\dfrac{2x-1}{6} - \dfrac{1}{3} = \dfrac{x+2}{4} - \dfrac{x}{2}$

25. $\dfrac{4x}{x-3} + \dfrac{12}{3-x} = \dfrac{1}{5}$

26. $\dfrac{4x}{2x+1} - \dfrac{3}{8} = -\dfrac{2}{2x+1}$

27. $(x-1)(x+3) = x(x-2) + 4$

28. $(x-2)(x-3) = 3 + (x-1)(x+2)$

29. $(3x-1)(x-2) = 3x(x-4) + 5$

30. $x(x+2) - 4 = x(x-5)$

c 31. $2.105x - 3.14 = 1.276x + 0.317$

c 32. $1.41x + 1.73 = 0.005x + 2.413$

c 33. $\dfrac{2.516}{x-3.018} = \dfrac{5.101}{x+4.021}$

c 34. $\dfrac{1}{0.05t} - 0.126 = 1.306 - \dfrac{1}{1.08t}$

c 35. $(5.3 \times 10^8)x - 3.15 \times 10^{-5} = (1.21 \times 10^7)x + 1.07 \times 10^{-6}$

c 36. $(1.316 \times 10^{-7})x + 1.105 \times 10^{-8} = 3.06 \times 10^{-3} - (6.72 \times 10^{-5})x$

In Exercises 37–50, formulas that occur in mathematics and the applied sciences are given. Solve for the indicated variable in terms of the others.

37. The relationship between a temperature measured in degrees Fahrenheit (F) and in degrees Celsius (C) is $F = (9/5)C + 32$. Solve for C.

38. Solve the equation $1/f = 1/d_1 + 1/d_2$ for d_1.

39. Boyle's law for gases states that $PV/T = C$, where P is the pressure, V is the volume, T is the temperature, and C is a constant. a) Solve for T. b) Solve for V.

40. The formula $A = (1/2)bh$ gives the area of a triangle of base b and height h. a) Solve for b. b) Find the length of the base if the area is equal to 150 cm² and the height is 35 cm.

41. The perimeter P of a rectangle with sides x and y is given by the formula $P = 2x + 2y$. i) Solve it for y. ii) If $x = 50$ cm and $P = 300$ cm, find y.

42. The surface area of a parallelepiped with sides a, b, and c is given by $S = 2ab + 2ac + 2bc$. Solve for a.

43. Solve the equation $p = p_0(1 + rt)$ for p_0 and for r.

44. In an electric circuit containing two capacitors C_1 and C_2 in series, the total capacitance C_T is given by

$$\frac{1}{C_T} = \frac{1}{C_1} + \frac{1}{C_2}.$$

Solve for C_1.

45. The lateral surface area of a right circular cylinder (think of a soda can *without* top and bottom) is equal to $S = 2\pi rh$, where r is the radius and h is the height of the cylinder. Solve this equation for r.

46. The total surface area of a right circular cylinder (think of a soda can *with* top and bottom) of radius r and height h is $S = 2\pi r^2 + 2\pi rh$. Solve the equation for h.

47. Let $F = G(m_1 m_2)/r^2$. Solve for m_1.

48. One of Newton's laws says that the force exerted on an object is the product of the object's mass by its acceleration: $F = ma$. Solve for a.

49. Ohm's law states that in an electrical circuit $V = RI$, where V is the potential (in volts), I the current (in amperes), and R the resistance (in ohms). Solve the equation for R.

50. The intelligence quotient (IQ) of an individual is obtained by dividing his/her mental age (MA) by his/her chronological age (CA) and multiplying the result by 100. Thus, IQ is given by the formula

$$IQ = (MA/CA)100.$$

Solve this equation for MA.

2.2   WORD PROBLEMS INVOLVING LINEAR EQUATIONS

In mathematics and the applied sciences we encounter a large variety of problems leading to linear equations. In such problems, known quantities are related to unknown quantities. By correctly expressing the relationship among them, we obtain an equation translating the problem into mathematical terms. The solution of the equation will give us the solution of the problem.

There are no fixed rules that lead us from a specific problem to a mathematical equation associated with the problem. Each problem has its own characteristics, and different problems may give rise to different equations. However, on many occasions, we are faced with similar problems, and knowing how to solve one of them might help us solve others.

Here are some suggestions for solving these problems:

1. Read each problem carefully several times, looking for all relevant information.
2. Whenever possible, draw a diagram to reorganize the information.
3. Identify known and unknown quantities.
4. Assign letters to unknown quantities.
5. Write down an equation relating known and unknown quantities according to the available information.
6. Solve the equation.
7. Check the solution in the original problem.

◆ **Example 1.**   Find three consecutive integers whose sum is 243.

Solution.   If x denotes the first number, then $x + 1$ denotes the second, and $x + 2$ denotes the third. The mathematical equation corresponding to this problem is then

$$x + (x + 1) + (x + 2) = 243.$$

Solving for x, we have

$$3x + 3 = 243,$$
$$3x = 240,$$
$$x = 80.$$

Thus, the three consecutive integers are: 80, 81, 82. We may now check our answer: $80 + 81 + 82 = 243$.

◇ **Practice Exercise 1.**   The sum of three consecutive integers is -45. Find the integers.

Answer. $-14, -15, -16$.

◆ **Example 2.**   A girl has 43 nickels and quarters in her piggy bank, corresponding to a total of $5.75. How many nickels and how many quarters does she have?

Solution.   Let x denote the number of nickels. Since the number of nickels and quarters is 43, it follows that the number of quarters is $43 - x$. Now, we could organize our work according to the following table:

	Number	*Value per coin*	*Total value*
Nickels	x	5¢	$5x$¢
Quarters	$43 - x$	25¢	$25(43 - x)$¢

Since the piggy bank total is the total value of nickels plus the total value of quarters, we obtain the equation

$$5x + 25(43 - x) = 575.$$

Solving for x, we get

$$5x + 1075 - 25x = 575$$
$$20x = 500$$
$$x = 25.$$

The girl has then 25 nickels and $43 - 25 = 18$ quarters.
Check: $25 \times 5 + 18 \times 25 = 125 + 450 = 575$¢ $= \$5.75$.

◇ **Practice Exercise 2.**   Sharon spent $5.90 buying pencils and notebooks. Each pencil cost 25¢ and each notebook cost $1.10. She bought 2 more pencils than notebooks. How many pencils and notebooks did she buy?

Answer.   6 pencils and 4 notebooks.

Percent Problems

The word *percent* is used to indicate a fraction whose denominator is 100. For example, $25/100 = 0.25$ is written as 25% (and read as 25 percent). To find r% of a number, multiply the number by $r/100$. For example, 25% of 120 is equal to

$$\frac{25}{100} \times 120 = 0.25 \times 120 = 30.$$

◆ **Example 3.**   If 30% of a number is 210, find the number.

Solution.   Let x be the number. If 30% of x is 210, we obtain the equation

$$0.3x = 210$$

which solved for x gives us

$$x = 700.$$

Check: 30% of $700 = 0.3 \times 700 = 210$.

◇ **Practice Exercise 3.**   If x% of 350 is 84, find x.

Answer. $x = 24$.

◆ **Example 4.**   Among 3200 college freshmen, 2080 are taking a calculus course. What percentage of the freshmen are taking calculus?

Solution.   The fraction of college freshmen taking calculus is

$$\frac{2080}{3200} = \frac{13}{20}.$$

Converted to a decimal, this is

$$\frac{13}{20} = 0.65.$$

Therefore, 65% of the freshmen are taking calculus.

◇ **Practice Exercise 4.**   In a small town, 210 people live in brick structures and 290 live in wooden structures. What percentage of the whole population live in brick structures?

Answer.   42%.

If a sum of money p_0 (the *initial principal*) is invested at a *simple interest rate* of $100r$ percent per year, the *amount of interest* at the end of t years is given by

$$(2.3) \quad i = p_0rt.$$

The *amount of principal* at the end of t years is equal to the initial principal p_0 plus the amount of interest:

$$p = p_0 + p_0rt$$

or

$$(2.4) \quad p = p_0(1 + rt).$$

◆ **Example 5.**   If you invest \$1,500 at 6% annual simple interest, find **a)** the amount of interest at the end of one year, **b)** the amount of principal at the end of two years.

Solution.   The interest rate is 6% per year, so the variable r appearing in formulas (2.3) and (2.4) is $r = 6/100 = 0.06$.

a) According to (2.3), the amount of interest at the end of one year ($t = 1$) is

$$1{,}500 \times 0.06 \times 1 = \$90.$$

b) At the end of two years the amount of principal is

$$p = 1{,}500(1 + 0.06 \times 2)$$
$$= 1{,}500(1.12)$$
$$= \$1{,}680.$$

◇ **Practice Exercise 5.**   How much money invested at a simple interest rate of 8% per year earns $192 in a year? What is the principal at the end of four years?

Answer. $2,400 and $3,168.

Mixture Problems

◆ **Example 6.**   A flask contains 120 milliliters (ml) of a saline solution whose salt concentration is 8%. How many ml of water should be added to decrease the salt concentration to 6%?

Solution.   Let x be the number of ml of water to be added to the 120 ml of saline solution. The volume of the new solution will be $(120 + x)$ ml.

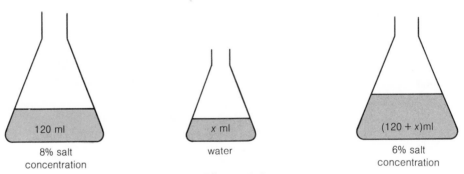

Figure 2.1

The volumes of the original and the new solutions are different, but the amount of salt remains the same. In the original solution the amount of salt is 8% of 120 ml, that is

$$(0.08) \times 120 = 9.6 \text{ ml},$$

while in the new solution the amount of salt is 6% of $(120 + x)$ ml, that is,

$$(0.06) \times (120 + x) \text{ ml}.$$

Since these two quantities are the same, we obtain the equation

$$(0.06)(120 + x) = (0.08) \times 120.$$

Solving for x, we have

$$7.2 + 0.06x = 9.6$$
$$0.06x = 2.4$$
$$6x = 240$$
$$x = 40 \text{ ml.}$$

Therefore, 40 ml of water should be added to decrease the salt concentration to 6%.

◇ **Practice Exercise 6.**   How many kilograms of water must be evaporated from 80 kg of a 6% saline solution to obtain a 15% solution?

Answer.   48 kg.

The mixing of stocks and bonds is often used by investors to improve the performance of their financial holdings.

◆ **Example 7.**   An investor owns $5,000 of stock A yielding 6% a year. He wants to buy shares of stock B yielding 8% per year. How much should he invest so that his total investment earns $6\frac{3}{4}$% per year?

Solution.   If x denotes the dollar amount used to buy shares of stock B, then the total investment is $5000 + x$. The yield over this amount is

$$6\tfrac{3}{4}\% \text{ of } (5000 + x) = 0.0675(5000 + x) \text{ dollars.}$$

The yield over the original $5,000 invested in stock A is

$$6\% \text{ of } 5000 = 0.06 \times 5000 = 300 \text{ dollars,}$$

and the yield over x dollars invested in stock B is

$$8\% \text{ of } x = 0.08x \text{ dollars.}$$

Thus, we obtain the equation

$$0.08x + 300 = 0.0675(5000 + x)$$
$$0.08x + 300 = 337.50 + 0.0675x$$
$$0.0125x = 37.50$$
$$125x = 375000$$
$$x = \$3,000.$$

◇ **Practice Exercise 7.**   Darlene invests part of $10,000 in a stock yielding 8% a year and the rest in another yielding 11% a year. If after one year she earned $938, how much did she invest in each stock?

Answer. $5,400 at 8% and $4,600 at 11%.

Rate Problems

If an object moves along a line with a constant *rate* (or *speed*) r, and covers a *distance d* in *time t*, then the quantities are related by the equation

$$(2.5) \quad d = rt.$$

◆ **Example 8.**   On a river, a boat traveled upstream from a point A to a point B at a rate of 15 miles per hour (mi/h). It returned to A, traveling downstream, at a rate of 20 mi/h. Knowing that the total traveling time was $2\frac{1}{3}$ h, find the distance from A to B.

Solution.   Let d be the distance from A to B. According to (2.5), if t_1 is the time it took the boat to travel upstream, then $d = 15t_1$, so $t_1 = d/15$. Similarly, if t_2 denotes the time it took to travel downstream, then $d = 20t_2$, so $t_2 = d/20$. With these facts we can make the following table:

	Distance	*Rate*	*Time*
Upstream	d	15	$d/15$
Downstream	d	20	$d/20$

Since the total traveling time was $2\frac{1}{3}$ hr, we have the equation

$$\frac{d}{15} + \frac{d}{20} = 2\frac{1}{3}.$$

Solving for d, we obtain

$$\frac{d}{15} + \frac{d}{20} = \frac{7}{3}$$
$$4d + 3d = 140$$
$$7d = 140$$
$$d = 20 \text{ miles.}$$

◇ **Practice Exercise 8.**   A single-engine plane flies from one city to another at a rate of 120 km/h. Flying back at a rate of 100 km/h, it takes the plane 1/2 hour longer to complete the trip. What is the distance between the two cities?

Answer.   300 km.

◆ **Example 9.**   A boy in a town 18.5 miles from home got a car ride for all but the last 3.5 miles, which he had to walk. The average speed of the car was 45 mi/h, and the whole trip took 1 hour and 30 minutes. Find the boy's walking rate.

Solution.   Again, this is a problem where you have to use formula (2.5). Let r be the boy's walking rate. Since he traveled 15 miles by car at a rate of 45 mi/h and walked 3.5 miles at a rate of r mi/h, you may use this information to complete the following table:

	Distance	Rate	Time
By car	15	45	15/45 = 1/3
Walking	3.5	r	3.5/r

The whole trip took 1 hour and 30 minutes ($= 3/2$ h), so write the equation

$$\frac{1}{3} + \frac{3.5}{r} = \frac{3}{2}.$$

Solving for r, obtain

$$2r + 21 = 9r$$
$$-7r = -21$$
$$r = \frac{21}{7}$$
$$= 3 \text{ mi/h}.$$

To check your solution, note that the car ride lasted $1/3$ h $= 20$ minutes, while the walking time was $3.5/3$ h $= 7/6$ h $= 1$ hour and 10 minutes. Thus, the total time was 1 hour and 30 minutes.

◇ **Practice Exercise 9.**   Janet can ride her bicycle 4 miles per hour faster than Peter. If Peter rides 12 miles in the same time that Janet rides 15 miles, find the rates of Janet and Peter.

Answer.   Janet: 20 mi/h, Peter: 16 mi/h.

◆ **Example 10.**   A water pipe fills a tank in 2 hours and another one fills it in 4 hours. How long will it take to fill the tank if both pipes are used?

Solution.   In one hour the first pipe fills $1/2$ of the tank, while the second pipe fills $1/4$ of the tank. Together, they fill $(1/2) + (1/4)$ of the tank in one hour.
If t denotes the time it takes for both pipes to fill the tank, then in one hour they fill $1/t$ of the tank. We thus obtain the following equation:

$$\frac{1}{t} = \frac{1}{2} + \frac{1}{4},$$

or

$$\frac{1}{t} = \frac{2 + 1}{4} = \frac{3}{4},$$

so

$$t = \frac{4}{3} \text{ h} = 1 \text{ h } 20 \text{ min}.$$

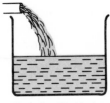

First pipe fills ½ of
the tank in 1 hr

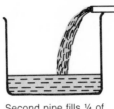

Second pipe fills ¼ of
the tank in 1 h

Both pipes fill ½ +
¼ =¾ of the tank in
1 h

Figure 2.2

◇ **Practice Exercise 10.**   Working alone, Ed can paint a garage in 2 days and Liz can paint it in 3 days.
How long will it take if both work together?

Answer.   6/5 days = 1.2 days.

 EXERCISES 2.2

1. Find four consecutive integers whose sum is 450.
2. Find three consecutive even numbers whose sum is 612.
3. If the difference of the squares of two consecutive natural numbers is 49, find the two numbers.
4. Find four consecutive numbers so that the sum of the last three equals 218 plus twice the first number.
5. The area of a rectangle is 112.5 square feet. Find its dimensions if its length is twice its width.
6. The length of a rectangle is three times its width. If the length is reduced by 4 m and the width is increased by 1 m, then the area is decreased by 12 m². Find the dimensions of the original rectangle.
7. If the perimeter of a rectangle is 98 cm and its length is 5 cm less than twice its width, find its dimensions.
8. A student has test scores of 77, 81, and 79. What should his next test score be in order to bring his average up to 80?
9. In an algebra course, a student took three tests and ended up with an average of 85. Knowing that her first and third tests scores were 90 and 87, find her second test score.
10. A manufacturer makes bicycles at a cost of $115 each. His fixed monthly costs are $2,575. How many bicycles has he made in a month if his total costs were $11,200?
11. The price tag of a tape recorder was reduced by $15. If this reduction corresponds to 12% of the original price, find the original price.
12. After getting a 20% discount, Peter bought a camera for $150. Find the original price of the camera.
13. What is the wholesale price of a tape recorder whose retail price of $116 is 45% above its wholesale price?
14. In a town of 8000 people, 4400 voted in the last election. What percentage of the people voted?
15. In a poll, 858 people preferred a new product as opposed to 35% who preferred another brand. How many people were polled?
16. A boy has 72 coins consisting of nickels and dimes, and totaling $4.90. How many nickels and dimes does he have?
17. A girl has $4.55 in nickels, dimes, and quarters. If she has four fewer dimes than quarters and seven more nickels than dimes, find how many of each type of coins she has.
18. Suppose that you invest $1,200 at a simple interest annual rate of $6\frac{1}{4}\%$ per year. How much money will you have at the end of the year?
19. A man puts 5.5% of his salary into a retirement plan. If his yearly salary is $21,520, how much will he put into the fund in one year?

20. Joan deposited $2,500 in a bank account earning simple interest. At the end of four years she had $3,175 in her account. Find the yearly rate of interest.

21. A businessman wishes to invest $50,000 in a tax-free fund and in a money market fund with annual yields of 7.1% and 14.5%, respectively. How much should he invest in each fund in order to realize a gain of $5,622 in one year?

22. A woman has $5,000 more invested at a simple interest rate of 8.5% a year than she has invested at 7%. If her total yearly interest is $2,285, how much money is invested at each rate?

23. A lab technician wants to prepare 720 grams of a compound by mixing 5 parts of chemical *A* with 3 parts of chemical *B*. How many grams of each chemical should be used?

24. An alloy contains 5 parts of copper, 3 parts of zinc, and 2 parts of nickel. How much of each metal is needed to make 300 kg of the alloy?

25. To make a certain amount of concrete, a builder mixes 3 parts of cement, 2 parts of sand, and 2 parts of stone. How many cubic feet of each ingredient should he use to prepare 875 cubic feet of concrete?

26. What volume of water should be evaporated from 600 liters of a 15% saline solution so that the remaining is a 25% saline solution?

27. A chemical manufacturer has two acid solutions, *A* and *B*. The first one is 20% acid and the second one is 40% acid. How many gallons of each should she mix to obtain 160 gallons of a 35% acid solution?

28. A lab assistant has two solutions of sulfuric acid: a 30% solution and a 70% solution. How many milliliters of each should he use to prepare 20 ml of a 60% solution?

29. On the moon's surface, a lunar rover traveled from the base of the lunar module to a nearby hill at a rate of 8 mi/h and then back at a rate of 12 mi/h. If the total traveling time was $1\frac{1}{4}$ h, find the total distance traveled.

30. At a river resort, you rent a motor boat for a sightseeing trip. You travel upstream at 9 mi/h and then return, traveling downstream at 12 mi/h. The whole trip lasts 3.5 h. How far from the resort were you when you turned back?

31. An airplane doing meteorological observation flies against a headwind, traveling from *A* to *B* at 250 mi/h. It flies back from *B* to *A*, with the wind, at 300 mi/h. If the whole trip lasted $5\frac{1}{2}$ h, find a) the distance from *A* to *B*, b) the time it took to fly from *A* to *B*.

32. An airplane traveled from one city to another at a cruising speed of 150 mi/h. On the return trip, traveling at 200 mi/h, the elapsed time was 1 hour less. What was the distance between the two cities?

33. On a trip to his summer cottage in the mountains, a man takes a ferryboat across a lake and then drives 180 mi at an average of 45 mi/h. The distance from the ferry dock to the cottage is 225 mi. If the total traveling time was 5 hours and 15 minutes, find the average cruising speed of the ferryboat.

34. A highway 200 miles long connects two cities *A* and *B*. A car leaves *A* traveling toward *B* at a uniform rate of 55 mi/h. At the same time, another car leaves *B* traveling toward *A* at a rate of 45 mi/h. How long will it take for the cars to pass each other? How far will they be from the respective cities from which they departed?

35. An automobile leaves a city at 9:00 AM traveling on a highway at a rate of 50 mi/h. Half an hour later, another automobile leaves the same city traveling in the same direction at a rate of 60 mi/h. How long will it take for the second automobile to overtake the first one? At what time will it overtake the first?

36. Three pipes are connected to a storage tank. The first one fills the tank in 3 hours, the second in 4 hours, and the third in 6 hours. How long will it take to fill the tank if all three pipes are used?

37. A swimming pool can be filled in 6 hours using water from an outlet. Another outlet fills the pool in 4 hours. If both outlets are used, how long will it take to fill the swimming pool?

38. A construction crew takes 30 days to rebuild a road. A second crew is able to rebuild it in 20 days. Working together, how long will it take for both crews to complete the job?

39. Pipes *A* and *B* can fill a tank in 2 h and 3 h, respectively. Pipe *C* can empty it in 4 h. If the three pipes are open simultaneously, how long will it take to fill the tank?

40. An electronic card sorter operates at a rate of 250 cards/min and a second one at a rate of 225 cards/min. Operating together, how long will it take them to sort a batch of 5700 cards?

2.3   THE ORDER RELATION IN ℝ

In Section 1.2, we described several properties of the real number system. It was mentioned that the set of all real numbers is closed under addition and multiplication and that these operations satisfy the commutative, associative, and distributive properties. Moreover, real numbers can be represented by points of a line, called a *coordinate* or *number line* and, vice-versa, points on a coordinate line are labeled by real numbers. Points to the right of the origin on the number line correspond to positive numbers, while points to the left of the origin correspond to negative real numbers. Besides these properties, the real number system has another important feature, which we now describe.

Order Relation

Given two real numbers a and b, a is said to be *greater than* b and we write $a > b$ if $a - b$ is *positive*. If the difference $a - b$ is *negative*, then a is said to be *smaller than* b and we write $a < b$. These statements define an *order relation* among real numbers.

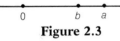

Figure 2.3

For example, $9 > 4$, because $9 - 4 = 5$, a positive number, and $-8 < -6$, because $(-8) - (-6) = -8 + 6 = -2$, a negative number.

If $a > b$, then the point corresponding to a on a coordinate line will be to the right of the point corresponding to b (Figure 2.3). For short, we say that a is *to the right* of b.

Similarly, if $a < b$ then a is to the left of b.

◆ **Example 1.**   Write in ascending numerical order the following numbers: **a)** -2 and -5, **b)** $\sqrt{2}$ and 1.41.

Solution.   **a)** Since $(-5) - (-2) = (-5) + 2 = -3$, then $-5 < -2$.
b) $\sqrt{2}$ is an irrational number and 1.4142136 is a decimal approximation of $\sqrt{2}$ (obtained with the help of a calculator). Comparing this approximation with the rational number 1.41, we see that $1.41 < \sqrt{2}$.

◇ **Practice Exercise 1.**   Locate on a coordinate line the numbers   **a)** 1/2 and 1/3,   **b)** $\sqrt{5}$ and 2.24.

Answer.

Inequalities

Given two real numbers a and b, we write $a \le b$ to indicate that a is *less than or equal* to b. Similarly, $a \ge b$ means that a is *greater than or equal* to b. Both statements are, sometimes, called *weak inequalities*.

The statements $a < b$ or $a > b$ as defined above are called *strict inequalities*.

The term inequality will be used for each of the statements $a < b$, $b > a$, $a \le b$, or $a \ge b$.

Observe that $a \le b$ consists of two *mutually exclusive* statements: *either $a < b$ or $a = b$*. For example, $2 \le 5$ is a true statement because $2 < 5$. Also $5 \le 5$ is a true statement because $5 = 5$.

Now we list the basic properties of inequalities. Together with properties I through VI of real numbers, discussed in Section 1.2, they will be used in the next section to solve inequalities.

Properties of Inequalities

1. If x and y are real numbers, then exactly one of the following statements is true: $x < y$, $x = y$, or $x > y$. This is the *trichotomy property*.
2. If $x \le y$ and $y \le z$, then $y \le z$. This is the *transitive property*.
3. If $x \le y$ and z is a real number, then $x + z \le y + z$.
4. If $x \le y$ and $z > 0$, then $xz \le yz$. However, if $z < 0$, then $xz \ge yz$.

Properties 1, 2, 3, and 4 are statements about strict inequalities as well as equalities. Property 3 asserts that a strict inequality or an equality is preserved if we add *any* number to both sides of the inequality. For example:

a) since $-3 < 5$, it follows that $-3 + 4 < 5 + 4$, or $1 < 9$,

b) since $-2 < 3$, it follows that $-2 + (-4) < 3 + (-4)$, or $-6 < -1$,

c) since $8 = 8$, it follows that $8 + (-2) = 8 + (-2)$, or $6 = 6$.

Property 4 says that if we multiply both sides of an inequality by a *positive* number, then the inequality is preserved. However, multiplication by a *negative* number reverses the sense of the inequality. Notice that an equality is always preserved if it is multiplied by either a positive or a negative number.

For example:

a) since $-4 < 3$, it follows that $(-4) \cdot 2 < 3 \cdot 2$, or $-8 < 6$,

b) since $-4 < 3$, it follows that $(-4) \cdot (-2) > 3 \cdot (-2)$, or $8 > -6$,

c) since $-5 = -5$, it follows that $(-5) \cdot (-1) = (-5) \cdot (-1)$, or $5 = 5$.

Intervals

The order relation $\leq$ and $<$ on real numbers allows us to define the notion of *intervals*.

Closed Intervals

If a and b are real numbers, with $a \leq b$, the closed interval $[a, b]$ is defined as the set of all real numbers x such that $a \leq x \leq b$.

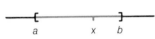

Figure 2.4

In set notation, the closed interval $[a, b]$ is denoted by

$$[a, b] = \{x \in \mathbb{R}: a \leq x \leq b\}.$$

Figure 2.4 shows the geometric interpretation of $[a, b]$ on the real line: it is a line segment with left endpoint a and right endpoint b. *Closed* means that both endpoints are included in it.

Observe that if $a = b$, then the closed interval reduces to a single point.

Open Intervals

If a and b are real numbers, with $a < b$, then the *open interval* (a, b) is defined as the set of all real numbers x such that $a < x < b$.

Figure 2.5

In set notation, the open interval (a, b) is denoted by

$$(a, b) = \{x \in \mathbb{R}: a < x < b\}$$

and illustrated on the number line in Figure 2.5.

Again the interval is a line segment. *Open* means that the interval does not include its endpoints.

Half-Open Intervals

If $a < b$, we define

$$[a, b) = \{x \in \mathbb{R}: a \leq x \leq b\} \quad \text{and} \quad (a, b] = \{x \in \mathbb{R}: a < x \leq b\}.$$

These intervals, illustrated in Figure 2.6, are called *half-open intervals*.

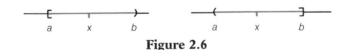

Figure 2.6

Note that "open" and "closed" as used in this special technical way are not really antonyms. Half-open intervals are not open, but they are not closed either.

Unbounded Intervals

The intervals defined above are called *bounded* intervals, in contrast to *unbounded* intervals, which we now define. A point P on a line divides the line into two parts, each of which is called a *half-line*. If the abscissa of P is a, then the set of all numbers x such that $x \geq a$ is said to be an *unbounded* (or *infinite*) *closed interval* (Figure 2.7). The set of all numbers x such that $x \leq a$ is also an unbounded closed interval. The set of all numbers x such that $x < a$ is said to be an *unbounded open interval* (Figure 2.8), as is the set of all numbers x such that $x > a$.

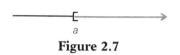

Figure 2.7

Figure 2.8

To represent unbounded intervals we use the *symbols* $+\infty$ and $-\infty$. These are *not* real numbers but just notational devices. In set notation, the interval illustrated in Figure 2.7 is denoted by

$$[a, +\infty) = \{x \in \mathbb{R} : x \geq a\}$$

and the interval shown in Figure 2.8 is denoted by

$$(-\infty, a) = \{x \in \mathbb{R} : x < a\}.$$

Analogously, we have the infinite intervals

$$(a, +\infty) = \{x \in \mathbb{R} : x > a\} \quad \text{and} \quad (-\infty, a] = \{x \in \mathbb{R} : x \leq a\}.$$

The real line itself can be viewed as an unbounded interval and denoted by $(-\infty, +\infty)$.

Absolute Values

If x is a real number, then its *absolute value* $|x|$ is defined by

$$|x| = \begin{cases} x & \text{if} \quad x \geq 0 \\ -x & \text{if} \quad x < 0. \end{cases}$$

This means that if a number is nonnegative, its absolute value is the number itself. If a number is negative, then its absolute value is the additive inverse of the number. For example, $|5| = 5$ and $|-8| = -(-8) = 8$. Loosely speaking, the absolute value of a signed number is *the number without the sign*.

◆ **Example 2.** Determine the set of numbers satisfying each of the following relations:
a) $|x| = |-4|$ **b)** $|2x| = 6$ **c)** $|x - 5| = 3$ **d)** $|1 - 3x| = 11$.

Solution. **a)** Since $|-4| = 4$, write the given relation as $|x| = 4$. According to the definition of absolute value, the last relation is the same as $x = 4$ or $-x = 4$. Hence, $x = 4$ and $x = -4$ are the two numbers satisfying the given relation.

b) Using the definition of absolute value, write $|2x| = 6$ as

$$2x = 6 \quad \text{or} \quad -2x = 6.$$

Thus,

$$x = 3 \quad \text{or} \quad x = -3.$$

c) If $|x - 5| = 3$, then

$$x - 5 = 3 \quad \text{or} \quad -(x - 5) = 3.$$

Solving for x in each equation, obtain $x = 8$ or $x = 2$.

d) The relation $|1 - 3x| = 11$ is equivalent to

$$1 - 3x = 11 \quad \text{or} \quad -(1 - 3x) = 11.$$

Solving for x, we get

$$x = -\frac{10}{3} \quad \text{or} \quad x = 4.$$

All statements in Example 2 correspond to *equations involving absolute values.*

◇ **Practice Exercise 2.**   Solve each of the following equations:
a) $|x| = |8|$ **b)** $|3x| = 12$ **c)** $|4 - x| = |-2|$ **d)** $|2x - 3| = 7$.

Answer.   **a)** -8 or 8 **b)** -4 or 4 **c)** 6 or -2 **d)** -2 or 5.

◆ **Example 3.**   By determining whether the expression inside the absolute value bars is non-negative or positive, write each expression without the absolute value bars.
a) $|-7 + 3|$ **b)** $|4 - \sqrt{7}|$
c) $|\pi - (22/7)|$ **d)** $|1 - a|$ if $a > 1$.

Solution.   **a)** $|-7 + 3| = |-4| = 4$
b) Since $\sqrt{7} < 4$ (Why?), we know that $4 - \sqrt{7} > 0$, so $|4 - \sqrt{7}| = 4 - \sqrt{7}$.
c) An approximation for π is 3.1416. Next, check that 3.1428 is an approximation for 22/7. Thus, $\pi < 22/7$ or $\pi - (22/7) < 0$. Hence,

$$|\pi - (22/7)| = -(\pi - (22/7)) = 22/7 - \pi.$$

d) If $a > 1$, then $a - 1 > 0$ and $1 - a < 0$, so $|1 - a| = -(1 - a) = a - 1$.

◇ **Practice Exercise 3.**   Write each expression without the absolute value bars.
a) $|-4 + 9|$ **b)** $|2 - \pi|$
c) $|\sqrt{2} - (7/5)|$ **d)** $|a - 1|$ if $a > 1$.

Answer.   **a)** 5 **b)** $\pi - 2$ **c)** $\sqrt{2} - 7/5$ **d)** $a - 1$.

Properties of the Absolute Value

The absolute value of a number is always nonnegative and satisfies the following properties:

$$|x| = |-x|$$
$$|xy| = |x||y|$$
$$-|x| \leq x \leq |x|$$

For example, if $x = -15$, then $-x = 15$ and $|-15| = |15| = 15$. If $x = -7$ and $y = 3$, then $xy = (-7) \cdot 3 = -21$ and $|-21| = |-7| \cdot |3|$. If $x = -4$, then $|-4| = 4$ and we have $-|-4| = -4 < 4$.

Using the notion of absolute value, we can define the *distance* between two points on a coordinate line.

Distance Between Two Points on an Axis

Let A and B be two points with coordinates a and b, respectively. The *distance between A and B*, denoted by $d(A, B)$ or AB, is defined by

$$d(A, B) = |b - a|.$$

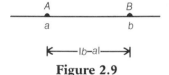

Figure 2.9

It follows that the distance $d(A, B)$ is always a nonnegative number. (Can you explain why?) Also, $d(A, B)$ is called the *length of the segment AB*, as shown in Figure 2.9.

Distance from a Point to the Origin

If P is a point with abscissa x, the distance between P and the origin is

$$d(P, O) = |x - 0| = |x|.$$

Thus, *the absolute value $|x|$ represents the distance from P to the origin.*

Properties of Distance

From the definition of absolute value, we can derive the following properties for the distance between two points:

$$d(A, B) = 0 \quad \text{if and only if} \quad A = B.$$
$$d(A, B) = d(B, A).$$

◆ **Example 4.**   Let A, B, C, and D be points with abscissas 5, 8, -4, and -7, respectively. Find $d(A, B)$, $d(C, A)$, and $d(D, C)$.

Solution.   See Figure 2.10.

$$d(A, B) = |8 - 5| = 3,$$
$$d(C, A) = |5 - (-4)| = |5 + 4| = 9,$$
$$d(D, C) = |-4 - (-7)| = |-4 + 7| = 3.$$

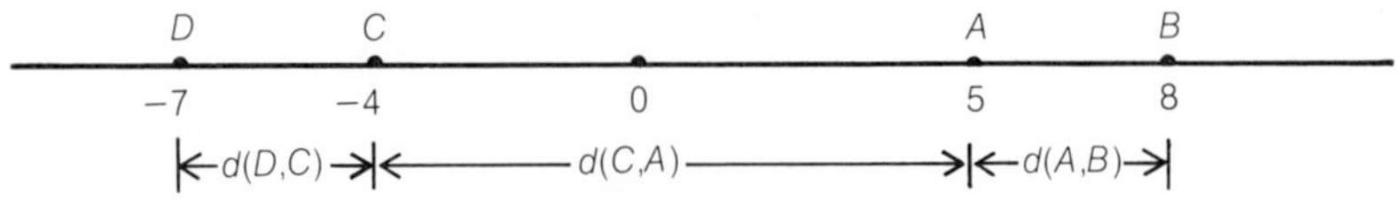

Figure 2.10

Note that $d(B, A) = |5 - 8| = |-3| = 3$, so $d(A, B) = d(B, A)$.

◇ **Practice Exercise 4.**   Let E, F, G, and H be points with abscissas -8, -5, -2 and 4, respectively. Find $d(E, G)$, $d(F, H)$, and $d(H, E)$.

Answer.   6, 9, and 12.

◻ **EXERCISES 2.3**

In Exercises 1–10, write the given numbers in ascending numerical order.

1. $-2, 2, \sqrt{3}$

2. $-3, 1, \sqrt{5}$

3. $-\dfrac{2}{3}, \dfrac{1}{2}, -\sqrt{2}$

4. $\dfrac{3}{5}, -\dfrac{2}{3}, -\sqrt{3}$

5. $\dfrac{1}{4}, \sqrt{5}, \sqrt{3}, 4$

6. $\dfrac{2}{3}, \sqrt{6}, \sqrt{2}, 3$

7. $-\sqrt{2}, 1, -\dfrac{5}{2}, \sqrt{6}$

8. $-\sqrt{5}, 2, \dfrac{6}{5}, \sqrt{7}$

9. $-2.45, -\sqrt{6}, \sqrt{2}, \dfrac{17}{12}$

10. $-\sqrt{2}, -\dfrac{41}{29}, \sqrt{7}, 2.63$

In Exercises 11–16, write each of the given intervals in set notation and sketch the interval.

11. $[-2, 3]$

12. $[-5, 0]$

13. $[2, 8)$

14. $(-2, 7)$

15. $(-\infty, 5)$

16. $[0, +\infty]$

In Exercises 17–22, write each inequality in interval notation and sketch the interval.

17. $-2 < x < 7$

18. $3 < x < 11$

19. $-3 < x < 3$

20. $-2 \geq x \geq -6$

21. $x \geq -2$

22. $7 > x > -7$

In Exercises 23–34, solve each equation.

23. $|x| = |-5|$

24. $|-x| = 4$

25. $|2x| = |-7|$

26. $|-3x| = |-2|$

27. $-x = |-2|$

28. $-x = |-8|$

29. $|x - 1| = 3$

30. $|2 - x| = |-5|$

31. $|3 - 2x| = 4$

32. $|2x - 1| = 7$

33. $|3x - 4| = |-2|$

34. $|4x - 1| = |-2|$

Find the value (without absolute value bars) of each of the following expressions.

35. $|-15|$

36. $-|-9|$

37. $|-9+4|$

38. $|5-12|$

39. $|-6-8|$

40. $|-10-3|$

41. $|\pi-(25/8)|$

42. $|\sqrt{10}-4|$

43. $|\sqrt{2}-\sqrt{3}|$

44. $|\sqrt{3}-\sqrt{5}|$

45. $|a-3|$ if $a<3$

46. $|5-b|$ if $b>5$

47. $|2x-10|$ if $x<5$

48. $|3x-9|$ if $x>3$

In Exercises 49–54, the abscissas of A, B, and C are given. Find $d(A, B)$, $d(A, C)$, and $d(B, C)$.

49. $-6, -4, -1$

50. $-8, -2, 5$

51. $-\dfrac{5}{2}, 0, \dfrac{7}{2}$

52. $-5, -\dfrac{1}{5}, 0$

53. $0.5, 1.26, 0.08$

54. $1.25, 2.11, 0.75$

Prove each of the following properties of the absolute value.

55. $|x| = |-x|$

56. $|xy| = |x||y|$ (*Hint*: First, consider the case when both x and y have the same sign; then prove the case when x and y have opposite signs.)

57. If $x \neq 0$, then $|1/x| = 1/|x|$. (*Hint*: Use the relation $x(1/x) = 1$.)

58. If $y \neq 0$, then $|x/y| = |x|/|y|$. (*Hint*: Use Exercises 56 and 57.)

59. $d(A, B) = 0$ if and only if $A = B$.

60. $d(A, B) = d(B, A)$.

2.4 LINEAR INEQUALITIES

> A *linear inequalities* in one variable is an expression of the form
>
> $$(2.6) \quad ax + b \leq 0$$
>
> where $a, b \in \mathbb{R}$ and $a \neq 0$.

To *solve* this inequality means to find all real numbers which, when substituted for x, make (2.6) a true statement. The set of all such numbers is the *solution set* of the inequality. In order to find the solution set of (2.6), we apply the properties of inequalities listed in the previous section and replace (2.6) by a chain of equivalent but simpler inequalities, ending up with an inequality whose solution set is easy to obtain.

◆ **Example 1.** Solve the inequality $(3/5)x + 2 \leq 0$ and sketch its solution set.

Solution. According to property 3 of inequalities (Section 2.3), adding -2 to both sides of the given inequality yields the equivalent inequality

$$\frac{3}{5}x \leq -2.$$

Multiplying both sides by 5/3 (a positive number) preserves the sense of the inequality (property 4), giving us

$$x \le \frac{5}{3}(-2)$$

or

$$x \le -\frac{10}{3}.$$

The solution set, illustrated in Figure 2.11, consists of all real numbers that are less than or equal to $-10/3$.

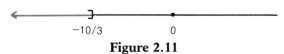

−10/3   0

Figure 2.11

It is important to notice how the solution set of the inequality $(3/5)x + 2 \le 0$ is related to the solution of the equation $(3/5)x + 2 = 0$. The latter has a *unique* solution $x = -10/3$. The point with abscissa $-10/3$ splits the real line into two half-lines. The inequality $(3/5)x + 2 \le 0$ is satisfied for all points on the half-line to the left of $-10/3$. (What inequality describes the points with abscissa greater than $-10/3$?)

◇ **Practice Exercise 1.** Find and sketch the solution set of the inequality $-4x + 2/3 \le 0$.

Answer. $x \ge \frac{1}{6}$

0   1/16

We can also consider inequalities of the form

$$ax + b \ge 0, \quad ax + b < 0, \quad \text{or} \quad ax + b > 0.$$

Inequalities such as $ax + b < 0$ or $ax + b > 0$ are called *strict inequalities*. Notice that multiplication by -1 changes the inequality $ax + b \ge 0$ into an equivalent inequality of the form (2.6).

◆ **Example 2.** Solve the inequality $1 - 2x > 7$, and graph the results.

Solution. We have

$$1 - 2x > 7 \qquad \text{[Add } -1 \text{ to both sides]}$$
$$-2x > 6 \qquad \text{[Multiply both sides by } -1/2, \text{ reversing the sense of the inequality]}$$

$$x < \left(-\frac{1}{2}\right)6$$

$$x < -3.$$

Thus, the solution set is $\{x \in \mathbb{R}: x < -3\}$.

−3   0

◇ **Practice Exercise 2.** Find the solution set of the inequality $2 - 3x < 9$, and graph it.

Answer. $\left\{ x \in \mathbb{R}: x > -\dfrac{7}{3} \right\}$

In many cases, algebraic operations must be performed in order to transform a given inequality into a linear inequality.

◆ **Example 3.** Solve $\dfrac{x-2}{2} \geq \dfrac{x}{5} - \dfrac{1}{4}$.

Solution. Multiplying both sides of the inequality by 20 in order to eliminate the denominators, we obtain

$$10(x - 2) \geq 4x - 5,$$
$$10x - 20 \geq 4x - 5,$$
$$6x \geq 15;$$

hence

$$x \geq \frac{15}{6} = \frac{5}{2}.$$

Thus, the solution set is $\{x \in \mathbb{R}: x \geq 5/2\}$.

◇ **Practice Exercise 3.** Solve the inequality

$$\frac{x}{3} < \frac{5}{6} - \frac{x-1}{4}.$$

Answer. $x < \dfrac{13}{7}$.

◆ **Example 4.** Find the solution set of the inequality $(2x + 1)(x - 2) \geq (x - 1)(2x + 3) + 5$.

Solution. First, compute the products and obtain

$$2x^2 - 3x - 2 \geq 2x^2 + x - 3 + 5.$$

Next, add $-2x^2$ to both sides and obtain the equivalent inequality

$$-3x - 2 \geq x + 2.$$

Solving for x, get

$$-4x \geq 4;$$

multiplying by $-1/4$ reverses the sense of the inequality, so

$$x \le -1.$$

Thus, the solution set is $\{x \in \mathbb{R}: x \le -1\}$.

◇ **Practice Exercise 4.**   Solve the inequality $(x + 4)(3x - 1) < 3(x + 1)^2 - 7$.

Answer. $x < 0$.

◆ **Example 5.**   Solve the inequality

$$\frac{3x}{6x - 1} > \frac{1}{2}.$$

Solution.   *Do not* multiply both sides by $2(6x - 1)$, trying to eliminate the denominators! If $x < 1/6$, then $2(6x - 1) < 0$ (can you see why?) and multiplication by a negative number reverses the sense of the inequality. Instead, add $-1/2$ to both sides of the given inequality to obtain the equivalent inequality

$$\frac{3x}{6x - 1} - \frac{1}{2} > 0.$$

Next, rewrite the two terms on the left-hand side with the same denominator

$$\frac{6x}{2(6x - 1)} - \frac{6x - 1}{2(6x - 1)} > 0$$

$$\frac{6x - (6x - 1)}{2(6x - 1)} > 0,$$

$$\frac{1}{2(6x - 1)} > 0.$$

Since the numerator is positive, in order for this rational expression to be positive, the denominator $2(6x - 1)$ must be positive. Thus, the last inequality is equivalent to

$$2(6x - 1) > 0,$$

or

$$6x - 1 > 0;$$

hence

$$x > \frac{1}{6}.$$

We conclude that the solution set of the given inequality is $\{x \in \mathbb{R}: x > 1/6\}$.

◇ **Practice Exercise 5.**   Find the solution set of the inequality $\dfrac{x}{4x - 3} < \dfrac{1}{4}$.

Answer. $\left\{ x \in \mathbb{R} : x < \dfrac{3}{4} \right\}$

Inequalities Involving Absolute Values

◆ **Example 6.**   Describe and sketch the sets of numbers such that   **a)** $|x| < 5$,   **b)** $|x| \geq 4$.

Solution.   **a)** As we already know (Section 2.3), $|x|$ represents the distance between a point with abscissa x and the origin. Since -5 and 5 are the coordinates of the only two points that are 5 units away from the origin, then every number x such that $|x| < 5$ must lie between -5 and 5, and vice versa (Figure 2.12). In other words, the statement "$|x| < 5$" is equivalent to the statement "$-5 < x < 5$." It follows that the set of all numbers satisfying the inequality $|x| < 5$ is the same as the open interval $(-5, 5)$.

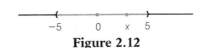

Figure 2.12

b) If $|x| \geq 4$, then the distance between the point with abscissa x and the origin is greater than or equal to 4. Such points lie outside the open interval $(-4, 4)$ (Figure 2.13). We can also say that the statement "$|x| \geq 4$" is equivalent to the statement "$x \leq -4$ or $x \geq 4$".

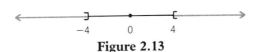

Figure 2.13

◇ **Practice Exercise 6.**   Determine all numbers satisfying the inequalities   **a)** $|x| > 2$,   **b)** $|x| \leq 3$.

Answer.   **a)** $\{ x \in \mathbb{R} : x < -2 \text{ or } x > 2 \}$ **b)** $\{ x \in \mathbb{R} : -3 \leq x \leq 3 \}$

Example 6 and Practice Exercise 6 illustrate two important properties of the absolute value that we now state in general terms.

The Two Basic Absolute Value Inequalities

If $b > 0$, then

$$(2.7) \quad |x| \leq b \quad \text{if and only if} \quad -b \leq x \leq b$$

and

$$(2.8) \quad |x| \geq b \quad \text{if and only if} \quad x \leq -b \text{ or } x \geq b.$$

◆ **Example 7.**   Solve the inequality $|x - 2| < 7$.

Solution.   According to (2.7) with $b = 7$ and x replaced by $x - 2$, we have

$$|x - 2| < 7 \quad \text{if and only if} \quad -7 < x - 2 < 7.$$

The last inequality can be written as a double inequality:

$$-7 < x - 2 \quad and \quad x - 2 < 7.$$

Solving both inequalities, we obtain

$$-5 < x \quad and \quad x < 9,$$

that is,

$$-5 < x < 9.$$

Thus, x satisfies $|x - 2| < 7$ if and only if $-5 < x < 9$. In other words, the solution set of the given inequality is the open interval $(-5, 9)$.

Figure 2.14

In this example, we could have worked as well with the inequality $-7 < x - 2 < 7$ instead of splitting it into two inequalities. By adding 2 to each of the three terms of the last inequality, we obtain

$$-7 + 2 < x - 2 + 2 < 7 + 2$$

or

$$-5 < x < 9.$$

◇ **Practice Exercise 7.**   Find the solution set of the inequality $|x + 3| \leq 5$.

Answer. $[-8, 2]$

◆ **Example 8.**   Find the solution set of the inequality $|2x + 1| \geq 5$.

Solution.   According to (2.8) with $b = 5$ and x replaced by $2x + 1$, we may write

$$|2x + 1| \geq 5 \quad \text{if and only if} \quad 2x + 1 \leq -5 \text{ or } 2x + 1 \geq 5.$$

Solving each of these inequalities, we obtain

$$2x \leq -6 \quad or \quad 2x \geq 4.$$

Hence

$$x \leq -3 \quad or \quad x \geq 2.$$

Thus, the solution set of $|2x + 1| \geq 5$ (Figure 2.15) consists of all real numbers lying outside the interval $(-3, 2)$.

Figure 2.15

◇ **Practice Exercise 8.**   Solve the inequality $|3x - 1| > 4$.

Answer. $x < -1$ or $x > \dfrac{5}{3}$

Applications

◆ **Example 9.**   In a chemical experiment the temperature must be kept between 104°F and 122°F. Find the corresponding range of temperature in the Celsius scale.

Solution.   The equation expressing Fahrenheit degrees in terms of Celsius degrees is

$$F = \left(\frac{9}{5}\right)C + 32.$$

Thus, we have to solve the inequalities

$$104 \le \left(\frac{9}{5}\right)C + 32 \le 122,$$

$$72 \le \left(\frac{9}{5}\right)C \le 90,$$

$$72\left(\frac{5}{9}\right) \le C \le 90\left(\frac{5}{9}\right);$$

hence

$$40° \le C \le 50°.$$

◇ **Practice Exercise 9.**   Assume that in Ohm's law, $V = RI$, the resistance R is kept constant and equal to 15 ohms. If the voltage V varies from 120 to 210 volts, find the range of variation of the intensity I of the current.

Answer. $8 \le I \le 14$.

◆ **Example 10.**   The owner of a movie theater estimates that his daily operating cost C is given by the equation $C = 2.5x + 100$, while his daily revenue R is given by $R = 3x$, where x is the number of tickets sold in a day. How many tickets a day must he sell to realize a profit?

Solution.   To make a profit the revenue R must be greater than the cost C, from which we get the inequality

$$3x > 2.5x + 100.$$

Solving for x, we obtain

$$0.5x > 100,$$

so

$$x > 200.$$

Thus, he must sell more than 200 tickets a day to realize a profit. This number is called the *break even* point; that is, by selling exactly 200 tickets, his revenue equals his cost, and there is no profit.

$$R = 3 \times 200 = 600 \quad \text{and} \quad C = 2.5 \times 200 + 100 = 600.$$

◇ **Practice Exercise 10.**   The daily cost of manufacturing x dozens of pencils is given by $C = 0.5x + 150$, and the revenue from selling them is given by $R = 3.5x$. How many dozens of pencils must be sold in order to realize a profit?

Answer. $x > 50$.

◇ **EXERCISES 2.4**

In Exercises 1–10, find and sketch the solution set of each inequality.

1. $6x - 3 > 8$

2. $3x - 4 < 10$

3. $5 - 2x \leq 7$

4. $4 - 5x \geq -8$

5. $\dfrac{3}{4}x - 2 \geq \dfrac{1}{3}$

6. $\dfrac{2}{3}x + \dfrac{1}{2} \geq \dfrac{1}{4}$

7. $3x - 2 \geq 4 - 2x$

8. $5x - 2 < 3 - 2x$

9. $\dfrac{2x - 3}{4} \geq \dfrac{3x}{2} - \dfrac{1}{3}$

10. $3 - \dfrac{x}{5} \leq \dfrac{x + 1}{2} - 2$

Solve each of the following inequalities.

11. $1 \leq \dfrac{2x - 5}{6} < 3$

12. $\dfrac{1}{3} \leq \dfrac{3x + 1}{2} \leq 2$

13. $5 > 2 - \dfrac{x}{3} \geq 0$

14. $3x \leq 4 - \dfrac{2x}{5} < 6$

15. $(3x + 2)(x - 4) < (x + 2)(3x - 1)$

16. $6x(2x - 8) \geq (4x - 1)(3x + 2)$

17. $(x + 1)^2 > (x - 2)(x - 1)$

18. $(2x - 3)(2x + 5) \leq (2x + 1)^2$

19. $\dfrac{2}{x + 3} > 0$

20. $\dfrac{3}{x + 5} < 0$

21. $\dfrac{x}{x + 4} > 1$

22. $\dfrac{2x}{3 - 2x} + 1 > 0$

23. $\dfrac{2x}{4x - 1} > \dfrac{1}{2}$

24. $\dfrac{2x}{2 - 3x} < -\dfrac{2}{3}$

In Exercises 25–40, solve each absolute value inequality and sketch the solution set.

25. $|x - 4| > 1$

26. $|x + 2| < 6$

27. $|3x + 1| < 9$

28. $|4x - 5| \leq 1$

29. $|3 - 6x| > 2$

30. $|1 - 2x| \leq 4$

31. $\left|\dfrac{3 - x}{4}\right| < 1$

32. $\left|\dfrac{5 - 2x}{3}\right| \geq 2$

33. $\left|\dfrac{2x-5}{3}\right| < 3$ **34.** $\left|\dfrac{3x-2}{2}\right| \le 2$ **35.** $\left|2x-\dfrac{1}{3}\right| < 4$ **36.** $\left|3x+\dfrac{2}{4}\right| \ge 1$

37. $\left|\dfrac{x}{3}-\dfrac{1}{2}\right| \ge 1$ **38.** $\left|\dfrac{3}{4}-\dfrac{5x}{2}\right| < .1$ **39.** $\left|\dfrac{3x}{4}-\dfrac{1}{3}\right| \ge 1$ **40.** $\left|\dfrac{3x}{5}-\dfrac{1}{2}\right| < 3$

41. If the temperature of a room must be kept between 20° and 25°C, find the range of temperatures in degrees Fahrenheit.

42. The temperature of an oven is kept between 284° and 347°F. What is the range of temperatures in degrees Celsius?

43. According to Ohm's law, $V = RI$. If V is kept constant and equal to 200 volts, find the range in resistance (measured in ohms) if the intensity I of the current varies from 10 to 20 amperes.

44. A student wants to keep her test average above 75. What is the minimum score that she should get in her fourth test if her previous scores were 70, 76, and 80?

45. Assume that in Boyle's law $PV/T = C$, the temperature T is kept at 100° and that $C = 1$. If the pressure P varies in the interval $25 \le P \le 75$, find the interval of variation for V.

46. Suppose that the force F (in pounds) required to stretch a spring x inches beyond its natural length is given by $F = 2.5x$. Find the range of F if the range of x is $10 \le x \le 12$.

47. If the range of the intelligence quotient of a group of 15-year-old children is $70 \le \text{IQ} \le 140$, find the range of their mental age. [See problem 50 of Section 2.1 for the IQ formula].

48. The daily cost equation for a jeans manufacturer is $C = 775 + 2.5x$ and his daily revenue equation is $R = 15x$, where x is the number of pairs of jeans sold per day. Find how many pairs of jeans he has to sell in a day to realize a profit.

49. The driver of a car wants to travel 150 mi in a time interval of 2.5 hr to 3 hr. What is the corresponding speed range?

50. A publisher estimates that to print a certain magazine, the fixed costs are $3,000 and the variable costs are 35 cents per copy. How many copies must he sell at 60 cents per copy to realize a profit?

2.5 SOLVING QUADRATIC EQUATIONS

A *quadratic equation* is an expression of the form

(2.9) $ax^2 + bx + c = 0$

where a, b, and c are real numbers with $a \ne 0$. A *solution* or *root* of the quadratic equation is a number r that, when substituted for x, makes (2.9) a true statement:

(2.10) $ar^2 + br + c = 0.$

In Section 2.1 we discussed linear equations and showed how the solution of a linear equation can easily be obtained by using properties of the real numbers system. The situation is different for quadratic equations.

First of all, a quadratic equation with real coefficients may have:

a) *two distinct real roots*;

b) *one real root* (called a *double root*); or

c) *no real roots*.

For example, the numbers 1 and 2 are two distinct roots of $x^2 - 3x + 2 = 0$. This can be seen if you substitute 1 and 2 for x into the given equation:

$$1^2 - 3 \cdot 1 + 2 = 0 \quad \text{and} \quad 2^2 - 3 \cdot 2 + 2 = 0.$$

The only solution (or double root) of $x^2 - 6x + 9 = 0$ is the number 3. Finally, the equation $x^2 + 1 = 0$ has no real roots.

Second, the solutions of a quadratic equation are not obtained as easily as the solution of a linear equation. Either we have to perform some algebraic operations to transform the quadratic equation into an equivalent but simpler equation, or we have to use the quadratic formula.

If a quadratic equation does not have real roots, we say that the equation is *not solvable over* the real numbers $\mathbb{R}$. As we shall see, quadratic equations that are not solvable over $\mathbb{R}$ always have *complex roots*, so although they are not solvable over $\mathbb{R}$ they are solvable over $\mathbb{C}$.

There are essentially two methods for solving a quadratic equation, by *factoring* and by *completing the square*. When the latter method is applied to the general equation (2.9), it yields the *quadratic formula*.

Solving By Factoring

The solutions of a quadratic equation $ax^2 + bx + c = 0$ are easily obtained whenever we can factor the quadratic polynomial $ax^2 + bx + c$. The method uses the multiplicative property of zero [Section 1.2]: if r and s are real numbers such that $rs = 0$, then either $r = 0$ or $s = 0$ (or both).

◆ **Example 1.**   Solve by factoring the quadratic equation $2x^2 - 7x - 4 = 0$.

Solution.   After factoring the quadratic polynomial, we can write the given equation in its equivalent form

$$(2x + 1)(x - 4) = 0.$$

Since the product is zero, it follows that either $2x + 1 = 0$ or $x - 4 = 0$. The first equation has solution $x = -1/2$ and the second has solution $x = 4$. To check that both numbers are solutions of the quadratic equation, substitute $-1/2$ and 4 for x into the equation:

$$2\left(-\frac{1}{2}\right)^2 - 7\left(-\frac{1}{2}\right) - 4 = \frac{1}{2} + \frac{7}{2} - 4 = 0;$$

$$2(4)^2 - 7(4) - 4 = 32 - 28 - 4 = 0.$$

◇ **Practice Exercise 1.**   Find the solutions of the equation $3x^2 + 11x - 4 = 0$.

Answer. $-4, \dfrac{1}{3}.$

◆ **Example 2.**   Solve by factoring $x^2 - 6x + 9 = 0$.

Solution.   The quadratic polynomial $x^2 - 6x + 9$ is a *perfect square*, that is, $x^2 - 6x + 9 = (x - 3)^2$. Thus the given equation is equivalent to

$$(x - 3)^2 = 0.$$

The latter has a unique solution $x = 3$, called a *double root*.

◇ **Practice Exercise 2.**   Solve the quadratic equation $4x^2 - 4x + 1 = 0$.

Answer. $\dfrac{1}{2}$ (double root).

Solving By Completing the Square

Suppose that we have a quadratic equation of the type

$$(2.11) \quad (x + p)^2 = q,$$

with $p, q \in \mathbb{R}$ and $q > 0$. To solve this equation, take square roots of both sides to obtain

$$x + p = \pm\sqrt{q}.$$

Next, add $-p$ to both sides, to get

$$x = -p \pm \sqrt{q}.$$

Thus, the two solutions of (2.11) are $x_1 = -p + \sqrt{q}$ and $x_2 = -p - \sqrt{q}$.

◆ **Example 3.**   Solve the equation $(x - 2)^2 - 8 = 0$.

Solution.   We have

$$(x - 2)^2 - 8 = 0,$$
$$(x - 2)^2 = 8,$$
$$x - 2 = \pm\sqrt{8} = \pm2\sqrt{2},$$
$$x = 2 \pm 2\sqrt{2}.$$

Hence the solutions are $x_1 = 2 + 2\sqrt{2}$ and $x_2 = 2 - 2\sqrt{2}$.

◇ **Practice Exercise 3.**   Find the solutions of $(2x + 1)^2 - 6 = 0$.

Answer. $\dfrac{-1 - \sqrt{6}}{2}, \dfrac{-1 + \sqrt{6}}{2}.$

The method of solving a quadratic equation by completing the square consists of transforming the given equation into an equation of the form (2.11) before solving it.

◆ **Example 4.**   Solve by completing the square $x^2 + 6x + 8 = 0$.

Solution.   First, observe that in a perfect square such as $(x + m)^2 = x^2 + 2mx + m^2$, the term m^2 is equal to *the square of one-half the coefficient of x.*
Next, write the given equation with its constant term on the right-hand side:

$$x^2 + 2 \cdot (3)x = -8.$$

(Note that we have written the coefficient of x as $2 \cdot 3$ so that it appears in the form $2 \cdot m$, with $m = 3$.) If we add $9 = 3^2$ to the left-hand side of the equations, then it will become a perfect square: $x^2 + 2 \cdot (3)x + 9 = (x + 3)^2$. In order to balance the equation, we must add 9 to the right-hand side of it so that we obtain

$$x^2 + 2 \cdot (3)x + 9 = -8 + 9,$$

or

$$(x + 3)^2 = 1,$$

which is an equation of the type (2.11). Taking square roots and solving for x yields

$$x + 3 = \pm\sqrt{1} = \pm 1,$$
$$x = -3 \pm 1.$$

Hence, the solutions are $x = -2$ and $x = -4$.

◇ **Practice Exercise 4.**   Use the method of completing the square to solve the equation $x^2 + 4x - 5 = 0$.

Answer. $-5, 1$.

◆ **Example 5.**   Solve by completing the square $2x^2 - 12x + 9 = 0$.

Solution.   Since $2x^2 - 12x + 9 = 2(x^2 - 6x + 9/2) = 0$, the given equation is equivalent to

$$x^2 - 6x + \frac{9}{2} = 0,$$

where the leading coefficient is now 1. Next, complete the square by arranging your work as follows:

$$x^2 - 6x \qquad\qquad = -\frac{9}{2},$$

$$x^2 + 2 \cdot (-3)x \qquad\qquad = -\frac{9}{2},$$

$$x^2 + 2 \cdot (-3)x + (-3)^2 = -\frac{9}{2} + (-3)^2.$$

Observe that, in order to complete the square, you have to add to both sides *the square of one-half the coefficient of x.* Now write

$$x^2 - 6x + 9 = -\frac{9}{2} + 9,$$

$$(x - 3)^2 = \frac{9}{2},$$

$$x - 3 = \pm\sqrt{\frac{9}{2}} = \pm\frac{3\sqrt{2}}{2}$$

$$x = 3 \pm \frac{3\sqrt{2}}{2} = \frac{6 \pm 3\sqrt{2}}{2}.$$

Thus, the solutions are

$$x_1 = \frac{6 + 3\sqrt{2}}{2} \quad \text{and} \quad x_2 = \frac{6 - 3\sqrt{2}}{2}.$$

◇ **Practice Exercise 5.** Find, by the method of completing the square, the solutions of $3x^2 - 6x - 3 = 0$.

Answer. $1 - \sqrt{2}, 1 + \sqrt{2}$.

The Quadratic Formula

From the method of completing the square, we now derive the quadratic formula for solving any quadratic equation. Consider the quadratic equation

$$ax^2 + bx + c = 0,$$

where a, b, $c \in \mathbb{R}$ and $a \neq 0$. By dividing both members by a, we obtain the equivalent equation

$$x^2 + \left(\frac{b}{a}\right)x + \frac{c}{a} = 0.$$

Shifting the constant term c/a to the right-hand side and writing b/a, the coefficient of x, as $2(b/2a)$, we get

$$x^2 + 2\left(\frac{b}{2a}\right)x = -\frac{c}{a}.$$

Now we complete the square, by adding $b^2/4a^2$, the square of $b/2a$, to both sides of the equation,

$$x^2 + 2\left(\frac{b}{2a}\right)x + \frac{b^2}{4a^2} = -\frac{c}{a} + \frac{b^2}{4a^2},$$

and rewrite the last expression as follows:

$$\left(x + \frac{b}{2a}\right)^2 = \frac{(b^2 - 4ac)}{4a^2}.$$

If the quantity $b^2 - 4ac$ is *nonnegative*, then we can take square roots and obtain

$$x + \frac{b}{2a} = \pm\sqrt{\frac{b^2 - 4ac}{4a^2}} = \pm\frac{\sqrt{b^2 - 4ac}}{2a},$$

$$x = -\frac{b}{2a} \pm \frac{\sqrt{b^2 - 4ac}}{2a},$$

$$x = \frac{-b \pm \sqrt{b^2 - 4ac}}{2a}.$$

HISTORICAL NOTE

Quadratic Equations
Evidence from tablets of burnt clay from the Babylonian empire (2000 BC) shows that the Babylonians knew how to solve quadratic equations in certain simple cases. The coefficients were then rational numbers and the discriminant a perfect square, so that the roots were rational numbers. Negative numbers or 0 were not known, and tablets of the time described in words how the solutions were obtained.

The Quadratic Formula

Let $ax^2 + bx + c = 0$, with a, b, $c \in \mathbb{R}$ and $a \neq 0$, be a quadratic equation. If the quantity $b^2 - 4ac$ is nonnegative, then the quadratic equation has *real roots* given by

$$(2.12) \quad x = \frac{-b \pm \sqrt{b^2 - 4ac}}{2a}.$$

If $b^2 - 4ac > 0$, then the equation has *two distinct real roots*. If $b^2 - 4ac = 0$, then the equation has *one real root* (called a *double root*).

◆ **Example 6.** Use the quadratic formula to solve $2x^2 - 5x - 3 = 0$.

Solution. Applying formula (2.12) with $a = 2$, $b = -5$, and $c = -3$ gives us

$$x = \frac{-(5) \pm \sqrt{(-5)^2 - 4 \cdot 2 \cdot (-3)}}{(2 \cdot 4)} = \frac{5 \pm \sqrt{25 + 24}}{8}$$

$$= \frac{5 \pm \sqrt{49}}{8} = \frac{5 \pm 7}{4}.$$

Hence, the solutions are

$$x_1 = \frac{5 + 7}{4} = 3 \quad \text{and} \quad x_2 = \frac{5 - 7}{4} = -\frac{2}{4} = -\frac{1}{2}.$$

◇ **Practice Exercise 6.** Solve the equation $3x^2 - x - 1 = 0$.

Answer. $\dfrac{1 - \sqrt{13}}{6}, \dfrac{1 + \sqrt{13}}{6}.$

Complex Roots

When the quantity $b^2 - 4ac$ is *negative*, the quadratic equation has *no real roots*. The solutions are complex numbers, as we now explain.

Principal Square Root of a Negative Number

Let $q > 0$ be a real number. The *principal square root* of $-q$ is the complex number $i\sqrt{q}$, where $\sqrt{q}$ is the principal square root of q. We write

$$\sqrt{-q} = i\sqrt{q}.$$

For example, $\sqrt{-4} = 2i$, $\sqrt{-9} = 3i$, and $\sqrt{-5} = i\sqrt{5}$. Now, consider a quadratic equation of the form

$$x^2 + q = 0,$$

where $q > 0$. This equation is equivalent to

$$x^2 = -q.$$

From the definition of the principal square root of a negative number, we conclude that the last equation has two complex solutions,

$$x_1 = i\sqrt{q} \quad \text{and} \quad x_2 = -i\sqrt{q}.$$

This can be checked by substituting $i\sqrt{q}$ and $-i\sqrt{q}$ for x into the equation and recalling that $i^2 = -1$ [Section 1.9]. We have

$$(i\sqrt{q})^2 + q = i^2(\sqrt{q})^2 + q = -q + q = 0$$

and

$$(-i\sqrt{q})^2 + q = i^2(\sqrt{q})^2 + q = -q + q = 0.$$

Reasoning in a similar way, you can check that the solutions of a quadratic equation of the form

$$(x + p)^2 + q = 0,$$

where $p, q \in \mathbb{R}$ and $q > 0$ are the complex numbers

$$x_1 = -p + i\sqrt{q} \quad \text{and} \quad x_2 = -p - i\sqrt{q}.$$

◆ **Example 7.** Solve the equation $(x - 2)^2 + 27 = 0$.

Solution. We have

$$(x - 2)^2 + 27 = 0,$$
$$(x - 2)^2 = -27,$$
$$x - 2 = \pm i\sqrt{27} = \pm i3\sqrt{3},$$
$$x = 2 \pm i3\sqrt{3}.$$

Hence $x_1 = 2 + i3\sqrt{3}$ and $x_2 = 2 - i3\sqrt{3}$ are the solutions of the equation.
Check: $((2 + i3\sqrt{3}) - 2)^2 + 27 = (i3\sqrt{3})^2 + 27 = -27 + 27 = 0,$
$((2 - i3\sqrt{3}) - 2)^2 + 27 = (-i3\sqrt{3})^2 + 27 = -27 - 27 = 0.$

◇ **Practice Exercise 7.**   Solve the equation $(x + 1)^2 + 16 = 0$.

Answer. $-1 - 4i, -1 + 4i$.

◆ **Example 8.**   Solve $x^2 - 6x + 13 = 0$ by completing the square.

Solution.   We have

$$x^2 - 6x + 13 = 0,$$
$$x^2 + 2(-3)x + (-3)^2 = -13 + (-3)^2,$$
$$x^2 + 2(-3)x + (-3)^2 = -13 + 9,$$
$$(x - 3)^2 = -4.$$

Taking square roots, we get

$$x - 3 = \pm\sqrt{-4} = \pm 2i$$

or

$$x = 3 \pm 2i.$$

Thus, the given equation has two complex roots, $x = 3 + 2i$ and $x = 3 - 2i$.

◇ **Practice Exercise 8.**   By completing the square, find the solutions of $x^2 - 4x + 13 = 0$.

Answer. $2 - 3i, 2 + 3i$.

If $b^2 - 4ac < 0$, then formula (2.12), together with the definition of the principal square root of a negative number, gives us two complex roots.

◆ **Example 9.**   Solve the quadratic equation $3x^2 - x + 1 = 0$.

Solution.   Apply the quadratic formula (2.12) with $a = 3$, $b = -1$, and $c = 1$:

$$x = \frac{-(-1) \pm \sqrt{(-1)^2 - 4 \cdot 3 \cdot 1}}{2 \cdot 3}$$
$$= \frac{1 \pm \sqrt{1 - 12}}{6}$$
$$= \frac{1 \pm \sqrt{-11}}{6}$$
$$= \frac{1 \pm i\sqrt{11}}{6}.$$

The two complex solutions are $x_1 = (1 + i\sqrt{11})/6$ and $x_2 = (1 - i\sqrt{11})/6$.

◇ **Practice Exercise 9.**   Use the quadratic formula to solve the equation $2x^2 - 3x + 2 = 0$.

Answer. $\dfrac{3 - i\sqrt{7}}{4}, \dfrac{3 + i\sqrt{7}}{4}.$

Now, we summarize the results of this section.

The Discriminant

The quantity $b^2 - 4ac$ appearing under the radical sign in formula (2.12) is called the *discriminant* of the quadratic polynomial $ax^2 + bx + c = 0$, and it is denoted by

$$(2.13) \quad \Delta = b^2 - 4ac.$$

The discriminant plays an important role in the study of quadratic polynomials and equations. If all coefficients of a quadratic equation are integers and the discriminant is a perfect square, then the roots are rational numbers. If the discriminant is a positive number, but not a perfect square, then the roots are irrational numbers. If the discriminant is equal to zero, then the quadratic equation has a double root. Finally, if the discriminant is negative, then the roots are complex numbers.

In general, we can say the following:

a) If $\Delta > 0$, then the quadratic equation has *two distinct real roots*

$$(2.14) \quad x_1 = \frac{-b + \sqrt{\Delta}}{2a} \quad \text{and} \quad x_2 = \frac{-b - \sqrt{\Delta}}{2a}.$$

b) If $\Delta = 0$, then the quadratic equation has a *double root*

$$(2.15) \quad x_1 = x_2 = -\frac{b}{2a}$$

c) If $\Delta < 0$, then the quadratic equation has *two complex roots*

$$(2.16) \quad x_1 = \frac{-b + i\sqrt{-\Delta}}{2a} \quad \text{and} \quad x_2 = \frac{-b - i\sqrt{-\Delta}}{2a}.$$

Note that since $\Delta < 0$, then $-\Delta > 0$, and we have $\sqrt{\Delta} = \sqrt{-(-\Delta)} = i\sqrt{-\Delta}$. Also, observe that the complex roots of a quadratic equation with *real coefficients* are always complex conjugates of each other. According to formulas (2.16) the real and imaginary parts of the complex roots are, respectively,

$$-\frac{b}{2a} \quad \text{and} \quad \frac{\sqrt{-\Delta}}{2a}.$$

Factoring Quadratic Polynomials

We have seen that whenever we can factor a quadratic polynomial as a product of two linear factors, then the solutions of the corresponding quadratic equation are readily obtained: they are the solutions of the linear factors. Conversely, if the solutions of a quadratic equation are known, then we can express the quadratic polynomial as a product of two linear factors.

◆ **Example 10.**   Solve each quadratic equation and use the solutions to factor the corresponding quadratic polynomial: **a)** $x^2 + 6x + 8 = 0$,   **b)** $2x^2 - 5x - 3 = 0$.

Solution.   **a)** We have seen (Example 4) that the solutions of the equation $x^2 + 6x + 8 = 0$ are $x = -2$ and $x = -4$. From these we obtain the linear factors $x + 2$ and $x + 4$. Next, it is easy to check that these are factors of the quadratic polynomial and that the factorization

$$x^2 + 6x + 8 = (x + 2)(x + 4)$$

holds.

b) The solutions of $2x^2 - 5x - 3 = 0$ are $x = -1/2$ and $x = 3$ (see Example 6). From these we obtain the linear factors $x + 1/2$ and $x - 3$. You can check that multiplication of these two factors *will not* reproduce the original polynomial: the leading coefficient of $2x^2 - 5x - 3$ is 2, while the leading coefficient of the product $(x + 1/2)(x - 3)$ is 1. To obtain the original polynomial we have to multiply this product by 2. Thus

$$2x^2 - 5x - 3 = 2(x + 1/2)(x - 3).$$

Since $2(x + 1/2) = 2x + 1$, the last expression can also be written as

$$2x^2 - 5x - 3 = (2x + 1)(x - 3).$$

◇ **Practice Exercise 10.**   Solve the quadratic equations   **a)** $x^2 + 2x - 3 = 0$,   **b)** $3x^2 + x - 2 = 0$. Use the solutions to factor each polynomial.

Answer.   **a)** $(x - 1)(x + 3)$,   **b)** $(3x - 2)(x + 1)$.

Factoring Quadratic Polynomials with Known Roots

It can be proved that if x_1 and x_2 are the roots (real or complex) of the quadratic equation

$$ax^2 + bx + c = 0,$$

then the following factorization holds:

$$ax^2 + bx + c = a(x - x_1)(x - x_2).$$

For more details, we refer you to Exercise 69.

 EXERCISES 2.5

In Exercises 1–16, solve each equation by moving the constant term to the right side and taking square roots.

1. $x^2 - 4 = 0$ **2.** $x^2 - 25 = 0$ **3.** $2x^2 - 32 = 0$ **4.** $3x^2 - 27 = 0$

5. $4x^2 - 18 = 0$ **6.** $5x^2 - 60 = 0$ **7.** $(x + 1)^2 - 2 = 0$ **8.** $(x - 2)^2 - 4 = 0$

9. $(5x - 2)^2 - 11 = 0$ **10.** $x^2 + 8 = 0$ **11.** $x^2 + 15 = 0$ **12.** $2x^2 + 18 = 0$
13. $(x + 3)^2 + 10 = 0$ **14.** $(x - 5)^2 + 24 = 0$ **15.** $(3x - 2)^2 + 20 = 0$ **16.** $(5x + 1)^2 + 32 = 0$

In Exercises 17–26, solve each equation by factoring.

17. $x^2 - 6x + 8 = 0$ **18.** $x^2 + 5x - 14 = 0$ **19.** $u^2 + 8u + 15 = 0$ **20.** $y^2 + 6y + 8 = 0$
21. $x^2 - 6x + 5 = 0$ **22.** $x^2 - 14x + 24 = 0$ **23.** $v^2 - 3v = 0$ **24.** $w^2 + 5w = 0$
25. $4x^2 - 5x - 6 = 0$ **26.** $5x^2 - 6x + 1 = 0$

In Exercises 27–42, complete the square and solve each equation.

27. $x^2 - 2x - 8 = 0$ **28.** $x^2 + 2x - 8 = 0$ **29.** $m^2 + 8m + 15 = 0$ **30.** $p^2 - 8p + 15 = 0$

31. $x^2 + 10x + 16 = 0$ **32.** $y^2 - 10y + 21 = 0$ **33.** $y^2 - y + \dfrac{2}{9} = 0$ **34.** $x^2 - \dfrac{2}{3}x - \dfrac{1}{3} = 0$

35. $2x^2 + 3x - \dfrac{1}{2} = 0$ **36.** $3x^2 - \dfrac{1}{2}x - \dfrac{1}{4} = 0$ **37.** $x^2 - 6x + 8 = 0$ **38.** $x^2 + 2x + 8 = 0$

39. $2x^2 - 3x + 2 = 0$ **40.** $z^2 - 5z + 7 = 0$ **41.** $3x^2 + 2x + \dfrac{1}{2} = 0$ **42.** $2x^2 - \dfrac{1}{2}x + \dfrac{1}{4} = 0$

In Exercises 43–56, use the quadratic formula to solve each equation.

43. $2x^2 + 4x - 5 = 0$ **44.** $2x^2 + 4x + 5 = 0$ **45.** $u^2 = 2 - u$ **46.** $3v^2 - 3 = -2v$

47. $p + 2p^2 = 3$ **48.** $q = 2 - 3q^2$ **49.** $x^2 + x = \dfrac{3}{2}$ **50.** $12 - 7x - x^2 = 0$

51. $5m^2 - 2m + 4 = 0$ **52.** $6r^2 - 11r + 7 = 0$ **53.** $x - 2 = x^2$ **54.** $3x^2 = 2x - 3$

55. $\dfrac{1}{2}x^2 - \dfrac{1}{3}x + 2 = 0$ **56.** $4x^2 - \dfrac{1}{4}x + 1 = 0$

c *In Exercises 57–60, use the quadratic formula and a calculator to solve each equation. Round off your answers to two decimal places.*

57. $0.6x^2 - 0.3x + 0.1 = 0$ **58.** $0.12x^2 + 0.2x + 0.18 = 0$
59. $1.31x^2 - 3.2x + 5.12 = 0$ **60.** $0.08x^2 - 0.01x + 0.03 = 0$

In Exercises 61–66, solve each quadratic equation and use the solutions to factor the corresponding quadratic polynomial.

61. $2x^2 - 18 = 0$ **62.** $3x^2 - 12 = 0$ **63.** $2x^2 + 3x - 2 = 0$
64. $3x^2 - x - 4 = 0$ **65.** $x^2 - 2x - 1 = 0$ **66.** $x^2 - 4x + 1 = 0$

In Exercises 67–70, x_1 and x_2 denote the roots (real or complex, equal or distinct) of the quadratic equation $ax^2 + bx + c = 0$.

● **67.** Prove that $x_1 + x_2 = -(b/a)$. (*Hint:* Use formula (2.12) and observe that the two radicals have opposite signs.)
● **68.** Prove that $x_1 \cdot x_2 = c/a$. (*Hint:* Use formula (2.12) and the special product $(A + B)(A - B) = A^2 - B^2$.)
● **69.** Using the results of Exercises 67 and 68, prove that

$$ax^2 + bx + c = a(x - x_1)(x - x_2).$$

● **70.** Suppose that the coefficients a, b, and c are real numbers and that the discriminant $\Delta = b^2 - 4ac$ is negative (case of complex conjugate roots). Let

$$h = -\frac{b}{2a} \quad \text{and} \quad k = \frac{\sqrt{-\Delta}}{2a}$$

be the real and imaginary parts of the roots. Prove that the quadratic polynomial can always be rewritten in the following form:

$$ax^2 + bx + c = a[(x - h)^2 + k^2].$$

2.6   EQUATIONS REDUCIBLE TO QUADRATIC FORM

Some equations that are not in quadratic form can be reduced to a quadratic equation. The new equation is then solved by the methods described in the previous sections and the solutions checked in the original equation. Sometimes, the new equation has solutions that are not solutions of the original one. Such solutions are called *extraneous solutions* and should be discarded.

◆ **Example 1.**   Solve $\sqrt{x + 5} = x - 1$.

Solution.   By squaring both sides, we eliminate the radical and transform the given equation into a quadratic equation,

$$x + 5 = (x - 1)^2,$$

or

$$x + 5 = x^2 - 2x + 1,$$

or

$$x^2 - 3x - 4 = 0.$$

The last equation has two solutions, $x = -1$ and $x = 4$, that must be checked into the original equation. Substituting -1 for x, we obtain

$$\sqrt{4} = -2$$

which is false. (Recall that $\sqrt{4}$ denotes the principal square root of 4, so $\sqrt{4} = +2$.) Substituting 4 for x in the original equation, we get

$$\sqrt{9} = 3$$

which is true. Hence, the solution of the original equation is $x = 4$.

◇ **Practice Exercise 1.**   Solve the equation $2x = \sqrt{10x - 4}$.

Answer. $\dfrac{1}{2}$, 2.

◆ **Example 2.**   Solve $\sqrt{3x - 2} - \sqrt{x + 3} = 1$.

Solution.   As a first step, we always rewrite the equation so that one radical appears alone on one side of the equation; this is called *isolating the radical*. Thus we rewrite the given equation as

$$\sqrt{3x - 2} = \sqrt{x + 3} + 1.$$

By squaring both sides, one of the radicals is eliminated:

$$3x - 2 = (\sqrt{x + 3} + 1)^2,$$
$$3x - 2 = x + 3 + 2\sqrt{x + 3} + 1,$$
$$3x - 2 = x + 4 + 2\sqrt{x + 3}.$$

Next, isolate the remaining radical to obtain

$$2x - 6 = 2\sqrt{x + 3},$$

or

$$x - 3 = \sqrt{x + 3}.$$

Squaring both sides of this equation again gives us

$$(x - 3)^2 = x + 3,$$
$$x^2 - 6x + 9 = x + 3,$$
$$x^2 - 7x + 6 = 0;$$

hence

$$(x - 1)(x - 6) = 0,$$

whose solutions are $x = 1$ and $x = 6$. By substituting into the original equation, you can check that 6 is a solution and that 1 is an extraneous solution. Thus, the only valid solution to the problem is $x = 6$.

◇ *Practice Exercise 2.*   Find all solutions of the equation $\sqrt{x + 5} - \sqrt{x - 3} = 2$.

Answer. $x = 4$.

◆ Example 3.   Solve the equation $\dfrac{1}{x} = \dfrac{1}{x^2}$.

Solution.   When $x \neq 0$, the given equation is equivalent to

$$x^2 = x \quad \text{or} \quad x^2 - x = 0.$$

The last equation has solutions $x = 0$ and $x = 1$. Since $x = 0$ cannot be a solution of the original equation, the only solution is $x = 1$.

◇ *Practice Exercise 3.*   Solve the equation $\dfrac{1}{x^2} + \dfrac{1}{3x} = 0$.

Answer. $x = -3$.

◆ **Example 4.**   Find all solutions of $x^4 + x^2 - 12 = 0$.

Solution.   By making the change of variable $x^2 = u$, so that $x^4 = u^2$, we obtain a quadratic equation in u:

$$u^2 + u - 12 = 0$$

or

$$(u + 4)(u - 3) = 0.$$

The solutions of this equation are $u = -4$ and $u = 3$. Since $x^2 = u$, we have

$$x^2 = -4 \quad \text{and} \quad x^2 = 3,$$

so

$$x = \pm 2i \quad \text{and} \quad x = \pm\sqrt{3}.$$

Substituting $2i$ for x into the given equation gives us

$$
\begin{aligned}
(2i)^4 + (2i)^2 - 12 &= 2^4 i^4 + 2^2 i^2 - 12 \\
&= 16 - 4 - 12 \qquad [i^4 = 1, \quad i^2 = -1] \\
&= 0.
\end{aligned}
$$

Thus, $2i$ is a solution. Substituting $\sqrt{3}$ for x, we get

$$(\sqrt{3})^4 + (\sqrt{3})^2 - 12 = 9 + 3 - 12 = 0.$$

Thus, $\sqrt{3}$ is another solution. We leave it to you to check that $-2i$ and $-\sqrt{3}$ are also solutions. Therefore, the given equation has four solutions $-2i$, $2i$, $-\sqrt{3}$, and $\sqrt{3}$.

◇ **Practice Exercise 4.**   Solve the equation $x^4 - 4 = 3x^2$.

Answer. $\pm 2, \pm i$.

◆ **Example 5.**   Find all solutions of the equation $x^{2/3} - 5x^{1/3} + 6 = 0$.

Solution.   If you set $u = x^{1/3}$, so that $u^2 = x^{2/3}$, then the given equation is transformed into the quadratic equation

$$u^2 - 5u + 6 = 0$$

or

$$(u - 2)(u - 3) = 0$$

which has solutions $u = 2$ and $u = 3$. Since $u = x^{1/3}$, then

$$x^{1/3} = 2 \quad \text{and} \quad x^{1/3} = 3.$$

Hence

$$(x^{1/3})^3 = 2^3 \quad \text{and} \quad (x^{1/3})^3 = 3^3,$$

so that

$$x = 8 \quad \text{and} \quad x = 27.$$

You can check that 8 and 27 are both solutions of the given equation.

◇ **Practice Exercise 5.**   Solve the equation $x + 2x^{1/2} - 8 = 0$.

Answer. $x = 4$.

___

📦 **EXERCISES 2.6**

In Exercises 1–16, solve each equation. Check for extraneous solutions.

1. $x = \sqrt{7x - 12}$

2. $x = \sqrt{3x + 10}$

3. $3x = \sqrt{9x - 2}$

4. $2x = \sqrt{4x + 3}$

5. $x + 1 = 2\sqrt{x}$

6. $2x + 1 = \sqrt{8x}$

7. $\sqrt{x - 2} = x - 8$

8. $\sqrt{x + 17} = x - 3$

9. $3\sqrt{x - 8} = x - 6$

10. $5\sqrt{2x + 1} = 4x - 1$

11. $\sqrt{3x + 7} - 1 = x$

12. $\sqrt{7x + 8} - x = 4$

13. $\sqrt{2x + 2} = \sqrt{x + 2} + 1$

14. $\sqrt{x - 3} + 3 = \sqrt{2x + 8}$

15. $\sqrt{2x - 3} + \sqrt{x + 2} = 3$

16. $\sqrt{2x + 14} - \sqrt{x + 3} = 2$

In Exercises 17–50, reduce each equation to a quadratic equation before solving it. Always check for extraneous solutions.

17. $1 + \dfrac{2}{x} - \dfrac{3}{x^2} = 0$

18. $3 + \dfrac{1}{y} + \dfrac{2}{y^2} = 0$

19. $u = \dfrac{2}{u} - 1$

20. $3 = \dfrac{3}{v^2} - \dfrac{2}{v}$

21. $x^{-1} - 4x^{-2} = 0$

22. $3x^{-1} - 5x^{-2} = 0$

23. $x^{-1} - 2x^{-2} = 1$

24. $4 - 4x^{-1} + x^{-2} = 0$

25. $\dfrac{3}{(x - 1)^2} - \dfrac{5}{x - 1} - 2 = 0$

26. $1 - \dfrac{5}{t - 1} + \dfrac{6}{(t - 1)^2} = 0$

27. $\dfrac{3}{(x^2 - 1)^2} - \dfrac{4}{x^2 - 1} + 1 = 0$

28. $2 + \dfrac{5}{t^2 - 1} - \dfrac{3}{(t^2 - 1)^2} = 0$

29. $\left(\dfrac{x}{x + 1}\right)^2 - \dfrac{x}{x + 1} - 2 = 0$

30. $2\left(\dfrac{t}{t - 1}\right)^2 - \dfrac{7t}{t - 1} + 6 = 0$

31. $u^4 - 6u^2 - 8 = 0$

32. $v^4 - 8v^2 + 15 = 0$

33. $4x^4 + 6x^2 - 1 = 0$

34. $12u^4 - 2u^2 - 1 = 0$

35. $2x^{-4} + 3x^{-2} - 2 = 0$

36. $3y^{-4} - 6y^{-2} - 1 = 0$

37. $x + 6\sqrt{x} - 16 = 0$

38. $y + y^{1/2} - 12 = 0$

39. $p - 15p^{1/2} + 26 = 0$

40. $m - 4\sqrt{m} = 0$

41. $y^{-1/2} - 5y^{-1/4} + 6 = 0$

42. $2x^{-1/2} - 11x^{-1/4} + 5 = 0$

43. $x^{2/3} - 6x^{1/3} + 8 = 0$

44. $y^{2/3} + 5y^{1/3} - 14 = 0$

45. $x^6 + 8x^3 + 15 = 0$

46. $r^6 - 14r^3 + 24 = 0$

47. $(x - 2)^2 - 4(x - 2) = 0$

48. $(x + 1)^2 + 7(x + 1) = 0$

49. $(2x + 1)^2 + 15 = 8(2x + 1)$

50. $(3x + 2)^2 + 21 = 10(3x + 2)$

2.7   WORD PROBLEMS INVOLVING QUADRATIC EQUATIONS

In many applications we encounter problems whose solution is given by one or both solutions of a quadratic equation. To handle such problems, you should follow the guidelines explained at the beginning of Section 2.2. As we already said, there are no fixed rules that tell you how to translate a given problem into a mathematical equation. The only way to develop and sharpen your problem-solving skills is to work out as large a number of problems as possible.

◆ **Example 1.**   Find two consecutive odd natural numbers whose product is 143.

Solution.   Any odd natural number can be written as $2n + 1$ where n is a suitable natural number. For example, $7 = 2 \cdot 3 + 1$, $11 = 2 \cdot 5 + 1$, and so on. Moreover, if $2n + 1$ is an odd number, then $2n + 3$ is the following odd number, so that $2n + 1$ and $2n + 3$ are two consecutive odd natural numbers. If their product is 143, we obtain the equation

$$(2n + 1)(2n + 3) = 143.$$

Solving for n, we have

$$4n^2 + 8n - 140 = 0,$$
$$n^2 + 2n - 35 = 0,$$
$$(n - 5)(n + 7) = 0,$$

so $n = 5$ and $n = -7$. Since n represents a natural number, the negative solution must be discarded. Thus the two consecutive odd numbers are $2 \cdot 5 + 1 = 11$ and $2 \cdot 5 + 2 = 13$.
    Check: $11 \times 13 = 143$.

◇ **Practice Exercise 1.**   Find two consecutive even natural numbers whose product is 2808.

Answer.   52 and 54.

◆ **Example 2.**   A rectangular plot of ground measuring 9 ft by 15 ft is to be used as a flower garden. An inside pavement surrounding the entire border of the plot is to be built so that an area of 72 sq ft is left for the flowers. How wide should the pavement be?

Solution.   If x denotes the width of the pavement, then $9 - 2x$ and $15 - 2x$ are the dimensions of the rectangular plot left for the flowers (Figure 2.16), whose area is 72 sq ft. Thus, we have the equation

$$(9 - 2x)(15 - 2x) = 72.$$

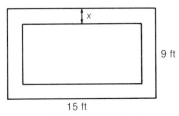

15 ft

Figure 2.16

Solving this equation for x, we get

$$4x^2 - 48x + 135 = 72$$
$$4x^2 - 48x + 63 = 0$$
$$x = \frac{48 \pm \sqrt{1296}}{8} = \frac{48 \pm 36}{8}.$$

Thus, $x = 10.5$ and $x = 1.5$ are the two values that might solve the problem. Since the width of the plot is only 9 ft, the value $x = 10.5$ must be discarded. Therefore, the pavement should be 1.5 ft wide.

Check: The dimensions of the inner rectangle are 6 ft and 12 ft, and its area is 72 sq ft.

◇ **Practice Exercise 2.**   Jane crochets a border of equal width around all sides of a rectangular tablecloth measuring 36 inches by 60 inches. How wide is the crocheted border if the area of the new rectangle is 2560 square inches?

Answer.   2 inches.

◆ **Example 3.**   On a river where the rate of the current was 2 mi/hr, a boat traveled 12 mi upstream and then returned to its departing point. If the round trip took 2.5 h, find the speed of the boat in still water.

Solution.   If x denotes the speed of the boat in still water, then $x - 2$ and $x + 2$ denote, respectively, the speeds of the boat traveling upstream and downstream. Since *distance* = *rate* × *time*, let us make the following table:

	Distance	Rate	Time
Upstream	12	$x - 2$	$12/(x - 2)$
Downstream	12	$x + 2$	$12/(x + 2)$

The total traveling time is the sum of the times to travel upstream and downstream, so

$$\frac{12}{x - 2} + \frac{12}{x + 2} = \frac{5}{2}.$$

Eliminating the denominators and solving for x, we obtain

$$24(x + 2) + 24(x - 2) = 5(x + 2)(x - 2)$$
$$24x + 48 + 24x - 48 = 5x^2 - 20$$
$$5x^2 - 48x - 20 = 0$$
$$(5x + 2)(x - 10) = 0,$$

so $x = -2/5$ and $x = 10$. Speed is always a positive quantity. Thus the speed of the boat in still water is 10 mi/h.

◇ **Practice Exercise 3.**   Fred can row a boat in still water at a rate of 6 miles per hour. On a certain stream, he rows 4 miles upstream and another 4 miles downstream to the departing point, in 1 hour and 30 minutes. What is the rate of the current?

Answer.   2 miles per hour.

◆ **Example 4.**   A travel agent estimates that 20 people will take a sightseeing tour at a cost of $100 per person. She also believes that each $5 reduction of the price will bring 2 additional people. To receive a total of $2250, how many people should take the tour?

Solution.   If n denotes the number of price reductions, then $20 + 2n$ is the number of people taking the tour at $100 - 5n$ dollars per person. Thus, the mathematical equation corresponding to this problem is

$$(20 + 2n)(100 - 5n) = 2250,$$
$$2000 + 100n - 10n^2 = 2250,$$
$$n^2 - 10n + 25 = 0,$$
$$(n - 5)^2 = 0.$$

The number of reductions is then $n = 5$, and the number of people taking the tour is $20 + 2 \cdot 5 = 30$.

◇ **Practice Exercise 4.**   A farmer has 30 orange trees planted in his orchard, and they produce an average of 600 oranges per tree per year. He estimates that for each additional tree planted in the orchard, the annual yield of each tree decreases by 10 oranges. How many additional trees should he plant in order to produce a total average of 20,250 oranges per year?

Answer.   15 trees.

Upward and Downward Motion of a Projectile

If air resistance is neglected, the quadratic equation

$$s = -16t^2 + v_0 t + s_0$$

gives the distance s (in feet) above the ground, at the time t seconds, of a projectile fired directly upward from a height of s_0 feet above the ground and with an initial velocity of v_0 feet per second.

◆ **Example 5.**   Assume that a projectile is fired vertically from ground level with an initial velocity of 88 ft/s. If air resistance is neglected, find when the projectile is 96 ft above the ground.

Solution.   Since $s_0 = 0$ and $v_0 = 88$, the quadratic equation giving the height of the pro-

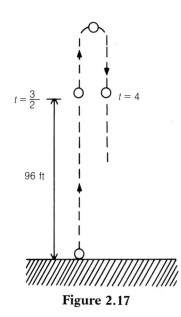

Figure 2.17

jectile is

$$s = -16t^2 + 88t.$$

Setting $s = 96$, we obtain

$$96 = -16t^2 + 88t.$$

Solving for t, we have

$$16t^2 - 88t + 96 = 0,$$
$$2t^2 - 11t + 12 = 0,$$
$$(2t - 3)(t - 4) = 0.$$

Thus, $t = 3/2$ seconds and $t = 4$ seconds. Both solutions are physically possible (Figure 2.17). Moving upward, the projectile takes 3/2 seconds to reach the altitude of 96 ft. Due to the Earth's gravitational pull, the projectile's velocity decreases continuously from the initial value of 88 m/s. When the velocity is zero, the projectile starts to fall back. The time of 4 seconds corresponds to how long it takes for the projectile to reach its maximum altitude and then fall back to a height of 96 ft.

◇ **Practice Exercise 5.** Assume that the projectile in Example 5 is fired upwards from 50 ft above ground level. When will it be 162 ft above the ground?

Answer. 2 seconds and 3.5 seconds.

 EXERCISES 2.7

1. The product of two consecutive natural numbers is 272. Find the numbers.
2. Find two consecutive positive integers whose product is 1260.
3. Find two consecutive even positive integers whose product is 624.
4. Find two consecutive odd natural numbers whose product is 675.
5. The sum of two numbers is 27 and their product is 162. Find the numbers.
6. Find two positive numbers whose difference is 7 and whose product is 120.
7. The sum of a number and its reciprocal is 26/5. Find the number.
8. The difference of a positive number and its reciprocal is 35/6. Find the number.
9. Find m so that one solution of the quadratic equation $x^2 - 16x - 4m = 0$ is 6. Determine the other solution.

10. One solution of the quadratic equation $2x^2 + kx - 40 = 0$ is -8. Find k and the other solution.
11. A rectangle has dimensions 3 in by 6 in. If each side is increased by the same amount, the area of the new rectangle is three times the area of the old one. Find the dimensions of the new rectangle.
12. If the sides of an 8-m by 12-m rectangle are decreased by the same amount, the area of the new rectangle is 5/8 the area of the original rectangle. What are the dimensions of the new rectangle?
13. The area of a rectangle is 96 sq ft and its length is 4 ft longer than its width. Find the dimensions of the rectangle.
14. What are the dimensions of a rectangle whose area is 65 m² and whose length is 3 m less than twice its width?

15. How long are the base and height of a triangle whose area is 45 cm² and whose height is 8 cm longer than twice the base?

16. The area of a triangle is 52 sq ft and its height is 5 ft longer than its base. Find the base and height of the triangle.

17. An open box is to be made from a square piece of cardboard, by cutting out a 4-in square from each corner and folding up the sides. If the volume of the box is to be 144 in³, find the dimensions of the square piece of cardboard.

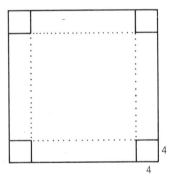

18. The length of a rectangular sheet of tin is twice its width. An open box is to be constructed by cutting out 2-in squares from each corner and folding up the sides. What are the dimensions of the rectangular sheet of tin if the volume of the box is 192 in³?

19. A pavement of uniform width is to be built around a rectangular plot 20 m by 30 m. If the area of the new rectangle is 704 m², find the width of the pavement.

20. The dimensions of a rectangular framed picture are 18 inches by 24 inches. The picture itself is a rectangle whose area is 352 square inches. If the frame has the same width on all four sides, what is its width?

21. The radius of a circle is 8 ft. What change in the radius will increase the area by 36π ft?

22. The surface area of a sphere of radius r is given by $S = 4\pi r^2$. If $r = 4$ cm, what change in the radius will double the sphere's surface area?

23. If p_0 dollars are invested at an interest rate of $100r$ percent compounded annually, then the amount of principal p at the end of t years is given by

$$p = p_0(1 + r)^t.$$

If $p_0 = \$1200$, $p = \$1323$, and $t = 2$ years, find the interest rate r.

24. At what annual rate of interest, compounded annually, should $5000 be invested so that at the end of two years it will increase to $5832?

25. To print 100 posters, a printer charges 20 cents per poster. For each increase of 50 posters, he decreases by 2 cents the price to print each poster. How many posters will he print for $36?

26. A landlady has 100 apartments for rent. She estimates that at $250 per month all apartments will be rented and that for each successive increase of $10 in monthly rent of each apartment, two apartments will remain vacant. If she wants to receive $27,000 per month from rentals, how much should she rent each apartment for?

27. The speed of a boat in still water is 14 mi/h. Knowing that the boat takes one hour longer to travel 48 miles up a river than it takes to return, find the rate of the current.

28. On a stream where the rate of the current was 3 mi/h, Janice rowed upstream for 5 miles and then returned to the starting point. She took $1\frac{3}{4}$ hours to complete the trip. How fast can Janice row in still water?

29. Working together, Peter and Mary can paint a garage in $1\frac{1}{5}$ days. Working alone, Mary can paint the garage in one day less time than Peter. How long does it take each of them to paint the garage working alone?

30. A drain can empty a tank in 40 minutes more time than it takes a pipe to fill it. When both the drain and the pipe are open, it takes 1/2 hour to fill the tank. If the drain is closed, how long will it take the pipe to fill the tank?

31. One bus travels 10 mi/h faster than another bus. If the faster bus takes 1/2 h less time to travel 150 mi, what are the rates of the two buses?

32. A plane doing meteorological observation flies from point A to point B and then back to A in 2.5 hours. The two points are 360 miles apart and the wind is blowing from A to B at a rate of 60 miles per hour. What is the speed of the plane in still air?

33. A projectile is fired vertically from ground level with an initial velocity of 960 ft/s.
   a) When will it be 8000 ft above ground?
   b) When will it hit the ground?

34. A projectile is fired straight upward from a height of 32 ft above the ground and with an initial velocity of 96 ft/s.
   a) When will the projectile be 112 ft above the ground?
   b) When will it hit the ground?

35. A piece of wire of length *l* inches is cut into two pieces, one of which is 16 in long. Each piece is bent into the shape of a square. If the sum of the areas of the squares is 272 in², find the length *l* of the original piece of wire.

36. A wire 28 cm long is cut into two pieces. One piece is bent into the shape of a rectangle whose length is three times its width. The other piece is bent into the shape of a square. If the areas of the rectangle and square add up to 21 cm², what are the dimensions of the rectangle?

● **37.** Suppose that in a certain town the daily demand equation for 6-volt batteries is $q = 4500/p$, where p is the price in dollars of each battery, and q is the quantity of batteries demanded in a day. Suppose also that the supply equation is $Q = 3000p - 1500$, where Q is the quantity of batteries that a manufacturer is willing to supply at p dollars per battery. At what price will $Q = q$; that is, at what price will supply equal demand?

● **38.** The cost equation for a manufacturer of digital watches is $C = x^2 - 18x + 125$, where x is the number of units produced daily. If the company sells each watch for $12, its daily revenue equation is $R = 12x$. Find the break-even points for the manufacturer.

● **c 39.** The Earth's mean radius is 6.371×10^6 m. How far is the horizon of the Earth from a balloon 2.056×10^3 m high?

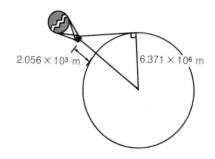

● **c 40.** The area of an equilateral triangle is 15.85 sq ft. Find the dimension of one of its sides.

2.8 QUADRATIC INEQUALITIES

Now, we turn our attention to the problem of solving quadratic inequalities such as $ax^2 + bx + c \geq 0$, $ax^2 + bx + c > 0$, $ax^2 + bx + c \leq 0$, or $ax^2 + bx + c < 0$.

◆ *Example 1.*   Solve the inequality $2x^2 - 5x - 3 \leq 0$ and sketch its solution set.

Solution.   By factoring the quadratic polynomial, notice that the given inequality is equivalent to $(2x + 1)(x - 3) \leq 0$. Before you attempt to solve this inequality, find the roots of the quadratic equation $2x^2 - 5x - 3 = 0$ or $(2x + 1)(x - 3) = 0$. The roots are $x = -1/2$ and $x = 3$, and they divide the real line into the three intervals $(-\infty, -1/2)$, $(-1/2, 3)$, and $(3, +\infty)$. As we already remarked when we solved linear inequalities (Section 2.4), the factor $2x + 1$ is negative whenever $x < -1/2$ and positive whenever $x > -1/2$. Similarly, $x - 3$ is negative for all values of $x < 3$ and positive for all values of $x > 3$. Thus, we have to combine the signs of these two factors in each of the three intervals, to determine the sign of the product $(2x + 1)(x - 3) = 2x^2 - 5x - 3$. This can be done by arranging our work according to the following table.

For x in each of the intervals

Sign of	$-\infty$	$-(1/2)$	3	$+\infty$
$2x + 1$	$-$	$+$		$+$
$x - 3$	$-$	$-$		$+$
$(2x + 1)(x - 3)$	$+$	$-$		$+$

The table shows the real line divided by the points $-1/2$ and 3 into three intervals. The signs of the factors $2x + 1$ and $x - 3$ are indicated in each of the intervals. Since, in the interval $(-1/2, 3)$, the sign of $2x + 1$ is *positive* while the sign of $x - 3$ is *negative*, it follows that the sign of the product $(2x + 1)(x - 3) = 2x^2 - 5x - 3$ is *negative*. At all points in the other two intervals, the product is positive, so the inequality is not satisfied there. At $x = -1/2$ and at $x = 3$, the quadratic polynomial is zero. Thus, the solution set is the closed interval $[-(1/2), 3]$.

◇ **Practice Exercise 1.**   Find and sketch the solution set of the inequality $2x^2 - 5x - 3 > 0$.

Answer.

◆ **Example 2.**   Solve the quadratic inequality $x^2 - 4x + 1 > 0$. Sketch the solution set.

Solution.   First, find the solutions of the quadratic equation $x^2 - 4x + 1 = 0$. By using the quadratic formula, you can check that the solutions are $x_1 = 2 - \sqrt{3}$ and $x_2 = 2 + \sqrt{3}$. According to our discussion about factoring of quadratic polynomials at the end of Section 2.5, factor the quadratic polynomial as follows:

$$x^2 - 4x + 1 = [x - (2 - \sqrt{3})][x - (2 + \sqrt{3})].$$

Now, to solve the given inequality, just determine the signs of each linear factor appearing in the right-hand side of the last expression. Proceeding as in the previous example, organize your work in tabular form as follows.

For each x in each of the intervals

Sign of	$-\infty$	$2 - \sqrt{3}$	$2 - \sqrt{3}$	$+\infty$
$x - (2 - \sqrt{3})$	$-$	$+$		$+$
$x - (2 + \sqrt{3})$	$-$	$-$		$+$
$x^2 - 4x + 1$	$+$	$-$		$+$

The table shows that the solution set of the given inequality is the union of the intervals $(-\infty, 2 - \sqrt{3})$ and $(2 + \sqrt{3}, +\infty)$.

Notice that since we are solving a strict inequality, the points $2 - \sqrt{3}$ and $2 + \sqrt{3}$ are not included in the solution set.

◇ **Practice Exercise 2.**   Solve the inequality $x^2 - 4x + 1 \leq 0$, and sketch the solution set.

Answer.

Certain other types of inequalities can be solved by the method described above.

◆ **Example 3.**   Solve the inequality $\dfrac{x - 7}{x - 5} \leq 2$.

Solution.   The given inequality is equivalent to $(x - 7)/(x - 5) - 2 \leq 0$. Next, write the left-hand side of this inequality as a single fraction:

$$\frac{x - 7}{x - 5} - \frac{2(x - 5)}{x - 5} \leq 0$$

$$\frac{x - 7 - 2x + 10}{x - 5} \leq 0$$

$$\frac{-x + 3}{x - 5} \leq 0$$

Multiplying both sides of the inequality by -1, so that the coefficient of x in the numerator becomes 1, we obtain $(x - 3)/(x - 5) \geq 0$, an inequality equivalent to the original one. (Note the reversal of the sense of the inequality.) Now, organize your work in tabular form.

| | *For each x in each of the intervals* | | |
Sign of	$-\infty$ 3	5	$+\infty$
$x - 3$	$-$ $+$		$+$
$x - 5$	$-$ $-$		$+$
$(x - 3)/(x - 5)$	$+$ $-$		$+$

The table shows that the rational expression $(x - 3)/(x - 5)$ is nonnegative in each of the intervals $(-\infty, 3]$ and $(5, +\infty)$. Thus, the solution set of the original inequality is the union of these two intervals. Note that 3 is included in the solution set, while 5 is not. (Do you know why?)

◇ **Practice Exercise 3.**   Find the solution set of the inequality $(x - 7)/(x - 5) \geq 2$.

Answer. $[3, 5)$

◆ **Example 4.**   Solve the inequality $(x^2 - x - 2)/(x - 3) > 0$.

Solution.   Factor the numerator, and the given inequality becomes

$$\frac{(x+1)(x-2)}{x-3} > 0,$$

where the left-hand side is now a product and quotient of linear terms. Next, notice that $x + 1 = 0$ when $x = -1$, $x - 2 = 0$ when $x = 2$, and $x - 3 = 0$ when $x = 3$. The points -1, 2, and 3 divide the real line into four intervals as shown in the table below. In each one of the intervals, determine the sign of each linear term. The sign of the rational expression $(x + 1)(x - 2)/(x - 3)$ is then obtained by the rule of signs for the product and quotient of numbers.

		For x in each of the intervals			
Sign of	$-\infty$	-1	2	3	$+\infty$
$x + 1$	$-$	$+$	$+$	$+$	
$x - 2$	$-$	$-$	$+$	$+$	
$x + 3$	$-$	$-$	$-$	$+$	
$(x + 1)(x - 2)/(x - 3)$	$-$	$+$	$-$	$+$	

Thus, the solution set of the given inequality is the union of the intervals $(-1, 2)$ and $(3, +\infty)$. Why are the points -1, 2, and 3 excluded from the solution set?

◇ **Practice Exercise 4.**   Find the solution set of the inequality $(x^2 - x - 2)/(x - 3) \leq 0$.

Answer.   Union of the intervals $(-1, 2)$ and $(3, +\infty)$.

Applications

◆ **Example 5.**   Assume that a projectile is fired vertically upward from ground level with an initial velocity of 88 ft/s. If air resistance is neglected, during what time interval will the projectile be above 96 ft?

Solution.   In Example 5 of Section 2.7, we explained that the equation $s = -16t^2 + 88t$ gives the height above ground of the projectile at time t. To find the time interval for which $s > 96$, we must solve the inequality

$$-16t^2 + 88t > 96,$$

or the equivalent one

$$16t^2 - 88t + 96 < 0.$$

This inequality can be simplified by dividing both sides by 8:

$$2t^2 - 11t + 12 < 0.$$

Factoring the quadratic polynomial, we obtain the equivalent inequality

$$(2t - 3)(t - 4) < 0.$$

Now, we organize our work in tabular form.

For t in each of the intervals

Sign of	$-\infty$		$3/2$		4		$+\infty$
$2t - 3$		$-$		$+$		$+$	
$t - 4$		$-$		$-$		$+$	
$(2t - 3)(t - 4)$		$+$		$-$		$+$	

Thus, for all t such that $3/2 < t < 4$, the projectile will be above 96 ft.

Note that when $t = 3/2$ and $t = 4$, the projectile will be exactly at the height of 96 ft (see Example 5 of Section 2.7).

◇ **Practice Exercise 5.** A travel agent estimates that 20 people will take a sightseeing tour at a cost of $100 per person. She also believes that each $5 reduction of the price will bring 2 additional people. How many reductions does she have to make to receive at least $2250? (*Hint*: See Example 4 of Section 2.7.)

Answer. Exactly 5 reductions.

EXERCISES 2.8

Solve the inequalities and sketch the solution set.

1. $(x + 2)(x - 5) > 0$
2. $(x - 3)(x + 3) \leq 0$
3. $(2x + 1)(x - 1) \geq 0$
4. $(3x - 1)(2x + 1) > 0$
5. $4 - x^2 < 0$
6. $9 - x^2 \geq 0$
7. $2x^2 - 3 < 0$
8. $3x^2 - 5 \leq 0$
9. $3x^2 + 1 > 0$
10. $2x^2 + 4 \leq 0$
11. $x^2 + 3x \geq 0$
12. $x^2 + 2x < 0$
13. $x^2 + 3x - 10 > 0$
14. $x^2 + x - 12 \geq 0$
15. $x^2 + 10 > 7x$
16. $x^2 + 2x < 15$
17. $x^2 - 4x + 1 \geq 0$
18. $5x^2 + 5x - 1 < 0$
19. $2x^2 + 2x < 7$
20. $x^2 + x \geq 1$
21. $x^2 - 2x - 15 < 0$
22. $2x^2 - x + 1 > 0$
23. $x^2 - 2x + 3 > 0$
24. $3x^2 + 2x + 1 > 0$

25. $\dfrac{x - 1}{x + 3} \leq 0$
26. $\dfrac{x + 4}{x - 5} > 0$
27. $\dfrac{x(x - 2)}{x - 4} \geq 0$
28. $\dfrac{x + 5}{x(x - 3)} < 0$

29. $\dfrac{8}{x + 3} < 1$
30. $\dfrac{2x}{x + 2} \geq 1$
31. $\dfrac{3x + 2}{x + 5} \geq 1$
32. $\dfrac{2x + 3}{x - 1} < 1$

33. $\dfrac{2}{x - 2} > \dfrac{1}{x + 1}$
34. $\dfrac{2}{x + 3} \geq \dfrac{3}{x + 4}$
35. $\dfrac{3}{x^2} > 1$
36. $\dfrac{2}{x^2} \leq 5$

37. $\dfrac{x^2 - 3x + 2}{x + 3} \leq 0$
38. $\dfrac{3x^2 - x}{x^2 - 2x - 8} < 0$
39. $\dfrac{2x^2 + 5x - 3}{x^2 - 3x - 4} > 0$
40. $\dfrac{2x^2 - 5x + 3}{x^2 - 9} \geq 0$

41. $\dfrac{x}{2x - 1} < \dfrac{1}{x + 2}$
42. $\dfrac{2x}{2x - 1} > \dfrac{3}{x + 2}$

43. A projectile is fired vertically upward from ground level with an initial velocity of 960 ft/s. During what time interval will the projectile be   a) above 4400 ft?   b) above 7200 ft?   c) below 5184 ft?

44. A projectile is fire straight upward from height of 32 ft above the ground with an initial velocity of 96 ft/s. During what time interval will the projectile be above 160 ft? Below 112 ft?

45. The weekly revenue for a manufacturer is $R = 600p - 4p^2$, where p is the price in dollars of each unit of the product being manufactured. When is $R > 0$?

46. Suppose that when the price of a product is p dollars each, consumers will demand $225 - p^2$ units of the product. For what prices is the demand positive?

47. In a culture, the number of bacteria is given by $N = 1000 + 50t - 5t^2$, where t is the time in hours. When will the number of bacteria be smaller than 1000?

● **48.** Show that $x^2 - 2ax + a^2 \geq 0$, for all $x \in \mathbb{R}$.

● **49.** If a, b, and c are rational numbers, prove that the equation $ax^2 + bx + c = 0$ cannot have both a rational root and an irrational root.

● **50.** What can you say about the relation between the solutions of $ax^2 + bx + c = 0$ and the solutions of $ax^2 - bx + c = 0$?

CHAPTER SUMMARY

In this chapter we have discussed methods to solve linear or quadratic equations and inequalities. This was done by performing a sequence of admissible operations that change a given equation or inequality into an equivalent, but simpler, equation or inequality whose solution is evident.

For a linear equation, such operations use only the properties of the real numbers listed in Chapter 1. Quadratic equations can be solved either by *factoring* or *completing the square*. This last method also gives us the *quadratic formula*.

Quadratic equations may have *two real and distinct roots, one real root* (called a *double root*), or *two complex conjugate roots*. It is the sign of the *discriminant* that determines the nature of the roots of a quadratic equation.

Linear inequalities are connected with the notion of *order relation* in $\mathbb{R}$. By correctly using the properties of inequalities listed in Section 2.3, we can reduce a given linear inequality to a simpler one whose solution set is easy to find.

It is important to remember that to solve a simple inequality such as $ax + b > 0$ (with a, b real numbers and $a \neq 0$), we should *first* solve the equation $ax + b = 0$. The unique solution $x = -b/a$ of this equation divides the line into two parts. In each of the half-lines that remain when the point $-b/a$ is removed, the sign of the expression $ax + b$ remains the same: always *positive* in one half-line and always *negative* in the other.

To solve a quadratic inequality, first find the solutions of the quadratic equation. Then, use the solutions to divide the real line into disjoint intervals, and study the sign of each factor of the quadratic polynomial in each one of the intervals. For that purpose, a table such as the ones used in Section 2.8 may greatly simplify your work. By combining the signs of the factors in each of the intervals, determine the sign variation of the quadratic polynomial on the whole line. This will give you the solution of the quadratic inequality. The same method can be used for inequalities involving rational expressions that can be written as product and quotient of linear polynomials.

A good part of this chapter is also devoted to the discussion of word problems involving linear or quadratic equations or inequalities. Follow the guidelines described in Section 2.2, and work as many exercises as you can if you wish to improve your skills in this area.

REVIEW EXERCISES

In Exercises 1–36, solve each equation

1. $2x - 3(x - 1) = 4$

2. $4x - 6(x - 2) = 1$

3. $\dfrac{3}{2(x + 2)} = \dfrac{1}{4(x + 3)}$

4. $\dfrac{1}{5(x - 3)} - 4 = \dfrac{1}{2(x - 6)}$

5. $(2x - 1)(x + 3) = 2x(x - 4) + 1$

6. $3x(x - 1) - 8 = (x + 1)(3x - 2)$

7. $\dfrac{4}{x - 1} = \dfrac{2}{x - 3}$

8. $\dfrac{6}{x + 3} = \dfrac{2}{x - 4}$

9. $\dfrac{2}{x} - \dfrac{1}{3} = \dfrac{1}{2x} - \dfrac{1}{4}$

10. $\dfrac{1}{x} - \dfrac{2}{3} = \dfrac{1}{4} - \dfrac{1}{2x}$

11. $\dfrac{x - 7}{x^2 - 4x + 4} = \dfrac{6}{x - 2}$

12. $\dfrac{x + 10}{x^2 - x - 12} = \dfrac{5}{x + 3}$

13. $x^2 - 7x + 12 = 0$

14. $x^2 + x - 30 = 0$

15. $6x^2 - x - 2 = 0$

16. $8x^2 + 2x - 15 = 0$

17. $3x + 1 = \sqrt{12x}$

18. $3x + 2 = 2\sqrt{6x}$

19. $\sqrt{3x - 5} + 3 = x$

20. $\sqrt{1 - 2x} = x + 7$

21. $\dfrac{x - 2}{3x + 4} = \dfrac{x + 1}{2x - 5}$

22. $\dfrac{x + 2}{x - 3} = \dfrac{4x - 1}{2x - 3}$

23. $3 - \dfrac{6}{x} + \dfrac{2}{x^2} = 0$

24. $9 + \dfrac{6}{x} + \dfrac{1}{x^2} = 0$

25. $\dfrac{4}{x} = \dfrac{3}{x^2}$

26. $\dfrac{2}{x} = -\dfrac{5}{x^2}$

27. $\dfrac{4}{x - 5} - 1 = \dfrac{9}{x + 5}$

28. $\dfrac{5}{y - 1} - \dfrac{3}{y} = \dfrac{3}{2}$

29. $2x = \dfrac{2}{x} - 3$

30. $x = \dfrac{10}{x - 3}$

31. $x^{1/2} - 5x^{1/4} + 6 = 0$

32. $2y^{1/2} - 5y^{1/4} + 2 = 0$

33. $\dfrac{2}{x - 1} + \dfrac{5}{x - 7} = \dfrac{2}{4 - x}$

34. $\dfrac{7}{x + 1} + \dfrac{2}{1 - x} = \dfrac{1}{x + 5}$

35. $3\sqrt{x^2 - 9} = 4(x - 2)$

36. $4\sqrt{y^2 - 13} = 3(y + 1)$

In Exercises 37–44, we list several formulas that occur in mathematics and the applied sciences. Solve each of them for the indicated variable.

37. The area of a trapezoid of bases b_1 and b_2 and height h is given by $A = \dfrac{1}{2}(b_1 + b_2)h$. Solve it for b_1.

38. The equation $v = at + v_0$ expresses the linear velocity of an object at time t with initial velocity v_0 and constant acceleration a. Solve it for t.

39. The length of a rod at temperature t is given by $l = l_0(1 + \alpha t)$, where l_0 is the length at the temperature $t = 0°$ and α is a physical constant. Solve it for t.

40. Solve for h: $V = (1/3)\pi r^2 h$.

41. The formula $s = (a/2)t^2 + v_0 t$ gives the distance s traveled by an object at a time t, where v_0 is the initial velocity and a is the acceleration. Solve it for t.

42. Solve for r: $F = G(m_1 m_2)/r^2$.

43. Solve for r: $S = 2\pi r(r + h)$.

44. Solve for r: $V = \pi r^2 h$.

In Exercises 45–56, solve each inequality.

45. $\dfrac{x}{5} - 2 > \dfrac{x-3}{4+2}$

46. $5 - \dfrac{x}{3} \le \dfrac{4-x}{4-2}$

47. $\dfrac{x-5}{9} - \dfrac{3}{4} \le \dfrac{7x}{12} + \dfrac{x-1}{6}$

48. $\dfrac{x}{3} - \dfrac{2x-1}{2} > \dfrac{3x}{4} - \dfrac{x+1}{6}$

49. $\dfrac{x-1}{x+3} < 1$

50. $\dfrac{x+2}{x-4} \ge 2$

51. $\dfrac{x}{x-5} \ge 3$

52. $\dfrac{2x-1}{x-2} < 5$

53. $\dfrac{(2x-1)(x-3)}{3x-5} > 0$

54. $\dfrac{(x+3)(2x-7)}{5x-1} < 0$

55. $\dfrac{3x^2 + 5x - 2}{x^2 - 10x + 21} < 0$

56. $\dfrac{6x^2 - 13x - 5}{4x^2 + 7x - 2} > 0$

57. Find four consecutive integers so that twice the sum of the first three is the sum of 102 and three times the last number.

58. A wire 84 inches long is bent into the shape of a rectangle. Find the dimensions of the rectangle if its length is 8 inches longer than its width.

59. In a school board election, the winning candidate was elected with 990 votes. If 90 voters abstained and the remaining 45% voted against him, find how many people took part in the election.

60. Peter bought a record player at a 15% discount. A few months later, he sold it to a friend at 20% discount over the discounted price. If Peter received $163.20, what was the original price of the record player?

61. If the total box office receipts of a theater were $2537, and 70 more tickets of $6 were sold than tickets of $8.50, how many tickets of $6 were sold?

62. A tank contains 1200 liters of a 3% saline solution. Water evaporation increases the salt concentration of the solution. How many liters of water will have to evaporate to have a 3.2% saline solution?

63. How many gallons of a 90% antifreeze solution should be added to 4 gallons of water to make a 30% antifreeze solution?

64. Two pipes fill a tank in 40 minutes. If one of the pipes takes 2 hours to fill the tank, how long does it take the other pipe to fill the tank?

65. Eight people will share equally the rental cost of a deep-sea fishing boat. If two more people decide to come along, then each share of the original eight will be reduced by $10. How much does it cost to rent the boat?

66. A company that makes digital watches has fixed costs of $2600 per month. The manufacturing cost of each watch is $25, and they can be sold for $35 each. How many watches must be manufactured per month so that the company has a profit of $2580?

67. If the area of a rectangle is 44 m² and its length is 3 m less than twice its width, find the dimensions of the rectangle.

68. Find the value of m so that one solution of the equation $2x^2 + mx - m^2 = 0$ is $x = 3/2$.

69. Find the length of the side of an equilateral triangle whose area is $16\sqrt{3}$ m².

70. The speed of a faster runner is 2 miles per hour more than the speed of a slower runner. If it takes the faster runner 5 minutes longer to run 6 miles than it takes the slower runner to run 4 miles, find the speed of each runner.

71. A shopowner purchased a shipment of glasses for $2400. His employees broke 10 glasses in the process of unpacking them. He sold the remaining glasses at a profit of $1.50 each and realized a total profit of $1765. How many glasses did he buy?

72. A bus company organizes a group excursion under the following conditions. If 150 people or fewer go, the ticket price is $50 per person. For each person in excess of 150, the ticket price will be reduced by 12 cents. If the total received by the bus company was $9225, find how many people took the excursion.

73. In a right triangle the hypotenuse and one leg are, respectively, 8 and 7 units longer than the other leg. Find the dimensions of the triangle.

74. Pipe B takes 20 minutes longer than pipe A to fill a tank. If both pipes can fill the tank in 24 minutes, how long does it take for pipe A to fill the tank?

75. The formula $I = c/d^2$ gives the intensity of illumination in foot-candles on a surface d feet from a light source of c candlepower. How far should a 90-candlepower light be placed from a surface to give the same intensity of illumination as a 40-candlepower light at 20 feet?

● c 76. The radius of a circle is 3.125 cm. What change in the radius will decrease the area by 8.76 cm^2?

● c 77. A sphere has radius 3.015 cm. What change in the radius will triple its surface area?

● c 78. Find the approximate radius of a circle whose area equals that of a square of side 4.25 m.

● c 79. If a search light has an intensity of 1.5×10^6 foot-candles at 55 feet, find its intensity at a distance of 1 mile (1 mile = 5280 feet).

● c 80. How many kiloliters each of a 20.5% acid solution and a 41.2% acid solution should be mixed to prepare 5 kl of a 35.6% acid solution?

3

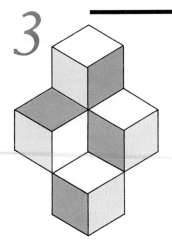

Coordinate Geometry

René Descartes (1596–1650), a French philosopher and mathematician, is considered the founder of analytic geometry, in which algebra and geometry are blended into a single discipline. Analytic geometry is based upon the notion of coordinate systems on the line, on the plane, and in three-dimensional space. Geometric objects are represented by equations, and equations are given geometric meanings. In the first section of this chapter, we introduce the Cartesian coordinate system in a plane and describe the midpoint and distance formulas. This is followed by a study of the straight line and linear relations. Next, we discuss graphs of simple equations and analyze different types of graph symmetry. Finally, the last part of this chapter is devoted to conic sections (circle, ellipse, parabola, and hyperbola), curves defined as intersections of a plane and a right circular cone.

3.1 CARTESIAN COORDINATE SYSTEM

In Section 1.3, we indicated how points on a line can be put into a one-to-one correspondence with real numbers. In this section, we briefly describe how points in a plane can be put into a one-to-one correspondence with ordered pairs of real numbers.

An *ordered pair* of real numbers is a pair of numbers, one of which is designated as the *first* number and the other the *second* number of the pair. If x is the first number and y is the second number, then the ordered pair is denoted (x, y). It is important to notice that the ordered pair $(1, 2)$ is different from the ordered pair $(2, 1)$, and both are different from the *set* $\{1, 2\}$ (which is the same as the set $\{2, 1\}$ because the order of elements in sets is not distinguished). In general, the pair (a, b) is different from the pair (b, a) unless $a = b$ (and from the set $\{a, b\}$ in all cases).

The correspondence between points in a plane and ordered pairs of real numbers is established as follows. In a plane, draw two perpendicular lines through a point O, called the *origin*, and define a coordinate system on each

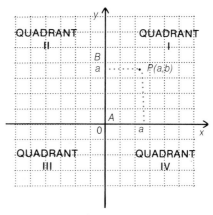

Figure 3.1

Analytical Geometry According to Descartes
In 1619, René Descartes, then 23 years old, had joined the forces of Maximilian, Duke of Bavaria. The Thirty Years' War was ravaging Europe.
   "Amid these campaigns, and especially in the long months when winter interrupted slaughter, Descartes continued his studies, especially of mathematics. One day (November 10, 1619), at Neuburg (near Ulm in Bavaria), he escaped the cold by shutting himself up in a 'stove' (probably an especially heated room). There, he tells us, he had three visions or dreams, in which he saw flashes of light and heard thunder; it seemed to him that some divine spirit was revealing to him a new philosophy. When he emerged from that 'stove' he had (he assures us) formulated analytical geometry, and had conceived the idea of applying the mathematical method to philosophy."
   From *The Story of Civilization*, volume VII, by Will and Ariel Durant.

line. It is customary to choose the same unit of length on both lines, to take one of the lines horizontal with positive direction to the right, and to take the other line vertical with positive direction upward. The horizontal line is often called the *x-axis* and the vertical line is the *y-axis*.

If P is a point in the plane, then the two lines passing through P and parallel to the axes will intersect the axes at the points A and B with coordinates a and b, respectively (Figure 3.1). We assign to the point P the ordered pair of real numbers (a, b), called the *coordinates* of P. The number a is the *first coordinate, x-coordinate,* or *abscissa* of P, and the number b is the *second coordinate, y-coordinate,* or *ordinate* of P.

Conversely, every ordered pair (a, b) determines a unique point in a plane with coordinate axes as follows. To the number a there corresponds a unique point A on the *x*-axis, and to the number b there corresponds a unique point B on the *y*-axis. The line through A parallel to the vertical axis intersects the line through B parallel to the horizontal axis at P, the point that corresponds to the pair (a, b).

A system of axes like the one we have described is called a *system of Cartesian coordinates* in the plane. The two axes divide the plane into four regions (Figure 3.1), each of which is called a *quadrant*. Points in the *first*

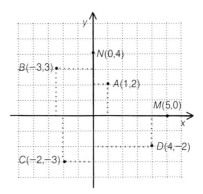

Figure 3.2

quadrant (I) have both coordinates positive; points in the *third quadrant* (III) have both coordinates negative. A point in the *second quadrant* (II) has negative abscissa and positive ordinate, while a point in the *fourth quadrant* (IV) has positive abscissa and negative ordinate. Finally, points on the *x*-axis have ordinates equal to zero, and points on the *y*-axis have abscissas equal to zero. Figure 3.2 shows several points plotted in a coordinate plane.

We now obtain a formula which gives the coordinates of the midpoint of a segment in terms of the coordinates of the endpoints of the segment.

Midpoint Formula

Let $P_1(x_1, y_1)$ and $P_2(x_2, y_2)$ be two points in a coordinate plane. Let M be the midpoint of segment P_1P_2. The coordinates (x, y) of M are given by

$$(3.1) \qquad x = \frac{x_1 + x_2}{2} \quad \text{and} \quad y = \frac{y_1 + y_2}{2}$$

To obtain this formula, observe that the lines through P_1 and P_2 parallel to the *y*-axis intersect the *x*-axis at the points $A_1(x_1, 0)$ and $A_2(x_2, 0)$, respectively. Assuming $x_2 > x_1$ (Figure 3.3), we have

$$d(A_1, A_2) = |x_2 - x_1| = x_2 - x_1.$$

From plane geometry, it follows that the line through M parallel to the *x*-axis intersects the *x*-axis at the midpoint M_1 of the segment A_1A_2. Thus,

$$d(M_1, A_1) = \frac{x_2 - x_1}{2}.$$

On the other hand, if x denotes the abscissa of M_1, then

$$d(M_1, A_1) = x - x_1.$$

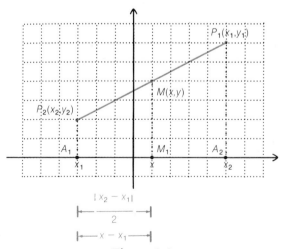

Figure 3.3

Equating the two expressions for $d(M_1, A_1)$ gives

$$x - x_1 = \frac{x_2 - x_1}{2},$$

so

$$x = x_1 + \frac{x_2 - x_1}{2}$$

$$= \frac{x_1 + x_2}{2}.$$

In the same way, the *y*-coordinate of *M* is

$$y = \frac{y_1 + y_2}{2}.$$

◆ **Example 1.**   Find the midpoint of the segment joining the points $A(2, 3)$ and $B(-4, 5)$ and plot the points.

Solution.   According to the midpoint formula, we get

$$x = \frac{2 + (-4)}{2} = \frac{2 - 4}{2} = \frac{-2}{2} = -1$$

$$y = \frac{3 + 5}{2} = \frac{8}{2} = 4.$$

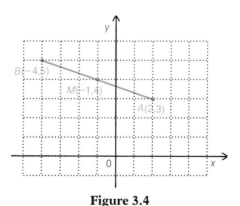

Figure 3.4

◇ **Practice Exercise 1.**   Find the midpoint *M* of the segment with endpoints $(-3, 5)$ and $(7, -2)$.

Answer. $(2, 3/2)$.

Next, we derive a formula that gives the distance between two points in terms of the coordinates of the points.

Distance Formula

If $P_1(x_1, y_1)$ and $P_2(x_2, y_2)$ are two points in a coordinate plane, then the *distance* between P_1 and P_2 is given by

$$(3.2) \quad d(P_1, P_2) = \sqrt{(x_1 - x_2)^2 + (y_1 - y_2)^2}$$

Assume that P_1 and P_2 are not on a horizontal or a vertical line. The triangle P_1AP_2 (Figure 3.5) is a right triangle whose hypotenuse P_1P_2 has length $d(P_1, P_2)$ and whose sides AP_2 and AP_1 have lengths $|x_1 - x_2|$ and $|y_1 - y_2|$, respectively. By the Pythagorean Theorem,

$$[d(P_1, P_2)]^2 = |x_1 - x_2|^2 + |y_1 - y_2|^2.$$

Taking square roots, we obtain

$$d(P_1, P_2) = \sqrt{(x_1 - x_2)^2 + (y_1 - y_2)^2}$$

which is formula (3.2).

If P_1 and P_2 lie on a horizontal line, then $y_1 = y_2$, and formula (3.2) becomes

$$d(P_1, P_2) = \sqrt{(x_1 - x_2)^2} = |x_1 - x_2|.$$

Similarly, if P_1 and P_2 lie on a vertical line, then $x_1 = x_2$ and

$$d(P_1, P_2) = |y_1 - y_2|.$$

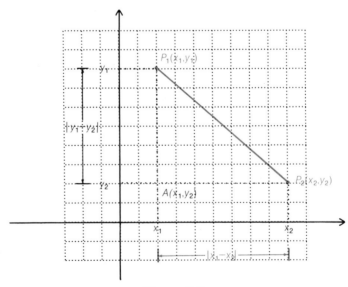

Figure 3.5

◆ **Example 2.** Find the distances between the points **a)** $P(-2, 3)$, $Q(-1, -4)$; **b)** $A(2, 0)$, $B(-3, 0)$.

Solution.   Using the distance formula, we have

a) $d(P, Q) = \sqrt{(-2 - (-1))^2 + (3 - (-4))^2}$
$= \sqrt{(-2 + 1)^2 + (3 + 4)^2}$
$= \sqrt{(-1)^2 + 7^2}$
$= \sqrt{1 + 49}$
$= \sqrt{50}$
$= 5\sqrt{2}$

b) $d(A, B) = \sqrt{(2 - (-3))^2 + (0 - 0)^2}$
$= \sqrt{(2 + 3)^2 - 0^2}$
$= \sqrt{25}$
$= 5$

Since A and B are points on the x-axis with x-coordinates 2 and -3, respectively, their distance could also be obtained by using the formula for the distance between two points on an axis (Section 2.3),

$$d(A, B) = |2 - (-3)| = |2 + 3| = 5.$$

◇ **Practice Exercise 2.**   Find the distances between the following pairs of points: **a)** $P(3, 4)$, $Q(-1, 6)$ **b)** $M(0, -3)$, $N(0, 8)$.

Answer.   **a)** $2\sqrt{5}$ **b)** 11

◆ **Example 3.**   Plot the points $A = (1, 3)$, $B = (2, 1)$, and $C = (6, 3)$ and show that the triangle ABC is a right triangle.

Solution.   By the distance formula, we have

$$d(A, B) = \sqrt{(1 - 2)^2 + (3 - 1)^2} = \sqrt{1 + 4} = \sqrt{5}$$
$$d(B, C) = \sqrt{(2 - 6)^2 + (1 - 3)^2} = \sqrt{16 + 4} = \sqrt{20}$$
$$d(A, C) = \sqrt{(1 - 6)^2 + (3 - 3)^2} = \sqrt{25 + 0} = \sqrt{25}$$

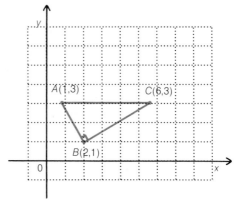

Figure 3.6

Now, to show that *ABC* is a right triangle, we use the converse of the Pythagorean Theorem: if the lengths *a*, *b*, and *c* of the sides of a triangle satisfy the relation $c^2 = a^2 + b^2$, then the triangle is a right triangle. Since

$$(\sqrt{25})^2 = (\sqrt{20})^2 + (\sqrt{5})^2,$$

then *ABC* is a right triangle with hypotenuse *AC*.

◇ **Practice Exercise 3.**   Do the points $X(4, -1)$, $Y(1, 2)$, and $Z(8, 3)$ determine a right triangle?

Answer.   Yes. (Obtain $d(X, Y) = 3\sqrt{2}$, $d(X, Z) = 4\sqrt{2}$, $d(Y, Z) = 5\sqrt{2}$ and use the converse of the Pythagorean Theorem.)

 EXERCISES 3.1

1. Plot the following points on a Cartesian coordinate system: $A(-1, 4)$, $B(-2, -5)$, $C(-1, -3)$, $M(2, 1)$, $N(-3, -2)$.
2. Plot the points $A(2, -4)$, $B(-4, 2)$, $C(1, 3)$, $D(3, 1)$ and draw the segments AB, AD, BC, and CD.

In Exercises 3–8, for the given points P and Q find the midpoint of the segment PQ.

3. $P(-1, 4)$, $Q(-2, 5)$
4. $P(2, 2)$, $Q(-3, -5)$
5. $P(-1, -3/2)$, $Q(1/4, 3)$
6. $P(2/3, -5)$, $Q(-2, 5/4)$
7. $P(2, 5)$, $Q(a, -5)$
8. $P(3, 4)$, $Q(-2, b)$

In Exercises 9–20, find the distance d(P, Q)

9. $P(1, 1)$, $Q(5, 5)$
10. $P(-2, -2)$, $Q(-4, 4)$
11. $P(6, -2)$, $Q(5, -6)$
12. $P(-1, -2)$, $Q(2, 3)$
13. $P(0, -2/3)$, $Q(0, 5/3)$
14. $P(1, 1/2)$, $Q(-1/4, 1)$
15. $P(2, 2)$, $Q(3, -3)$
16. $P(0, 3)$, $Q(-5, 0)$
17. $P(1, b)$, $Q(3, -b)$
18. $P(b, -1)$, $Q(-1, b)$
19. $P(t, t+2)$, $Q(t+3, t-2)$
20. $P(t-1, s)$, $Q(t, s-1)$

c *In Exercises 21–22, use a calculator to find the distance between the two given points. Round off your answers to four significant digits.*

21. $A(21.05, 1.314)$, $B(-3.114, 21.26)$
22. $M(\sqrt{2}, 3/8)$, $N(121/4, \sqrt{3})$
23. Plot the points $A(7, 2)$, $B(4, 6)$, and $C(-4, 0)$ and show that the triangle *ABC* is a right triangle.
24. Show that the triangle with vertices $P(0, 0)$, $Q(2, 0)$, and $R(2, 6)$ is a right triangle.
25. Find the area of the triangle in Exercise 23.
26. Find the area of the triangle *PQR* in Exercise 24.
27. Show that the triangle with vertices $A(1, -2)$, $B(-4, 2)$, and $C(1, 6)$ is an isosceles triangle.
28. Show that the triangle with vertices $P(1, 3)$, $Q(3, 1)$, and $R(4, 4)$ is an isosceles triangle.
29. Plot the points $A(-1, -2)$, $B(4, 3)$, $C(-1, 8)$, and $D(-6, 3)$ and show that they are the vertices of a square.

30. Show that the points $P(-1, -5)$, $Q(1, -2)$, $R(-1, 4)$, and $S(-3, 1)$ are the vertices of a parallelogram.

31. Determine b so that the triangle with vertices $O(0, 0)$, $A(2, 3)$, and $B(3, b)$ is a right triangle with right angle at O.

32. Given the points $A(2, 7)$ and $M(-3, -5)$, find the coordinates of B so that M is the midpoint of AB.

33. Given $A(3, 1)$ and $B(7, 5)$, find the coordinates of the point M on the segment AB such that $d(A, M) = (1/4)d(A, B)$. [*Hint*: If x is the abscissa of M, then $x - 3 = (1/4)(7 - 3)$.]

34. Given $A(1, 2)$ and $B(-3, 3)$, find the coordinates of the point on the segment AB that is 2/3 of the way from A to B.

35. Show that the point $P(4, 2)$ is on the perpendicular bisector of AB, where $A(-1, 3)$ and $B(3, -3)$. [*Hint*: If P is on the perpendicular bisector of AB, then $d(A, P) = d(B, P)$.]

36. Find the formula that expresses the fact that $P(x, y)$ is on the perpendicular bisector of AB, where $A(-1, 3)$ and $B(3, -3)$.

In Exercises 37–40 determine, by checking the relation $d(P, Q) = d(P, R) + d(R, Q)$, whether or not the point R lies on the line segment PQ. It can be shown that R lies on the segment PQ exactly when $d(P, Q) = d(P, R) + d(R, Q)$.

37. $P(-2, -3)$, $Q(4, 3)$, $R(2, 1)$

38. $P(0, 2/5)$, $Q(2, 2)$, $R(-3, -2)$

39. $P(-3, 1/2)$, $Q(1/3, 2)$, $R(1, 1)$

40. $P(2/3, 0)$, $Q(-1, -2)$, $R(-2, 1)$

3.2 THE STRAIGHT LINE

A line is a geometric object. It is known from Euclidean geometry that two distinct points determine a unique line. If the two points lie on a Cartesian coordinate plane, the line also lies in the same plane and it can be described by an algebraic equation that relates the coordinates x and y of an arbitrary point on the line to the coordinates of the two points. The line itself is called the *graph of the equation.*

In order to derive the equation of a line, we introduce the notion of *slope,* a number that measures the steepness of a line relative to the x-axis.

The Slope of a Line

> Let $P_1(x_1, y_1)$ and $P_2(x_2, y_2)$ be two distinct points on a coordinate plane such that $x_1 \neq x_2$. The *slope m of the line l determined by these two points* is defined by
>
> $$(3.3) \quad m = \frac{y_2 - y_1}{x_2 - x_1}.$$

If $x_1 = x_2$, the line l is *vertical* (Figure 3.7a) and, in this case, the slope is *not defined*. Referring to Figure 3.7b, we may say that the differences $y_2 - y_1$

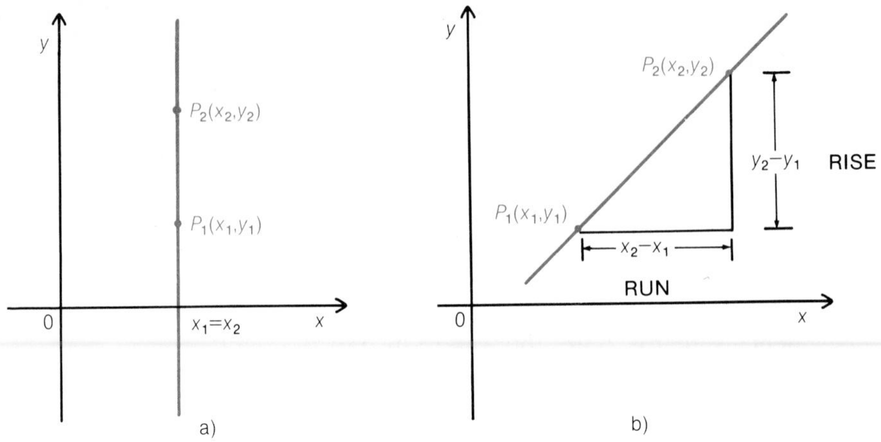

Figure 3.7

and $x_2 - x_1$ are the *rise* and *run* along l from P_1 to P_2 and write

$$\text{slope} = \frac{\text{rise}}{\text{run}}.$$

If the labeling of the points is reversed, the slope remains the same, since

$$m = \frac{y_2 - y_1}{x_2 - x_1} = \frac{y_1 - y_2}{x_1 - x_2}.$$

The slope of a line can be *any* real number: *positive, negative,* or *zero.* This corresponds to the fact that infinitely many lines can pass through a fixed point (Figure 3.8). If $m = 0$, then the line is parallel to the x-axis, so it is a *horizontal* line.

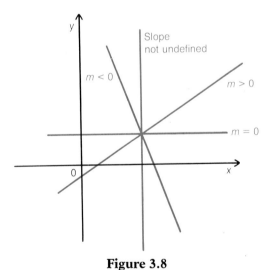

Figure 3.8

◆ **Example 1.** If possible, find the slopes of the lines determined by the following pairs of points: **a)** $A(-1, 2)$, $B(2, 5)$ **b)** $C(-1, -3)$, $D(-1, 2)$

Solution.   **a)** According to (3.3), we have

$$m = \frac{5 - 2}{2 - (-1)} = \frac{3}{3} = 1$$

b) Since C and D have the same x-coordinate (-1), the slope is undefined and the line through C and D is vertical.

◇ **Practice Exercise 1.**   Find the slopes of the following pairs of points: **a)** $M(-3, 2)$, $N(4, 2)$
b) $P(-3, 4)$, $Q(2, 1)$

Answer.   **a)** $m = 0$ **b)** $m = -\dfrac{3}{5}$.

Equations of a Line

Let l be the line determined by two distinct points $P_1(x_1, y_1)$ and $P_2(x_2, y_2)$, with $x_1 \neq x_2$, and let $P(x, y)$ be an arbitrary point on l (Figure 3.9). We want to find how the coordinates x and y of P are related to the coordinates of P_1 and P_2. By doing so we shall obtain an equation in x and y, called *equation of the line l.*

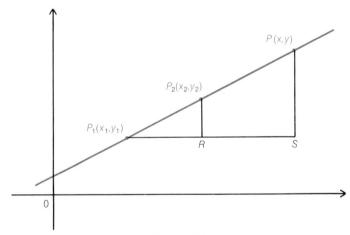

Figure 3.9

Point Slope Form

Since the triangles $P_1 R P_2$ and $P_1 S P$ are similar, the slope of the line determined by P_1 and P_2 is equal to the slope of the line determined by P_1 and P, which we express as

$$\frac{y - y_1}{x - x_1} = \frac{y_2 - y_1}{x_2 - y_1} = m$$

or

$$\frac{y - y_1}{x - x_1} = m.$$

Hence, we obtain

$$(3.4) \quad y - y_1 = m(x - x_1)$$

which is the *point-slope form* of the equation of the line *l*.

Notice that in this equation, x_1, x_2, and m are *known* quantities while x and y are *variables*. Actually, the ordered pair (x, y) represents the coordinates of an arbitrary point P on *l*. We can say that *l* is the set of all points $P(x, y)$ in the plane which satisfy the equation (3.4) and write

$$l = \{P(x, y): y - y_1 = m(x - x_1)\}.$$

Such a set is also called the *graph* of the equation (3.4). (Graphs of equations will be discussed in detail in the next section.)

◆ **Example 2.**   Find an equation of the line with slope $-1/3$ that passes through the point $A(1, -2)$.

Solution.   Using the point-slope form with $m = -1/3$, $x_1 = 1$, and $y_1 = -2$, we get

$$y - (-2) = -\frac{1}{3}(x - 1)$$

or

$$y + 2 = -\frac{1}{3}(x - 1).$$

We will write this equation in two equivalent forms. First, we solve for y and simplify, to obtain

$$y = -\frac{1}{3}x - \frac{5}{3}.$$

Next, we multiply by 3 and rearrange the terms, to get

$$x + 3y + 5 = 0.$$

◇ **Practice Exercise 2.**   Find the point-slope equation of the line through $P(-3, 7)$ and with slope $5/3$.

Answer. $y - 7 = \dfrac{5}{3}(x + 3).$

◆ **Example 3.**   Find an equation of the line through the points $A(-1, 2)$ and $B(3, 5)$.

Solution.   First, determine the slope of the line:

$$m = \frac{5 - 2}{3 - (-1)} = \frac{3}{4}.$$

Next, use equation (3.4) with $m = 3/4$, $x_1 = -1$, and $y_1 = 2$, to get

$$y - 2 = \frac{3}{4}(x + 1).$$

This is the point-slope equation of the line with slope 3/4 passing through $A(-1, 2)$. The last equation can also be written in the following equivalent forms:

$$y = \frac{3}{4}x + \frac{11}{4}$$

or

$$3x - 4y + 11 = 0.$$

As an exercise, you can check that the last two equations can be obtained by writing the point-slope equation of the line through $B(3, 5)$ with slope 3/4.

◇ **Practice Exercise 3.**   Find an equation of the line through $P(1, -2)$ and $Q(-5, 3)$.

Answer. $y + 2 = -\frac{5}{6}(x - 1)$ or $5x + 6y + 7 = 0$.

Equation of a Vertical Line

If $x_1 = x_2$, then the line l passing through $P_1(x_1, y_1)$ and $P_2(x_2, y_2)$ is vertical. A point $P(x, y)$ lies on l if and only if

$$(3.5) \quad x = x_1.$$

This equation represents a vertical line passing through the point $P_1(x_1, y_1)$. Such a line can be viewed as the set of points in a Cartesian plane whose x-coordinates are equal to x_1 and whose y-coordinates are arbitrary numbers. We may write

$$l = \{P(x, y): x = x_1\}.$$

Slope-Intercept Form

If we solve (3.4) for y, we obtain

$$y = mx + (y_1 - mx_1).$$

Next, setting $b = y_1 - mx_1$ gives us

$$(3.6) \quad y = mx + b.$$

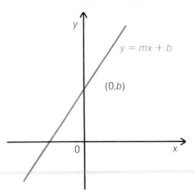

Figure 3.10

This is the *slope-intercept* form of the equation of the line *l*. Notice that in this form, the slope of the line is the *coefficient* of *x*. Moreover, if 0 is substituted for *x* in equation (3.6), then *y* = *b*. This tells us that the point (0, *b*) lies on the line and also on the *y*-axis. Thus, the line crosses the *y*-axis at the point (0, *b*) (Figure 3.10). The number *b* is called the *y-intercept* of the line.

◆ **Example 4.**   Find the slope-intercept equation of the line determined by the points $A(-1, 2)$ and $B(3, 5)$.

Solution.   We saw, in Example 3, that a point-slope equation of this line is

$$y - 2 = \frac{3}{4}(x + 1).$$

Solving for *y* and simplifying gives us

$$y = \frac{3}{4}x + \frac{11}{4},$$

the slope-intercept equation of the line (Figure 3.11). The slope is 3/4, the coefficient of *x*, and the *y*-intercept is 11/4.

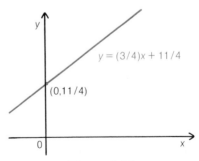

Figure 3.11

◇ **Practice Exercise 4.**   Find the slope-intercept equation of the line through $P(1, -2)$ and $Q(-5, 3)$.

Answer. $y = -\dfrac{5}{6}x - \dfrac{7}{6}.$

It is important to observe that every nonvertical line has a *unique* slope-intercept equation. From the slope-intercept equation of a line, we can read off sufficient geometric data to characterize the line: the coefficient of x measures the steepness of the line relative to the x-axis, and the constant term is the y-coordinate of the point of intersection of the line and the y-axis.

Linear Equations

An equation of the form

$$(3.7) \quad Ax + By + C = 0$$

with A and B not both zero is called a *linear equation in x and y.*

Every equation of a straight line can be put into the form (3.7). For example, we saw that the equation

$$y = \frac{3}{4}x + \frac{11}{4}$$

can be written as $3x - 4y + 11 = 0$, which is of the form (3.7) with $A = 3$, $B = -4$, and $C = 11$. Similarly, the equation

$$x = \frac{2}{3}$$

representing the vertical line through the point $(2/3, 0)$ can be written as $3x - 2 = 0$. This is also an equation of the form (3.7) with $A = 3$, $B = 0$, and $C = 2$.

Conversely, every equation of the form (3.7) with A and B not both zero represents a line. We say that (3.7) is an equation of a line in *standard form.*

◆ **Example 5.**   Write the equation $6x - 4y + 3 = 0$ in slope-intercept form. Find its slope and the x- and y-intercepts of the corresponding line, and sketch it.

Solution.   The slope-intercept form is obtained by solving the given equation for y:

$$-4y = -6x - 3,$$
$$y = \frac{3}{2}x + \frac{3}{4}.$$

Thus, the line has slope $3/2$ and y-intercept $3/4$.

The x-intercept is obtained by finding the point where the line crosses the x-axis. At this point, the y-coordinate is equal to zero. Thus, setting $y = 0$ in the equation $6x - 4y + 3 = 0$ and solving for x, we get

$$6x - 4 \cdot 0 + 3 = 0, \quad \text{or} \quad x = -\frac{3}{6} = -\frac{1}{2}.$$

Thus, $-1/2$ is the x-intercept.

To graph the line (Figure 3.12), just plot the x- and y-intercepts and join them by a straight line.

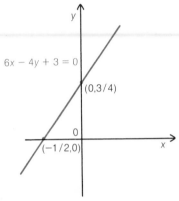

Figure 3.12

◇ **Practice Exercise 5.**   Given the line $2x + 5y - 3 = 0$, find its slope and its x- and y-intercepts. Sketch the line.

Answer.   Slope: $-2/5$, x-intercept: $3/2$, y-intercept: $3/5$.

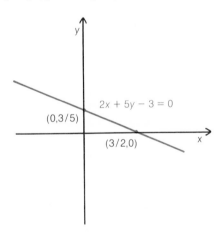

◆ **Example 6.**   What lines have equations   **a)** $6x - 5 = 0$ **b)** $5y + 7 = 0$?

Solution.   **a)** This is an equation of the form (3.7) with $A = 6$, $B = 0$, and $C = -5$. Solving for x, we get

$$x = \frac{5}{6}.$$

Thus, the given equation represents a vertical line through the point $(5/6, 0)$.

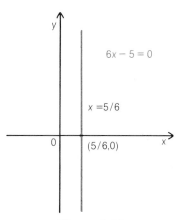

Figure 3.13

b) Here $A = 0$, $B = 5$, and $C = 7$. Solving for y, we obtain

$$y = -\frac{7}{5}.$$

In this case, the given equation represents a horizontal line (i.e., one that is parallel to the x-axis) through $(0, -7/5)$.

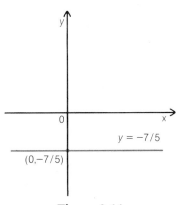

Figure 3.14

◇ **Practice Exercise 6.**   Find the slopes of the following lines: **a)** $3x - 12 = 0$ **b)** $7y - 10 = 0$.

Solution.   **a)** No slope, vertical line through $(4, 0)$ **b)** $m = 0$, horizontal line through $(0, 10/7)$.

Terminology

From now on and when no confusion is possible we shall say *the line* $Ax + By + C = 0$, meaning *the line with equation* $Ax + By + C = 0$.

Parallel Lines

Two lines are said to be parallel if they lie in the same plane and do not intersect each other. Since the slope of a line is a number that measures how steep the line is in relation to the *x*-axis, it seems clear that parallel lines have the same slope. This is the *algebraic condition for parallel lines* which we now state.

Condition for Parallel Lines

Let l and l' be two lines with slopes m and m', respectively (Figure 3.15). The lines are parallel if and only if

$$(3.8) \quad m = m'.$$

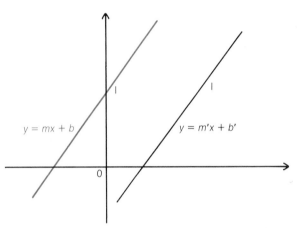

Figure 3.15

◆ **Example 7.**   Find an equation of the line parallel to $2x - y + 6 = 0$ and passing through the point $P(1, 5)$. Sketch it.

Solution.   First, solve the given equation for y and find its slope:

$$y = 2x + 6.$$

Hence, the slope of the given line is $m = 2$. Since the line through $P(1, 5)$ is parallel to the given line, its slope is also 2. Thus, the point-slope form of the line through $P(1, 5)$ and parallel to $2x - y + 6 = 0$ is

$$y - 5 = 2(x - 1).$$

From this equation you may derive the point-intercept equation

$$y = 2x + 3,$$

which is graphed in Figure 3.16.

Note that in the slope-intercept equation of two parallel lines only the *constant* terms differ from each other. The linear terms (i.e., containing *x*) are the same since the lines have the same slope.

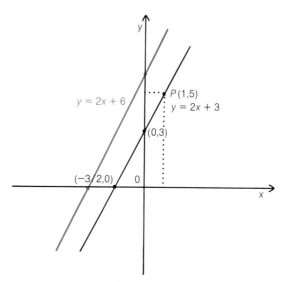

Figure 3.16

◇ **Practice Exercise 7.**   Find an equation of the line passing through the point $P(-2, 4)$ and parallel to the line determined by the points $A(2, 0)$ and $B(3, 5)$.

Answer. $y - 4 = 5(x + 2)$.

Perpendicular Lines

Two lines are perpendicular to each other if they intersect at a right angle. The *algebraic condition for perpendicular lines* (not as obvious as the condition for parallel lines) can be stated as follows.

Condition for Perpendicular Lines

Let l and l' be two lines with slopes m and m' both different from zero. The two lines are perpendicular if and only if

$$(3.9) \quad m' = -\frac{1}{m}.$$

Thus two lines (neither of which is vertical) are perpendicular if and only if *the slope of one line is the negative of the reciprocal of the other slope.*

◆ **Example 8.**   Find an equation of the line perpendicular to the line $2x - 3y - 6 = 0$ and passing through $P(-1, 3)$. Sketch it.

Solution.   First, find the slope of the given line:

$$2x - 3y - 6 = 0,$$
$$-3y = -2x + 6,$$
$$y = \frac{2}{3}x - 2,$$

so

$$m = \frac{2}{3}.$$

According to (3.9), every line perpendicular to the line $2x - 3y - 6 = 0$ has slope

$$m' = -\frac{1}{2/3} = -\frac{3}{2}.$$

Next, write the point-slope equation of the line through $P(-1, 3)$ with slope $-3/2$:

$$y - 3 = -\frac{3}{2}(x + 1).$$

This equation can also be written as

$$y = -\frac{3}{2}x + \frac{3}{2} \qquad \text{(slope-intercept form)}$$

or

$$3x + 2y - 3 = 0 \qquad \text{(standard form)}$$

The line is graphed in Figure 3.17.

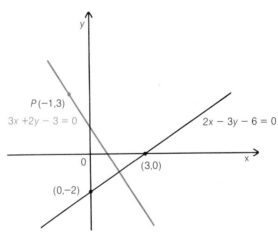

Figure 3.17

◇ **Practice Exercise 8.**   Find an equation of the line perpendicular to the line $3x + 2y + 7 = 0$ and passing through $(-1, -2)$.

Answer. $y + 2 = \frac{2}{3}(x + 1).$

Now we show how to derive the perpendicularity condition (3.9).

The Perpendicularity Condition

Let l and l' be two lines (neither of which is vertical) through the origin with equations $y = mx$ and $y = m'x$ (Figure 3.18). Let $x = 1$ be the vertical line through the point $(1, 0)$. This line intersects l and l' at the points $A(1, m)$ and $B(1, m')$. (Can you see why?) Now, the lines l and l' are perpendicular if and only if the triangle AOB is a right triangle with right angle at the vertex O. By the Pythagorean Theorem,

$$d(A, B)^2 = d(O, A)^2 + d(O, B)^2$$

According to the distance formula (3.2), we have

$$(m - m')^2 = (1 + m^2) + (1 + m'^2).$$

Squaring and simplifying give us

$$m^2 - 2mm' + m'^2 = 1 + m^2 + 1 + m'^2,$$
$$-2mm' = 2,$$
$$mm' = -1.$$

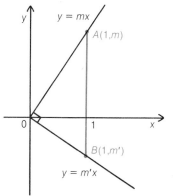

Figure 3.18

Hence

$$m' = -\frac{1}{m},$$

which is condition (3.9).

If the lines l and l' intersect at a point other than the origin, we can apply the same reasoning to the two lines through the origin and parallel to l and l'.

Linear Relations

Two variables x and y that satisfy a linear equation $Ax + By + C = 0$ are said to be *linearly related*.

For example, the relationship between degrees Fahrenheit (F) and degrees Celsius (C) is given by the equation

$$F = \frac{9}{5}C + 32.$$

This equation can be written as $9C - 5F + 160 = 0$, which is an equation of the form (3.7) with $x = C$, $y = F$, $A = 9$, $B = -5$, and $C = 160$. Thus the variables C and F are linearly related.

Notice that in the (C, F) coordinate plane (Figure 3.19), the given equation represents a line whose slope is 9/5 and whose y-intercept is 32.

◆ **Example 9.** Suppose that a variable s (space) is linearly related to a variable t (time). Find an equation relating s and t, knowing that $s = 6$ when $t = 3$, and that $s = 8$ when $t = 6$. Also, find s when $t = 10$.

$$F = (9/5)C + 32$$

(0,32)

Figure 3.19

Solution.   First, make a table of values as follows:

t	s
3	6
6	8

Next, observe that if t and s are linearly related, they must satisfy a linear equation in t and s. This means that the equation represents the line, in the (t, s) coordinate plane, passing through the points $(3, 6)$ and $(6, 8)$. To find the equation of this line, compute its slope:

$$m = \frac{8 - 6}{6 - 3} = \frac{2}{3}.$$

Then write the point-slope equation

$$s - 6 = \frac{2}{3}(t - 3),$$

$$s = \frac{2}{3}t + 4,$$

or, in standard form,

$$2t - 3s + 4 = 0.$$

This is a linear equation relating t and s.

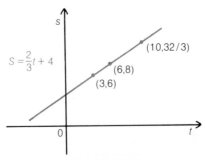

$S = \frac{2}{3}t + 4$

(10,32/3)

(6,8)

(3,6)

Figure 3.20

Setting $t = 10$, we obtain

$$s = \frac{2}{3} \cdot 10 + 4$$

$$= \frac{32}{3}.$$

◇ **Practice Exercise 9.**   Find a linear equation relating x and y if you know that $y = 2$ when $x = -1$, and that $y = -3$ when $x = 2$. Also, find y when $x = 0$.

Answer. $5x + 3y - 1 = 0$, $y = \dfrac{1}{3}$.

 EXERCISES 3.2

In Exercises 1–4, find the slope of the line passing through the given points.

1. $A(2, 2)$, $B(-1/2, -1/2)$
2. $A(1, 3)$, $B(-2/3, -2)$
3. $P(-2, -1/4)$, $Q(3, -2/3)$
4. $U(5, -1/4)$, $V(-5, 1/4)$

In Exercises 5–8, use slopes to determine whether or not the points P, Q, and R lie on a straight line.

5. $P(1, 2)$, $Q(3, 5)$, $R(5, 8)$
6. $P(-1, 3)$, $Q(2, -2)$, $R(0, 1)$
7. $P(2, 2)$, $Q(-2, 4)$, $R(0, 3)$
8. $P(0, 1)$, $Q(1/2, 0)$, $R(2, -3)$

In Exercises 9–16, write an equation of the line having the given properties.

9. Vertical and passing through $(-2, 5)$.
10. Horizontal and passing through $(-1, -3)$.
11. Having slope 3 and y-intercept -2.
12. Having slope $-1/3$ and y-intercept 3.
13. Passing through the origin and having slope $-2/5$.
14. Having slope $-2/3$ and x-intercept 3.
15. Passing through $(-6, -2)$ and having slope 0.
16. Passing through $(-2, 0)$ and having slope $3/4$.

In Exercises 17–24, write each of the lines in slope-intercept form. Find the slope, y-intercept, and x-intercept. Sketch the line.

17. $3x - 4y + 5 = 0$
18. $2x - 3y - 9 = 0$
19. $x = \dfrac{3}{5}y + 3$
20. $\dfrac{2}{3}x + \dfrac{2}{5}y - 4 = 0$

21. $2y + 3 = 0$
22. $4x - 3y = 2$
23. $\dfrac{x}{2} + \dfrac{y}{3} = 1$
24. $\dfrac{x}{4} - \dfrac{x}{2} = 1$

In Exercises 25–28, determine whether the given lines are parallel, perpendicular, or neither. Sketch the lines.

25. $2x + 7y - 2 = 0$, $7x - 2y + 1 = 0$
26. $y = \dfrac{3}{5}x - \dfrac{1}{4}$, $y = -\dfrac{5}{3}x + 2$

27. $x + y + 1 = 0$, $x - y = 0$
28. $\dfrac{x}{2} + \dfrac{y}{3} - 2 = 0$, $\dfrac{2}{3}x + y - 1 = 0$

In Exercises 29–36, find an equation of the line satisfying each of the given conditions. Write the equation in standard form. Sketch it.

29. Parallel to the line $y = 3x - 2$ and passing through $(0, 0)$.

30. Perpendicular to the line $y = 2x - 5$ and passing through $(-1, -2)$.

31. Parallel to the line $2x - 5y - 3 = 0$ and passing through $(-2, 3)$.

32. Perpendicular to the line $6x - 2y + 5 = 0$ and passing through $(-1, -2)$.

33. Perpendicular to the main diagonal (the line $y = x$) and passing through $(1, -4)$.

34. Parallel to the main diagonal and passing through $(-5, -7)$.

35. Parallel to $(1/2)x + (2/3)y - 3/4 = 0$ and passing through $P(3, -3)$.

36. Perpendicular to $(3/5)x - (2/3)y + 1/2 = 0$ and passing through $Q(-1/3, 4/5)$.

37. Find a so that the line $ax + 3y - 5 = 0$ is parallel to the line $x - 4y + 1 = 0$.

38. Find a so that the line $ax - 4y + 3 = 0$ is perpendicular to the line $3x - 2y + 5 = 0$.

39. Find b so that the line $5x - by - 3 = 0$ is perpendicular to the line $2x + 2y - 5 = 0$.

40. Find an equation of the line through $A(-1, 3)$ and parallel to the line determined by the points $B(2, 5)$ and $C(-3, -4)$.

41. Find an equation of the line through $P(-2, -5)$ and perpendicular to the line determined by the points $Q(-2, 9)$ and $R(3, -10)$.

42. By computing the slope of each side, show that the triangle with vertices $A(1, 3)$, $B(2, 1)$, and $C(8, 4)$ is a right triangle.

43. By writing the slope of each side, determine b so that the triangle with vertices $O(0, 0)$, $A(2, 3)$, and $B(3, b)$ is a right triangle with right angle at A.

44. Write an equation for the perpendicular bisector of the segment AB with $A(-1, 2)$ and $B(3, 1)$.

45. The variables u and v are linearly related. When $u = 5$ we have $v = 10$, and when $u = 8$ we have $v = 15$. Find an equation that relates u and v. Also, find v when $u = 2$.

46. Suppose that X and Y are linearly related in such a way that if $X = 2$ then $Y = 10$, and if $X = 5$ then $Y = 25$. Find a) a linear equation relating X and Y, and b) X when $Y = 0$.

47. Find a such that the point $A(1, 2)$ lies on the line $ax - 3y + 4 = 0$.

48. Find b such that the line $3x + by + 2 = 0$ has y-intercept $1/2$.

49. A w-lb weight suspended on a spring causes it to stretch s inches. Hooke's law states that the weight w and the stretch s are linearly related. If a 5-lb weight stretches the spring 1 in and no weight causes no stretch, find a) a linear equation relating s and w and b) the stretch of the spring for a 2-lb weight.

50. A 4-lb weight stretches a spring 1 in and an 8-lb weight stretches it 2 in. Assuming that Hooke's law (Exercise 49) holds, find a linear equation relating a w-lb weight to an s-inch stretch.

51. One of the depreciation methods used for tax purposes is the *straight line method*, which apportions the deductions evenly in each year. Assume that a typewriter purchased for $1,000 is being depreciated linearly for a period of 5 years. (This means that after 5 years the value of the typewriter is $0.) Find

a) the linear equation giving the typewriter's value in t years;

b) its value after 3 years.

52. If the total cost y to manufacture a certain product is linearly related to the number x of units manufactured, find an equation relating x and y from the information that it costs \$90 to manufacture 12 units and \$105 to manufacture 15 units.

53. When air resistance is neglected, the velocity of a projectile fired upward with initial velocity v_0 (in ft/s) is given by the linear equation $v = -32t + v_0$. Assuming that $v_0 = 192$ ft/s, find the velocity of the projectile after 3 seconds. When will the velocity be zero?

54. Suppose that in Exercise 53 the initial velocity is $v_0 = 240$ ft/s. Find the velocity after 5 seconds. When will the velocity be 120 ft/s?

● **55.** Show that the lines $ax + by + c = 0$ and $bx - ay + d = 0$ are perpendicular to each other.

● **56.** Show that the line passing through the point (x_0, y_0) and parallel to the line $ax + by + c = 0$ has equation $a(x - x_0) + b(y - y_0) = 0$. Write an equation of the line that passes through the point $P(7, -3)$ and parallel to the line $3x - 8y + 10 = 0$.

● **57.** Show that an equation of the line through the point (x_0, y_0) and perpendicular to the line $ax + by + c = 0$ is $b(x - x_0) - a(y - y_0) = 0$.

● **58.** If the x- and y-intercepts of a straight line are p and q, respectively, show that an equation of the line is $x/p + y/q = 1$.

● **59.** If both the points $P_1(x_1, y_1)$ and $P_2(x_2, y_2)$, with $x_1 \neq x_2$, lie on the graph of the line $y = mx + b$, prove that $m = (y_2 - y_1)(x_2 - x_1)$ and $b = (y_1 x_2 - y_2 x_1)/(x_2 - x_1)$.

● **60.** Show that if three distinct points (x_1, y_1), (x_2, y_2), and (x_3, y_3) lie on the same line and if $x_1 \neq x_2$, then $x_1 \neq x_3$, $x_2 \neq x_3$, and

$$\frac{y_1 - y_2}{x_1 - x_2} = \frac{y_1 - y_3}{x_1 - x_3} = \frac{y_2 - y_3}{x_2 - x_3}.$$

3.3 GRAPHS OF EQUATIONS

To every set S of ordered pairs (x, y) of real numbers, there is a corresponding set G of points $P(x, y)$ in a coordinate plane. The set G is called the *graph of S.*

In this section, we discuss graphs of simple equations in two variables, such as $x + y = 2$, $y = x^2$, $x = y^2$, $y = x^3$, and so on. Given an equation in two variables x and y, the *solution set* consists of all ordered pairs (x, y) that satisfy the equation. The corresponding set of points $P(x, y)$ in a coordinate plane is the *graph of the equation.*

Following an intuitive approach, we are going to sketch graphs of equations by *plotting points.* This is a rather crude and imprecise method that requires a certain amount of guessing and intuition. Nevertheless, the method provides good practice and we encourage you to graph a large number of equations. As we advance, better graphing techniques will be explained. As we shall see, a careful analysis of a given equation may unveil certain important features that minimize the plotting of points.

Graph Sketching

To sketch the graph of an equation, proceed as follows:
1. Examine the equation and, by reasoning, try to determine the general nature of the curve.
2. Construct a table of numerical values of x and y that satisfy the equation.
3. In a coordinate plane, plot the points (x, y) from the table.
4. Join the plotted points with a smooth curve.

◆ **Example 1.**   Sketch the graph of the set $S = \{(x, y): x + y = 2\}$.

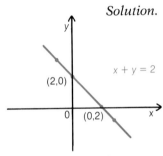

Solution. S is the solution set of the linear equation $x + y = 2$. As we already know, this equation describes a line. To sketch the line, we only have to plot two points such as $(2, 0)$ and $(0, 2)$, the x- and y-intercepts of the line. In this case, a table of values is not really necessary. Figure 3.21 illustrates the graph of the equation $x + y = 2$.

Figure 3.21

◇ **Practice Exercise 1.**   Graph the equation $3x - 2y = 1$.

Answer.

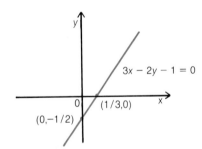

◆ **Example 2.**   Sketch the graph of the equation $y = x^2$.

Solution.   Choosing several values of x and finding the corresponding values of y, make a table such as

x	-3	-2	-1	0	1	2	3
y	9	4	1	0	1	4	9

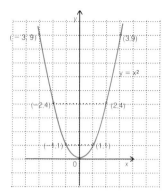

Figure 3.22

Plot the ordered pairs obtained from this table and connect them with a smooth curve as illustrated in Figure 3.22. Since $y = x^2 \geq 0$ for all values of x, the graph is located in the upper half-plane. The curve is called a *parabola* and the point $(0, 0)$ is its *vertex*.

The parabola opens *upward*, and the vertex is its lowest point. Notice that if a pair (x, y) satisfies the equation $y = x^2$, then the pair $(-x, y)$ also satisfies the equation. This means that the parabola is *symmetric with respect to the y-axis*, and the y-axis is called the *axis of symmetry*. If the coordinate plane were folded along the axis of symmetry, the point $(1, 1)$ would coincide with $(-1, 1)$, $(2, 4)$ with $(-2, 4)$, and so on.

◇ **Practice Exercise 2.**   Sketch the graph of the equation $y = -x^2$.

Answer.

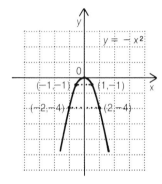

◆ **Example 3.**   Graph the equation $x = y^2$.

Solution.   In this case it is easier to assign values to y and find the corresponding x. Thus, make the following table:

y	-3	-2	-1	0	1	2	3
x	9	4	1	0	1	4	9

Plot the ordered pairs (x, y) obtained from this table and join them by a smooth curve as shown in Figure 3.23.

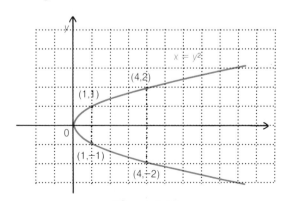

Figure 3.23

The graph is a parabola *opening to the right* and with vertex at the origin. Notice that if a point (x, y) lies on the graph, the point $(x, -y)$ also lies on the graph. In this case we have *symmetry with respect to the x-axis*. If the coordinate plane were folded along the *x*-axis, the point $(1, 1)$ would coincide with $(1, -1)$, $(4, 2)$ with $(4, -2)$, and so on.

◇ **Practice Exercise 3.**   Graph the equation $x = -y^2$.

Answer.

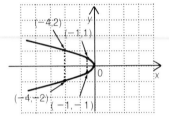

A parabola is a particular *conic section*, that is, a curve defined by the intersection of a plane and a right circular cone. Conic sections will be discussed in detail in Section 3.4.

◆ **Example 4.**   Graph the equation $y = x^3$.

Solution.   You may assign any value to x. The corresponding y is the cube of x. So, you will obtain a table such as

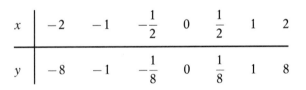

x	-2	-1	$-\dfrac{1}{2}$	0	$\dfrac{1}{2}$	1	2
y	-8	-1	$-\dfrac{1}{8}$	0	$\dfrac{1}{8}$	1	8

By plotting the corresponding pairs and connecting them with a smooth curve, obtain the curve illustrated in Figure 3.24. Now observe that the equation $y = x^3$ is equivalent to the equation $(-y) = (-x)^3$. (Why?) Thus, if (x, y) belongs to the graph, then $(-x, -y)$ also belongs to the graph. This means that the graph is *symmetric with respect to the origin*. If we rotate the graph $180°$ about the origin, the point $(1, 1)$ will coincide with $(-1, -1)$, $(2, 8)$ with $(-2, -8)$, etc., so that the graph will coincide with itself.

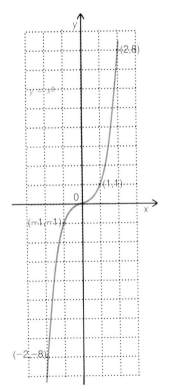

Figure 3.24

◇ **Practice Exercise 4.** Graph $y = -x^3$.

Answer.

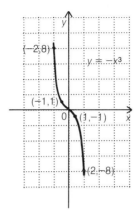

The Equation of a Circle

As an application of the distance formula (3.2), we now show that each circle in the plane is the graph of an equation, called an *equation of the circle*. We shall also show, conversely, that all the graphs of equations of a certain form are circles. Let C be a point in the plane and let r be a nonnegative real number.

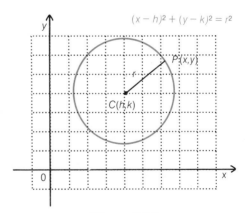

Figure 3.25

The *circle with center C and radius r* is defined, geometrically, to be the set of all points P in the plane whose distances from C are equal to r, that is, all points that satisfy

$$(3.10) \quad d(P, C) = r.$$

If C has coordinates (h, k) and P has coordinates (x, y) (Figure 3.25), then from (3.2) we have

$$d(P, C) = \sqrt{(x - h)^2 + (y - k)^2}$$

and (3.10) becomes

$$\sqrt{(x - h)^2 + (y - k)^2} = r.$$

Squaring both sides yields

$$(3.11) \quad (x - h)^2 + (y - k)^2 = r^2.$$

This is called the *standard form* of the equation of a circle, because it displays the coordinates (h, k) of the center and the radius r. When the center C coincides with the origin, equation (3.11) becomes

$$(3.12) \quad x^2 + y^2 = r^2.$$

◆ **Example 5.** Find an equation of the circle with center $C(1, 2)$ and radius 4.

Solution. By substituting $h = 1$, $k = 2$, and $r = 4$ into the formula (3.11), we obtain

$$(x - 1)^2 + (y - 2)^2 = 16,$$

which is the desired equation.

   Squaring and simplifying the last equation gives us

$$x^2 - 2x + 1 + y^2 - 4y + 4 = 16,$$
$$x^2 + y^2 - 2x - 4y - 11 = 0,$$

which is another form of the equation of the circle centered at $C(1, 2)$ with radius 4.

◇ **Practice Exercise 5.** Find an equation of the circle with center $C(-2, 3)$ and radius 2.

Answer. $(x + 2)^2 + (y - 3)^2 = 4$ or $x^2 + y^2 + 4x - 6y + 9 = 0$.

   If we expand the squares in (3.11), we obtain

$$x^2 - 2hx + h^2 + y^2 - 2ky + k^2 = r^2,$$

or

$$x^2 + y^2 - 2hx - 2ky + h^2 + k^2 - r^2 = 0,$$

which is an equation of the form

$$(3.13) \quad x^2 + y^2 + ax + by + c = 0,$$

where $a = -2h$, $b = -2k$, and $c = h^2 + k^2 - r^2$.

◆ **Example 6.** Find the center and radius of the circle with the equation

$$x^2 + y^2 + 2x + 6y - 6 = 0.$$

Graph this equation.

Solution. Our aim is to replace the given equation with an equivalent one in standard form. This is achieved by using the method of *completing the square*. First, write together the terms containing x and the terms containing y, arranging your work as follows:

$$(x^2 + 2x \quad) + (y^2 + 6y \quad) = 6.$$

Next, complete the squares by adding 1 to the terms inside the first parentheses and 9 to the terms inside the second parentheses:

$$(x^2 + 2x + 1) + (y^2 + 6y + 9) = 6 + 1 + 9.$$

Notice that, in order to balance the equation, it is necessary to add 1 and 9 to the right-hand side of it. The last equation can be rewritten as

$$(x + 1)^2 + (y + 3)^2 = 16.$$

Comparing this equation with (3.11), we see that it represents a circle with center $C(-1, -3)$ and radius 4 (Figure 3.26).

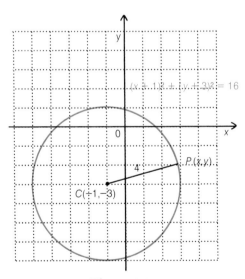

Figure 3.26

◇ **Practice Exercise 6.** Write the equation $x^2 + y^2 - 8x + 2y + 15 = 0$ in standard form.

Answer. $(x - 4)^2 + (y + 1)^2 = 2.$

◆ **Example 7.** Determine whether or not the following equations represent circles:
a) $x^2 + y^2 - 2x - 6y + 10 = 0$, and **b)** $x^2 + y^2 + 4x - 2y + 8 = 0$.

Solution.   **a)** Completing the square in the x and y terms, we obtain

$$(x^2 - 2x + 1) + (y^2 - 6y + 9) = -10 + 1 + 9$$
$$(x - 1)^2 + (y - 3)^2 = 0,$$

which is an equation of the form (3.11) with $h = 1$, $k = 3$, and $r = 0$. Since $x = 1$ and $y = 3$ are the only values that satisfy the last equation, its graph reduces to the *single point* $C(1, 3)$.

b) Completing the squares, we have

$$(x^2 + 4x + 4) + (y^2 - 2y + 1) = -8 + 4 + 1,$$
$$(x + 2)^2 + (y - 1)^2 = -3.$$

Since the right-hand side is negative, while the left-hand side is always non-negative, there are *no* real numbers that can be substituted for x and y to satisfy the last equation. Thus the given equation does not represent a circle; its graph is the *empty set*.

◇ **Practice Exercise 7.**   Do the equations **a)** $x^2 + y^2 + 2x + 3 = 0$ and **b)** $x^2 + y^2 - 4x + 6y + 13 = 0$ represent circles?

Answer.   **a)** No   **b)** No, the point $(2, -3)$ is its only solution.

As we have seen in Examples 6 and 7, an equation of the form (3.13) may represent a *circle*, a *point*, or the *empty set*.

◆ **Example 8.**   Find an equation of a circle passing through the origin and having center $C(-2, 3)$. Graph the equation.

Solution.   First, determine the radius of the circle. Since the origin is on the circle, the distance from the origin to the center $C(-2, 3)$ of the circle must equal the radius of the circle:

$$r = d(C, O) = \sqrt{(-2 - 0)^2 + (3 - 0)^2} = \sqrt{13}.$$

According to (3.11), we obtain the equation

$$(x + 2)^2 + (y - 3)^2 = 13,$$

whose graph is illustrated in Figure 3.27.

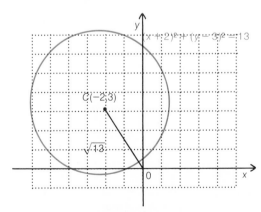

Figure 3.27

◇ **Practice Exercise 8.** Find the standard form equation of the circle with center at $(3, -1)$ and passing through $(1, 0)$.

Answer. $(x - 3)^2 + (y + 1)^2 = 5$.

Circles are also conic sections of a particular kind. You will find more information about conic sections in Section 3.4.

Symmetry

In Examples 2, 3, and 4, we have encountered different types of symmetry of graphs of equations. Now, we discuss the notion of symmetry in general.

Symmetry with Respect to a Line

A curve C is said to be *symmetric with respect to a line l* if to each point P in C there corresponds a point P' in C, such that the line l is the perpendicular bisector of the segment PP' (Figure 3.28).

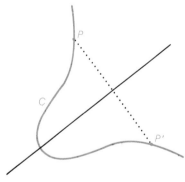

Figure 3.28

We say that l is a *line of symmetry* (or *axis of symmetry*) of the curve C and the points P and P' are *reflections* of each other with respect to the line l.

Symmetry with Respect to a Point

A curve C is *symmetric with respect to a point M* if to each point P in C there corresponds a point P' in C, such that M is the midpoint of the segment PP' (Figure 3.29).

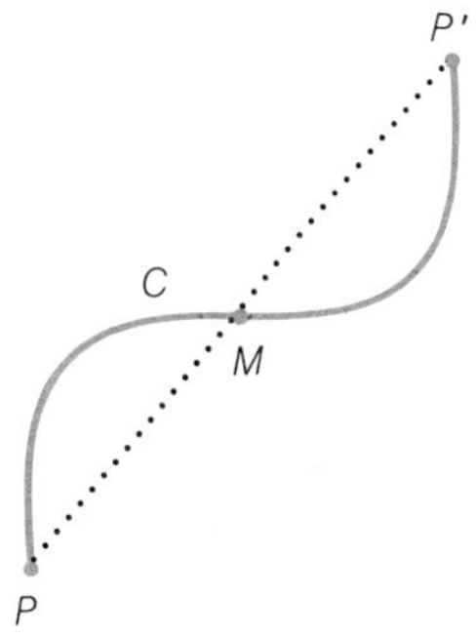

Figure 3.29

The point M is called the *center of symmetry* of the curve C. If we rotate the curve $180°$ about the point M, then the point P will coincide with P'.

Symmetry in Terms of Coordinates

In a coordinate plane, the various notions of symmetry just defined can be translated in terms of coordinates.

---

1. **A curve C is symmetric with respect to the y-axis if, whenever $(x, y) \in C$, then $(-x, y) \in C$.**

For example, the graph of $y = x^2$ (Figure 3.22) is symmetric with respect to the y-axis. Other examples are illustrated in Figure 3.30.

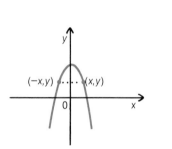 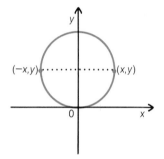

Figure 3.30

---

2. **A curve C is symmetric with respect to the x-axis if, whenever $(x, y) \in C$, then $(x, -y) \in C$.**

Recall that the curve of equation $x = y^2$ (Figure 3.23) is symmetric with respect to the x-axis. Figure 3.31 shows two other examples of symmetry with respect to the x-axis.

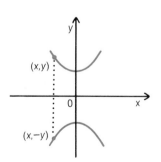 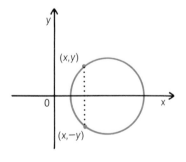

Figure 3.31

> 3. A curve is symmetric with respect to the origin if, whenever
> $(x, y) \in C$, then $(-x, -y) \in C$.

The graph of $y = x^3$ (Figure 3.24) is symmetric with respect to the origin. More examples of symmetry about the origin are shown in Figure 3.32.

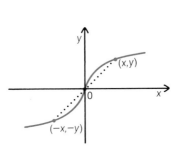 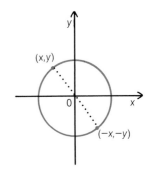

Figure 3.32

Symmetry with Respect to the Line $y = x$

Another important type of symmetry is *symmetry with respect to the line $y = x$.*

> 4. A curve C is symmetric with respect to the line $y = x$ if,
> whenever $(a, b) \in C$, then $(b, a) \in C$.

For example, the curves illustrated in Figure 3.33 are symmetric with respect to the line $y = x$. It is easy to see that the reflection of point (a, b) with respect to the line $y = x$ is the point (b, a) obtained by interchanging the coordinates.

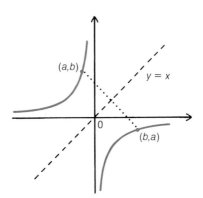 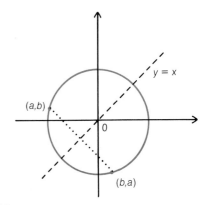

Figure 3.33

EXERCISES 3.3

In Exercises 1–8, plot the given points. Also, plot the points that are symmetric to each point with respect to a) *the x-axis,* b) *the y-axis,* c) *the origin, and* d) *the line y = x.*

1. $(2, -5)$ **2.** $(-3, 1)$ **3.** $(4, 0)$ **4.** $(-2, -2)$

5. $(2, 6)$ **6.** $(3, -8)$ **7.** $\left(\frac{3}{2}, -4\right)$ **8.** $\left(-6, \frac{5}{2}\right)$

In Exercises 9–18, graph each equation and check for symmetry.

9. $2x + y = 3$ **10.** $2x + 2y = 1$ **11.** $2y - 3x^2 = 0$ **12.** $y = 1 - x^2$
13. $x - y^2 - 4 = 0$ **14.** $x + y^2 + 3 = 0$ **15.** $y + x^3 = 0$ **16.** $y = -5x^3$
17. $y = 2 - x^3$ **18.** $x = y^3$

In Exercises 19–28, write an equation of the circle having the indicated properties. Graph it.

19. Center $(1, 1)$; radius $\sqrt{5}$.
20. Center $(-1, -3)$; radius 5.
21. Center at the origin and passing through $(2, 2)$.
22. Center $C(-1, -3)$ and passing through the origin.
23. Center $C(3, 5)$ and tangent to the x-axis.
24. Center $C(-2, -6)$ and tangent to the y-axis.
25. The points $P(-1, 3)$ and $Q(2, 5)$ lie on the circle and the segment PQ is a diameter.
26. The points $A(1, 0)$ and $B(0, 5)$ are endpoints of a diameter.
27. Center in the first quadrant, tangent to the coordinate axes and radius 3.
28. Center in the fourth quadrant, tangent to the coordinate axes and radius 5.

In Exercises 29–40, determine whether each equation represents a circle. If possible, write each equation in the form $(x - h)^2 + (y - k)^2 = r^2$ and state the center and radius. Graph each equation.

29. $x^2 + y^2 - 2y - 5 = 0$ **30.** $x^2 + 2x + y^2 - 3 = 0$
31. $2x^2 + 2y^2 - 18 = 0$ **32.** $3x^2 + 3y^2 - 75 = 0$
33. $x^2 + y^2 - 6x - 4y + 9 = 0$ **34.** $x^2 + y^2 - 2x + 10y + 18 = 0$
35. $x^2 + y^2 - 4y - 5 = 0$ **36.** $x^2 + y^2 - x + 3y - 2 = 0$
37. $x^2 + y^2 + 6x + 4y + 13 = 0$ **38.** $x^2 + y^2 - 3x - 5y - 1 = 0$
39. $3x^2 + 3y^2 + 9x + 12y - 6 = 0$ **40.** $2x^2 + 2y^2 + 4x - 8y - 6 = 0$

3.4 CONIC SECTIONS (Optional)

Conic sections are curves defined by the intersections of a plane and a right circular cone. They were studied by Apollonius of Perga (262–190 BC), a Greek mathematician. His treatise "Conics" is still considered one of the most important works on the subject. The study of conic sections is of contemporary importance: the orbits of planets, satellites, and comets are elliptical; parabolic mirrors are used in telescopes; certain navigational systems are based on properties of hyperbolas.

Since conic sections are plane curves, it follows that they can be defined using only two-dimensional concepts. Moreover, in a Cartesian coordinate system the conic sections can be described by quadratic equations in two variables. The study of such equations and their graphs is the main objective of this section.

Figure 3.34 illustrates four conic sections (parabola, ellipse, circle, and hyperbola) obtained as intersections of a plane and a cone.

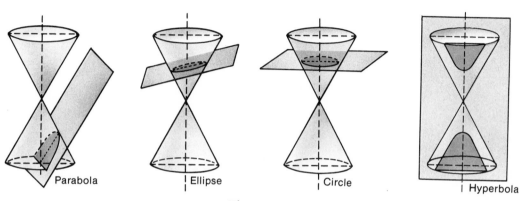

Parabola   Ellipse   Circle   Hyperbola

Figure 3.34

Parabolas

Definition

A *parabola* is the set of points in a plane equidistant from a fixed point and a fixed line. The fixed point is called the *focus* of the parabola, and the fixed line is called the *directrix*.

The line through *F* and perpendicular to the directrix is called the *principal axis* of the parabola (Figure 3.35). The point of intersection *V* of the curve and the principal axis is called the *vertex* of the parabola. It follows from the definition that the vertex *V* is midway between the focus and the directrix. The parabola is symmetric with respect to the principal axis which, for this reason, is also called the *axis of symmetry*.

Parabolic forms are frequently encountered in the physical world, as well as in art and architectural design. If air resistance is neglected, the path described by a projectile under the force of gravity is a parabola. When a parabola is rotated about its principal axis, it generates a surface called a *paraboloid of revolution*. A parabolic mirror (a paraboloid of revolution) has a very important physical property: when a light ray parallel to the principal axis reaches the mirror, it is reflected to the focus (Figure 3.36). For this reason, reflector telescopes use parabolic mirrors. Also, the antenna of a radio telescope has the shape of a parabolic dish. Rays of light emitted from the focus are reflected off the surface of a parabolic mirror in rays parallel to the axis, as in automobile headlights.

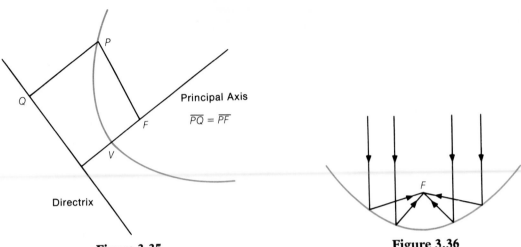

Figure 3.35

Figure 3.36

The Equation of a Parabola

In a Cartesian system of coordinates, a parabola is in *standard position* if its vertex coincides with the origin and its axis coincides with one of the coordinate axes.

To derive the equation of a parabola in standard position, consider the parabola illustrated in Figure 3.37, whose principal axis is the x-axis and whose focus F has coordinates $(p, 0)$ with $p > 0$. The directrix is the vertical line $x = -p$. If $P(x, y)$ lies on the parabola, then by definition,

$$d(P, F) = d(P, Q)$$

where $Q(-p, y)$ is the foot of the perpendicular from P to the directrix.

Since

$$d(P, Q) = |x + p|$$

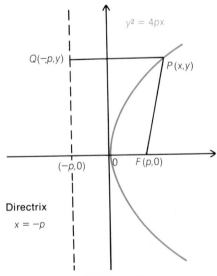

Figure 3.37

and

$$d(P, F) = \sqrt{(x - p)^2 + y^2},$$

we have

$$\sqrt{(x - p)^2 + y^2} = |x + p|.$$

Squaring and simplifying, we obtain

$$(x - p)^2 + y^2 = (x + p)^2$$
$$\cancel{x^2} - 2px + \cancel{p^2} + y^2 = \cancel{x^2} + 2px + \cancel{p^2}.$$

Hence

$$\text{(3.14)} \quad y^2 = 4px.$$

Thus the coordinates (x, y) of a point P on the parabola satisfy the equation (3.14). Since all the algebraic steps are reversible, any point whose coordinates satisfy (3.14) must lie on the parabola.

Equation (3.14) is the *standard equation* of the parabola with focus $F(p, 0)$, $p > 0$, and directrix $x = -p$. Observe that the curve opens to the right.

If the focus is the point $F(-p, 0)$, with $p > 0$, and the directrix is the vertical line $x = p$, then by reasoning in the same manner as above, we get the standard equation

$$\text{(3.15)} \quad y^2 = -4px.$$

This parabola, illustrated in Figure 3.38, opens to the left.

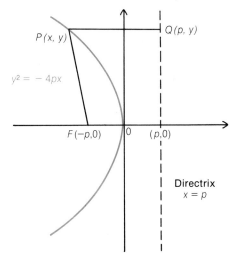

Figure 3.38

If the principal axis of the parabola coincides with the y-axis, then we obtain two other equations, as illustrated in Figure 3.39.

$$\text{(3.16)} \quad x^2 = 4py \qquad x^2 = -4py$$

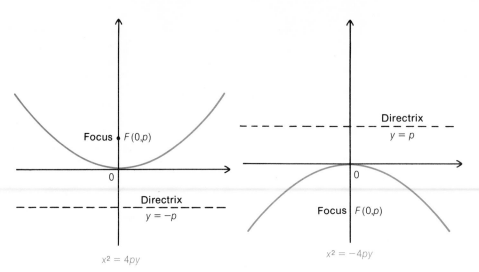

Figure 3.39

In Example 2 of Section 3.3, we indicated by plotting points that the graph of $y = x^2$ is a parabola. This equation can be rewritten as $x^2 = 4(1/4)y$, which is an equation of the form (3.16). Its graph is then a parabola with vertex at the origin whose directrix is the horizontal line $y = 1/4$. A similar observation applies to the equation $y^2 = x$ discussed in Example 3 of Section 3.3.

◆ **Example 1.**   Determine the vertex, focus, principal axis, and directrix of the parabola defined by $y^2 = 8x$. Graph this equation.

Solution.   This is an equation of the form $y^2 = 4px$. Setting $4p = 8$, we get $p = 2$. Thus, the focus is $F(2, 0)$ and the directrix is the vertical line $x = -2$ (Figure 3.40). Also, the parabola has vertex at the origin and the principal axis is the x-axis.

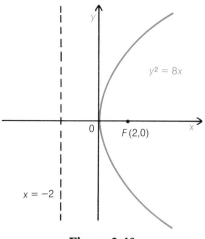

Figure 3.40

◇ **Practice Exercise 1.**   Find the vertex, focus, principal axis, and directrix of the parabola $y^2 = -6x$. Graph this equation.

Answer.   Vertex (0, 0), focus (−3/2, 0), principal axis $y = 0$, directrix $x = 3/2$.

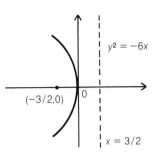

◆ **Example 2.**   Sketch the parabola $y = -(1/6)x^2$. Locate the vertex, focus, principal axis, and directrix.

Solution.   Rewrite the given equation as $x^2 = -6y$. This is now an equation of the form $x^2 = -4py$, with $p = 3/2$. It follows that the parabola has vertex at the origin and focus at $(0, -(3/2))$. The principal axis is the y-axis and the directrix is the horizontal line $y = 3/2$.

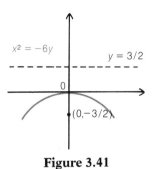

Figure 3.41

◇ **Practice Exercise 2.**   Sketch the parabola $y = (1/8)x^2$, locating the vertex, focus, principal axis, and directrix.

Answer.   Vertex (0, 0), focus (0, 2), principal axis $x = 0$, directrix $y = -2$.

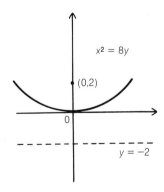

Parabola with Vertex at (h, k)

Four new equations can be derived for parabolas whose vertices are located at a point $V(h, k)$ and whose principal axes are parallel to one of the coordinate

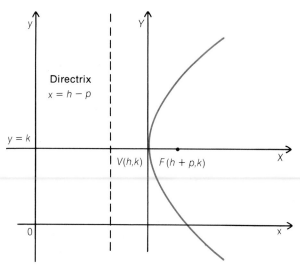

Figure 3.42

axes. For the parabola illustrated in Figure 3.42, the equation is

$$(y - k)^2 = 4p(x - h)$$

The other equations of parabolas are

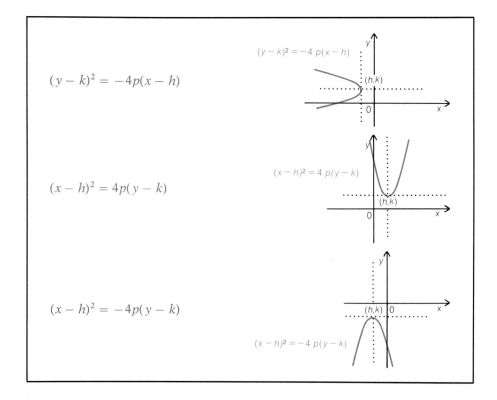

$$(y - k)^2 = -4p(x - h)$$

$$(x - h)^2 = 4p(y - k)$$

$$(x - h)^2 = -4p(y - k)$$

◆ **Example 3.**   Find an equation of the parabola with vertex $V(-2, -1)$ and directrix $y = 1$.

Solution.   On a coordinate plane, plot the vertex $V(-2, -1)$ and the directrix $y = 1$ (Figure 3.43). The axis of symmetry is the vertical line through the vertex, or $x = -2$. Since the vertex is midway between the directrix and the focus, it follows that the coordinates of the focus are $(-2, -3)$.

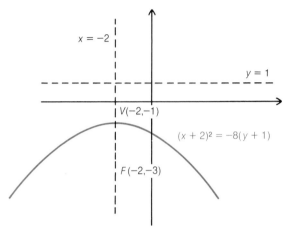

Figure 3.43

Thus, $p = 2$ and the equation is

$$(x - (-2))^2 = -4 \cdot 2(y - (-1))$$

or

$$(x + 2)^2 = -8(y + 1).$$

Squaring and simplifying give us the equivalent equation

$$x^2 + 4x + 8y + 12 = 0.$$

◇ **Practice Exercise 3.**   Find an equation of the parabola with vertex $V(2, 0)$ and focus $(5, 0)$.

Answer. $y^2 = 12(x - 2)$ or $y^2 - 12x + 24 = 0$.

◆ **Example 4.**   Show that $y = x^2 - 2x + 3$ is an equation of a parabola. Find its vertex, focus, principal axis, and directrix. Graph it.

Solution.   Writing the equation as

$$y - 3 = x^2 - 2x$$

and completing the square relative to x, we get

$$y - 3 + 1 = x^2 - 2x + 1$$

or

$$y - 2 = (x - 1)^2.$$

Now, rewrite the last equation as follows:

$$(x - 1)^2 = 4\left(\frac{1}{4}\right)(y - 2)$$

which is an equation of the form $(x - h)^2 = 4p(y - k)$ with $h = 1$, $k = 2$, and $p = 1/4$. This equation is equivalent to the original one, and it is the equation of a parabola with vertex $(1, 2)$ and principal axis $x = 1$ (Figure 3.44). The curve opens upward, since $y - 2$ is equal to a square and is thus always nonnegative. The focus, located on the principal axis, is $1/4$ unit above the vertex; thus $F = (1, 9/4)$. The directrix, parallel to the x-axis, is placed $1/4$ unit below the vertex. Its equation is $y = 7/4$.

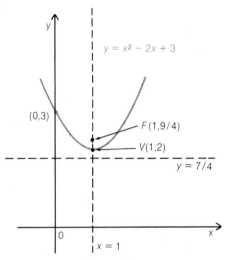

Figure 3.44

◇ **Practice Exercise 4.** Graph the parabola $y^2 + 4x + 2y - 11 = 0$. Locate the vertex, focus, principal axis, and directrix.

Answer. Vertex $(3, -1)$, focus $(2, -1)$, principal axis $y = -1$, directrix $x = 4$.

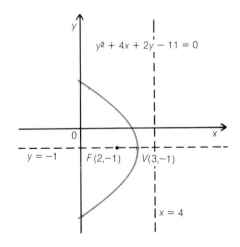

HISTORICAL NOTE

"No one, in Greek times, supposed that conic sections had any utility; at last, in the seventeenth century, Galileo discovered that projectiles move in parabolas, and Kepler discovered that planets move in ellipses. Suddenly, the work that the Greeks had done from pure love of theory became the key to warfare and astronomy."

From *A History of Western Philosophy*, by Bertrand Russell.

Ellipses

Definition

An *ellipse* is the set of points in a plane the sum of whose distances from two fixed points is constant. The two fixed points are called *foci*.

If we denote the foci by F_1 and F_2, the constant by $2a$, and a point on the ellipse by P, then by definition

$$d(P, F_1) + d(P, F_2) = 2a.$$

The line passing through the foci is called the *principal axis* of the ellipse. The curve is *symmetric* with respect to the principal axis. The points V_1 and V_2 (Figure 3.45) where the ellipse intersects the principal axis are called *vertices*. The *center* of the ellipse is the midpoint of the segment $V_1 V_2$.

A very simple method of drawing an ellipse is derived directly from the definition (Figure 3.46). Hold the two ends of a string of length $2a$ fixed at the foci F_1 and F_2. Trace the curve with a pencil, holding it taut against the string.

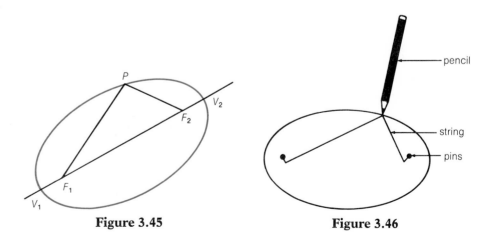

Figure 3.45   **Figure 3.46**

Ellipses are frequently encountered in the physical world. The orbits of planets, some comets, and satellites are elliptical. Some gears and cams, and the domes of some buildings have elliptical forms.

Equations for Ellipses

In a Cartesian coordinate plane, an ellipse is in *standard position* when its principal axis coincides with one of the coordinate axes and its center coincides with the origin.

Consider an ellipse in standard position and such that its principal axis coincides with the *x*-axis (Figure 3.47). The foci are then symmetrically located with respect to the origin, and we may assume that they have coordinates $F_1(-c, 0)$ and $F_2(c, 0)$. If $P(x, y)$ belongs to the ellipse, then

$$d(P, F_1) + d(P, F_2) = 2a$$

or, according to the distance formula,

$$\sqrt{(x + c)^2 + y^2} + \sqrt{(x - c)^2 + y^2} = 2a.$$

Transposing one of the radicals, squaring and simplifying, we obtain

$$a\sqrt{(x - c)^2 + y^2} = a^2 - cx.$$

Squaring and simplifying again, we have

$$(a^2 - c^2)x^2 + a^2y^2 = a^2(a^2 - c^2),$$

so

$$\frac{x^2}{a^2} + \frac{y^2}{a^2 - c^2} = 1.$$

If we consider the triangle PF_1F_2 of Figure 3.47 and remember that the sum of the lengths of any two sides of a triangle is greater than the third side, we see that

$$d(P, F_1) + d(P, F_2) = 2a > 2c = d(F_1, F_2)$$

which means that $a > c$, so $a^2 - c^2 > 0$. This lets us define the following quantity.

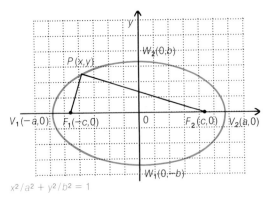

$x^2/a^2 + y^2/b^2 = 1$

Figure 3.47

For an ellipse, there is a number $b > 0$ such that

$$b^2 = a^2 - c^2$$

Replacing $a^2 - c^2$ in the equation above, we obtain the equation of the ellipse in *standard form*:

$$(3.17) \quad \frac{x^2}{a^2} + \frac{y^2}{b^2} = 1 \quad (a > b)$$

Setting $y = 0$ gives us the x-intercepts of the ellipse:

$$\frac{x^2}{a^2} + \frac{0}{b^2} = 1$$

$$\frac{x^2}{a^2} = 1$$

$$x^2 = a^2,$$

hence

$$x = \pm a.$$

Thus, the vertices V_1 and V_2 (Figure 3.47) have coordinates $(a, 0)$ and $(-a, 0)$. Similarly, if we set $x = 0$, we obtain the y-intercepts $\pm b$.

The segment $V_1 V_2$ of length $2a$ is called the *major axis* of the ellipse. The segment $W_1 W_2$ is the *minor axis*; it has length $2b$. Since $a > b$, the major axis is always greater than the minor axis.

Until now we have assumed that the principal axis coincided with the x-axis. If the principal axis is the y-axis, then the roles of the x- and y-coordinates are interchanged. Proceeding as before, we obtain the equation in standard form

$$(3.18) \quad \frac{x^2}{b^2} + \frac{y^2}{a^2} = 1 \quad (a > b)$$

◆ **Example 5.** Graph the equation $16x^2 + 25y^2 = 400$.

Solution. Divide both sides by 400 and get $x^2/25 + y^2/16 = 1$, the standard equation of an ellipse. Setting $y = 0$, obtain the x-intercepts ± 5. Setting $x = 0$, obtain the y-intercepts ± 4. Thus the major axis coincides with the x-axis (Figure 3.48).

Since $c^2 = a^2 - b^2 = 25 - 16 = 9$, it follows that $c = 3$. Thus, the foci are $F_1(-3, 0)$ and $F_2(3, 0)$.

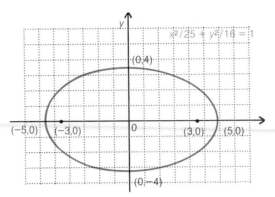

Figure 3.48

◇ **Practice Exercise 5.**   Graph the equation $4x^2 + y^2 = 16$. Determine the vertices, foci, and major axis.

Answer. $V(0, \pm 4)$, $F(0, \pm 2\sqrt{3})$, major axis vertical.

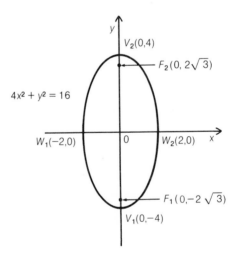

Ellipses with Center at (h, k)

For ellipses centered at a point $C(h, k) \neq O(0, 0)$ and major and minor axes parallel to the coordinate axes, the equations are

$\dfrac{(x - h)^2}{a^2} + \dfrac{(y - k)^2}{b^2} = 1$	$(a > b)$	major axis parallel to the x-axis
$\dfrac{(x - h)^2}{b^2} + \dfrac{(y - k)^2}{a^2} = 1$	$(a > b)$	major axis parallel to the y-axis

◆ **Example 6.**   Find an equation of the ellipse with center $C(2, 1)$, focus $F(0, 1)$, and semi-major axis $a = 3$.

Solution.   The center and focus lie on the horizontal line $y = 1$, as shown in Figure 3.49. The other focus is then $F_2(4, 1)$ and the vertices are $V_1(-1, 1)$ and $V_2(5, 1)$. Since $a = 3$ and $c = 2$ (Why?), we have

$$b^2 = a^2 - c^2$$
$$= 3^2 - 2^2$$
$$= 9 - 4$$
$$= 5.$$

Hence, the semiminor axis is $b = \sqrt{5}$. Thus, the equation of the ellipse in standard form is

$$\frac{(x - 2)^2}{9} + \frac{(y - 1)^2}{5} = 1.$$

By eliminating the denominators, squaring, and simplifying, we get

$$5(x - 2)^2 + 9(y - 1)^2 = 45,$$
$$5(x^2 - 4x + 4) + 9(y^2 - 2y + 1) = 45,$$
$$5x^2 + 9y^2 - 20x - 18y - 16 = 0.$$

The latter is another equation for the ellipse.

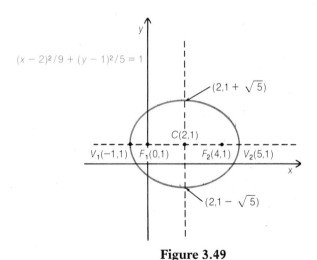

Figure 3.49

◇ **Practice Exercise 6.**   Find an equation of the ellipse with center at $(0, 2)$, vertex at $(0, 6)$, and semiminor axis $b = 3$. Graph it.

Answer. $\dfrac{x^2}{9} + \dfrac{(y - 2)^2}{16} = 1.$

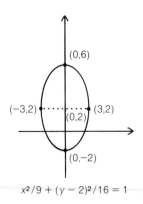

$$x^2/9 + (y - 2)^2/16 = 1$$

◆ Example 7.   Show that the equation $4x^2 + 9y^2 - 8x + 36y + 4 = 0$ represents an ellipse. Find its center and its major and minor axes.

Solution.   The method of solution is the same as the one used to find centers and radii of circles. First, complete the square on x and y, arranging your work as follows:

$$4(x^2 - 2x \quad) + 9(y^2 + 4y \quad) = -4$$
$$4(x^2 - 2x + 1) + 9(y^2 + 4y + 4) = -4 + 4 + 36.$$

Observe that we have added 1 inside the first parentheses and 4 inside the second parentheses. This has the effect of adding 4 and 36 to the left-hand side of the equation. To balance the equation, we have added 4 and 36 to the right-hand side. Next, obtain

$$4(x - 1)^2 + 9(y + 2)^2 = 36.$$

Dividing both sides by 36, we get

$$\frac{(x - 1)^2}{9} + \frac{(y + 2)^2}{4} = 1.$$

This is a standard equation of an ellipse with center at $(1, -2)$, semimajor axis $a = 3$, and semiminor axis $b = 2$. The graph is shown in Figure 3.50.

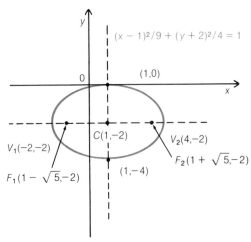

Figure 3.50

◇ **Practice Exercise 7.** Find the center and the major and minor axes of the ellipse

$$4x^2 + y^2 + 16x - 2y + 13 = 0.$$

Answer. $C(-2, 1)$, $a = 2$, $b = 1$, major axis parallel to the y-axis.

Eccentricity of an Ellipse

Definition
The ratio $$e = \frac{c}{a}$$ is called the *eccentricity* of an ellipse.

Since for an ellipse $0 < c < a$, we always have $0 < e < 1$. That is, the eccentricity of an ellipse is a positive number less than 1.

◆ **Example 8.** Find the eccentricity of the ellipse $\dfrac{x^2}{25} + \dfrac{y^2}{16} = 1$.

Solution. As we have seen in Example 5, $a = 5$, $b = 4$, and $c = 3$. Thus,

$$e = \frac{3}{5} = 0.6.$$

◇ **Practice Exercise 8.** What is the eccentricity of the ellipse $\dfrac{x^2}{1} + \dfrac{y^2}{4} = 1$?

Answer. $e = \dfrac{\sqrt{3}}{2}$.

The eccentricity measures how much an ellipse differs from a circle. If a is fixed and c varies from 0 to a, then the corresponding ellipses vary in shape as illustrated in Figure 3.51. If c were equal to zero, the eccentricity e would equal 0, and the ellipse would become a circle. On the other hand, if c were equal to a, in which case the eccentricity would equal 1, the ellipse would reduce to a line segment.

For example, the eccentricity of the orbit of the Earth is approximately 0.017, a small number. Thus the orbit of the Earth is almost a circle. On the other hand, Halley's comet (due to pay us another visit in 1986) has a very elongated orbit. Its eccentricity is approximately 0.98, a number very close to 1!

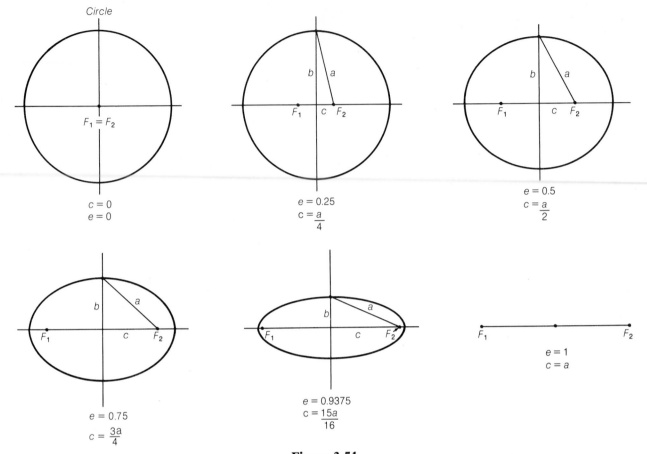

Figure 3.51

Hyperbolas

Definition

A *hyperbola* is the set of points in a plane the difference of whose distances from two fixed points is constant. The two fixed points are called *foci*.

If F_1 and F_2 are the foci, P is a point on the hyperbola such that $d(P, F_1) > d(P, F_2)$ and $2a$ is the constant, then by definition we have

$$d(P, F_1) - d(P, F_2) = 2a.$$

The line through the foci is the *principal axis* of the hyperbola (Figure 3.52). The points of intersection of the curve and the principal axis, V_1 and V_2, are called *vertices* of the hyperbola. The midpoint of the segment $V_1 V_2$ is the *center* of the hyperbola. It can be shown that the curve is symmetric with respect to the principal axis.

As with other conic sections, hyperbolas are encountered in the physical world. Under the action of an electric field, the path described by certain atomic

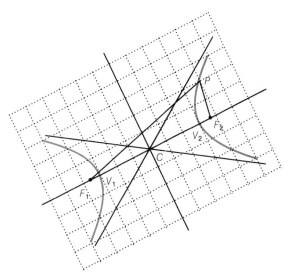

Figure 3.52

particles is a branch of a hyperbola. Some comets are believed to move along hyperbolic orbits. Hyperbolic forms are found in optics and some modern architectural structures.

Equations of Hyperbolas

In a rectangular coordinate system, a hyperbola is said to be in *standard position* if its center is the origin and its principal axis coincides with one of the coordinate axes.

We derive the equation of a hyperbola in standard position, assuming that the principal axis coincides with the x-axis and that the foci are $F_1(-c, 0)$ and $F_2(c, 0)$, with $c > 0$ (Figure 3.53).

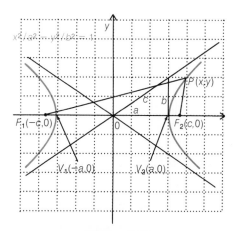

Figure 3.53

Let $P(x, y)$ be a point on the curve and assume that $d(P, F_1) > d(P, F_2)$ (the same reasoning applies if $d(P, F_2) > d(P, F_1)$). We have, by definition,

$$d(P, F_1) - d(P, F_2) = 2a$$

which, in terms of coordinates, can be written as

$$\sqrt{(x + c)^2 + y^2} - \sqrt{(x - c)^2 + y^2} = 2a.$$

Transposing the second radical to the right-hand side, squaring and simplifying, we get

$$cx - a^2 = a\sqrt{(x - c)^2 + y^2}.$$

Squaring again and simplifying give us

$$(c^2 - a^2)x^2 - a^2 y^2 = a^2(c^2 - a^2);$$

so

$$\frac{x^2}{a^2} - \frac{y^2}{c^2 - a^2} = 1.$$

In the triangle $PF_1 F_2$ (Figure 3.53), the difference between the lengths of the sides PF_1 and PF_2 is smaller than the length of the side $F_1 F_2$. Thus,

$$d(P, F_1) - d(P, F_2) = 2a < 2c = d(F_1, F_2)$$

which means that $c > a$, so $c^2 - a^2 > 0$. This lets us define the following quantity.

For a hyperbola, there is a number $b > 0$ such that

$$b^2 = c^2 - a^2.$$

Replacing $c^2 - a^2$ in the equation above, we get the equation in *standard form*

$$(3.19) \qquad \frac{x^2}{a^2} - \frac{y^2}{b^2} = 1$$

If the principal axis of the hyperbola coincides with the y-axis and the foci are $F_1(0, -c)$ and $F_2(0, c)$, then with the same reasoning as above we obtain the equation

---

$$(3.20) \qquad \frac{y^2}{a^2} - \frac{x^2}{b^2} = 1.$$

◆ **Example 9.** Find an equation of the hyperbola with foci $F_1(-5, 0)$ and $F_2(5, 0)$, and $a = 3$.

Solution. Since $a = 3$ and $c = 5$, then

$$b^2 = c^2 - a^2$$
$$= 5^2 - 3^2$$
$$= 16.$$

Hence, $b = 4$, and the equation of the hyperbola in standard form is

$$\frac{x^2}{9} - \frac{y^2}{16} = 1.$$

If we multiply both sides by $9 \cdot 16 = 144$, we obtain the equivalent equation

$$16x^2 - 9y^2 = 144.$$

◇ **Practice Exercise 9.**   Find an equation of the hyperbola with foci $F_1(0, -\sqrt{5})$ and $F_2(0, \sqrt{5})$, and $a = 2$.

Answer. $y^2/4 - x^2/1 = 1$.

Ellipses versus Hyperbolas

You have probably noticed the minus sign that differentiates the standard equation of a hyperbola from that of an ellipse. More importantly, the relations among the numbers a, b, and c are different. For an ellipse $a > c$ and $b^2 = a^2 - c^2$. For a hyperbola $c > a$ and $b^2 = c^2 - a^2$.

Transverse Axis

If we set $y = 0$ in the equation

$$\frac{x^2}{a^2} - \frac{y^2}{b^2} = 1,$$

we get $x = \pm a$, so the coordinates of the vertices are $V_1(-a, 0)$ and $V_2(a, 0)$ (Figure 3.53). The segment $V_1 V_2$ joining the vertices is called the *transverse axis* of the hyperbola; its length is $2a$.

Conjugate Axis

If we now set $x = 0$ in the equation

$$\frac{x^2}{a^2} - \frac{y^2}{b^2} = 1,$$

we obtain $y^2 = -b^2$, which has no real solution. Therefore, the curve does not intersect the y-axis. Nevertheless, we may consider the points $W_1(0, -b)$ and $W_2(0, b)$ on the y-axis and call the segment $W_1 W_2$ the *conjugate axis* of the hyperbola; its length is $2b$.

The Fundamental Rectangle

When sketching the graph of a hyperbola, it is very useful to consider the rectangle with sides parallel to the coordinate axes and with vertices (a, b), $(-a, b)$, $(-a, -b)$, and $(a, -b)$. This rectangle is called the *fundamental rectangle* (Figure 3.54). To locate the foci of the hyperbola, draw a circle with center at the origin and radius c (the *semidiagonal* of the fundamental rectangle). The points where the circle intersects the principal axis are the foci (because $c^2 = a^2 + b^2$).

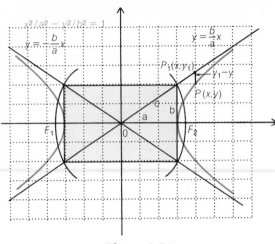

Figure 3.54

Asymptotes

The two lines $y = \pm(b/a)x$, extensions of the diagonals of the fundamental rectangle, are called *asymptotes* of the hyperbola. They have a very important geometric meaning. If $P(x, y)$ lies on the hyperbola and $P_1(x, y_1)$ is the point on the asymptote with the same x-coordinate as P, as shown in Figure 3.54, then the vertical distance $y_1 - y$ between P and P_1 becomes smaller and smaller as x becomes larger and larger.

The fundamental rectangle and the asymptotes are very helpful when graphing a hyperbola.

◆ **Example 10.**   Graph the hyperbola $\dfrac{x^2}{16} - \dfrac{y^2}{9} = 1$.

Solution.   In this case, $a = 4$ and $b = 3$. It follows that the x-intercepts (vertices) are at the points $(-4, 0)$ and $(4, 0)$. Also the vertices on the conjugate axis are $(0, -3)$ and

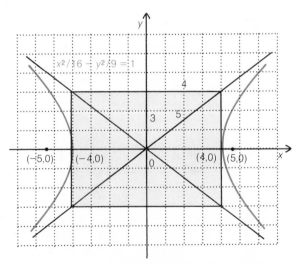

Figure 3.55

(0, 3). By plotting these points, we can draw the fundamental rectangle as shown in Figure 3.55. Also, from the relation $b^2 = c^2 - a^2$, we get

$$c^2 = a^2 + b^2$$
$$= 4^2 + 3^2$$
$$= 16 + 9$$
$$= 25.$$

Thus, $c = 5$ is the length of the semidiagonal of the fundamental rectangle. It follows that the foci are $F_1(-5, 0)$ and $F_2(5, 0)$. By extending the diagonals of the fundamental rectangle, we have the two asymptotes and we can now draw a rough sketch of the graph. Additional points may be plotted for precision.

◇ **Practice Exercise 10.** Sketch the graph of the hyperbola $\dfrac{y^2}{16} - \dfrac{x^2}{9} = 1$.

Answer.

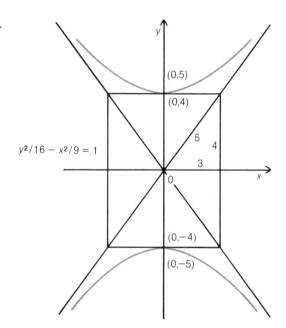

Hyperbolas Centered at (h, k)

If the center of a hyperbola is $C(h, k) \neq O(0, 0)$ and the principal axis is parallel to one of the coordinate axes, the equations are of the following form:

$\dfrac{(x - h)^2}{a^2} - \dfrac{(y - k)^2}{b^2} = 1$	Principal axis parallel to the x-axis
$\dfrac{(y - k)^2}{a^2} - \dfrac{(x - h)^2}{b^2} = 1$	Principal axis parallel to the y-axis

◆ **Example 11.**   Find an equation of the hyperbola with center at $(1, 2)$, a vertex at $(1, 6)$, and $b = 3$.

Solution.   The center and the vertex lie on the vertical line $x = 1$, four units away from each other. Thus $a = 4$. Since $b = 3$, we may write

$$\frac{(y - 2)^2}{16} - \frac{(x - 1)^2}{9} = 1.$$

The other vertex is at $(1, -2)$. The foci are $F_1(1, -3)$ and $F_2(1, 7)$. (Why?)

◇ **Practice Exercise 11.**   Find an equation of the hyperbola with center at $(-2, 1)$, vertex at $(-7, 1)$, and length of conjugate axis 8.

Answer. $\dfrac{(x + 2)^2}{25} - \dfrac{(y - 1)^2}{16} = 1.$

◆ **Example 12.**   Show that the equation $x^2 - 4y^2 - 2x - 8y - 7 = 0$ represents a hyperbola.

Solution.   Write the equation as follows:

$$(x^2 - 2x \quad) - 4(y^2 + 2y \quad) = 7$$

and complete the square on x and y:

$$(x^2 - 2x + 1) - 4(y^2 + 2y + 1) = 7 + 1 - 4$$
$$(x - 1)^2 - 4(y + 1)^2 = 4.$$

Dividing both sides by 4, we get

$$\frac{(x - 1)^2}{4} - \frac{(y + 1)^2}{1} = 1.$$

This is an equation of a hyperbola with center $C(1, -1)$, $a = 2$, and $b = 1$.

◇ **Practice Exercise 12.**   Does the equation $y^2 - 3x^2 - 2y - 12x - 14 = 0$ represent a hyperbola?

Answer.   Yes: $\dfrac{(y - 1)^2}{3} - \dfrac{(x + 2)^2}{1} = 1$; $C(-2, 1)$, $a = \sqrt{3}$, $b = 1$.

⬡ **EXERCISES 3.4**

Determine the vertex, focus, principal axis, and directrix of each of the following parabolas. Graph each one.

1. $y = 4x^2$

4. $y^2 - 3x - 6 = 0$

2. $y = -2x^2$

5. $(x - 1)^2 = y + 1$

3. $y = 3x^2 + 1$

6. $(y + 2)^2 = 2(x - 1)$

In Exercises 7–14, find an equation of the parabola corresponding to the given information. Sketch the graph.

7. $V(0, 0)$, $F(3/2, 0)$
8. $V(0, 0)$, $F(0, -1/4)$
9. $V(1, 2)$, $F(2, 2)$
10. $V(-2, 1)$, $F(-2, 3)$
11. $V(3, 0)$, directrix $x = 1$
12. $V(0, -2)$, directrix $y = -3$
13. $V(-1, -1)$, directrix $y = -1/2$
14. $V(2, 0)$, directrix $x = 1/3$

In Exercises 15–18, discuss each equation.

15. $2x^2 - 8y + 6 = 0$ **16.** $y^2 + x + 1 = 0$ **17.** $y^2 - 2y - x = 0$ **18.** $x^2 - 4x + 4y = 0$

19. Find the values of c for which the parabola $y = x^2 + 4x + c$ has two x-intercepts.

20. For which values of b does the parabola $x = y^2 + by + 9$ have two y-intercepts?

21. Find the points of intersection of the line $y = 2x + 1$ and the parabola $y = 3x^2 + 5x - 5$. [*Hint*: Set $2x + 1 = 3x^2 + 5x - 5$ and solve for x.]

22. Find the points of intersection of the parabolas $y = x^2 - 20$ and $y = 12 - x^2$.

In Exercises 23–27, determine the center, vertices, foci, and major and minor axes of each ellipse. Graph each one.

23. $\dfrac{x^2}{16} + \dfrac{y^2}{9} = 1$
24. $\dfrac{x^2}{36} + \dfrac{y^2}{64} = 1$
25. $4x^2 + 9y^2 = 36$

26. $25x^2 + 16y^2 = 400$
27. $\dfrac{(x - 1)^2}{25} + \dfrac{(y + 1)^2}{4} = 1$
28. $\dfrac{(x + 2)^2}{9} + \dfrac{(y - 3)^2}{16} = 1$

In each of Exercises 29–38, find an equation of the ellipse corresponding to the given data.

29. Center at $(0, 0)$, a focus at $(-2, 0)$, length of the major axis 8.
30. Center at $(0, 0)$, a focus at $(3, 0)$, length of the minor axis 8.
31. Center at $(1, 2)$ and vertices at $(6, 2)$ and $(1, 6)$.
32. Center at $(-3, -1)$ and vertices at $(-3, 4)$ and $(-1, -1)$.
33. Foci at $(0, 0)$ and $(6, 0)$ and vertices at $(-2, 0)$ and $(8, 0)$.
34. Foci at $(0, 1)$ and $(0, 9)$ and vertices at $(0, 0)$ and $(0, 10)$.
35. A focus at $(-5, 0)$, center at $(0, 0)$, and eccentricity $1/3$.
36. A focus at $(0, 6)$, center at $(0, 0)$, and eccentricity $3/4$.
37. A focus at $(2, 0)$, center at $(0, 0)$, and a vertex at $(3, 0)$.
38. A focus at $(0, -3)$, center at $(0, 0)$, and a vertex at $(0, 4)$.

Show that each equation in Exercises 39–42 represents an ellipse. Determine the center and the major and minor axes.

39. $9x^2 + 4y^2 - 18x + 8y - 23 = 0$
40. $x^2 + 4y^2 + 2y = 0$
41. $4x^2 - 9y^2 - 16x + 18y - 11 = 0$
42. $9x^2 + 25y^2 - 6x + 10y - 223 = 0$

According to Kepler's first law, the orbit of each planet is an ellipse with the sun at one focus. The vertex of the orbit closest to the sun is called perihelion *and the vertex of the orbit farthest from the sun is called* aphelion.

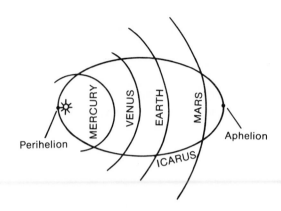

c 43. The eccentricity of the orbit of the Earth is 1.670×10^{-2} and the major axis has length 1.858×10^8 miles. How close is the Earth to the sun at perihelion? How far is it at aphelion?

c 44. The asteroid Icarus has perihelion of 1.7×10^7 miles and aphelion of 1.83×10^8 miles. What is the eccentricity of its elliptical orbit?

In Exercises 45–48, determine the center, vertices, and foci of the hyperbola.

45. $\dfrac{x^2}{64} - \dfrac{y^2}{36} = 1$

46. $\dfrac{x^2}{25} - \dfrac{y^2}{144} = 1$

47. $\dfrac{(x-2)^2}{81} - \dfrac{(y+3)^2}{144} = 1$

48. $\dfrac{(x+5)^2}{64} - \dfrac{(y+7)^2}{36} = 1$

In Exercises 49–54, find the equation of the hyperbola corresponding to the given data.

49. Center at $(0, 0)$, focus at $(4, 0)$, $a = 2$.

50. Center at $(2, 0)$, focus at $(4, 0)$, $a = 1$.

51. Vertices at $(-2, 1)$ and $(6, 1)$, foci at $(-3, 1)$ and $(7, 1)$.

52. Vertices at $(1, \pm 3)$ and foci at $(1, \pm 5)$.

53. Vertices at $(\pm 1, 0)$ and asymptotes $y = \pm 2x$.

54. Vertices at $(\pm 3, 0)$ and asymptotes $y = \pm(4/3)x$.

55. Foci at $(-4, 1)$ and $(6, 1)$, $a = 4$.

56. Foci at $(-6, -1)$ and $(10, -1)$, $a = 6$.

Show that the equations in Exercises 57–60 represent hyperbolas. Determine the center and vertices.

57. $2y^2 + 4y - x^2 = 0$

58. $3x^2 - 6x - y^2 = 0$

59. $4x^2 - 9y^2 - 16x - 18y - 29 = 0$

60. $2x^2 + 12x - y^2 + 10y + 1 = 0$

CHAPTER SUMMARY

In a *Cartesian coordinate system* a correspondence between points in a plane and ordered pairs of real numbers is established. To each point P corresponds an ordered pair (x, y) of real numbers called the *coordinates* of P. The number x is the *abscissa* or *x-coordinate* of P and the number y is the *ordinate* or *y-coordinate* of *P*. Conversely, every ordered pair (x, y) of real numbers determines a unique point in the plane.

As a first application, two simple formulas are derived: the *midpoint formula* and the *distance formula*. The first gives the coordinates of the midpoint of a segment in terms of the coordinates of the endpoints of the segment. The second, obtained from the Pythagorean Theorem, gives the distance between two points in terms of the coordinates of the points. Of these two formulas, the distance formula is more important and has wider applications.

Once a coordinate system is established, geometric objects are represented by equations and equations can be given a geometric meaning. As a first example we discuss the *straight line*. Two distinct points determine a unique line. Every line that is not *vertical* has a *slope* defined as the ratio of the difference between the *y*-coordinates and the difference between the *x*-coordinates of any pair of distinct points on a line. The slope of a vertical line is *not defined*. *Any* real number can be the slope of a line, and this corresponds to the fact that there are infinitely many lines passing through a point. If the slope is equal to zero, then the line is *horizontal*.

Among the equations that represent a given line, the following are particularly important. Given a nonvertical line, the *point-slope form*

$$y - y_1 = m(x - x_1)$$

represents a line in terms of the slope m and the coordinates (x_1, y_1) of a point on the line. In the *slope-intercept* form

$$y = mx + b,$$

the coefficient of x is the slope of the line and the number b is the *y-intercept* of the line. The equation of every line can be written in *standard form as* $Ax + By + C = 0$, with A and B not simultaneously zero. This is a *linear equation in x and y*. Conversely, the graph of any linear equation in x and y is a straight line.

Two lines of slopes m and m' are *parallel* if $m = m'$. Two lines of slopes m and m' both different from zero are *perpendicular* if $m' = -1/m$.

If two variables x and y satisfy a linear equation, such as $Ax + By + C = 0$, then they are said to be *linearly related*.

The *solution set* of an equation in two variables x and y consists of all ordered pairs (x, y) that satisfy the equation. The corresponding set of points $P(x, y)$ in a coordinate plane is the *graph of the equation*. Graphs of equations can be sketched by *plotting points*. Since plotting points is a rather crude and tedious method to graph an equation, we should always analyze the equation and look for certain features that reduce to a minimum the plotting of points. One of them is *symmetry*. A curve C is *symmetric with respect to the y-axis* if whenever $(x, y) \in C$, then $(-x, y) \in C$. The curve is *symmetric with respect to the x-axis* if whenever $(x, y) \in C$, then $(x, -y) \in C$. *Symmetry with respect to the origin* means that whenever (x, y) belongs to the curve, $(-x, -y)$ also belongs to the curve. Finally, *symmetry with respect to the line* $y = x$ means that if (a, b) lies on the curve, then (b, a) also lies on the curve.

Conic sections are curves defined by the intersections of a plane and a right circular cone. In a Cartesian coordinate system, conic sections are described by quadratic equations in two variables.

Of all conic sections the *circle* is the simplest one. It is defined as the set of points in a plane whose distances to a fixed point, the *center* of the circle, are equal to a given number, the *radius* of the circle.

A *parabola* is the set of points in a plane equidistant from a fixed point, the *focus*, and a fixed line, the *directrix*.

An *ellipse* is the set of all points in a plane the sum of whose distances from two fixed points, the *foci*, is constant. An ellipse is related to a circle, and the *eccentricity* of an ellipse (a number between 0 and 1) measures by the amount by which the curve deviates from a circle. Ellipses with eccentricity near zero approach circles; ellipses with eccentricity near 1 are very elongated curves.

A *hyperbola* is the set of points in a plane the difference of whose distances from two fixed points, the *foci*, is constant.

Conic sections are frequently encountered in the physical sciences and engineering, as well as in art and architectural designs. The path of a projectile is a parabola; planets and comets move around the sun in elliptical orbits; satellites have been placed into a circular orbit around the earth; and, under certain conditions, atomic particles move along one of the branches of a hyperbola.

REVIEW EXERCISES

In Exercises 1–4, the coordinates of P and Q are given. Find the coordinates of the midpoint of the segment PQ.

1. $P(2, 8)$, $Q(3, -5)$ **2.** $P(-3, -7)$, $Q(-2, -5)$

3. $P(1/3, -1/2)$, $Q(3/4, 1/5)$ **4.** $P(2/9, 1/4)$, $Q(3/5, 2/7)$

5. If $A(2, -3)$ and $M(4, -1)$, find the coordinates of B so that M is the midpoint of the segment AB.

6. If $B(-1, -3)$ and $M(-6, 2)$, find the coordinates of A so that M is the midpoint of the segment AB.

7. Find the points on the x-axis that are at a distance of 6 units from the point $P(2, -5)$.

8. Find the points on the main diagonal (i.e., the line $y = x$) that are at a distance of 5 units from the point $A(7, 6)$.

9. Show that the triangle with vertices $A(-4, 0)$, $B(1, 10)$, and $C(4, 6)$ is a right triangle. Find its area.

10. Plot the points $A(5, -6)$, $B(8, 5)$, and $C(1, -2)$. Show that ABC is a right triangle and find its area.

11. Show that the points $A(-3, -4)$, $B(1, -8)$, $C(7, -2)$, and $D(3, 2)$ are the vertices of a parallelogram.

12. Plot the points $A(-2, 1)$, $B(3, 7)$, $C(9, 2)$, and $D(4, -4)$ and show that they are the vertices of a square.

13. Find b so that the triangle with vertices $O(0, 0)$, $A(2, 3)$, and $B(3, b)$ is a right triangle with right angle at A.

14. Given $Q(0, -2)$ and $R(0, 3)$, find a point P on the positive x-axis so that QPR is a right triangle with right angle at P.

In Exercises 15–18, graph each equation and check for symmetry.

15. $3x = -5y^3$ **16.** $2y + 7x^3 = 0$ **17.** $y = x^2 - 4x$ **18.** $x = y^2 - 6y$

In Exercises 19–22, find the slope of the line containing the given points.

19. $A(-3, -5)$, $B(4, 7)$

c 21. $M(2.157, -3.019)$, $N(0.026, -3.107)$

20. $P(2, 1/3)$, $Q(1/4, 3)$

c 22. $R(\sqrt{3}, 1.414)$, $Q(-\sqrt{5}, 3.141)$

In Exercises 23–32, find an equation for each of the following lines. Write your answer in the form $Ax + By + C = 0$.

23. Passing through $(-2, -6)$ with slope $-1/3$.

24. Passing through $(3, -8)$ with slope $3/11$.

25. With y-intercept $3/2$ and slope $5/3$.

26. With x-intercept -5 and slope $-1/2$.

27. With x-intercept $3/5$ and y-intercept 6.

28. With x-intercept -3 and y-intercept -8.

c 29. Passing through $(2.35, -7.21)$ and $(1.07, -2.15)$.

c 30. Passing through $(-1.18, 3.26)$ and $(4.03, -1.38)$.

31. Parallel to $x/2 + y/3 = 1$ and passing through $(1, 6)$.

32. Perpendicular to $x/3 - y/3 = 1$ and passing through $(1, 3)$.

33. Find b so that the line $y = 5x + b$ contains the point $(-1, 3)$.

34. Find m so that the line $y = mx + 7$ contains the point $(-2, -5)$.

35. Find the slope and the y-intercept of the line $5.17x - 3.21y + 1.72 = 0$.

c 36. Find the slope and the x-intercept of the line $-3.371x + 1.016y - 2.371 = 0$.

37. A line passes through the point $(3, 5)$ and has equal x- and y-intercepts. Find its equation.

38. The y-intercept of a line is the opposite of its x-intercept. If the line passes through $(1, -4)$, find its equation.

39. The variable v (velocity) is linearly related to the variable t (time). Assume that $v = 15$ when $t = 0$, and $v = 45$ when $t = 1.5$. Find an equation relating v and t. Also, find t when $v = 85$.

40. The variables C (degrees Celsius) and F (degrees Fahrenheit) are linearly related. Let $F = 32$ when $C = 0$, and $F = 212$ when $C = 100$. Find an equation relating both variables.

41. The value of a stamp collection is appraised at \$20,000. If it appreciates linearly at a rate of 6% per year, find its value V in t years.

42. A farmer buys a piece of equipment for \$36,000 and estimates that in 15 years its scrap value will be \$3,600. Assuming that the depreciation is linear, find a formula for the value V of the equipment in t years.

Find an equation of the circle satisfying the stated conditions.

43. Center at the origin and passing through $(-1, 3)$.

44. Center at $(3, 5)$ and passing through the origin.

45. Center at $(-1, 1)$ and passing through $(2, -3)$.

46. Center in the second quadrant, tangent to both axes and of radius 2.

47. Center in the third quadrant, tangent to both axes and of radius $\sqrt{10}$.

48. Diameter AB with $A(1, -3)$ and $B(-2, 5)$.

In Exercises 49–54, find the center and radius of each circle.

49. $\dfrac{x^2}{2} + \dfrac{y^2}{2} = 8$

50. $\dfrac{x^2}{3} + \dfrac{y^2}{3} = 125$

51. $x^2 + y^2 - 3y - 1 = 0$

52. $x^2 + y^2 - x - 3 = 0$

53. $x^2 + y^2 + 3x + 5y - \dfrac{1}{2} = 0$

54. $x^2 + y^2 - 6x + 8y = 0$

In Exercises 55–60, find an equation of the parabola corresponding to the given data.

55. Focus at $(0, 3)$ and directrix $y = -3$.
56. Focus at $(2, 0)$ and directrix $x = -2$.
57. Vertex at $(0, 0)$, passing through the point $(2, 4)$, principal axis horizontal.
58. Vertex at the origin, passing through $(-1, -2)$, principal axis vertical.
59. Principal axis vertical, passing through $(0, 4)$, vertex at $(2, 1)$.
60. Principal axis horizontal, passing through $(2, -2)$, vertex at $(-1, 2)$.
61. Find the points of intersection of the parabola $y = x^2$ with the line $3x - 2y - 1 = 0$.
62. Find the points of intersection of the parabola $y = (x - 1)^2$ and the line $x + y = 3$.

In Exercises 63–68, find an equation of the ellipse corresponding to the given data.

63. Center at $(-1, 0)$, focus at $(-1, 2)$, $a = 3$.
64. Foci at $(\pm 5, 0)$ and $b = 2$.
65. Foci at $(-2, 3)$ and $(4, 3)$, and semimajor axis 5.
66. Foci at $(0, \pm 4)$ and eccentricity $1/10$.
67. Vertices $V_1(-5/3, 0)$ and $V_2(5/3, 0)$, and passing through the point $(1, 1)$.
68. The point $(\sqrt{3}, 2)$ lies on the curve, and the vertices are $V_1(-3, 0)$ and $V_2(3, 0)$.

In Exercises 69–74, find an equation of the hyperbola corresponding to the given data.

69. Foci at $(0, \pm 5)$ and vertices at $(0, \pm 3)$.
70. Foci at $(0, \pm 6)$ and conjugate axis 2.
71. Focus at $(-5, 0)$, vertex at $(-2, 0)$, and center at the origin.
72. Vertices at $(\pm 4, 0)$ and asymptotes $y = \pm \frac{1}{2} x$.
73. A focus at $(10, 0)$, center at the origin, and eccentricity $5/4$.
74. Conjugate axis 6, vertices at $(\pm 5, 0)$.

Discuss each of the equations in Exercises 75–82.

75. $x^2 + y^2 - 4x - 6y + 4 = 0$
76. $x^2 + y^2 - 6x - 7 = 0$
77. $9x^2 + 4y^2 + 18x + 8y - 12 = 0$
78. $x^2 + 3y^2 + 6y - 6 = 0$
79. $y^2 - 6y - 3x = 0$
80. $2x^2 - 2x + y - 2 = 0$
81. $x^2 - 4y^2 - 2x - 8y - 12 = 0$
82. $x^2 - 4y^2 + 2y = 0$.

83. For which values of b does the parabola $y = -2x^2 + bx - 5$ have only one x-intercept?
84. Find a so that the parabola $y = ax^2 - 6x + 1$ has only one x-intercept.
85. Find the points of intersection of the parabolas $y = 4 - (x - 1)^2$ and $y = x^2 - 2x - 3$.
86. Find the points of intersection of parabolas $y = x^2 + 2x - 15$ and $y = 9 - (x + 2)^2$.
87. Find an equation of the circle passing through the points $A(1, 4)$ and $B(-5, 2)$ and whose center is the midpoint of the segment AB.

88. Find an equation of the circle passing through the points $(0, 0)$, $(8, 0)$, and $(0, -6)$.

c 89. The major axis of the orbit of Mars has length 2.834×10^8 miles, and the eccentricity of the orbit is 9.3×10^{-2}. How close is Mars to the sun at perihelion and how distant is it at aphelion?

c 90. In April of 1983, the astronauts aboard the space shuttle Challenger launched into orbit the first of three global communications satellites. The satellite was to be placed in a circular orbit over the equator, at a distance of approximately 22,300 miles above the earth. In such an orbit, the satellite would be synchronized with the rotation of the earth so that, seen from the earth, it would remain at the same point. Due to equipment failure, the satellite was placed instead in an elliptical orbit whose closest point to the earth's surface was 13,540 miles and whose farthest point was 21,960 miles. Find the eccentricity of that elliptical orbit. (Radius of the earth = 3,959 miles).

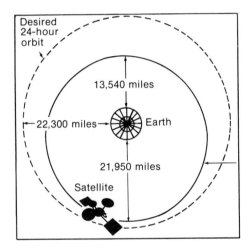

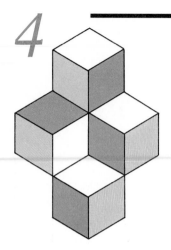

4

Functions _____

In this chapter, we introduce the concept of functions, one of the most important and useful concepts in mathematics. First, we discuss the notions of domain and range, dependent and independent variables, and function values. Next, the graph of a function is defined and the graphs of some simple functions are described. In Section 4.2, we discuss algebraic properties of functions and some examples of functions derived from the applied sciences. In Section 4.3, several techniques to graph functions are described; even and odd functions and increasing and decreasing functions are defined. Finally, in the last section we discuss the notion of variation of one variable with to others.

 HISTORICAL NOTE

Functions

The concept of function was first introduced by Leonhard Euler (1707–1783), a Swiss mathematician. It was also Euler who created the notation $f(x)$ and made functions the basis of calculus. Euler was among the most prolific mathematicians of his time, having made outstanding contributions to analysis, number theory, calculus of variations, astronomy, mechanics, optics, and hydrodynamics.

4.1 FUNCTIONS

The notion of *function*, one of the most important concepts in mathematics, is linked to the notation of *related variables*, which we frequently encounter in the applied sciences and everyday life.

Here are some examples of related variables. At the end of the previous section, we saw that the relationship between degrees Fahrenheit F and degrees Celsius C is given by the linear equation

$$F = \frac{9}{5}C + 32.$$

The variables F and C are linearly related as defined in Chapter 3, page 145. To each value assigned to C there corresponds a unique value of F and vice-versa.

If an object moves along a line at a *constant rate* (or *speed*) r, then the *distance d* covered by the object is related to the *time t* by the equation

$$d = rt.$$

To each assigned value of the time t there corresponds a unique value of the distance d.

The formula

$$A = l^2$$

gives the area of a square in terms of the length l of the side of the square. It shows that to each positive value l there corresponds a unique value of the area A.

According to Boyle's law, the pressure P of a gas enclosed in a container and kept at a constant temperature is related to the volume of the gas by the formula

$$PV = C,$$

where C is a constant. This formula can be written as

$$P = \frac{C}{V},$$

showing that to each value of the volume V there corresponds a unique value of the pressure P. This formula indicates that pressure and volume are in *inverse* proportion. That is, a decrease in volume corresponds to an increase in pressure, and an increase in volume corresponds to a decrease in pressure.

There is a relationship between the cost of producing a unit of a certain item and the number of units produced. Within certain limits, the more you produce the less it costs to produce it.

These are examples of an important kind of relationship between two variables that characterizes the notion of *function*.

Definition

A function f is a rule that assigns to each element x of a set X, called the *domain* of the function, a unique element y of a set Y. The element y is called the *image* of x under f, and the set of all images of elements of X is the *range* of the function.

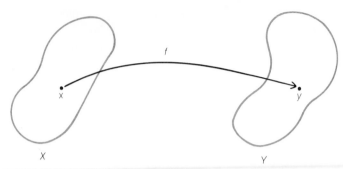

Figure 4.1

Functions are sometimes illustrated by diagrams such as Figure 4.1. The arrow indicates that the element *y* is the image of *x* under *f*.

As an example of how general the function concept is, suppose that *X* is the set of all counties in the State of New Jersey and that *Y* denotes the set of all natural numbers. If to each county *x* we assign the number *y* of registered voters in the county, a function *f* from *X* into *Y* is defined. The domain of this function is the set of all counties in New Jersey and the range is a subset of the set of natural numbers.

As another example, suppose that to every individual we assign his or her age in years and months, with the number of months converted into a fraction of the year. This defines a function from the set of all individuals into the set of rational numbers.

It is important to observe that, in the definition of function, to each element *x* in *X* there corresponds a *unique* element *y* in *Y*. However, the same *y* may be assigned to different elements in *X*. For instance, two different counties may have the same number of registered voters, and several individuals may have the same age.

Constant Function

When the range of a function consists of a *single element*, the function is said to be a *constant function*.

Notations

The domain of a function *f* is denoted by D_f or, when no confusion is possible, by *D*.

If *f* is a function and *x* an element in the domain of *f*, the image *y* of *x* under *f* is denoted by $y = f(x)$, and we read "*y* is equal to *f* of *x*." We also say that $y = f(x)$ is a function of *x*. The notations

$$f: x \to f(x) \quad \text{or} \quad f: x \to y$$

are also used to indicate that the image of *x* under *f* is *f(x)* or *y*.

Real-Valued Functions

> A *real-valued function* is a rule assigning to each element of the domain of the function a unique real number.

For example, the two functions described above are real-valued functions.

Function Values

For the most part, throughout this book we shall consider only *real-valued functions* whose domains are *sets of real numbers*.

For example, the correspondence that assigns to every real number x its square x^2 defines a real-valued function denoted by

$$f(x) = x^2 \quad \text{or} \quad f: x \to x^2.$$

Also, the correspondence that assigns to every nonnegative number x its principal square root $\sqrt{x}$ defines the *square root* function denoted by

$$g(x) = \sqrt{x} \quad \text{or} \quad g: x \to \sqrt{x}.$$

If f is a real-valued function and x is an element in the domain of f, then the number $f(x)$ is called the *function value* of f at x.

◆ **Example 1.** Find the domain and range of the function $f(x) = x^2$. Also, find the function values $f(-5)$, $f(3.5)$, $f(12.1)$, and $f(2/3)$.

Solution. For any given number x, we can always compute x^2, so the domain of the function $f(x) = x^2$ is the set of all real numbers. Since every nonnegative number has a square root, it follows that the range of f is the set of all nonnegative numbers. Next, we have

$$f(-5) = (-5)^2 = 25,$$
$$f(3.5) = (3.5)^2 = 12.25,$$
$$f(12.1) = (12.1)^2 = 146.41,$$
$$f(2/3) = \left(\frac{2}{3}\right)^2 = \frac{4}{9}.$$

◇ **Practice Exercise 1.** Find the domain and range of the function $F(x) = x^2 + 3$. Compute the function values $F(-1.2)$, $F(0)$, $F(10)$, $F(1/4)$.

Answer. Domain: all real numbers. Range: all numbers ≥ 3. $F(-1.2) = 4.44$, $F(0) = 3$, $F(10) = 103$, $F(1/4) = 49/16$.

◆ **Example 2.** Find the domain and range of the function $g(x) = \sqrt{x}$. Can you find the function values $g(0)$, $g(16)$, and $g(-5)$?

Solution.   The principal square root $\sqrt{x}$ is defined only when x is a nonnegative number. So, the domain of g is the set of all nonnegative numbers. On the other hand, let y be a nonnegative number. If we set $x = y^2$, then x is nonnegative and the principal root of x is y, that is, $y = \sqrt{x}$. Thus, the range of g is also the set of all nonnegative numbers. Next, we have

$$g(0) = \sqrt{0} = 0 \quad \text{and} \quad g(16) = \sqrt{16} = 4.$$

Since -5 is negative, we see that $g(-5)$ is *not* defined.

◇ **Practice Exercise 2.**   Find the domain and range of $G(x) = \sqrt{x} + 2$. Can you find the function values $G(3)$, $G(9)$, and $G(-1)$?

Answer.   Domain: $x \geq 0$. Range: all numbers ≥ 2. $G(3) = \sqrt{3} + 2$, $G(9) = 5$, $G(-1)$ is undefined.

Independent and Dependent Variables

In the definition of a function f, the roles of the two variables are conceptually different: x denotes a freely chosen element in the domain of f, while y is the number that corresponds to x by f. For this reason, x is called the *independent variable* and y the *dependent variable*.

For example, consider the square root function $y = g(x) = \sqrt{x}$. If we take

$$x = \frac{4}{9}, \quad x = 6, \quad \text{and} \quad x = 25$$

as our choices for the independent variable, then the corresponding values of the dependent variable are

$$y = \frac{2}{3}, \quad y = \sqrt{6}, \quad \text{and} \quad y = 5.$$

The independent variable is also called the *input* and the dependent variable is called the *output* of a function. For the square root function, if the input is 100, the output is 10.

Graphs of Functions

Let $y = f(x)$ be a real-valued function defined on D, a subset of the real numbers. The *graph* of f is the set of points in a Cartesian plane whose coordinates are $(x, f(x))$, for all $x \in D$.

By the definition of function, to each element a in the domain D there corresponds a unique element $f(a)$ in the range of f. Thus *every vertical line $x = a$, with $a \in D$, intersects the graph in a single point.* Given a function $y = f(x)$, to find the function value of f at $x = a$, start at the point a and move vertically until the graph is reached, and then move horizontally to the y-axis (Figure 4.2).

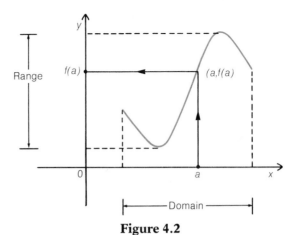

Figure 4.2

The Vertical Line Test

In a coordinate plane, a given set of points is the graph of a function exactly when every vertical line intersects the set in *at most* one point.

For example, consider the parabola of equation $y = x^2$ (Figure 4.3a). Since every vertical line intersects the curve in exactly one point, the curve is the graph of a function that we may denote by $f(x) = x^2$. This function is also said to be defined by the equation $y = x^2$.

By contrast, there are vertical lines that intersect the graph of the equation $x = y^2$ in more than one point (Figure 4.3b). For example, the vertical line $x = 4$

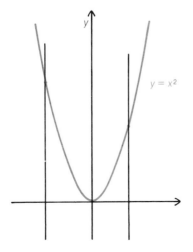
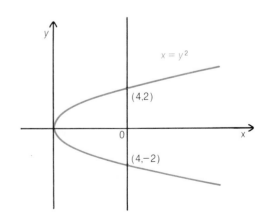

Figure 4.3

intersects the graph at the points $(4, 2)$ and $(4, -2)$. In fact, for each nonnegative number x, there are two numbers y and $-y$ so that $y^2 = (-y)^2 = x$. Thus, the parabola illustrated in Figure 4.3b *cannot* be the graph of a function.

The Graph of $g(x) = \sqrt{x}$

The equations $x = y^2$ *does not* define y as a function of x because to each nonnegative number x there correspond *two* numbers: y, the principal square root of x, and $-y$, the negative square root of x. The equation itself does not tell us which of the two numbers to assign to x.

However, as we have already said, the function $g(x) = \sqrt{x}$ assigns to each x the principal square root of x. If we set $y = g(x) = \sqrt{x}$, then $y \geq 0$ and $y^2 = x$. It follows that the function $g(x) = \sqrt{x}$ is defined by the equation $y^2 = x$ with $y \geq 0$. Thus the graph of g (Figure 4.4) consists of all points on the graph $x = y^2$ lying above or on the x-axis.

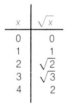

x	$\sqrt{x}$
0	0
1	1
2	$\sqrt{2}$
3	$\sqrt{3}$
4	2

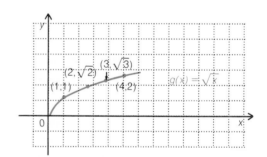

Figure 4.4

Graph Sketching

Graphs of functions can be sketched, by plotting points, in the same way as graphs of equations (Section 3.3). The method is not foolproof and, in many cases, requires guesswork and intuition. Nevertheless, in simple situations, a lot of relevant information concerning the graph of a function can be obtained through simple analysis of the function and without having to rely on more advanced techniques. By using this information, the plotting of points can be reduced to a minimum.

◆ **Example 3.**   Graph the function $f(x) = 2x + 3$.

Solution.   If we set $y = f(x)$, then the graph of the given function is the same as the graph of the linear equation $y = 2x + 3$, which represents a straight line with slope 2 and y-intercept 3. As we already said, to sketch the graph we only have to plot two points, such as the x- and y-intercepts, and join them by a straight line (Figure 4.5).

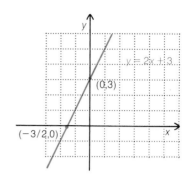

Figure 4.5

◇ **Practice Exercise 3.**   Sketch the graph of the function $f(x) = -3x + 6$.

Answer.

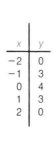

◆ **Example 4.**   Sketch the graph of the function $f(x) = 4 - x^2$.

Solution.   Setting $y = f(x)$, we see that the given function is defined by the equation $y = 4 - x^2$. The graph, illustrated in Figure 4.6, is a parabola obtained by adding 4 to the ordinate of each point on the graph of $y = -x^2$. (See Practice Exercise 3 of Section 3.3.) Or, if we write $y - 4 = -x^2$ and use the results of Section 3.4, we see that the graph is a parabola opening downward, with vertex at $(0, 4)$ and principal axis the y-axis. To double check our work, a table of values accompanies the graph of $f(x) = 4 - x^2$.

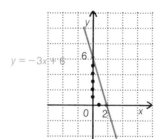

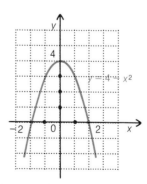

Figure 4.6

◇ **Practice Exercise 4.**   Sketch the graph of the function $f(x) = x^2 - 1$.

Answer.

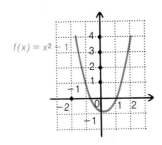

◆ **Example 5.** Find the domain of $f(x) = \sqrt{x + 4}$ and sketch the graph of the function.

Solution. The function f is defined whenever $x + 4 \geq 0$, that is, for all $x \geq -4$. Thus, the domain of f is the unbounded interval $[-4, +\infty)$. We could make a table of values and plot a few points as shown in Figure 4.7, to sketch the graph of f.

However, it is better to proceed in a different manner. If we write $y = \sqrt{x + 4}$, then we can see that the graph is the part of the parabola $x + 4 = y^2$ which lies above the x-axis. (Why?)

x	y
-4	0
-3	1
0	2
5	3

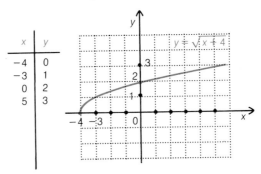

Figure 4.7

◇ **Practice Exercise 5.** What is the domain of $f(x) = \sqrt{x - 2}$? Graph this function.

Answer. Domain: $[2, +\infty)$.

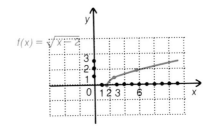

📦 **EXERCISES 4.1**

Find the domain and range of each of the following functions:

1. $f(x) = 3x - 4$

2. $g(x) = 4x - 1$

3. $F(x) = \dfrac{2x}{3} + \dfrac{1}{5}$

4. $G(x) = \dfrac{x}{5} - \dfrac{.3}{4}$

5. $H(x) = 4 - x^2$ **6.** $h(x) = 1 - x^2$ **7.** $r(x) = 3x^2 - 1$ **8.** $s(x) = 4x^2 + 5$

9. $u(x) = \sqrt{x} - 3$ **10.** $v(x) = \sqrt{x} + 4$ **11.** $l(x) = 2\sqrt{x} + 1$ **12.** $m(x) = 3\sqrt{x} - 2$

13. $p(x) = \sqrt{x - \dfrac{1}{2}}$ **14.** $g(x) = \sqrt{x + \dfrac{2}{3}}$

For each function in Exercises 15–22, find the indicated function values. When necessary, use your calculator.

15. $f(x) = 3x - 4$; $f(-2)$, $f(0)$, $f(2/3)$, and $f(\sqrt{1.5})$

16. $g(x) = 4x - 1$; $g(-3/8)$, $g(0)$, $g(4)$, and $g(\sqrt{2})$

17. $F(x) = 4 - x^2$; $F(-3)$, $F(-2)$, $F(1)$, and $F(\sqrt{3})$

18. $G(x) = 1 - x^2$; $G(-\sqrt{5})$, $G(-3)$, $G(1)$, and $G(5)$

19. $h(x) = 3 - \sqrt{x}$; $h(2)$, $h(16)$, $h(3.5)$, $h(-8)$, $h(1/4)$

20. $H(x) = \sqrt{x} + 5$; $H(0)$, $H(-1)$, $H(4)$, $H(5.8)$, $H(9/16)$

21. $k(x) = \sqrt{x - 4}$; $k(0)$, $k(5)$, $k(12)$, $k(7.25)$, $k(17/4)$

22. $K(x) = \sqrt{x + 1}$; $K(0)$, $K(5)$, $K(-12)$, $K(5.5)$, $K(-7/16)$

In Exercises 23–40, find the domain of each function. Sketch the graph.

23. $f(x) = 4x - 3$ **24.** $f(x) = 5x - 4$ **25.** $f(x) = 2 - \dfrac{3}{5}x$

26. $f(x) = 1 - \dfrac{3}{7}x$ **27.** $f(x) = x^2 - 4$ **28.** $f(x) = x^2 - 6$

29. $f(x) = 3 - x^2$ **30.** $f(x) = 9 - x^2$ **31.** $f(x) = \sqrt{x - 1}$

32. $f(x) = \sqrt{x + 3}$ **33.** $f(x) = \sqrt{4 - x}$ **34.** $f(x) = \sqrt{2 - x}$

35. $f(x) = \sqrt{x - 2}$ **36.** $f(x) = \sqrt{x + 4}$ **37.** $f(x) = \sqrt{x} - 3$

38. $f(x) = \sqrt{x} + 4$ **39.** $f(x) = 1 - \sqrt{x}$ **40.** $f(x) = 3 - \sqrt{x}$

4.2 ALGEBRA OF FUNCTIONS

Functions can be added, subtracted, multiplied, and divided to generate new functions.

Definitions

Let f and g be two functions with domains D_f and D_g, respectively. The *sum* $f + g$, *difference* $f - g$, *product* $f \cdot g$, and *quotient* f/g are functions defined by

$$(f + g)(x) = f(x) + g(x)$$
$$(f - g)(x) = f(x) - g(x)$$
$$(f \cdot g)(x) = f(x) \cdot g(x)$$
$$(f/g)(x) = f(x)/g(x), \quad \text{provided that } g(x) \neq 0.$$

In these definitions, the domain of $f + g$, $f - g$, and $f \cdot g$ is the set of all points that belong to the domains D_f and D_g; that is, the domain of the three functions is the *intersection* of the domains D_f and D_g. The domain of f/g is the intersection of the domains, *except* the points for which $g(x) = 0$.

◆ **Example 1.**   Find the sum, difference, product, and quotient of the functions $f(x) = 3x - 2$ and $g(x) = x^2 - 4$, and determine the domains of definition.

Solution.   Since f and g are defined for all real numbers, the functions $f + g$, $f - g$, and $f \cdot g$ are also defined for all real numbers. We have

$$(f + g)(x) = f(x) + g(x) = (3x - 2) + (x^2 - 4) = x^2 + 3x - 6,$$
$$(f - g)(x) = f(x) - g(x) = (3x - 2) - (x^2 - 4) = -x^2 + 3x + 2,$$

and

$$(f \cdot g)(x) = f(x) \cdot g(x) = (3x - 2)(x^2 - 4) = 3x^3 - 2x^2 - 12x + 8.$$

Since the polynomial $x^2 - 4$ is zero when $x = -2$ and $x = 2$, the domain of the quotient

$$(f/g)(x) = f(x)/g(x) = \frac{3x - 2}{x^2 - 4}$$

is the set $\mathbb{R}$ of all real numbers, except the numbers -2 and 2. Thus, the domain is the union of the open intervals $(-\infty, -2)$, $(-2, 2)$, and $(2, +\infty)$.

◇ **Practice Exercise 1.**   Let $F(x) = 2x - 5$ and $G(x) = 1 - x^2$. Find $F + G$, $F - G$, $F \cdot G$, and F/G. Determine the domains of these functions.

Answer. $(F + G)(x) = -x^2 + 2x - 4$, defined for all real numbers;
$(F - G)(x) = x^2 + 2x - 6$, for all real numbers;
$(F \cdot G)(x) = -2x^3 + 5x^2 + 2x - 5$, for all real numbers;
$(F/G)(x) = \dfrac{2x - 5}{1 - x^2}$, for all real numbers except -1 and 1.

◆ **Example 2.**   If $F(x) = 2x - 8$ and $G(x) = \sqrt{x + 3}$, find $(F/G)(x)$ and its domain of definition.

Solution.   The function F is defined for all real numbers, while G is defined only for all real numbers x such that $x \geq -3$. Since $G(3) = \sqrt{-3 + 3} = \sqrt{0} = 0$, the quotient

$$(F/G)(x) = \frac{F(x)}{G(x)} = \frac{2x - 8}{\sqrt{x + 3}}$$

is defined for all numbers x such that $x > -3$. That is, the domain of F/G is the unbounded interval $(-3, +\infty)$.

◇ **Practice Exercise 2.**   Given $f(x) = \sqrt{x - 4}$ and $g(x) = 3x - 8$, find f/g and its domain.

Answer. $(f/g)(x) = \dfrac{f(x)}{g(x)} = \dfrac{\sqrt{x - 4}}{3x - 8}$, defined for $x \geq 4$.

Polynomial Functions

Among the simplest functions that we encounter are the functions defined by polynomials.

If n is a nonnegative integer, a *polynomial function of degree n* is a function defined by an expression of the form

$$f(x) = a_n x^n + a_{n-1} x^{n-1} + \cdots + a_1 x + a_0$$

where $a_0, a_1, \ldots, a_{n-1}, a_n$ are real numbers and $a_n \neq 0$. The domain of definition of a polynomial function is the set $\mathbb{R}$ of all real numbers.

Constant Functions

If $n = 0$, then $f(x) = a_0$, and f is called a *constant function*.

Linear Functions

A polynomial function of degree 1 is said to be a *linear function*.

◆ **Example 3.** Let $f(x) = 2x + 1$ be a linear function. Find the function values $f(-3)$, $f(0)$, and $f(1)$. Also, if a and b are real numbers, find $f(a) + f(b)$ and $f(a + b)$.

Solution. We have

$$f(-3) = 2(-3) + 1 = -6 + 1 = -5,$$
$$f(0) = 2 \cdot 0 + 1 = 1,$$

and

$$f(1) = 2 \cdot 1 + 1 = 3.$$

Next, if a and b are real numbers, $f(a)$ and $f(b)$ are the values of the linear polynomial $2x + 1$ at $x = a$ and $x = b$. So,

$$f(a) = 2a + 1 \quad \text{and} \quad f(b) = 2b + 1.$$

It follows that

$$f(a) + f(b) = (2a + 1) + (2b + 1)$$
$$= 2a + 2b + 2.$$

Now, $f(a + b)$ is the function value of f at $x = a + b$, that is,

$$f(a + b) = 2(a + b) + 1$$
$$= 2a + 2b + 1.$$

Compare $f(a + b)$ and $f(a) + f(b)$, and notice that $f(a + b) \neq f(a) + f(b)$.

◇ **Practice Exercise 3.** For $g(x) = 3x + 5$, find $g(-2)$, $g(4/3)$, and $g(5)$. Also find $g(a - b)$ and $g(a) - g(b)$.

Answer. $g(-2) = -1$, $g(4/3) = 9$, $g(5) = 20$, $g(a - b) = 3a - 3b + 5$, $g(a) - g(b) = 3a - 3b$.

Quadratic Functions

A *quadratic function* is a polynomial function of degree 2.

◆ **Example 4.**   Let $F(x) = x^2 - 5x + 6$ be a quadratic function. Find the function values $F(-2)$, $F(0)$, and $F(2)$. Also, find $2F(a)$ and $F(2a)$.

Solution.   We have

$$F(-2) = (-2)^2 - 5(-2) + 6 = 20,$$
$$F(0) = 0^2 - 5 \cdot 0 + 6 = 6,$$
$$F(2) = 2^2 - 5 \cdot 2 + 6 = 0.$$

Since $F(a) = a^2 - 5a + 6$, then

$$2F(a) = 2(a^2 - 5a + 6)$$
$$= 2a^2 - 10a + 12.$$

On the other hand,

$$F(2a) = (2a)^2 - 5(2a) + 6$$
$$= 4a^2 - 10a + 6.$$

Are there numbers for which the relation $F(2a) = 2F(a)$ holds?

◇ **Practice Exercise 4.**   If $G(x) = 3x^2 - 2x - 1$, find $G(-1/3)$, $G(0.5)$, $G(4)$. Also, find $G(b/3)$ and $G(b)/3$.

Answer. $G(-1/3) = 0$, $G(0.5) = -1.25$, $G(4) = 39$, $G(b/3) = b^2/3 - 2b/3 - 1$, $G(b)/3 = (3b^2 - 2b - 1)/3$.

◆ **Example 5.**   If $F(x) = x^2 - 5x + 6$ and a and h are real numbers, with $h \neq 0$, find

$$\frac{F(a + h) - F(a)}{h}.$$

Solution.   We have

$$F(a + h) = (a + h)^2 - 5(a + h) + 6$$

and

$$F(a) = a^2 - 5a + 6.$$

Thus,

$$\frac{F(a + h) - F(a)}{h} = \frac{[(a + h)^2 - 5(a + h) + 6] - [a^2 - 5a + 6]}{h}.$$

Simplifying the right-hand side of the last expression, we obtain

$$\frac{F(a + h) - F(a)}{h} = \frac{(a^2 + 2ah + h^2) - 5(a + h) + 6 - a^2 + 5a - 6}{h}$$
$$= \frac{a^2 + 2ah + h^2 - 5a - 5h + 6 - a^2 + 5a - 6}{h}$$
$$= \frac{h(2a + h - 5)}{h}$$
$$= 2a + h - 5.$$

◇ **Practice Exercise 5.**   Let $G(x) = 3x^2 - 2x - 1$ and let b and t be real numbers, with $t \neq 0$. Find

$$\frac{G(b + t) - G(b)}{t}.$$

Answer. $6b - 2 + 3t$.

Computations of the kind discussed in Example 5 are important. They form the foundation on which differential calculus is built.

Rational Functions

> The quotient of two polynomials $f(x)$ and $g(x)$ defines a *rational function*
>
> $$r(x) = \frac{f(x)}{g(x)},$$
>
> whose domain is the set of all real numbers x for which the denominator $g(x)$ is different from zero.

◆ **Example 6.**   Find the domain of the function

$$f(x) = \frac{x + 2}{x + 5}.$$

If a, b, and $a - b$ are in the domain of f, find $f(a) - f(b)$ and $f(a - b)$.

Solution.   Since f is the quotient of two polynomials $x + 2$ and $x + 5$, the domain D of f is the set of all real numbers for which $x + 5 \neq 0$, that is, $x \neq -5$. We can also say that D is the union of the intervals $(-\infty, -5)$ and $(-5, +\infty)$. If a, $b \in D$, then

$$f(a) = \frac{a + 2}{a + 5}, \quad f(b) = \frac{b + 2}{b + 5},$$

and we have

$$f(a) - f(b) = \frac{a + 2}{a + 5} - \frac{b + 2}{b + 5}$$

$$= \frac{(a + 2)(b + 5) - (b + 2)(a + 5)}{(a + 5)(b + 5)}$$

$$= \frac{ab + 5a + 2b + 10 - ab - 5b - 2a - 10}{(a + 5)(b + 5)}$$

$$= \frac{3a - 3b}{(a + 5)(b + 5)}.$$

If $a - b \in D$, then

$$f(a - b) = \frac{(a - b) + 2}{(a - b) + 5}.$$

Notice that, in most cases, $f(a - b) \neq f(a) - f(b)$. (What happens if $a = -2$ and $b = 1$?)

◇ **Practice Exercise 6.**   Let

$$g(x) = \frac{3x - 5}{4x + 1}.$$

What is the domain of g? Find $g(a)$ and $g(1/a)$, assuming that these two function values are defined.

Answer.   Domain: All real numbers $\neq -1/4$.

$$g(a) = \frac{3a - 5}{4a + 1}, \quad g\left(\frac{1}{a}\right) = \frac{3 - 5a}{4 + a}.$$

The operations of sum, difference, product, and quotient of functions can be combined with root extraction to produce new functions.

◆ **Example 7.**   What is the domain of the function $G(x) = \sqrt{2x + 1}$? Find the function values $G(-1/2)$, $G(4)$, and $G(8)$. Assuming that a, b, and $ab \in D_G$, find $G(a) \cdot G(b)$ and $G(ab)$.

Solution.   In order for G to be defined, the linear expression $2x + 1$ must be nonnegative. Thus, the domain of G consists of all numbers x satisfying the inequality

$$2x + 1 \geq 0,$$

that is, all numbers x so that $x \geq -1/2$.

Next, we have

$$G\left(-\frac{1}{2}\right) = \sqrt{2\left(-\frac{1}{2}\right) + 1} = \sqrt{-1 + 1} = \sqrt{0} = 0,$$

$$G(4) = \sqrt{2 \cdot 4 + 1} = \sqrt{9} = 3,$$

and

$$G(8) = \sqrt{2 \cdot 8 + 1} = \sqrt{17} \simeq 4.123.$$

Assuming that a and b belong to D_G, we have

$$G(a) = \sqrt{2a + 1} \quad \text{and} \quad G(b) = \sqrt{2b + 1},$$

so

$$G(a) \cdot G(b) = \sqrt{2a + 1} \cdot \sqrt{2b + 1}$$
$$= \sqrt{(2a + 1)(2b + 1)}$$
$$= \sqrt{4ab + 2a + 2b + 1}.$$

If $ab \in D_G$, then

$$G(ab) = \sqrt{2ab + 1}.$$

Notice that, in general, $G(ab) \neq G(a) \cdot G(b)$. Can you find a pair a, b for which $G(ab) = G(a) \cdot G(b)$?

◇ **Practice Exercise 7.**   Find the domain of $g(x) = \sqrt{2 - 3x}$. Find the function values $g(-1)$, $g(-2/3)$, $g(1)$. Also, find $g(a)/g(b)$ and $g(a/b)$, assuming that these function values are defined.

Answer.   Domain: $x \leq 2/3$; $g(-1) = \sqrt{5}$, $g(-2/3) = 2$, $g(1)$ not defined;

$$\frac{g(a)}{g(b)} = \sqrt{\frac{2 - 3a}{2 - 3b}}; \quad g\left(\frac{a}{b}\right) = \sqrt{\frac{2b - 3a}{b}}.$$

Functions Defined By Formulas

Many formulas used in mathematics and the applied sciences provide examples of functions.

One of the greatest scientific discoveries of the Renaissance was Galileo's law of falling bodies. The law states that the distance s covered by an object falling freely from rest is proportional to the square of time; that is, the formula

$$s(t) = ct^2$$

expresses s as a function of t.

 HISTORICAL NOTE

Galileo Galilei (1564–1642)

was an Italian astronomer, mathematician, and physicist. He discovered the law of falling bodies: when a body is falling freely and the resistance of the air is neglected, its acceleration is constant; furthermore, the acceleration is the same for all bodies. Galileo constructed telescopes with which he made important astronomical discoveries; he observed the phases of Venus, found out that the Milky Way consists of numberless stars, and discovered the satellites of Jupiter. He ardently adopted Copernicus' theory that the earth and the planets move around the sun. Brought to trial by the Inquisition, he was forced to renounce all his beliefs and writings holding that the sun was the center of the planetary system. Galileo's persistent investigations of natural laws laid the foundations for modern experimental science.

◆ **Example 8.** When the distance s is measured in *feet* and the time t in *seconds*, the constant c in the preceding formula is approximately equal to 16 feet per square second. How many feet does an object fall in **a)** 2 seconds, **b)** 3 seconds, **c)** 6.5 seconds?

Solution. We have $s(t) = 16t^2$. So,

$$s(2) = 16 \cdot 2^2 = 64 \text{ ft,}$$
$$s(3) = 16 \cdot 3^2 = 144 \text{ ft,}$$
$$s(6.5) = 16 \cdot (6.5)^2 = 676 \text{ ft.}$$

◇ **Practice Exercise 8.** The area A of a circle of radius r is equal to π times the square of the radius. Write a formula expressing A as a function of r. Also, find the area if $r = \sqrt{2}$, 4, and 6.2 units of length.

Answer. $A(r) = \pi r^2$, $A(\sqrt{2}) = 2\pi$, $A(4) = 16\pi$, $A(6.2) = 38.44\pi$.

◆ **Example 9.** A manufacturer can produce a maximum of 60 car radios per day, at a cost of $35 per radio. If his daily operating expenses are $575, write a formula for the total cost of producing x radios per day. What is the domain of definition of this function?

Solution. If $C(x)$ denotes the total cost to produce x radios per day, then

$$C(x) = 575 + 35x$$

with x a nonnegative integer less than or equal to 60. Thus, the domain of C is the set $\{x: 0 \le x \le 60,\ x \text{ integer}\}$.

◇ **Practice Exercise 9.** It costs $25 per day plus 25¢ per mile to rent a car. Find the cost per day, $C(x)$, as a function of x miles.

Answer. $C(x) = 25 + 0.25x$.

▢ EXERCISES 4.2

In Exercises 1–10, find the domain of each function.

1. $f(x) = \dfrac{3}{4 - 2x}$

2. $f(x) = \dfrac{-2}{3x - 9}$

3. $f(x) = \sqrt{4 - 3x}$

4. $f(x) = \sqrt{2x + 6}$

5. $f(x) = \dfrac{x}{(x - 1)^2}$

6. $f(x) = \dfrac{2x - 1}{(x + 2)^2}$

7. $f(x) = \dfrac{2x + 1}{x^2 - 3x + 2}$

8. $f(x) = \dfrac{5x - 3}{x^2 - 3x - 4}$

9. $f(x) = \sqrt{1 - x^2}$

10. $f(x) = \sqrt{x^2 + 2x}$

For each function in Exercises 11–18, find the indicated function values. When necessary, use your calculator.

11. $f(x) = 3x + 6$; $f(1)$, $f(0)$, $f(2/3)$, $f(\sqrt{1.5})$

12. $g(x) = 3 - 5x$; $g(-2)$, $g(0)$, $g(4/5)$, $g(\sqrt{6})$

13. $F(x) = -x^2 - x - 1$; $F(-1)$, $F(0)$, $F(1)$, $F(\sqrt{2})$

14. $G(x) = x^2 - 3x + 4$; $G(-4)$, $G(-1/2)$, $G(0)$, $G(3)$

15. $h(x) = \sqrt{2x - 4}$; $h(2)$, $h(10)$, $h(20/9)$, $h(\sqrt{3})$

16. $H(x) = \sqrt{2x + 2}$; $H(-2/3)$, $H(5)$, $H(-34/74)$, $H(14/3)$

17. $k(x) = \dfrac{x}{x + 3}$; $k(-4)$, $k(-2)$, $k(0)$, $k(3.512)$

18. $K(x) = \dfrac{2x}{4 - x}$; $K(-1)$, $K(0)$, $K(3)$, $K(-6.418)$

In Exercises 19–24, for each of the given functions, find a) $f(a)$, b) $f(-a)$, c) $-f(a)$, d) $f(a) + f(b)$, and e) $f(a + b)$.

19. $f(x) = 5 - 3x$

20. $f(x) = 2 - 7x$

21. $f(x) = x^2 - 4$

22. $f(x) = x^2 + 1$

23. $f(x) = \dfrac{x - 1}{x + 1}$

24. $f(x) = \dfrac{2 - x}{x + 2}$

In Exercises 25–30, for each of the given functions, find a) $h(1/a)$, b) $1/h(a)$, c) $h(a) + h(1/a)$, and d) $h\left(a + \dfrac{1}{a}\right)$.

25. $h(x) = \dfrac{1}{x - 5}$

26. $h(x) = \dfrac{1}{x + 3}$

27. $h(x) = \dfrac{1}{2x - 1}$

28. $h(x) = \dfrac{3x}{x - 1}$

29. $h(x) = 1 - x^2$

30. $h(x) = x^2 + 2$

In Exercises 31–36, find a) $(f + g)(x)$, b) $(f - g)(x)$, c) $(f \cdot g)(x)$, and d) $(f/g)(x)$. State the domain of each function.

31. $f(x) = 3x + 5$, $g(x) = x^2 - 3$

32. $f(x) = x^2$, $g(x) = 2x - 5$

33. $f(x) = \dfrac{3x}{2x - 1}$, $g(x) = \dfrac{x + 1}{5x + 3}$

34. $f(x) = \dfrac{1}{x + 7}$, $g(x) = x + 7$

35. $f(x) = 1 - \dfrac{3}{x}$, $g(x) = 1 + \dfrac{3}{x}$

36. $f(x) = x + \dfrac{2}{x}$, $g(x) = x - \dfrac{2}{x}$

In Exercises 37–44, find $\dfrac{f(a + h) - f(a)}{h}$ and simplify your answer.

37. $f(x) = 8$

38. $f(x) = -6$

39. $f(x) = 3x + 5$

40. $f(x) = 5 - 2x$

41. $f(x) = x^2 + 4$

42. $f(x) = x^2 - 1$

43. $f(x) = 2x^2 - 3x - 1$

44. $f(x) = x^2 - x + 1$

45. The length of a rectangle is 4 cm longer than twice its width. Write a formula for the area of the rectangle as a function of the width.

46. The height of a triangle is 5 ft longer than three times its base. Express the area as a function of the base.

47. Express the area of a square as a function of its perimeter.

48. Write a formula for the area of an equilateral triangle as a function of the length of a side.

49. The length of a rectangular box is four times the width and the height is half the width. What is the volume of the box as a function of the width?

50. The height of a cylinder is five times the radius. What is the volume of the cylinder as a function of the radius?

51. The volume of a cylindrical tin can is 16π cubic centimeters. Write a formula for $L(h)$, the lateral surface area of the can, as a function of the height h. (Recall that $L = 2\pi rh$.)

52. For the tin can in Exercise 51, write a formula for $S(h)$, the total surface area, as a function of the radius. (Recall that $S = \pi r^2 + 2\pi rh$.)

53. If $1,000 is invested at a simple interest rate of 6% per year, express the principal p as a function of t years.

54. Suppose that if you invest p_0 dollars today, at a simple interest rate, you'll receive $5,000 in 10 years. Write a formula giving p_0 as a function of the interest rate.

55. Suppose that in the simple lens equation

$$\frac{1}{f} = \frac{1}{d_1} + \frac{1}{d_2},$$

the object distance d_1 is five times the image distance d_2. Express the focal length f as a function of d_2.

● **56.** A bus company can organize an excursion at a cost of $35 per person if 80 people decide to take the tour. Moreover, the company will reduce the cost at a rate of 12¢ per person in excess of 80. If x denotes the number of people taking the tour, write the total cost, C, as a function of x.

● **57.** Margie has a total of A dollars in nickels, dimes, and quarters. If she has five fewer dimes than quarters and ten more dimes than nickels, write A as a function of the number x of nickels.

● **58.** An open box is to be made from a rectangular piece of cardboard having dimensions 24 inches by 36 inches, by cutting a square of side x inches from each corner and turning up the sides. Express the volume $V(x)$ of the box in terms of x. What is the domain of this function?

● **59.** The height h of a cylinder is 3 times the radius.

a) Write a formula expressing $L(h)$, the lateral surface area of the cylinder, as a function of h.

b) Find the function values $L(0.25)$, $L(1)$, $L(1.25)$, and $L(2.012)$. Round off your answers to two decimal places.

● **60.** For a simple pendulum, the period T, in seconds, is a function of the length l, in feet, given by

$$T(l) = 2\pi \sqrt{\frac{l}{32}}.$$

Find the function values $T(3/4)$, $T(1.5)$, and $T(2.25)$. Round off your answers to three decimal places.

4.3   GRAPH SKETCHING

In this section, we discuss several graphing techniques, such as the *stretching* and *shrinking* of graphs, and *horizontal* and *vertical translations* of graphs. Combined with the plotting of points, these techniques considerably simplify the task of graphing a function.

◆ **Example 1.**   Graph the functions

a) $f(x) = 3x^2$,   **b)** $f(x) = \dfrac{1}{4}x^2$.

Solution.   **a)** The graph of $f(x) = 3x^2$ may be obtained multiplying by 3 the ordinates of all points on the graph of the parabola $y = x^2$. Figure 4.8 illustrates the graphs of $f(x) = 3x^2$ and $y = x^2$ (dotted line). The graph of $f(x) = 3x^2$ is also a parabola with vertex at the origin and axis of symmetry the y-axis.

b) If we multiply by 1/4 the ordinates of points on the graph of $y = x^2$, we obtain the graph of $f(x) = (1/4)x^2$, shown in Figure 4.9.

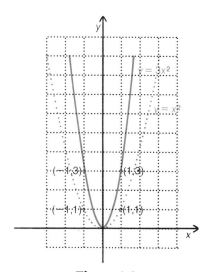

Figure 4.8

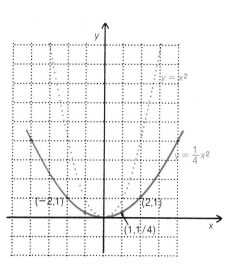

Figure 4.9

◇ **Practice Exercise 1.**   Graph the functions

a) $f(x) = 2\sqrt{x}$,   **b)** $f(x) = \dfrac{1}{2}\sqrt{x}$

Answer.

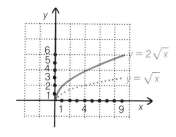

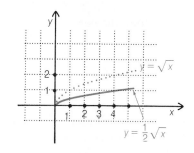

Example 1 and Practice Exercise 1 illustrate the following general fact.

> **Stretching and Shrinking Graphs**
>
> To obtain the graph of $y = af(x)$, multiply by a the ordinates of points on the graph of $y = f(x)$. If $a > 1$, the graph of $y = f(x)$ is *stretched* vertically by a factor of a. If $0 < a < 1$, the graph is *shrunk* (or *squeezed*) vertically. If $a = -1$, the graph is reflected across the x-axis. For example, compare the graphs of $y = x^2$ and $y = -x^2$.

Now we discuss vertical translations of graphs.

◆ **Example 2.** Graph the functions
a) $f(x) = \sqrt{x} + 1$, **b)** $f(x) = \sqrt{x} - 2$.

Solution. The graphs of both functions together with the graph of $y = \sqrt{x}$ (dotted lines) are illustrated in Figure 4.10.

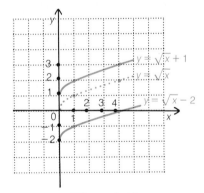

Figure 4.10

To obtain the graph of $y = \sqrt{x} + 1$, we increase by 1 the ordinate of each point on the graph of $y = \sqrt{x}$. In other words, the graph of $y = \sqrt{x} + 1$ is obtained by shifting the graph of $y = \sqrt{x}$ one unit up. If we shift the graph of $y = \sqrt{x}$ two units down, the result will be the graph of $y = \sqrt{x} - 2$.

Notice that the x- and y-intercepts were changed as a result of this shifting. For example, the origin is both the x- and y-intercepts of the graph of $y = \sqrt{x}$. However, the x-intercept of the graph of $y = \sqrt{x} - 2$ is 4, and the y-intercept is -2.

◇ **Practice Exercise 2.** Graph the functions
a) $f(x) = x^2 + 1$, **b)** $f(x) = x^2 - 1$.

Answer.

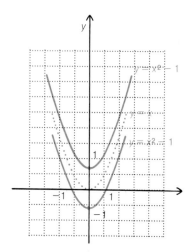

The technique of shifting that we considered in Example 2, called *vertical translation*, applies to any function and can be summarized as follows.

Vertical Translations of Graphs

The graph of

$$y = f(x) + k$$

may be obtained by shifting the graph of $y = f(x)$ as shown in Figure 4.11, k units up if $k > 0$, or $-k$ units down if $k < 0$.

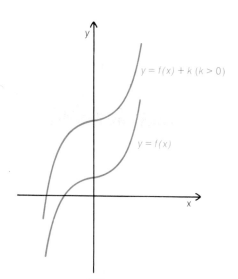

Figure 4.11

Next, we discuss the notion of horizontal translation of graphs.

◆ **Example 3.** Graph the functions
a) $f(x) = (x - 1)^2$ **b)** $f(x) = (x + 3)^2$.

Solution. **a)** Let $g(x) = x^2$. Since

$$f(a) = (a - 1)^2 = g(a - 1),$$

we see that the point with abscissa a on the graph of f has the same ordinate as the point with abscissa $a - 1$ on the graph of g. The graph of $f(x) = (x - 1)^2$ is then obtained by shifting the graph of $g(x) = x^2$ one unit to the *right* (Figure 4.12). The graph is a parabola whose vertex is the point $(1, 0)$ and whose axis of symmetry is the vertical line $x = 1$.

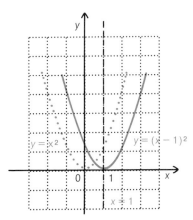

Figure 4.12

b) Again let $g(x) = x^2$. The relation

$$f(a) = (a + 3)^2 = g(a + 3)$$

shows that the point of abscissa a on the graph of f has the same ordinate as the point of abscissa $a + 3$ on the graph of g. Thus the graph of $f(x) = (x + 3)^2$ is a parabola obtained by shifting the graph of $g(x) = x^2$ three units to the *left*

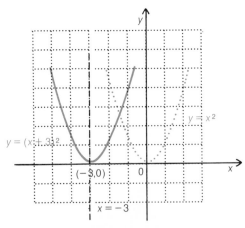

Figure 4.13

(Figure 4.13). The point $(-3, 0)$ is the vertex of the parabola, and the line $x = -3$ is the axis of symmetry.

◇ **Practice Exercise 3.**   Graph the functions

a) $f(x) = \sqrt{x - 2}$,   **b)** $f(x) = \sqrt{x + 4}$

Answer.

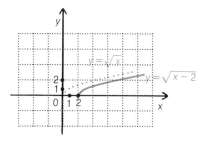
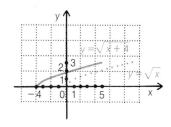

The technique described in Example 3, called *horizontal translation*, applies to any function. It can be summarized as follows.

Horizontal Translation of Graphs

The graph of

$$y = f(x - h)$$

may be obtained from the graph of $y = f(x)$ by a *horizontal translation* as shown in Figure 4.14. If $h > 0$, the graph of f is shifted h units to the right; if $h < 0$, the graph of f is shifted $-h$ units to the left.

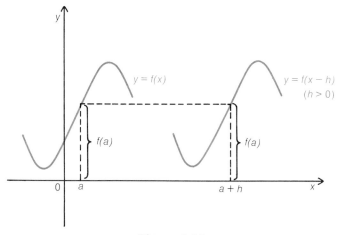

Figure 4.14

Sometimes we encounter functions defined in special ways such as in the following example.

◆ **Example 4.**   Graph the function

$$f(x) = \begin{cases} x + 2 & \text{if } x \le 0 \\ 4 - x^2 & \text{if } 0 < x \le 2 \\ 3 & \text{if } x > 2. \end{cases}$$

Solution.   The given function is defined for all real values of x. The graph, illustrated in Figure 4.15, consists of three distinct parts. When $x \le 0$, $f(x) = x + 2$, thus the graph is part of the line $y = x + 2$. On the interval $(0, 2)$ the graph is part of the parabola $y = 4 - x^2$. Finally, when $x > 2$, $f(x)$ is constant and equal to   , and the graph is part of the horizontal line $y = 3$.

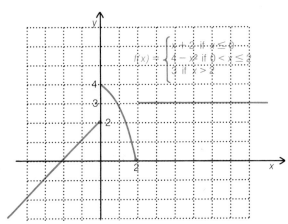

Figure 4.15

◇ **Practice Exercise 4.**   Graph the function defined as follows:

$$f(x) = \begin{cases} -x - 1 & \text{if } x < 0 \\ x^2 - 4 & \text{if } 0 < x \le 2 \\ 2 & \text{if } x > 2. \end{cases}$$

Answer.

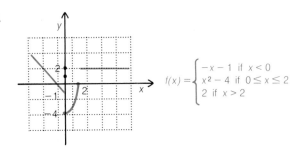

The Absolute Value Function

Consider the function

$$f(x) = |x|$$

called the *absolute value* function. Its domain of definition is the set of all real numbers and its range consists of the set of all nonnegative numbers. If $x > 0$, then $f(x) = |x| = x$, so the graph of the absolute value function coincides with the graph of $y = x$. If $x < 0$, then $f(x) = |x| = -x$ and the graph of f coincides with the graph of $y = -x$.

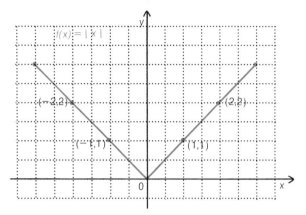

Figure 4.16

Notice that the graph is symmetric with respect to the x-axis.

The Greatest Integer Function

If x is a real number, denote by $[x]$ the *greatest integer which is less than or equal to x*. For example, the greatest integer less than or equal to 1/2 is 0, the greatest integer less than or equal to 15 is 15, and the greatest integer less than or equal to $\sqrt{2}$ is 1. Thus we write

$$\left[\frac{1}{2}\right] = 0, \quad [15] = 15, \quad \text{and} \quad [\sqrt{2}] = 1.$$

We also have

$$\left[-\frac{2}{3}\right] = 1, \quad \left[\frac{25}{3}\right] = 8, \quad \text{and} \quad [\pi] = 3.$$

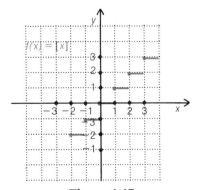

Figure 4.17

The rule that assigns to every real number x the integer $[x]$ defines a function called the *greatest integer function* and denoted by $f(x) = [x]$. The domain of definition of f is the set of all real numbers and the range is the set of all integers.

Even and Odd Functions

A function f with domain D is said to be *even* if $f(x) = f(-x)$ for all x in D.

    The function f is said to be *odd* if $f(-x) = -f(x)$, for all $x \in D$.

The definition says that the value of an even function does not change when $-x$ is substituted for x. However, the function value of an odd function changes sign if $-x$ is substituted for x.

The two simplest examples of odd and even functions are, respectively, $f(x) = x$ and $f(x) = x^2$. The reason for the names *even* and *odd functions* comes from the fact that *even powers* of x define even functions, while *odd powers* of x define odd functions.

In Figure 4.18, we illustrate the graphs of the even functions $f(x) = x^2$ and $f(x) = x^4$. Observe that both graphs are symmetric with respect to the y-axis.

x	0	1/2	1/3	1	3/2	2
x^2	0	1/4	1/9	1	9/4	4
x^4	0	1/16	1/81	1	81/16	16

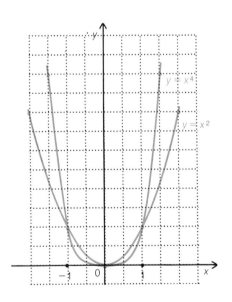

Figure 4.18

Figure 4.19 illustrates the graphs of $f(x) = x$ and $f(x) = x^3$, two odd functions. The graphs are symmetric with respect to the origin.

x	0	1/2	1	2
x^3	0	1/8	1	8

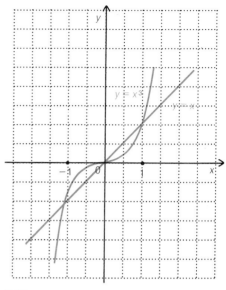

Figure 4.19

As an exercise, you can check that the graph of any even function is symmetric with respect to the *y*-axis, while the graph of any odd function is symmetric with respect to the origin.

Increasing and Decreasing Functions

Let *f* be a function and let *I* be a subset of the domain of *f*. If x_1 and x_2 are in *I*, then:
  f is *increasing* on *I* if $f(x_1) < f(x_2)$ whenever $x_1 < x_2$, and
  f is *decreasing* on *I* if $f(x_1) > f(x_2)$ whenever $x_1 < x_2$.

For example, we already know that the graph of a linear function

$$f(x) = mx + b$$

is a line with slope *m*. If $m > 0$, the *f* is an increasing function on $\mathbb{R}$, and if $m < 0$, then *f* is a decreasing function on $\mathbb{R}$ (Figure 4.20).

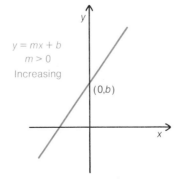

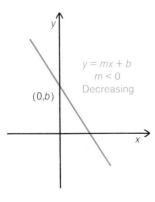

Figure 4.20

As another example, the function $f(x) = x^2$ is decreasing on the interval $(-\infty, 0]$ and increasing on the interval $[0, +\infty)$ as shown in Figure 4.21.

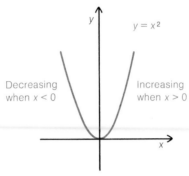

Figure 4.21

A constant function $f(x) = b$ is neither increasing nor decreasing (Figure 4.22).

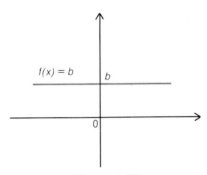

Figure 4.22

 EXERCISES 4.3

In Exercises 1–22, graph each function. (Make use of stretching, shrinking, horizontal and vertical shifts.)

1. $f(x) = \dfrac{x^2}{2}$

2. $f(x) = 4x^2$

3. $f(x) = -3x^2$

4. $f(x) = -\dfrac{x^2}{4}$

5. $f(x) = 3\sqrt{x}$

6. $f(x) = \dfrac{\sqrt{x}}{2}$

7. $f(x) = -2\sqrt{x}$

8. $f(x) = \dfrac{-3\sqrt{x}}{4}$

9. $f(x) = x^2 - 4$

10. $f(x) = 1 - x^2$

11. $f(x) = \sqrt{x} + 3$

12. $f(x) = \sqrt{x} - 1$

13. $f(x) = 2 - \sqrt{x}$

14. $f(x) = 3 - \sqrt{x}$

15. $f(x) = (x - 3)^2$

16. $f(x) = (x + 4)^2$

17. $f(x) = -(x + 1)^2$

18. $f(x) = -(x - 2)^2$

19. $f(x) = \sqrt{x + 5}$

20. $f(x) = \sqrt{x - 1}$

21. $f(x) = (x - 1)^2 + 4$

22. $f(x) = (x + 4)^2 - 1$

Graph each function.

23. $f(x) = \begin{cases} -2x - 1 & \text{if } -2 \le x < -1 \\ x^2 & \text{if } -1 \le x \le 1 \\ \dfrac{x+2}{3} & \text{if } 1 < x \le 2 \end{cases}$

24. $f(x) = \begin{cases} 3x + 4 & \text{if } x \le 1 \\ x^2 - 1 & \text{if } x > 1 \end{cases}$

In Exercises 25–34, combine the definition of the absolute value function and the various graphing techniques to graph each function.

25. $f(x) = 3|x|$

26. $f(x) = \dfrac{|x|}{4}$

27. $f(x) = |x| - 2$

28. $f(x) = |x| + 3$

29. $f(x) = 4 - |x|$

30. $f(x) = 1 - |x|$

31. $f(x) = |x - 1|$

32. $f(x) = |x + 2|$

33. $f(x) = |x + 1| - 2$

34. $f(x) = |x - 3| + 1$

Graph each of the functions in Exercises 35–40. Use the definition of the greatest integer function and the various techniques of graphing discussed in Section 4.3.

35. $f(x) = 2[x]$

36. $f(x) = \dfrac{[x]}{4}$

37. $f(x) = [x] + 1$

38. $f(x) = [x] - 2$

39. $f(x) = [x - 1]$

40. $f(x) = [x - 2]$

In Exercises 41–54, determine whether each function is even, odd, or neither.

41. $f(x) = 3x^4 + x^2$

42. $f(x) = x^2 + 4x$

43. $F(x) = 2x^3 - 6x$

44. $F(x) = -x^4 - 3x^2$

45. $g(x) = x^6 - 2x^2 - 4$

46. $g(x) = x^5 - 1$

47. $G(x) = |x| + 2$

48. $G(x) = 4 - |x|$

49. $h(x) = |x|^3$

50. $h(x) = -|x|^5$

51. $H(x) = 5/x$

52. $H(x) = -3/x^2$

53. $k(x) = \dfrac{1}{x^2 - 1}$

54. $k(x) = \dfrac{2}{x - 1}$

Find intervals (each as large as you can find) in which each of the functions in Exercises 55–62 is increasing or decreasing. First, graph each function.

55. $f(x) = -5x + 3$

56. $f(x) = 4x - 2$

57. $f(x) = x^2 - 4$

58. $f(x) = 9 - x^2$

59. $f(x) = -|x - 3|$

60. $f(x) = |x + 5|$

61. $f(x) = x - |x|$

62. $f(x) = x + |x|$

63. A coin collection is worth \$16,000 now and its value appreciates linearly at a rate of 8% per year. Find the function $V(x)$ giving the value of the coin collection in x years. Graph this function.

64. A piece of equipment purchased for \$48,000 is expected to have a scrap value of \$6,000 after 15 years. Assuming that the depreciation is linear, what will be the value $V(t)$ of the equipment after t years? Graph this function.

65. To mail a first-class letter costs 20¢ for the first ounce or fraction, plus 17¢ for each additional ounce or fraction thereof. Graph the function $f(x)$ that gives the cost of mailing an x-ounce letter.

66. A car rental company charges $15 per day plus $1.50 for each additional 10 miles or fraction of 10 miles. Graph the function $C(x)$ that expresses the daily cost of renting and driving a car for x miles.

67. Let $D(p) = 125 - \dfrac{p^2}{2}$ be the demand function for a certain item, where p is the price of the item. Graph this function.

68. Suppose that the cost to produce x units of a certain item is

$$C(x) = \frac{x^2}{4} + 25, \quad 0 \le x \le 10.$$

Sketch the graph.

● 69. Show that for any function $f(x)$, the function

$$F(x) = \frac{1}{2}[f(x) + f(-x)]$$

is even, and that the function

$$G(x) = \frac{1}{2}[f(x) - f(-x)]$$

is odd.

● 70. Prove that any function $f(x)$ can always be written as the sum of an even and an odd function. [*Hint*: Use Exercise 69.]

4.4 VARIATION

In the beginning of Section 4.1, we described two simple, but important, examples of related variables. In the first one,

$$d = r \times t \quad \text{(distance = rate × time)}$$

if the rate r remains constant, the distance d *varies directly* as t; that is, d increases (or decreases) as t increases (or decreases). In the second one,

$$P = \frac{C}{V} \quad \text{(Boyle's law)},$$

the pressure P of a gas enclosed in a container and kept at a constant temperature increases if the volume decreases, and decreases if the volume increases. This is an example of *inverse variation*.

In mathematics and the applied sciences we often have to deal with relations involving a dependent variable and one or more independent variables. This leads us to the notion of variation.

Direct Variation

A variable y *varies directly* as x, or y is *directly proportional* to x, if there is a real number k such that

$$y = kx.$$

The number k is called the *constant of proportionality*.

For example, if a car is travelling at 55 mi/h, the distance d (in miles) traveled by the car in t hours is given by $d = 55t$. Thus, d is directly proportional to t, and 55 is the constant of proportionality.

Observe that in direct variation, the dependent variable increases (or decreases) numerically as the independent variable increases (or decreases).

Sometimes, a variable is directly proportional to a power of another variable. For example, in the formula $s = 16t^2$ giving the distance in feet that an object falls in t seconds, s *varies directly as the square* of the time t. The constant of proportionality is $k = 16$.

Next, we describe *inverse variation*.

Inverse Variation

A variable y *varies inversely* as x, or y is said to be *inversely proportional* to x, if

$$y = \frac{k}{x}$$

for some fixed number k, called the *constant of proportionality.*

If y varies inversely as x, then the product xy is constant and equal to the constant of proportionality k.

◆ *Example 1.* When the volume of air inside a bicycle pump is 54 in³, the air pressure is 15 lb/in². Find the air pressure when the volume is equal to 18 in³.

Solution. According to Boyle's law, the pressure P is inversely proportional to the volume V and we may write $PV = C$. If the volume changes to V', the corresponding pressure P' is such that $P'V' = C$. Hence,

$$P'V' = PV.$$

Setting $V = 54$, $P = 15$, and $V' = 18$ in the last equation and solving it for P', we obtain

$$P' \times 18 = 15 \times 54,$$

or

$$P' = \frac{15 \times 54}{18} = 45 \text{ lb/in}^2,$$

which is the air pressure when the volume is 18 in³.

In inverse variation, the dependent variable increases (or decreases) numerically as the independent variable decreases (or increases).

Sometimes y varies inversely as a power of x. For example, the intensity of illumination, I, produced by a light source varies inversely as the square of the distance, d, from the source; that is, $I = c/d^2$, where c is the constant of proportionality.

◇ **Practice Exercise 1.**　A light source at 20 feet produces an illumination of 60 foot-candles. What is the intensity of illumination produced at 10 feet?

Answer.　240 foot-candles.

In certain cases, the dependent variable may depend on more than one independent variable.

Joint Variation

A variable y is said to *vary jointly* as x and z if there is a number k (constant of proportionality) such that

$$y = kxz.$$

For example, in the formula $d = rt$, the distance d varies jointly as the speed r and the time t. A box of dimensions x, y, and z has volume $V = xyz$. Thus, the volume of a box varies jointly as the dimensions of its sides. In both examples, the constant of proportionality is 1.

In certain cases, a variable may vary jointly as powers of other variables. For instance, the volume V of a right cylinder of radius r and height h is given by the formula

$$V = \pi r^2 h,$$

which shows that V varies jointly as the square of the radius and the height of the cylinder. The constant of proportionality is π.

We can also have combined variation, such as in the formula

$$F = G\frac{mm'}{r^2}$$

[Example 5 of Section 1.3] giving the magnitude of the force of attraction between two masses m and m' placed at a distance r from each other; G is a physical constant.

As a side remark, we mention that joint variation provides examples of *functions of several variables*, a topic studied in calculus. For instance, we may say that the volume V of a cylinder is a function of two variables: the radius r and the height h. The force F is a function of three variables: the masses m and m', and the distance r separating these masses.

◆ **Example 2.**　Suppose that w varies directly as u and inversely as the square of v. If $w = 2$ when $u = 16$ and $v = 2$, find an explicit formula for w. Also, find w when $u = 8$ and $v = 2$.

Solution.　The first statement translates into

$$w = k\frac{u}{v^2}.$$

To evaluate the constant of proportionality, substitute the given values for u, v, and w in the last equation and solve it for k:

$$2 = k\frac{16}{2^2},$$

$$2 = 4k,$$

$$k = \frac{2}{4} = \frac{1}{2}.$$

Therefore, the explicit formula for w is

$$w = \frac{1}{2}\frac{u}{v^2}.$$

Now, substituting 8 for u and 2 for v, we obtain a numerical value for w:

$$w = \frac{1}{2} \cdot \frac{8}{2^2} = \frac{1}{2} \cdot \frac{8}{4} = 1.$$

◇ **Practice Exercise 2.** A variable m varies directly as the square of n and inversely as p. Find **a)** the constant of proportionality if $m = 3$ when $n = 3$ and $p = 2$, and **b)** m when $n = 4$ and $p = 4$.

Answer. **a)** $k = \dfrac{2}{3}$, **b)** $m = \dfrac{8}{3}$.

c ◆ **Example 3.** Empirical observation indicates that the shoe size of a normal man varies directly as the 3/2 power of his height. On the average, a man 6 feet tall wears a size 11 shoe. What is the approximate shoe size of a basketball player 7 feet and 3 inches tall?

Solution. Let s denote the shoe size and let h denote the height. Since s varies directly as the 3/2 power of h, we have

$$s = ch^{3/2}.$$

To determine the constant of proportionality c, substitute 6 for h and 11 for s, obtaining

$$11 = c6^{3/2}.$$

Using a calculator we have

$$11 \simeq c(14.7)$$

hence

$$c \simeq 0.75.$$

Thus, the equation

$$s = 0.75h^{3/2}$$

relates the shoe size s of a normal man to his height h. Since $7'3'' = 7.25$ ft, substituting 7.25 for h in the last equation yields

$$s = 0.75(7.25)^{3/2}.$$

Now, performing the following keystrokes in your calculator

$$0.75 \boxed{\times} 7.25 \boxed{y^x} 1.5 \boxed{=}$$

obtain

$$s \simeq 14.64 \simeq 15.$$

Alternate solution. We can also find the desired value of s without finding the constant of proportionality. Start with the equation $s = ch^{3/2}$. When $h = 6$, then $s = 11$, so that

$$11 = c6^{3/2}.$$

When $h = 7.25$, we have

$$s = c(7.25)^{3/2}.$$

Dividing this equation by the preceding one, we obtain

$$\frac{s}{11} = \frac{(7.25)^{3/2}}{6^{3/2}}$$

$$s = \frac{11 \times (7.25)^{3/2}}{6^{3/2}}$$

$$\simeq \frac{11 \times 19.52}{14.70}$$

$$\simeq 14.61 \simeq 15.$$

◇ **Practice Exercise 3.**   The frequency of vibration of a guitar string of constant length is proportional to the square root of the tension. If a tension of 40 units produces a frequency of 330 hertz (Hz), what is the frequency when the tension is **doubled**?

Answer.   467 Hz.

 EXERCISES 4.4

In Exercises 1–4, find an explicit formula for the dependent variable.

1. u varies directly as v, and $u = 18$ when $v = 6$.
2. p varies directly as the square root of q, and $p = 2$ when $q = 2$.
3. y varies inversely as x^2, and $y = 27$ when $x = 1/9$.
4. f varies jointly as p and q and inversely as r^2, and $f = 1$ when $p = 2$, $q = 3$, and $r = 3$.

Solve each of the following problems.

5. If w varies jointly as x and y, and $w = 6$ when $x = 3$ and $y = 4$, find w when $x = 1$ and $y = 6$.
6. If a varies directly as b and inversely as c, and

$a = 15/2$ when $b = 5$ and $c = 2$, find a when $b = 2$ and $c = 7$.

7. Let p be directly proportional to the square of q and inversely proportional to r. If $p = 9/4$ when $q = 3$ and $r = 2$, find p when $q = 2$ and $r = 3$.

8. Suppose that y is directly proportional to m and inversely proportional to n and r^2. If $y = 1/4$ when $m = 3$, $n = 1$, and $r = 2$, find y when $m = 2$, $n = 3$, and $r = 4$.

9. Hooke's law states that the force F required to stretch a spring a length l beyond its original length is directly proportional to the additional length. If a force of 6 pounds stretches a certain

spring 2 inches, how much will a force of 28 pounds stretch the spring?

10. The pressure exerted by a liquid at a point is proportional to the depth of the point below the surface of the liquid. If the pressure at 18 meters below the surface of the liquid is 60 kg/cm^2, what is the pressure at 26 meters below the surface?

11. The intensity of illumination at a given point is inversely proportional to the square of the distance from the point to the source. If a light source at 20 feet produces an illumination of 40 foot-candles, what is the illumination 8 feet from the light source?

12. In an artery, the resistance r to blood flow is directly proportional to the length l of the artery and inversely proportional to the fourth power of its diameter d. Write a formula expressing r in terms of l and d. What happens if l is increased by a factor of 4 and d is doubled?

13. The length of the skid marks caused by braking a car varies directly as the square of the initial velocity. Assume that the skid marks are 80 feet long when the initial velocity is 40 mi/h. Find the length of the skid marks when the velocity is 60 mi/h.

14. The frequency of vibration of a guitar string is proportional to the square root of the tension and inversely proportional to the length of the string. What is the effect on the frequency if both the tension and the length are increased by a factor of 4?

15. The volume of a sphere is directly proportional to the cube of its radius. If a sphere of radius 3 in has a volume of 36π in^3, find the formula for the volume of a sphere. **c** Use this formula to find the volume of a sphere of radius $r = 2.25$ in.

c 16. The weight of an object above the surface of the earth varies inversely with the square of its distance from the center of the earth. If an astronaut weighs 175 lb at the surface of the earth, what will be the weight of the astronaut 150 miles above the surface of the earth? (The radius of the earth is 3959 mi.)

17. The kinetic energy K of a particle varies directly as its mass m and the square of its velocity v. A particle of mass 4 g moving at 10 cm/sec has a kinetic energy of 200 ergs.
   a) Find the formula for kinetic energy.
   c b) Find the kinetic energy of a particle of mass 6.80 g moving at 18.5 cm/sec.

c 18. The electrical resistance of a wire varies inversely as the square of the diameter of the wire. If a wire of diameter 2×10^{-2} cm has a resistance of 10^{-1} ohm, what is the resistance of a wire of diameter 1.25×10^{-2} cm?

● 19. The period of a pendulum varies directly as the square root of its length and inversely as the square root of the acceleration due to gravity.
   a) Find an explicit formula for the period, knowing that at a place where the acceleration due to gravity is g, a pendulum of length g/π^2 has period 2.
   c b) Find the period of a pendulum 50 cm long at a place where $g = 9.8$ m/sec^2.

● 20. According to Kepler's third law, the time required for a planet to make a complete revolution around the sun varies directly with the 3/2 power of its average (mean) distance from the sun. The earth-sun mean distance is 9.29×10^7 mi. Find how long it takes for Jupiter to complete one revolution around the sun if the Jupiter-sun mean distance is 4.83×10^8 mi.

CHAPTER SUMMARY

A *function f* is a rule that assigns to each element x of a set a *unique* element y of another set. The element x is the *independent variable* or *input* of f, and the element y is the *dependent variable* or *output* of f. The *domain* of f is the set where f is defined and the *range* is the set of *images* under f of all domain elements.

If f is a *real-valued function* and x is a number in the domain of f, the number $f(x)$ is the *function value* of f at x.

The *graph* of a function f with domain D is the set of points in a Cartesian coordinate plane with coordinates $(x, f(x))$ for all $x \in D$. If $a \in D$, every vertical line through $(a, 0)$ intersects the graph of f at a single point. Conversely, a set of points in a coordinate plane is the graph of a function if every vertical line intersects the set in *at most* one point.

Functions can be added, subtracted, multiplied, and divided, generating new functions. Polynomial functions are the simplest functions in mathematics. A polynomial function of degree 0 is the same as a *constant function. Linear functions* are polynomial functions of degree 1. *Quadratic functions* are polynomial functions of degree 2. A *rational function* is the quotient of two polynomials.

Several equations in mathematics and many formulas in the applied sciences provide important examples of functions.

The task of graphing a function can be considerably simplified if the *plotting of points* is combined with the techniques of *stretching, shrinking,* and *translating* graphs.

A function f is *even* if $f(x) = f(-x)$, for all x in the domain of f. A function f is *odd* if $f(-x) = -f(x)$ for all x in the domain of f.

A function f *increases* on a set I if $f(x_1) < f(x_2)$ whenever $x_1 < x_2$ in I. It *decreases* on I if $f(x_1) > f(x_2)$ whenever $x_1 < x_2$.

The notion of *variation* tells us how a variable y may depend on a single variable or on several independent variables. We say that y is *directly proportional* to x if there is a number k, called *constant of proportionality*, such that $y = kx$. It is *inversely proportional* to x if $y = k/x$. In both cases, we may say that y is a function of x. We may also have direct or inverse variation of y as a power of x. Finally, y may vary directly as powers of several variables and inversely as powers of other variables.

REVIEW EXERCISES

Find the domain of definition of each of the following functions.

1. $f(x) = \dfrac{3x + 1}{5 - 4x}$

2. $f(x) = \dfrac{-2x - 3}{3x - 7}$

3. $f(x) = \sqrt{6 - 9x}$

4. $f(x) = \sqrt{4x + 10}$

5. $f(x) = \dfrac{x + 1}{(x - 2)(3x + 5)}$

6. $f(x) = \dfrac{2x - 3}{(x + 4)(2x - 1)}$

7. $f(x) = \dfrac{3x - 8}{x^2 - x - 2}$

8. $f(x) = \dfrac{x - 4}{2x^2 - 5x + 3}$

9. $f(x) = \sqrt{2x^2 - 18}$

10. $f(x) = \sqrt{48 - 3x^2}$

11. $f(x) = \sqrt{\dfrac{x - 1}{x + 2}}$

12. $f(x) = \sqrt{\dfrac{2x + 3}{x - 4}}$

13. $f(x) = \sqrt{\dfrac{-x}{x^2 + 1}}$

14. $f(x) = \sqrt{\dfrac{x^2}{1 - x^2}}$

In Exercises 15–24, find the indicated function values. When necessary, use your calculator.

15. $f(x) = 5x - 1$, $f(-7/10)$, $f(2/5)$, $f(3)$, $f(3.12)$

16. $f(x) = 2 - 3x$, $f(-4/9)$, $f(5/3)$, $f(-8)$, $f(1.45)$

17. $f(x) = 3 - 2x^2$, $f(-5)$, $f(0)$, $f(5)$, $f(\sqrt{10})$

18. $f(x) = 4 - 3x^2$, $f(-2)$, $f(1)$, $f(3)$, $f(\sqrt{8})$

19. $f(x) = \sqrt{4 - 3x}$, $f(-7)$, $f(-15)$, $f(11/12)$, $f(2.5)$

20. $f(x) = \sqrt{5 - 2x}$, $f(-10)$, $f(-31/2)$, $f(2/9)$, $f(4.1)$

21. $f(x) = \dfrac{-2x}{x^2 + 1}$, $f(-3)$, $f(0)$, $f(3)$, $f(4.25)$ **22.** $f(x) = \dfrac{x - 1}{x^2 - 2}$, $f(-2)$, $f(10)$, $f(2)$, $f(0.05)$

23. $f(x) = \sqrt{\dfrac{x - 1}{x - 3}}$, $f(1)$, $f(-10)$, $f(7/2)$, $f(9.5)$ **24.** $f(x) = \sqrt{\dfrac{2 - x}{x - 5}}$, $f(2)$, $f(3/5)$, $f(4)$, $f(2.75)$

In Exercises 25–32, find $\dfrac{f(x + h) - f(x)}{h}$ *and simplify your answer.*

25. $f(x) = 3 - 7x$ **26.** $f(x) = 6x + 2$ **27.** $f(x) = 5x^2 - x$

28. $f(x) = 3x^2 - 2x + 4$ **29.** $f(x) = \dfrac{2}{x}$ **30.** $f(x) = \dfrac{1}{x - 1}$

31. $f(x) = \dfrac{x - 1}{x + 1}$ **32.** $f(x) = \dfrac{1 - x}{2x + 1}$

33. If $f(x) = \dfrac{x + a}{x - a}$, find $f(a + b)$, $f\left(\dfrac{1}{a}\right) + f\left(\dfrac{1}{b}\right)$ and $f\left(\dfrac{1}{a} + \dfrac{1}{b}\right)$.

34. If $g(x) = x + \dfrac{1}{x}$, find $g(a) + \dfrac{1}{g(a)}$ and $g\left(a + \dfrac{1}{a}\right)$.

● **35.** If $f(x) = x$, show that $(f(a + b))^2 - (f(a - b))^2 = 4f(a)f(b)$.

● **36.** If $f(x) = \dfrac{1}{x}$, show that $f(a + 1) - f(a - 1) = -2f(a + 1)f(a - 1)$.

● **37.** Express the area of a triangle as a function of its height h, when its base is one-third of h.

● **38.** Let $S = 2ab + 2ac + 2bc$ be the lateral surface area of a box with sides a, b, and c. Express S as a function of a, when b is twice a, and c is one-half of a.

● **39.** The height of a right circular cylinder is five times its radius. Express the lateral surface area of the cylinder as a function of the radius.

● **40.** Express the total surface area of the cylinder in Exercise 39 as a function of the radius.

In Exercises 41–54, use the techniques of stretching, shrinking, and shifting to graph each function.

41. $f(x) = 8 - 2x^2$ **42.** $f(x) = 3x^2 - 27$ **43.** $f(x) = (x - \frac{1}{2})^2$

44. $f(x) = -(x + 4)^2$ **45.** $f(x) = -(2x + 3)^2$ **46.** $f(x) = (3x - 4)^2$

47. $f(x) = \sqrt{x - \frac{3}{5}}$ **48.** $f(x) = \sqrt{x + 5}$ **49.** $f(x) = \sqrt{4x - 2}$

50. $f(x) = \sqrt{9x + 6}$ **51.** $f(x) = \dfrac{|x - 2|}{3}$ **52.** $f(x) = 3|x + 5|$

53. $f(x) = 1 - |x + 3|$ **54.** $f(x) = 2 - |x + 1|$

Determine whether each function in Exercises 55–62 is even, odd, or neither.

55. $f(x) = x^3 - 6x$ **56.** $f(x) = x^2 - 20$ **57.** $f(x) = x(x^2 + 2)$

58. $f(x) = x(x^3 + 2x)$ **59.** $f(x) = \dfrac{x^2}{2} + |x|^3$ **60.** $f(x) = -3x^2 + 2|x|^5$

61. $f(x) = \dfrac{|x|}{x^2 - 1}$ **62.** $f(x) = \dfrac{|x|}{x^3 - 1}$

Find intervals (each as large as you can find) in which each of the functions in Exercises 63–70 is increasing or decreasing.

63. $f(x) = \dfrac{1}{2}x - 3$ **64.** $f(x) = 2 - \dfrac{5}{7}x$ **65.** $f(x) = 2x^2 - 5$ **66.** $f(x) = 3 - 4x^2$

67. $f(x) = -|x + 3|$ **68.** $f(x) = 2|x - 4|$ **69.** $f(x) = \sqrt{x - 4}$ **70.** $f(x) = -\sqrt{x + 2}$

Express each statement in Exercises 71–74 as an equation. Use k for the constant of proportionality.

71. The pressure in the ears of an underwater swimmer is directly proportional to the depth at which she is swimming.

72. At sea level, the distance that a person can see to the horizon is directly proportional to the square root of the height of the person's eyes above sea level.

73. The weight of a solid cylinder varies jointly as the height and the square of its radius.

74. The electric current I in a wire is directly proportional to the electromotive force E and inversely proportional to the resistance R.

Solve each of the following problems.

75. If y varies directly as the square root of x, and $y = 3$ when $x = 4$, find y when $x = 12$.

76. If a varies directly as b and inversely as the square root of c, and $a = 2$ when $b = 2$ and $c = 16$, find a when $b = 4$ and $c = 8$.

● **77.** The striking force F of a moving car is proportional to its weight w and the square of its velocity v. By how much would the striking force increase if the velocity were doubled?

● **78.** A manufacturer estimates that the number of units u of a certain item sold per week is inversely proportional to the quantity $p + 35$, where p is the price per unit in dollars. Suppose that at a price of \$15 per unit, 300 units are sold per week. How many units per week would be sold if the price were \$25 per unit?

● **79.** At a constant speed, the number of gallons of gasoline used by a car is directly proportional to the length of time it travels. If the car uses 4 gallons in 1 hour and 20 minutes of travelling, how many gallons will it use in 4.5 hours?

● **80.** The maximum safe load L that can be supported by a wooden beam of rectangular cross-section varies jointly as the width w and the square of the height h of the cross-section, and inversely as the length l of the beam. A beam 6 ft long with cross-section of width 2 inches and height 4 inches will safely support 450 pounds. How much weight can be safely supported by a beam of the same material which is 12 feet long and has a cross-section of width 4 inches and height 8 inches?

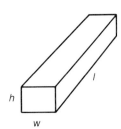

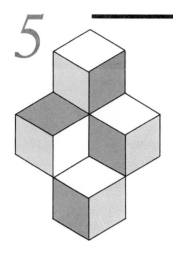

5 Polynomial and Rational Functions; Composite and Inverse Functions

Polynomial and rational functions deserve our special attention, in view of their many applications. As we have already mentioned, polynomial functions are among the simplest functions. In Chapter 4 we discussed polynomial functions of degrees 0 and 1. In this chapter we shall study polynomial functions of degrees 2 and higher. First we discuss quadratic functions, whose graphs are parabolas. As we shall see, all relevant properties of the graph are derived from algebraic properties of the quadratic polynomial. To a certain extent, this is also true for graphs of other polynomials, particularly those that can be factored as a product of linear functions. Rational functions are more difficult to graph. In simple cases, though, a lot of information about the graph of a rational function (such as locations of vertical and horizontal asymptotes) can be derived from algebra. In the last section of this chapter we discuss composition of functions and study the important concept of inverse functions.

5.1 QUADRATIC FUNCTIONS

Let f be a quadratic function given by

$$(5.1) \quad f(x) = ax^2 + bx + c,$$

where a, b, and c are real numbers and $a \neq 0$. The method of completing the square (Section 2.5 of Chapter 2) allows us to rewrite f as follows:

$$(5.2) \quad f(x) = a(x - h)^2 + k$$

where h and k are real numbers.

From this new expression and using horizontal and vertical translations of graphs as explained in Section 4.3 of Chapter 4, we can easily obtain the graph of f.

First, shift the graph of $y = ax^2$ horizontally $|h|$ units, to the right if $h > 0$ or to the left if $h < 0$. The result is the graph of $y = a(x - h)^2$. Next, shift the graph of $y = a(x - h)^2$ vertically $|k|$ units, upward if $k > 0$ or downward if $k < 0$.

The following examples show how powerful this simple method is.

◆ **Example 1.** Graph $f(x) = 2x^2 - 12x + 10$.

Solution. Our aim is to write f in the form (5.2). This can be done by arranging our work as follows:

$$f(x) = 2(x^2 - 6x \quad) + 10.$$

Next we convert $x^2 - 6x$ into a perfect square by adding 9 inside the parentheses and obtaining $x^2 - 6x + 9 = (x - 3)^2$. Since 9 was added inside the parentheses, $18 = 2 \cdot 9$ must be subtracted outside the parentheses so that f remains unchanged.

$$f(x) = 2(x^2 - 6x + 9) + 10 - 18$$
$$= 2(x - 3)^2 - 8.$$

The last expression is of the form (5.2) with $a = 2$, $h = 3$, and $k = -8$. The graph, shown in Figure 5.1, is a parabola obtained by shifting the graph of $y = 2x^2$ three units to the right and eight units down. The parabola opens upward, and its vertex has coordinates $(3, -8)$. The minimum value of the function f is -8, corresponding to the y-coordinate of the vertex. Note that the minimum value is obtained by substituting 3 for x in the new expression

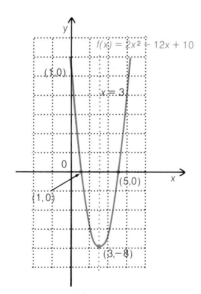

Figure 5.1

for f.

$$f(3) = 2(3 - 3)^2 - 8 = -8.$$

Also observe that the same result is obtained if 3 is substituted for x in the original form of f. Thus, comparing the results of the two substitutions is a good check to determine whether the arithmetic was done correctly.

The axis of symmetry of the curve is the line $x = 3$. Additional points on the graph are found from the intercepts. Setting $x = 0$, $f(0) = 2 \cdot 0^2 - 12 \cdot 0 + 10 = 10$. Thus the graph crosses the y-axis at the point $(0, 10)$. To obtain the x-intercepts, solve the quadratic equation $2x^2 - 12x + 10 = 0$. This equation is equivalent to $(2x - 2)(x - 5) = 0$, so the solutions are $x = 1$ and $x = 5$. Thus the x-intercepts are 1 and 5.

◇ **Practice Exercise 1.**   Graph the function $f(x) = 2x^2 + 4x - 6$.

Answer.

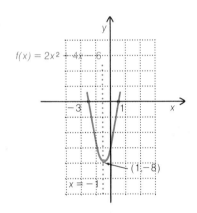

◆ **Example 2.**   Graph the function $f(x) = -x^2 - x + 2$.

Solution.   We begin by completing the square as follows:

$$f(x) = -(x^2 + x \qquad) + 2$$
$$= -\left(x^2 + 2\left(\frac{1}{2}\right)x \qquad\right) + 2$$
$$= -\left(x^2 + 2\left(\frac{1}{2}\right)x + \frac{1}{4}\right) + 2 + \frac{1}{4}$$
$$= -\left(x + \frac{1}{2}\right)^2 + \frac{9}{4}.$$

The function f is now of the form (5.2) with $a = -1$, $h = -1/2$, and $k = 9/4$. Since $a < 0$, the graph opens downward. The vertex $(-1/2, 9/4)$ is the highest point of the graph, and the maximum value of f is $f(-1/2) = 9/4$. The line $x = -1/2$ is the axis of symmetry. The graph is illustrated in Figure 5.2. The y-intercept 2 is obtained by setting $x = 0$ in the equation of f: $f(0) = 2$. The x-intercepts -2 and 1 are the solutions of the quadratic equation $-x^2 - x + 2 = 0$.

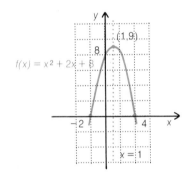

Figure 5.2

◇ **Practice Exercise 2.**   Graph the function $f(x) = -x^2 + 2x + 8$.

Answer.

The results of Examples 1 and 2 can be summarized as follows:

The Graph of $y = ax^2 + bx + c$

To graph a quadratic function

$$f(x) = ax^2 + bx + c,$$

complete the square and write f in the form

$$f(x) = a(x - h)^2 + k.$$

The graph is a parabola with vertex at (h, k) and axis of symmetry $x = h$.

If $a > 0$, the parabola opens upward and the vertex is its lowest point. The function f has a *minimum*, obtained by evaluating f at $x = h$, that is, $f(h) = k$.

If $a < 0$, the parabola opens downward and the vertex is the highest point. The function f has a *maximum* equal to $f(h) = k$.

To find the y-intercept, evaluate $f(x)$ at $x = 0$.

To find the x-intercepts (if any), solve the quadratic equation $ax^2 + bx + c = 0$.

The *x*-Intercepts of the Parabola $y = ax^2 + bx + c$

When graphing a quadratic function

$$f(x) = ax^2 + bx + c,$$

it is helpful to find the *x*-intercepts (if any) of the graph. They are the solutions of the quadratic equation

$$ax^2 + bx + c = 0.$$

As we said in Section 2.5 of Chapter 2, this equation has two real roots if the discriminant $\Delta = b^2 - 4ac$ is positive; it has only one root if $\Delta = 0$; and it has no real root if Δ is negative.

Examples 1 and 2 illustrate the case $\Delta > 0$. We consider the other two cases in the examples that follow.

◆ **Example 3.** Graph the function $f(x) = 4x^2 - 4x + 1$.

Solution. You can check that the quadratic polynomial is a perfect square: $4x^2 - 4x + 1 = (2x - 1)^2$. Thus

$$f(x) = (2x - 1)^2.$$

Since $2x - 1 = 2(x - 1/2)$, we can write f as follows:

$$f(x) = 4\left(x - \frac{1}{2}\right)^2.$$

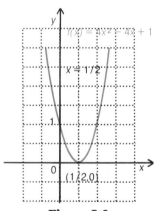

Figure 5.3

The graph of f, illustrated in Figure 5.3, can be obtained by translating the parabola $y = 4x^2$ horizontally 1/2 unit to the right. Note that the graph just touches the *x*-axis at the point $(1/2, 0)$, the vertex of the parabola. When this happens, we say that the parabola is *tangent* to the *x*-axis.

In this example the discriminant of the quadratic polynomial is

$$\Delta = b^2 - 4ac = 4^2 - 4 \cdot 4 \cdot 1 = 0.$$

The only solution of the equation $4x^2 - 4x + 1 = 0$ is $x = 1/2$. (Thus the *x*-intercept is 1/2.) The *y*-intercept is 1 and the axis of symmetry is $x = 1/2$.

◇ **Practice Exercise 3.** Graph $f(x) = -2x^2 - 4x - 2$.

Answer.

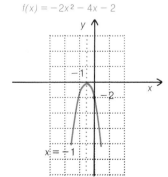

◆ **Example 4.**   Graph the function $f(x) = -2x^2 + 4x - 5$.

Solution.   Here $a = -2$, $b = 4$, $c = -5$, and $\Delta = 4^2 - 4(-2)(-5) = 16 - 40 < 0$. The quadratic equation has no real root, and the graph of f has no x-intercept. Since $a = -2 < 0$, the parabola opens downward and the vertex is its highest point. To find the coordinates of the vertex as well as the axis of symmetry, write f in the form (5.2):

$$
\begin{aligned}
f(x) &= -2x^2 + 4x - 5 \\
&= -2(x^2 - 2x \qquad) - 5 \\
&= -2(x^2 - 2x + 1) - 5 + 2 \\
&= -2(x - 1)^2 - 3.
\end{aligned}
$$

Thus the vertex has coordinates $(1, -3)$ and the axis of symmetry is $x = 1$. The graph is shown in Figure 5.4.

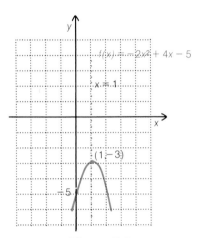

Figure 5.4

◇ **Practice Exercise 4.**   Graph $f(x) = x^2 + 2x + 5$.

Answer.

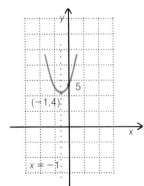

We can now summarize our results.

Let $f(x) = ax^2 + bx + c$ be a quadratic function and let $\Delta = b^2 - 4ac$ be the discriminant of the quadratic polynomial.
1. If $\Delta > 0$, the graph of f has two x-intercepts.
2. If $\Delta = 0$, the graph of f has one x-intercept, and the graph is said to be *tangent* to the x-axis.
3. If $\Delta < 0$, the graph has no x-intercept.

◆ **Example 5.**   Mrs. Robinson owns a pizza parlor. By analyzing her costs, she estimates that the daily cost, $C(x)$, of operating her business is given by

$$C(x) = 2x^2 - 240x + 7380,$$

where x is the number of pizza pies she sells per day. Find the number of pies that she must sell to minimize her costs. What is the minimum cost?

Solution.   This problem is equivalent to that of finding the minimum point on the graph of $C(x)$. By completing the square, we have

$$
\begin{aligned}
C(x) &= 2(x^2 - 120x \qquad) + 7380 \\
&= 2(x^2 - 2 \cdot 60x + 3600) + 7380 - 7200 \\
&= 2(x - 60)^2 + 180.
\end{aligned}
$$

Thus the minimum point on the graph of $C(x)$ is $(60, 180)$, as shown in Figure 5.5.

Mrs. Robinson must sell 60 pizza pies in order to minimize her costs. The minimum cost is $C(60) = 180$.

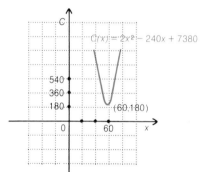

Figure 5.5

Nicole Oresme

Nicole Oresme (1320–1382) was a chaplain to King Charles V of France. He later became the Bishop of Liseux in Normandy. He is interesting to us in his role as a mathematician because, in a work written about 1360, he graphed a function (of course, he did not use the modern notation). He called the vertical coordinate *latitude* and the horizontal coordinate *longitude* in analogy with the use of those terms in map-making, which had been practiced since ancient times. This work was reprinted as late as 1515. Thus while it was Descartes who first completely saw that coordinatization could turn geometry into algebra and conversely, the method of graphical representation that underlies analytic geometry was no less than 130 years older than Descartes.

◇ **Practice Exercise 5.** The profit function for a manufacturer is

$$P(x) = 300x - 2x^2$$

where x is the number of units produced. How many units must be produced in order to maximize profit? What is the maximum profit?

Answer. 75 units, $11.250.

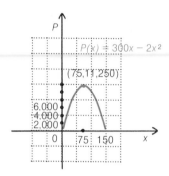

The following example and practice exercise are related to Example 4 and Practice Exercise 4 of Section 2.7 of Chapter 2.

◆ **Example 6.** A travel agent estimates that 20 people will take a sightseeing tour at a cost of $100 per person. She also believes that each $5 reduction of the price will bring 2 additional people. How many price reductions would maximize her revenue? What is the maximum revenue?

Solution. If x denotes the number of price reductions, then $10 + 2x$ is the number of people taking the tour at $100 - 5x$ dollars per person. Thus the revenue is

$$R(x) = (20 + 2x)(100 - 5x)$$
$$= 2000 + 100x - 10x^2.$$

By completing the square, we rewrite the quadratic function R as follows:

$$R(x) = -10(x - 5)^2 + 2250.$$

We see that the corresponding parabola has a maximum point with coordinates $(5, 2250)$. Thus $x = 5$ is the number of price reductions that maximizes the revenue. The maximum revenue is $R(5) = 2250$.

◇ **Practice Exercise 6.** A farmer has 30 orange trees in his orchard that produce an average of 600 oranges per tree per year. He estimates that for each additional tree planted in the orchard, the annual yield of each tree decreases by 10 oranges. How many additional trees should he plant to maximize the total yield of oranges? What is the maximum yield?

Answer. 15 trees, 20,250 oranges.

EXERCISES 5.1

1. Graph $f(x) = ax^2$ if

   a) $a = 3$ b) $a = -\dfrac{1}{2}$

   c) $a = -4$ d) $a = -2$

2. Given $f(x) = ax^2 + c$, graph f if

   a) $a = -3$, $c = 1$ b) $a = 2$, $c = -1$

   c) $a = \dfrac{1}{4}$, $c = -2$ d) $a = -\dfrac{1}{2}$, $c = \dfrac{3}{5}$

3. Given $f(x) = (x - h)^2$, graph f if

   a) $h = 2$ b) $h = -3$

   c) $h = -5$ d) $h = 4$

4. Graph $f(x) = a(x - h)^2 + k$, if

   a) $a = 2$, $h = 1$, $k = 3$

   b) $a = -2$, $h = -3$, $k = 1$

   c) $a = -1$, $h = 2$, $k = -1$

   d) $a = 3$, $h = 0$, $k = -3$

Calculate the discriminant $\Delta = b^2 - 4ac$ of each quadratic function in Exercises 5–12. What can you say about the number of x-intercepts of the graph? Does the parabola open upward or downward?

5. $f(x) = x^2 - 8x + 20$

6. $f(x) = -x^2 + 6x - 2$

7. $f(x) = -2x^2 + 7x + 4$

8. $f(x) = 3x^2 - 5x + 1$

9. $f(x) = 9x^2 - 12x + 4$

10. $f(x) = -4x^2 + 20x - 25$

11. $f(x) = -x^2 - 2x - 3$

12. $f(x) = x^2 - x + 6$

In Exercises 13–24, graph each of the given quadratic functions. Give the vertex and the axis of each parabola.

13. $f(x) = x^2 - 9$

14. $f(x) = 16 - 4x^2$

15. $f(x) = x^2 - 6x + 8$

16. $f(x) = x^2 - 8x + 12$

17. $f(x) = x^2 - 6x + 9$

18. $f(x) = x^2 + 8x + 16$

19. $f(x) = 2x^2 + 1$

20. $f(x) = 3x^2 + x + 1$

21. $f(x) = -x^2 + 2x$

22. $f(x) = -3 - x^2$

23. $f(x) = -2x^2 + 8x - 11$

24. $f(x) = 3x^2 - 6x + 4$

25. For what value of c is the graph of $f(x) = x^2 + 8x + c$ tangent to the x-axis?

26. For what value of k is the graph of $f(x) = x^2 + kx + k$ tangent to the x-axis?

27. Find all values of b for which the graph of $f(x) = 2x^2 + bx + 2$ does not cross the x-axis.

28. Find all values of c for which the graph of $f(x) = -x^2 + 3x - c$ does not intersect the x-axis.

29. Find two positive numbers whose sum is 36 and whose product is maximum.

30. Find two numbers whose difference is 48 and whose product is minimum.

31. What is the area of the largest rectangular field that can be enclosed with 1260 m of fencing?

32. A woman wants to make a vegetable garden on a rectangular plot bordered on one side by her house. She has 36 ft of fencing to fence off the other three sides of the plot. What should be the dimensions of the plot if the enclosed area is to be a maximum?

33. Peter operates a hot-dog stand. If his profit $P(x)$ is given by the function

$$P(x) = 0.8x - 0.002x^2$$

where x is the number of hot dogs, how many hot dogs must Peter sell to maximize his profit? What is the maximum profit?

34. The demand function for a manufacturer's product is $p = 300 - 2x$, where p is the price (in dollars) per unit when x units are demanded. The revenue function $R(x)$ is defined by

$$R(x) = (300 - 2x)x.$$

Find the maximum revenue and the price corresponding to the maximum revenue.

35. If an object is thrown upward from the ground level with an initial velocity of 88 feet per second, then its height s (in feet) above the ground after t seconds is given by

$$s = -16t^2 + 88t.$$

Find the maximum height reached by the projectile. After how many seconds will the object hit the ground?

36. After t hours the number of bacteria in a certain culture is given by

$$N = 1000 + 50t - 5t^2.$$

When will the number of bacteria be a maximum? What is the maximum number?

● **37.** To print 100 posters, a printer charges 20¢ per poster. For each increase of 50 posters, he decreases by 2¢ the price of printing each poster. How many posters should he print to maximize the revenue?

● **38.** A merchant sells 80 pounds of coffee daily at $3.00 per pound. Each 10¢ increase in price decreases sales by 5 pounds. If his costs are $2.00 per pound, how much should he charge per pound of coffee in order to maximize profit?

● **39.** Show that every quadratic function

$$f(x) = ax^2 + bx + c$$

can be written as

$$f(x) = a(x - h)^2 + k,$$

where

$$h = -\frac{b}{2a} \quad \text{and} \quad k = -\frac{\Delta}{4a} = -\frac{b^2 - 4ac}{4a}.$$

[*Hint*: Complete the square as shown in Example 1.]

● **40.** Show that the graph of

$$f(x) = ax^2 + bx + c$$

is a parabola with vertex at the point

$$\left(-\frac{b}{2a}, -\frac{\Delta}{4a} \right)$$

and symmetry axis $x = -\frac{b}{2a}$.

5.2 POLYNOMIAL FUNCTIONS

We already mentioned that a function of the form

$$f(x) = a_n x^n + a_{n-1} x^{n-1} + \cdots + a_1 x + a_0,$$

where $a_0, a_1, \ldots, a_n$ are real numbers and $a_n \neq 0$, is called a *polynomial function of degree n*.

We know from Chapter 4 that the graph of a linear polynomial $f(x) = ax + b$ is a straight line. In Section 5.1 we saw that the graph of a quadratic polynomial $f(x) = ax^2 + bx + c$ is always a parabola.

The graphs of most polynomial functions of degree greater than 2 are difficult to describe without using certain techniques studied in calculus. Certainly, it is always possible to obtain an approximate graph of a polynomial by plotting points; the more points we plot, the more accurate is the graph. However, our aim is to describe several graphing techniques that reduce to a minimum the plotting of points and still enable us to draw a reasonable sketch of the graph of a polynomial. First we discuss some simple cases.

◆ **Example 1.** Graph the function $f(x) = \frac{1}{2} x^3$.

Solution. Using the technique of shrinking (see Section 4.3), we see that the graph can be obtained by multiplying by 1/2 the ordinates of all points on the graph of $y = x^3$. (Recall that the graph of $y = x^3$ was discussed in Example 4 of Section 3.3.) Figure 5.6 illustrates for comparison purposes the graphs of $y = x^3$ (dotted line) and $y = (1/2)x^3$ (solid line.)

x	-2	-1	0	1	2
y	-4	$-\dfrac{1}{2}$	0	$\dfrac{1}{2}$	4

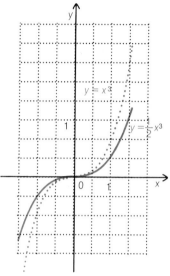

Figure 5.6

Note that f is an odd function; thus the graph is symmetric with respect to the origin. Also, since $y = (1/2)x^3 > 0$ when $x > 0$, that part of the graph lies above the x-axis. It lies below the x-axis when $x < 0$.

◇ **Practice Exercise 1.** Graph the function $f(x) = (x - 1)^3$.

Answer.

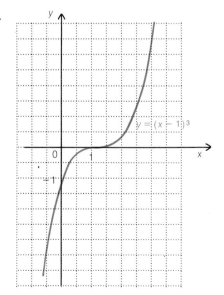

◆ **Example 2.**   Graph $f(x) = x^4 - 1$.

Solution.   Using the technique of vertical translation (Section 4.3), it follows that the graph is obtained by vertically shifting the graph of $y = x^4$ one unit down.

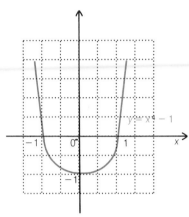

Figure 5.7

◇ **Practice Exercise 2.**   Graph $f(x) = (x + 1)^4$.

Answer.

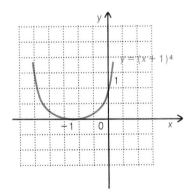

Polynomials in Factored Form

The task of graphing a polynomial is considerably simplified when the polynomial is written as a product of linear factors.

First, it can be shown in advanced courses that the graph of any polynomial is always a curve that contains no gaps, corners, jumps, or missing points.

Second, the graph is unbounded as $|x|$ increases without bound, that is, the graph cannot level off as $|x|$ becomes very large.

Third, for a polynomial $f(x)$ the x-intercepts are the real solutions of the equation $f(x) = 0$. These are easily obtained when the polynomial is written as a product of linear factors.

Fourth, the *x*-intercepts divide the real line into a finite number of intervals, in each of which the sign of the polynomial function remains constant. To determine the sign of the polynomial in any particular such interval, we may use the method for solving inequalities explained in Section 2.8. Alternatively, we can determine the sign by simply evaluating the polynomial at one conveniently chosen test point in the interval.

◆ **Example 3.** Graph the function $f(x) = x^3 + x^2 - 2x$.

Solution. First factor the polynomial as follows:

$$f(x) = x(x^2 + x - 2) = x(x - 1)(x + 2).$$

Next observe that at the points $x = -2$, $x = 0$, and $x = 1$, the polynomial function is equal to zero. Thus, -2, 0, and 1 are the *x*-intercepts of the graph. They divide the line into four intervals: $(-\infty, -2), (-2, 0), (0, 1)$, and $(1, +\infty)$. By using the method described in Section 2.8, we can determine the sign changes of $f(x)$ as *x* varies on the real line.

Sign of	For x in each of the intervals			
	$-\infty$ -2		0 1	$+\infty$
x	$-$	$-$	$+$	$+$
$x - 1$	$-$	$-$	$-$	$+$
$x + 2$	$-$	$+$	$+$	$+$
$f(x) = x(x - 1)(x + 2)$	$-$	$+$	$-$	$+$

The table shows that *f* is negative in the intervals $(-\infty, -2)$ and $(0, 1)$, so the graph of *f* lies below the *x*-axis there. In the intervals $(-2, 0)$ and $(1, +\infty)$, *f* is positive, so the graph lies above the *x*-axis in those intervals. Using this information and plotting a few points, we obtain the curve illustrated in Figure 5.8.

It is often simpler to determine the sign changes of the polynomial function *f* by evaluating it at some arbitrarily chosen point inside each of the intervals between *x*-intercepts. As shown in Figure 5.8, $f(-1) = 2 > 0$ and $f(1/2) = -5/8 < 0$. As an exercise, check that $f(-3) = -12 < 0$ and $f(2) = 8 > 0$.

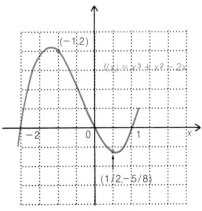

Figure 5.8

◇ **Practice Exercise 3.**   Graph $f(x) = -x^3 + 2x^2 + 3x$.

Answer.

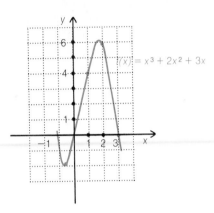

High and Low Points of Polynomial Functions

As we already know, the graph of a quadratic function always has a maximum (high point) or a minimum (low point). These are *turning points* of the graph. At a maximum, the function changes from increasing to decreasing; at a minimum it changes from decreasing to increasing.

Polynomials of degree >2 may also have high and low points. It can be proved that a polynomial of degree n has *at most* $n-1$ turning points. For example, the graph of $f(x) = x^3 + x^2 - 2x$ (Figure 5.8) indicates that the function $f(x)$ attains a maximum at some point in the interval $(-2, 0)$, while similarly $f(x)$ attains a minimum in the interval $(0, 1)$. However, the problem of finding the maximum and minimum points of a polynomial function of degree ≥ 3 is difficult; there is no analogue of the simple method of completing the square that we used with quadratic functions. Historically, differential calculus was developed in order to solve this problem. The appropriate methods to handle the general problem thus lie beyond the scope of this book.

The examples and exercises considered in this section are fairly simple. If you determine the x-intercepts and plot a reasonable number of points you will be able to determine the approximate location of the maxima and minima of the polynomial function.

◆ **Example 4.**   Graph the function $f(x) = -x^3 - x^2 + 4x + 4$.

Solution.   Factoring the polynomial, we get

$$f(x) = -x^2(x+1) + 4(x+1)$$
$$= -(x^2 - 4)(x+1)$$
$$= -(x+2)(x-2)(x+1).$$

It follows that the polynomial function $f(x)$ is zero at the points $x = -2$, $x = -1$, and $x = 2$. These points divide the real line into four intervals: $(-\infty, -2)$, $(-2, -1)$, $(-1, 2)$, and $(2, +\infty)$. We could now determine the sign changes of f by making a table such as in Example 3. Instead, we are going to select a point in each interval and substitute it into the function to determine whether the function is positive or negative in that interval. Here is a selection of points.

Interval	Selected point	Function value
$(-\infty, -2)$	-3	$f(-3) = 10 > 0$
$(-2, -1)$	$-\dfrac{3}{2}$	$f\left(-\dfrac{3}{2}\right) = -\dfrac{7}{8} < 0$
$(-1, 2)$	0	$f(0) = 4 > 0$
$(2, +\infty)$	3	$f(3) = -20 < 0$

Using this information and plotting the points, we obtain the graph in Figure 5.9. Note that 4 is the y-intercept of the graph because $f(0) = 4$. Also, f has a minimum in the interval $(-2, -1)$ and a maximum in the interval $(-1, 2)$.

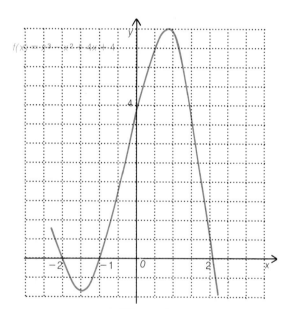

Figure 5.9

◇ **Practice Exercise 4.**   Graph the function $f(x) = x^3 - 2x^2 - x + 2$.

Answer.

EXERCISES 5.2

Graph the following functions:

1. $f(x) = (x + 3)^4$
2. $f(x) = (2 - x)^3$
3. $f(x) = x^2(x - 1)$
4. $f(x) = 2x(x - 1)^2$
5. $f(x) = x(x + 3)^2$
6. $f(x) = -x^2(x - 4)$
7. $f(x) = x^3 - 4x$
8. $f(x) = (x - 1)(x + 1)(x + 2)$
9. $f(x) = x(x - 2)(x - 4)$
10. $f(x) = x^2(x + 1)(x + 3)$
11. $f(x) = x^4 - 2x^2 - 8$
12. $f(x) = x^4 - 8x^2 - 9$
13. $f(x) = x^3 + x^2 - 4x - 4$
14. $f(x) = x^3 + 3x^2 - x - 3$
15. $f(x) = 2x^3 + 3x^2 - 2x - 3$
16. $f(x) = x^3 + 2x^2 - 9x - 18$

In Exercises 17–26, the graph of each polynomial function is symmetric about a line or a point. Graph each function and name the line or point of symmetry.

17. $f(x) = 3x^4$

18. $f(x) = \dfrac{1}{2} x^5$

19. $f(x) = 2x^3 - \dfrac{1}{3}$

20. $f(x) = \dfrac{1}{2} x^6 + 1$

21. $f(x) = x^3 - 4$

22. $f(x) = 2 - x^4$

23. $f(x) = \dfrac{1}{2} (x + 1)^3$

24. $f(x) = -2(x - 3)^4$

25. $f(x) = (x - 2)^4 + 1$

26. $f(x) = (x - 1)^3 + 4$

c *In Exercises 27–30, use a calculator to determine several points on the graph of each polynomial. Sketch the graph.*

27. $f(x) = x^3 + x - 1$
28. $f(x) = x^3 - x^2 + 3x - 3$
29. $f(x) = 1 - x - x^3$
30. $f(x) = x^2 - x^3$

5.3 RATIONAL FUNCTIONS

Graphing a rational function

$$f(x) = \frac{P(x)}{Q(x)}$$

where $P(x) = a_n x^n + \cdots + a_1 x + a_0$ and $Q(x) = b_m x^m x + \cdots + b_1 x + b_0$ are polynomial functions, is not a simple matter, particularly if the degrees of the polynomials are large. Nevertheless, in simple cases, by combining the plotting of points with certain important properties of rational functions, it is possible to sketch the graph of a rational function fairly accurately.

As we already know, a rational function is defined for all real numbers x, except the numbers for which the denominator is equal to zero.

Two Initial Steps

When dealing with a rational function $f(x)$, we have to analyze, among other things, two important features.
1. The function values of f at x near the numbers for which the denominator is equal to zero.
2. The function values of f when $|x|$ becomes very large.

This can be illustrated by studying the graph of

$$f(x) = \frac{1}{x}$$

which is the simplest nontrivial rational functions. First, f is defined for all real numbers $x \neq 0$. Next note that the graph of f is the same as the graph of the equation

$$y = \frac{1}{x}.$$

This equation is equivalent to the equation $x = 1/y$ or the equation $xy = 1$. We can now derive the following information about the coordinates (x, y) of a point on the graph of $y = 1/x$:

a) x and y are multiplicative inverses of each other;

b) neither x nor y can be zero, so the graph does not have x- or y-intercepts;

c) x and y are both positive or both negative. This tells us that the graph lies on the first and third quadrants.

Now we make a table of values such as

x	-3	-2	-1	$-\dfrac{1}{2}$	$\dfrac{1}{2}$	1	2	3
y	$-\dfrac{1}{3}$	$-\dfrac{1}{2}$	-1	-2	2	1	$\dfrac{1}{2}$	$\dfrac{1}{3}$

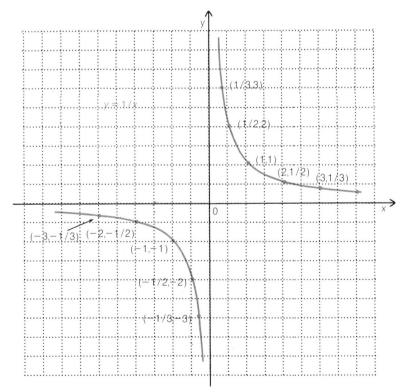

Figure 5.10

Plotting the points and connecting them with a smooth curve, we obtain the graph illustrated in Figure 5.10. You can check that the graph is symmetric with respect to the origin and with respect to the main diagonal.

Now we analyze what happens to the function $f(x) = 1/x$ as x approaches 0. If x is a small positive number, then $f(x) = 1/x$ is a large positive number. For instance, if $x = 0.1$, 0.01, and 0.001, then $y = 10$, 10^2, and 10^3. On the other hand, if x is a negative number with small absolute value, then y is a negative number with large absolute value. For instance, if $x = -0.1$, -0.01, and -0.001, then $y = -10$, -10^2, and -10^3. In both cases we can say that as x approaches 0 (*from the left or right*), *the absolute value of $y = 1/x$ becomes large and increases without bound* (Figure 5.11). For this reason, the vertical line $x = 0$ (i.e., the y-axis) is said to be a *vertical asymptote* of the graph of $f(x) = 1/x$.

Next let us see what happens to f as $|x|$ becomes very large. If x is a large positive number, then $y = 1/x$ is a small positive number. Moreover, as x increases without bound, y decreases and approaches 0. For instance, if $x = 10^4$, 10^9, and 10^{12}, then $y = 10^{-4}$, 10^{-9}, and 10^{-12}. We leave it to you to check that if x is negative with a large absolute value, then y is negative with a small absolute value. Summarizing these results, we can say that as $|x|$ increases without bound, the values of $y = 1/x$ approach 0 (Figure 5.12). Thus we say

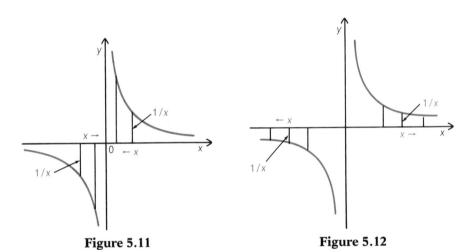

Figure 5.11   **Figure 5.12**

that the horizontal line $y = 0$ (i.e., the x-axis) is a *horizontal asymptote* of the graph of $y = 1/x$.

The discussion above motivates the following definitions.

Vertical and Horizontal Asymptotes

A line $x = a$ is a *vertical asymptote* of the graph of a function $y = f(x)$ if $|f(x)|$ increases without bound as x approaches a from the left or right.

A line $y = b$ is a *horizontal asymptote* of the graph of a function $y = f(x)$ if the values $f(x)$ approach b as $|x|$ increases without bound.

◆ **Example 1.** Sketch the graph of $f(x) = \dfrac{2}{x-5}$.

Solution. Using horizontal translation and stretching (see Section 4.3), we can obtain the graph of f by drawing the graph of $y = 1/x$, shifting it 5 units to the right, and then multiplying each ordinate by 2 (Figure 5.13). Thus the graph is merely a translation of the graph of $y = 1/x$ followed by a stretch. Note that when $x = 0$, we have $f(0) = -2/5$; so $-2/5$ is the y-intercept of the graph.

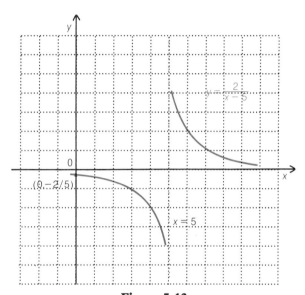

Figure 5.13

The graph indicates that the line $x = 5$ is a vertical asymptote and the line $y = 0$ is a horizontal asymptote.

To see that $x = 5$ is a vertical asymptote, note that f is not defined at 5, because at this number the denominator is zero. As x approaches 5 (from the left or right), the difference $x - 5$ approaches 0 so that $|f(x)| = 2/|x - 5|$ becomes large and increases without bound.

To see that the line $y = 0$ is a horizontal asymptote, we have to proceed in a different manner. If we divide the numerator and denominator of f by x, we can write

$$f(x) = \frac{\dfrac{2}{x}}{1 - \dfrac{5}{x}}.$$

Now we study the behavior of f as $|x|$ becomes very large. We already know that as $|x|$ increases without bound, then $1/x$ approaches 0. It follows that the two terms $2/x = 2 \cdot 1/x$ and $5/x = 5 \cdot 1/x$ also approach 0. Thus the numerator of f approaches 0 while the denominator approaches 1. Therefore f approaches 0 as $|x|$ increases without bound, and the line $y = 0$ is a horizontal asymptote.

◇ **Practice Exercise 1.**   Sketch the graph of $f(x) = \dfrac{3}{x+1}$.

Answer.

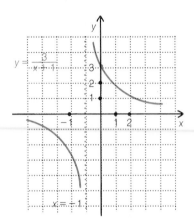

Vertical Asymptotes of the Graph of a Rational Function

The following is a general result concerning rational functions. Let

$$f(x) = \frac{P(x)}{Q(x)}$$

be a rational function. If a is a number such that $Q(a) = 0$, but $P(a) \neq 0$, then the line $x = a$ is a *vertical asymptote* of the graph of f.

◆ **Example 2.**   Discuss and sketch the graph of the function

$$f(x) = \frac{2x + 1}{x - 4}.$$

Solution.   The domain of definition of this function is the set of all numbers $x \neq 4$. When $x = 4$, the denominator is equal to zero, but the numerator is different from zero. Thus, the line $x = 4$ is a vertical asymptote of the graph.

In order to determine a horizontal asymptote, divide (as in Example 1) both numerator and denominator of f by x:

$$f(x) = \frac{2x + 1}{x - 4} = \frac{2 + \dfrac{1}{x}}{1 - \dfrac{4}{x}}.$$

and study what happens to f for large values of $|x|$. As $|x|$ increases without bound, both $1/x$ and $4/x = 4 \cdot 1/x$ approach zero. It follows that the numerator of f approaches 2, while the denominator approaches 1. Thus f approaches $2/1 = 2$ as $|x|$ increases without bound and, therefore, the line $y = 2$ is a horizontal asymptote.

Now determine the *x*- and *y*-intercepts. If $x = 0$, then $f(0) = -1/4$, so $-1/4$ is the *y*-intercept of the graph. To find the *x*-intercepts we must solve $f(x) = 0$, so we set

$$\frac{2x + 1}{x - 4} = 0.$$

For a fraction to be zero, the numerator must be zero but the denominator different from zero. Thus $2x + 1 = 0$, which gives us $x = -1/2$. Therefore $-1/2$ is the *x*-intercept of the graph.

Using all the above information and plotting a few points, we obtain the graph shown in Figure 5.14.

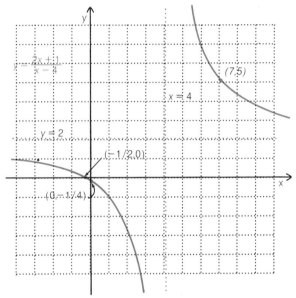

Figure 5.14

◇ **Practice Exercise 2.**   Discuss and graph the function $f(x) = \dfrac{x - 1}{2x + 4}$.

Answer.

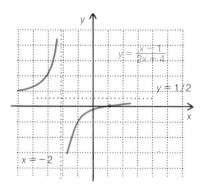

Reciprocals of Powers of x

When considering rational functions, it is important to study the reciprocals of powers of x. Here, we discuss the reciprocal of the square function. Consider the function

$$y = f(x) = \frac{1}{x^2}$$

whose domain is the set of all numbers $x \neq 0$. The line $x = 0$ is a vertical asymptote of the graph of $f(x) = 1/x^2$, because the denominator of f is zero when $x = 0$, while the numerator is always $\neq 0$. The function f is even, thus the graph is symmetric with respect to the y-axis. Moreover, since $f(x)$ is always positive, the graph is located above the x-axis.

Next, we discuss what happens to $y = 1/x^2$ as $|x|$ increases without bound. If $|x|$ is a large number, then $y = 1/x^2$ is a small number. For instance, if $x = \pm 10$, $\pm 10^2$, and $\pm 10^6$, then $y = 10^{-2}$, 10^{-4}, and 10^{-12}, which are very small numbers.

At this point, we could compare the values of $y = 1/x^2$ with the values of $y = 1/x$. If $|x|$ is a large number, then $1/x$ is a small number and $1/x^2$ is even smaller. For example, if $x = 10^4$, then $1/x = 10^{-4}$ and $1/x^2 = 10^{-8}$, which is smaller. Thus we can say that as $|x|$ increases without bound, $1/x^2$ approaches 0 *faster* than $1/x$.

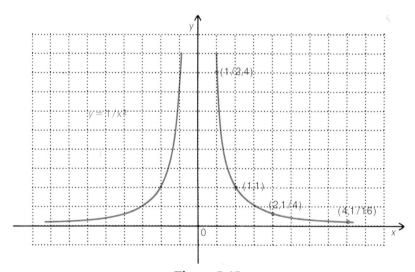

Figure 5.15

Let us make a final remark about the graph of a rational function of the form

$$f(x) = \frac{1}{x^n}.$$

If n is an *odd* number, the graph of f resembles that of $y = 1/x$. If n is an *even* number, the graph of f resembles the graph of $y = 1/x^2$.

In the next example we discuss the case of a rational function whose numerator and denominator are two simple quadratic polynomials.

◆ **Example 3.** Discuss and sketch the graph of $f(x) = \dfrac{x^2}{x^2 - 1}$.

Solution. The function f is defined for all real numbers, except $x = -1$ and $x = 1$. Since the denominator is zero at $x = -1$ and $x = 1$, but the numerator is different from zero, the lines $x = -1$ and $x = 1$ are vertical asymptotes.

To find horizontal asymptotes, divide both numerator and denominator by x^2 (the highest power of x in the expression defining f) and write

$$f(x) = \frac{x^2}{x^2 - 1} = \frac{1}{1 - \dfrac{1}{x^2}}.$$

As $|x|$ becomes large, the term $1/x^2$ approaches 0 and $f(x)$ approaches 1. Thus the line $y = 1$ is a horizontal asymptote.

Next check that 0 is both the x- and y-intercept of the graph of f.

The table below gives us a few points on the graph. We have selected only positive values of x because f is an even function ($f(x) = f(-x)$) and, as a consequence, the graph is symmetric with respect to the y-axis.

x	$\dfrac{1}{2}$	$\dfrac{2}{3}$	$\dfrac{3}{2}$	2
$f(x)$	$-\dfrac{1}{3}$	$-\dfrac{4}{5}$	$\dfrac{9}{5}$	$\dfrac{4}{3}$

You should plot some additional points and check the graph illustrated in Figure 5.16.

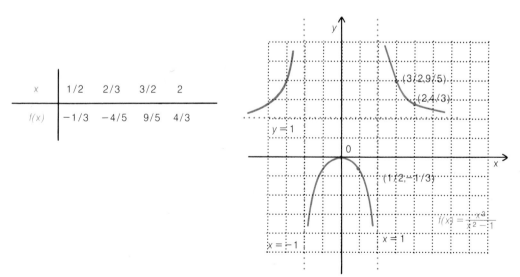

x	1/2	2/3	3/2	2
$f(x)$	$-1/3$	$-4/5$	$9/5$	$4/3$

Figure 5.16

◇ **Practice Exercise 3.** Discuss and sketch the graph of $f(x) = \dfrac{x^2 - 1}{x^2}$.

Answer.

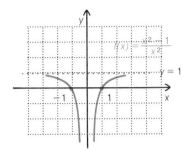

Horizontal Asymptotes of the Graph of a Rational Function

We have seen in Example 2 that the horizontal asymptote of the graph of the function

$$f(x) = \frac{2x + 1}{x - 4}$$

is the line $y = 2$. Note that 2 is the quotient of 2 and 1, the leading coefficients in the numerator and denominator. Analogously, the horizontal asymptote of the graph of

$$f(x) = \frac{x^2}{x^2 - 1}$$

(Example 3) is $y = 1$. Also note that 1 is the quotient of the leading coefficients in the numerator and denominator.

These are no chance events; they are consequences of a general result about rational functions, which we now explain.

Method of Determining the Horizontal Asymptote

Let

$$f(x) = \frac{a_n x^n + \cdots + a_1 x + a_0}{b_m x^m + \cdots + b_1 x + b_0}$$

be a rational function. In order to find the horizontal asymptote of f, we have to consider three cases.

1. $n = m$. If the numerator and denominator have the *same* degree, then the line $y = a_n/b_m$ is a horizontal asymptote. Observe that a_n/b_m is the quotient of the leading coefficients of the two polynomials.
2. $n < m$. If the degree of the numerator is *smaller* than the degree of the denominator, then the line $y = 0$ is a horizontal asymptote. For example, $y = 0$ is the horizontal asymptote of the graph of $f(x) = 2/(x - 5)$.
3. $n > m$. If the degree of the numerator is *greater* than the degree of the denominator, then the graph of f has no horizontal asymptote.

These properties tell us that to find horizontal asymptotes of rational functions, we just have to compare the degrees of the numerator and denominator, and if they are equal, then look at their leading coefficients.

◆ **Example 4.** Determine the horizontal asymptote of each function:

a) $f(x) = \dfrac{2x^2 - 5}{3x^2 - 2x - 1}$, **b)** $f(x) = \dfrac{6x^2 - 1}{x^3 - 4x - 2}$

Solution. **a)** Both polynomials have the same degree so that, according to item 1) above, $y = 2/3$ is the horizontal asymptote of the graph.
b) In this case, the degree of the numerator is smaller than the degree of the denominator. According to item 2) above, $y = 0$ is the horizontal asymptote.

◇ **Practice Exercise 4.** Find the horizontal asymptotes of each function:

a) $f(x) = \dfrac{2x^3 - 1}{x^2 - x + 4}$ **b)** $f(x) = \dfrac{6x^2 - 7}{3x^2 - 4x - 2}$

Answer. **a)** No horizontal asymptote   **b)** $y = 2$.

◆ **Example 5.** Discuss and sketch the graph of

$$f(x) = \frac{x^2}{x^2 + 2x - 8}.$$

Solution. Since $x^2 + 2x - 8 = (x - 2)(x + 4)$, the function f is defined for all real numbers except $x = 2$ and $x = -4$. The two lines $x = 2$ and $x = -4$ are vertical asymptotes. Since the numerator and denominator have the same degree, the line $y = 1/1 = 1$ is a horizontal asymptote.
Next, we determine x- and y-intercepts. If $x = 0$, then $f(0) = 0$, so 0 is the y-intercept. Setting $f(x) = 0$ and solving for x we obtain $x = 0$, so 0 is also the x-intercept.
Now it is helpful to have an idea of the location of the graph of f. This can be done by investigating the sign of $f(x)$ as x varies in the domain of definition. In the intervals where $f(x) < 0$, the graph lies below the x-axis. To determine the sign changes of f we can use the method of solving inequalities explained in Section 2.8 of Chapter 2. The first step is to factor the denominator of f and write

$$f(x) = \frac{x^2}{(x - 2)(x + 4)}.$$

We must now keep track of the signs of the various factors. For this purpose, we organize your work in tabular form as follows.

For x in each of the intervals

Sign of	$-\infty$		-4		2		$+\infty$
x^2		$+$		$+$		$+$	
$x - 2$		$-$		$-$		$+$	
$x + 4$		$-$		$+$		$+$	
$x^2/(x - 2)(x + 4)$		$+$		$-$		$+$	

The table shows that the graph of f lies above the x-axis for all values of x in the intervals $(-\infty, -4)$ and $(2, +\infty)$, and below the x-axis for all values of x in the interval $(-4, 2)$.

Using all the available information and plotting a few points, we can sketch the graph of f shown in Figure 5.17.

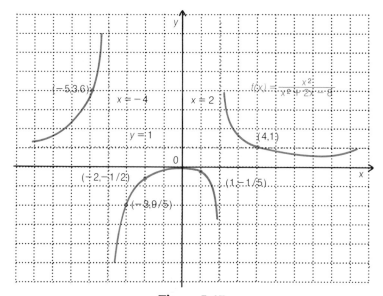

Figure 5.17

The graph illustrates a new important feature of certain rational functions, which we now describe. If f is a rational function, its graph *never crosses a vertical asymptote*. (Why?) However, the graph *may cross a horizontal asymptote*. To find the points where the graph crosses the horizontal asymptote, set $f(x)$ equal to the number corresponding to the asymptote and solve the equation for x. For example, $y = 1$ is the horizontal asymptote of

$$f(x) = \frac{x^2}{x^2 + 2x - 8}.$$

If we set $f(x) = 1$, that is,

$$\frac{x^2}{x^2 + 2x - 8} = 1,$$

and solve for x, we obtain $x = 4$. Clearly $f(4) = 1$. Thus $(4, 1)$ is a point that lies both on the graph of f and on the line $y = 1$.

◇ **Practice Exercise 5.** Discuss and sketch the graph of

$$f(x) = \frac{x^2}{x^2 - 2x - 3}.$$

Answer.

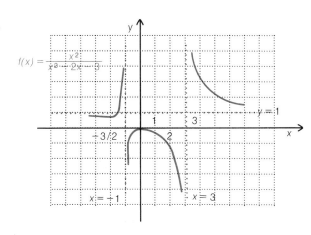

We end this section with a summary of all of our results concerning the graphing of rational functions.

Steps in Sketching the Graph of a Rational Function

To sketch the graph of a rational function $f(x)$ proceed as follows.
1. Determine the domain of the function.
2. Find any vertical and horizontal asymptotes.
3. Find all x- and y-intercepts of the graph.
4. Study the sign variation of f as x varies in the domain of definition.
5. Look for points where the graph may cross a horizontal asymptote.
6. Make a table of function values and plot the corresponding points.
7. Join the points with a smooth curve.

 EXERCISES 5.3

In Exercises 1–10, find vertical and horizontal asymptotes of each function.

1. $f(x) = \dfrac{1}{2x + 1}$

2. $f(x) = \dfrac{-2}{3x - 2}$

3. $f(x) = \dfrac{2x - 4}{3x + 2}$

4. $f(x) = \dfrac{x}{4x - 5}$

5. $f(x) = \dfrac{2x^2 + 3}{x^2 - 4x + 1}$

6. $f(x) = \dfrac{3x^2 - 2x + 1}{x^2 + 4}$

7. $f(x) = \dfrac{2x + 3}{x^2 - 3x + 5}$

8. $f(x) = \dfrac{5x - 3}{3x^2 - 4x - 1}$

9. $f(x) = \dfrac{3x^2 - 4x + 1}{5x - 3}$

10. $f(x) = \dfrac{4x^3 - 2}{x^2 - 4x + 3}$

In Exercises 11–20, use the techniques of stretching, shrinking, reflecting, and/or shifting graphs to sketch the graphs of the given functions.

11. $f(x) = \dfrac{5}{x}$

12. $f(x) = -\dfrac{3}{x}$

13. $f(x) = \dfrac{-3}{x+5}$

14. $f(x) = \dfrac{2}{4-x}$

15. $f(x) = 2 - \dfrac{3}{x}$

16. $f(x) = \dfrac{4}{x} + 3$

17. $f(x) = \dfrac{1}{x^2} + 3$

18. $f(x) = 2 - \dfrac{1}{x^2}$

19. $f(x) = \dfrac{2}{(x-1)^2} + 3$

20. $f(x) = 4 - \dfrac{1}{(x+1)^2}$

In Exercises 21–40, sketch the graph of each rational function. Find horizontal asymptotes and x- and y-intercepts. In order to improve the accuracy of your graph, you may use a calculator to determine more points on the graph.

21. $f(x) = \dfrac{-3}{x-5}$

22. $f(x) = \dfrac{5}{1-x}$

23. $f(x) = \dfrac{3}{1-2x}$

24. $f(x) = \dfrac{4}{3x-2}$

25. $f(x) = \dfrac{x-4}{2x+1}$

26. $f(x) = \dfrac{2x-1}{x+3}$

27. $f(x) = \dfrac{x^2}{x^2-4}$

28. $f(x) = \dfrac{2x^2}{x^2-1}$

29. $f(x) = \dfrac{x^2}{x^2+1}$

30. $f(x) = \dfrac{-x^2}{x^2+1}$

31. $f(x) = \dfrac{3x}{x^2-4}$

32. $f(x) = \dfrac{4x}{x^2-1}$

33. $f(x) = \dfrac{3x-1}{x^2-4}$

34. $f(x) = \dfrac{x+3}{x^2-1}$

35. $f(x) = \dfrac{2x+1}{(x-1)(x+3)}$

36. $f(x) = \dfrac{3x-2}{(x-2)(x+1)}$

37. $f(x) = \dfrac{2x^2}{x^2-x-6}$

38. $f(x) = \dfrac{3x^2}{x^2-x+2}$

39. $f(x) = \dfrac{x^2}{x^2-2x+1}$

40. $f(x) = \dfrac{x}{x^2-2x+1}$

● **41.** Sketch the graph of $F(x) = 1/x^3$ and compare it with the graph of $f(x) = 1/x$.

● **42.** Sketch the graph of $G(x) = 1/x^4$ and compare it with the graph of $g(x) = 1/x^2$.

Find the horizontal asymptote of the graph of each function in Exercises 43–48.

43. $f(x) = \dfrac{3x^2-1}{5x^2+3x+2}$

44. $f(x) = \dfrac{4x^2+x-4}{3x^2+5}$

45. $f(x) = \dfrac{x^2+1}{3x^4-5}$

46. $f(x) = \dfrac{x^3-1}{3x^5-4x^2+1}$

47. $f(x) = \dfrac{3x^{99}-1}{x^{100}+x^{50}}$

48. $f(x) = \dfrac{5x^{200}-4x^{100}+3}{6x^{200}+x^2}$

● **49.** Let

$$f(x) = \dfrac{x^n}{3x^6-5x+1}.$$

Find the horizontal asymptote of f if   **a)** $n = 6$,   **b)** $n < 6$.

50. Let n be a positive integer and let

$$f(x) = \frac{2x^n + 5}{5x^{2n} + 6x^n + 1}.$$

Find the horizontal asymptote of f.

5.4 COMPOSITE AND INVERSE FUNCTIONS

In Chapter 4 we discussed ways of combining functions, and we defined the sum, difference, product, and quotient of functions. There is another important way of combining functions, called *composition of functions*, which we now explain. Composition of functions leads us to *inverse functions*, a concept of great significance in mathematics and its applications.

Composition of Functions

Consider the function

$$h(x) = \sqrt{3x - 2}$$

defined for all $x \geq 2/3$. The function h can be obtained from the functions $f(x) = 3x - 2$ and $g(x) = \sqrt{x}$ in the following manner. First, the function f assigns to each number x the value $3x - 2$. Next, whenever $3x - 2$ is nonnegative, we can take its square root and obtain the value of h at x. Symbolically, we may write

$$x \xrightarrow{f} 3x - 2 \xrightarrow{g} \sqrt{3x - 2}$$
$$\underset{h}{\underline{\hspace{3cm}}\uparrow}$$

and say that the function $h(x)$ is the *composition* of the functions $f(x)$ and $g(x)$.

With the help of a calculator we can illustrate the composition of these two functions. For example, to find the function value $h(4)$, we perform the following keystrokes:

$$4\;\boxed{\times}\;3\;\boxed{-}\;2\;\boxed{=}\;\boxed{\sqrt{}}$$

and get the final display 3.162277. You can see that in the sequence of keystrokes $4\;\boxed{\times}\;3\;\boxed{-}\;2\;\boxed{=}$, we have entered the number 4, multiplied it by 3, and subtracted 2 to obtain 10, which is the function value of $f(x)$ at $x = 4$. The last keystroke $\boxed{\sqrt{}}$ gives us the square root of 10, that is, $\sqrt{10} \simeq 3.162277$. Note that the keystrokes correspond exactly to the above diagram indicating the composition of f and g.

A word of caution is in order about the domain of the *composite function* h. The function f is defined for all real numbers, while g is defined only for nonnegative numbers. Thus the function h is defined for all real numbers x satisfying the inequality

$$3x - 2 \geq 0,$$

that is, for all $x \geq 2/3$. That is, the composition of f and g *is defined only for the values of x for which f(x) lies in the domain of g.*

Composite Functions

Given two functions f and g, the *composite function* $g \circ f$ is the function defined by

$$(g \circ f)(x) = g(f(x))$$

for all x in the domain of f for which $f(x)$ lies in the domain of g (Figure 5.18).

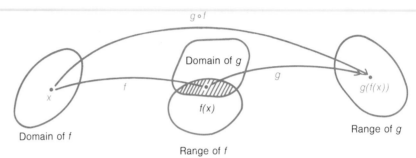

Figure 5.18

When we define the composite functions, the order in which the functions are combined may affect the final result. For example, if $f(x) = 3x - 2$ and $g(x) = \sqrt{x}$, then the composite function $k = f \circ g$ is

$$k(x) = (f \circ g)(x) = f(g(x)) = 3\sqrt{x} - 2,$$

which is defined for all $x \geq 0$. The two composite functions

$$h(x) = \sqrt{3x - 2} \quad \text{and} \quad k(x) = 3\sqrt{x} - 2$$

are clearly distinct. For example, $k(4) = 4 \neq h(4) = \sqrt{10}$.

◆ **Example 1.**   **a)** If $f(x) = x^2 - 4$ and $g(x) = 1/x$, find $h(x) = (g \circ f)(x)$ and $k(x) = (f \circ g)(x)$, and determine the domain of definition of each function.
   b) Evaluate $(g \circ f)(-3)$ and $(f \circ g)(2)$.

Solution.   **a)** We have

$$h(x) = g(f(x)) = g(x^2 - 4) = \frac{1}{x^2 - 4}$$

and the domain of definition of h is the set $\{x \in \mathbb{R}: x \neq -2, \, x \neq 2\}$.
   On the other hand,

$$k(x) = f(g(x)) = f\left(\frac{1}{x}\right) = \left(\frac{1}{x}\right)^2 - 4 = \frac{1}{x^2} - 4,$$

defined for all numbers $x \neq 0$.
   b) Since $(g \circ f)(x) = h(x) = 1/(x^2 - 4)$, we see that

$$(g \circ f)(-3) = \frac{1}{(-3)^2 - 4} = \frac{1}{9 - 4} = \frac{1}{5}.$$

Since $(f \circ g)(x) = k(x) = (1/x^2) - 4$, it follows that

$$(f \circ g)(2) = \frac{1}{2^2} - 4 = \frac{1}{4} - 4 = -\frac{15}{4}.$$

◇ Practice Exercise 1. **a)** Let $f(x) = 2x - 4$ and $g(x) = \sqrt{x + 3}$. Find $(g \circ f)(x)$ and $(f \circ g)(x)$, and determine their domains of definition.
   b) Evaluate $(g \circ f)(5)$ and $(f \circ g)(6)$.

Answer. **a)** $(g \circ f)(x) = \sqrt{2x - 1}$ for $x \geq 1/2$ and $(f \circ g)(x) = 2\sqrt{x + 3} - 4$ for $x \geq -3$;   **b)** 5, 2.

One-to-One Functions

We have seen that a function f assigns to each x in its domain a unique value $y = f(x)$, the image of x under f. However, different elements in the domain of f may have the same image. For example, if $f(x) = x^2$, then $x = -2$ and $x = 2$ have the same image: $f(-2) = f(2) = 4$.

For some functions, however, each element in the range is the image of *exactly one* element in the domain. These are said to be *one-to-one functions*.

Definition

A function $y = f(x)$ is one-to-one if whenever $f(x_1) = f(x_2)$, then $x_1 = x_2$.
   In other words, $f(x)$ is one-to-one if $x_1 \neq x_2$ implies $f(x_1) \neq f(x_2)$.

For example, the function $f(x) = x^3$ is one-to-one on $\mathbb{R}$, while the function $f(x) = x^2$ is not. This can be seen by looking at the graphs of both functions shown in Figure 5.19. We have already remarked that every vertical line through

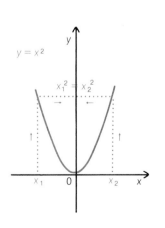
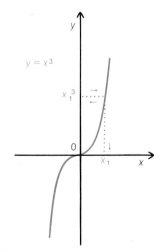

Figure 5.19

a point in the domain of a function intersects the graph at a single point. However, a horizontal line may intersect the graph in more than one point. Figure 5.19a shows a horizontal line intersecting the graph of $f(x) = x^2$ in two points. The x-coordinates of these points, x_1 and x_2, are such that $x_1^2 = x_2^2$; thus $f(x) = x^2$ is not a one-to-one function. Figure 5.19b indicates that every horizontal line intersects the graph of $f(x) = x^3$ in a single point. Thus $f(x) = x^3$ is a one-to-one function.

We now give the *horizontal line test*, which tells us whether or not a function is one-to-one.

Horizontal Line Test

A function is one-to-one if each horizontal line intersects its graph in at most one point (Figure 5.20).

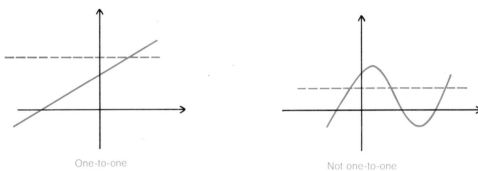

One-to-one   Not one-to-one

Figure 5.20

It is easy to see that every *increasing* or *decreasing* function is one-to-one.

◆ **Example 2.**   Determine whether each of the following functions is one-to-one:
a) $f(x) = 4x - 1$ b) $g(x) = 5 - 3x^2$.

Solution.   a) We may apply the definition of one-to-one function. If $x_1 \neq x_2$, then $4x_1 \neq 4x_2$ and $4x_1 - 1 \neq 4x_2 - 1$. Thus $x_1 \neq x_2$ implies $f(x_1) \neq f(x_2)$, so that f is one-to-one. Note that $f(x) = 4x - 1$ is an increasing function.

   b) For $x = -1$ and $x = 1$, we have

$$g(-1) = 5 - 3 \cdot (-1)^2 = 2 = 5 - 3 \cdot (1)^2 = g(1);$$

that is, the function values of g at $x = -1$ and $x = 1$ are the same. Thus g is not one-to-one.

   You could also apply the horizontal line test to show that f is one-to-one, but g is not.

◇ **Practice Exercise 2.**   Which of the following functions is one-to-one?
a) $f(x) = 3 - 4x$ b) $g(x) = 3x^2 + 2$.

Answer.   a) f is one-to-one (decreasing)   b) g is not one-to-one.

Inverse Functions

Let f be a function with domain X and range Y. If f is a one-to-one function, then every element y in Y is the image under f of only one element x in X. Thus we can define a new function g by assigning to each y in Y the unique x in X such that $y = f(x)$, as shown by Figure 5.21. This new function g with domain Y and range X is called the *inverse* of f.

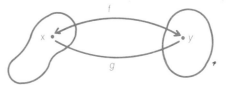

Figure 5.21

According to this definition, we can say that

$$y = f(x) \quad \text{if and only if} \quad x = g(y).$$

In other words, if the function f assigns y in Y to any x in X, then the function g assigns x to this element y. Conversely, if y is any element in Y and x its image under g, then the image of x under f is y.

We can summarize these results as follows:

Two Properties Concerning a Function and its Inverse

If f is a one-to-one function with domain X and range Y and g is the inverse of f, we have

$$(g \circ f)(x) = x \quad \text{for every } x \in X$$
$$(f \circ g)(y) = y \quad \text{for every } y \in Y.$$

These two relations say that if two functions are inverses of each other, then each function "undoes" what the other "does."

Notation

If g is the inverse of f, we often denote g by f^{-1} and read "f-inverse." With this notation, we may rewrite the two properties above as follows:

$$(f^{-1} \circ f)(x) = x \quad \text{for every } x \in X$$
$$(f \circ f^{-1})(y) = y \quad \text{for every } y \in Y.$$

In certain cases, we can find the inverse of a function by solving an equation associated with the function.

◆ **Example 3.**   Find the inverse of the function $f(x) = 4x - 1$.

Solution.   As we already know [Example 2], f is a one-to-one function, so it has an inverse function g. To find the inverse, set $y = f(x)$, obtaining the equation

$$y = 4x - 1.$$

Solve this equation for x:

$$4x - 1 = y$$
$$4x = y + 1$$
$$x = \frac{y + 1}{4}.$$

The function

$$g(y) = \frac{y + 1}{4}$$

is the inverse of f. This can be verified by checking the two relations between a function and its inverse. We have

$$(g \circ f)(x) = g(f(x)) = \frac{f(x) + 1}{4} = \frac{4x - 1 + 1}{4} = x$$

and

$$(f \circ g)(y) = f(g(y)) = 4g(y) - 1 = 4\left(\frac{y + 1}{4}\right) - 1 = y.$$

Since it is customary to use x as the independent variable of a function, we write g as follows:

$$g(x) = \frac{x + 1}{4}.$$

◇ **Practice Exercise 3.**   Find the inverse of the function $f(x) = \dfrac{3x + 1}{5}$.

Answer. $g(x) = \dfrac{5x - 1}{3}.$

We now summarize the method described in Example 3.

Algebraic Method for Finding the Inverse

Let $f(x)$ be a one-to-one function. To find the inverse g of f, proceed as follows.

1. Set $y = f(x)$.
2. Solve this equation for x and get $x = g(y)$.
3. Interchange x and y so that the last equation becomes $y = g(x)$.
4. Check that $(g \circ f)(x) = x$ and $(f \circ g)(x) = x$.

◆ **Example 4.** Show that the functions $f(x) = 3 - 4x$ and $g(x) = (3 - x)/4$ are inverses of each other and sketch their graphs on the same coordinate plane.

Solution. If f and g are inverses of each other, then we must have

$$(g \circ f)(x) = x \quad \text{and} \quad (f \circ g)(x) = x$$

for all x, and conversely. To verify this, we have

$$(g \circ f)(x) = g(f(x)) = \frac{3 - f(x)}{4} = \frac{3 - (3 - 4x)}{4} = x$$

and

$$(f \circ g)(x) = f(g(x)) = 3 - 4g(x) = 3 - 4\left(\frac{3 - x}{4}\right) = x.$$

Thus f and g are inverses of each other. Now we sketch the graph of both functions in Figure 5.22.

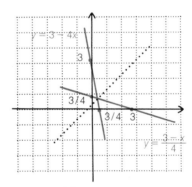

Figure 5.22

◇ **Practice Exercise 4.** Show that $f(x) = 2x - 4$ and $g(x) = (x/2) + 2$ are inverses of each other. Graph these two functions on the same coordinate plane.

Answer.

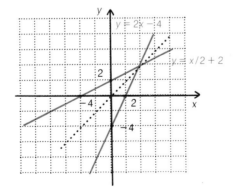

◆ **Example 5.** Show that the functions $f(x) = x^2$, for $x \geq 0$, and $g(x) = \sqrt{x}$ are inverses of each other. Graph these functions on the same coordinate plane.

Solution.   We have

$$(g \circ f)(x) = g(f(x)) = \sqrt{x^2} = x \quad \text{for all } x \geq 0$$

and

$$(f \circ g)(x) = f(g(x)) = (\sqrt{x})^2 = x \quad \text{for all } x \geq 0.$$

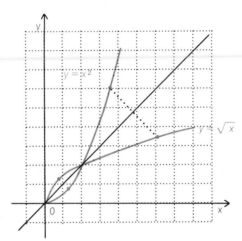

Figure 5.23

◇ **Practice Exercise 5.**   Show that the functions $f(x) = \sqrt{x - 1}$, for $x \geq 1$, and $g(x) = x^2 + 1$, for $x \geq 0$, are inverses of each other. Graph both functions on the same coordinate plane.

Answer.

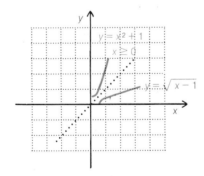

The Graph of an Inverse Function

Consider the two functions $f(x) = 3 - 4x$ and $g(x) = (3 - x)/4$ of Example 4. Their graphs appear to be symmetric to each other with respect to the line $y = x$ (Figure 5.22). The same can be said of the graphs of $f(x) = x^2$ for $x \geq 0$, and $g(x) = \sqrt{x}$ for $x \geq 0$, shown in Figure 5.23.

The two examples illustrate a general fact about the graphs of two functions that are inverses of each other. Suppose that f^{-1} is the inverse of f and let (a, b) be an arbitrary point belonging to the graph of f. This means that $b = f(a)$. Since f^{-1} is the inverse of f, it follows that $a = f^{-1}(b)$ and that the point (b, a) belongs

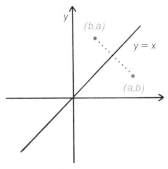

Figure 5.24

to the graph of f^{-1}. Now, as shown in Figure 5.24, the points (a, b) and (b, a) are symmetric to each other with respect to the line $y = x$. This implies that the graph of f^{-1} can be obtained by *reflecting the graph of f about the line* $y = x$.

We may state the following result.

Symmetry About the Line $y = x$

The graphs of f and f^{-1} are symmetric to each other about the line $y = x$.

◆ **Example 6.** Find the inverse f^{-1} of the function $f(x) = 4 - x^2$ for $x \geq 0$. Graph f and f^{-1} on the same coordinate plane.

Solution. Set $y = f(x)$. The function f is defined for $x \geq 0$ and its range is the set of all numbers $y \leq 4$. (Can you see why?) Next solve the equation for x:

$$y = 4 - x^2,$$
$$x^2 = 4 - y,$$
$$x = \sqrt{4 - y}$$

where $\sqrt{}$ denotes the principal square root. Thus

$$f^{-1}(y) = \sqrt{4 - y}, \quad y \leq 4,$$

is the inverse of f. Changing y into x, write the inverse as

$$y = f^{-1}(x) = \sqrt{4 - x}, \quad x \leq 4,$$

Now graph the two functions. Start with $f(x) = 4 - x^2$, $x \geq 0$, whose graph is part of a parabola (Figure 5.25). You may obtain the graph of f^{-1} by reflecting the graph of f about the line $y = x$.

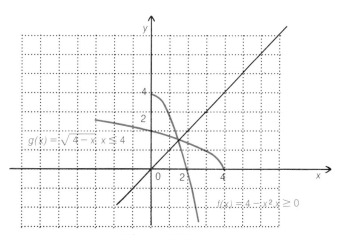

Figure 5.25

◇ **Practice Exercise 6.**   Find the inverse $f^{-1}(x)$ of the function $f(x) = \sqrt{4 + x}$, for $x \ge -4$. Graph both functions on the same coordinate plane.

Answer.

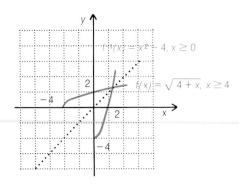

 EXERCISES 5.4

In Exercises 1–8, find $(f \circ g)(x)$, $(f \circ g)(x)$, and $(f \circ f)(x)$.

1. $f(x) = 2x + 3$, $g(x) = 3x - 4$

2. $f(x) = 4x^2 - 3x$, $g(x) = 5x - 4$

3. $f(x) = 3x + 1$, $g(x) = \sqrt{x + 5}$

4. $f(x) = x^2 - 9$, $g(x) = \dfrac{1}{x + 1}$

5. $f(x) = 2x - 5$, $g(x) = \dfrac{1}{x^2 + 2}$

6. $f(x) = 2x^2 - 1$, $g(x) = \sqrt{2x + 1}$

7. $f(x) = x^3 + 1$, $g(x) = \sqrt[3]{x - 1}$

8. $f(x) = 2x - 3$, $g(x) = \dfrac{3x}{x - 4}$

In Exercises 9–14, let $f(x) = \sqrt{x + 2}$ and $g(x) = x^2 - 4$. Find the indicated function values.

9. $(f \circ g)(\sqrt{6})$

c 10. $(f \circ g)(3.5)$

c 11. $(g \circ f)(\sqrt{3.25})$

12. $(g \circ f)(100)$

c 13. $(f \circ f)(23)$

14. $(g \circ g)(2)$

In Exercises 15–20, express each function as a composition of two functions chosen from $f(x) = \sqrt{x}$, $g(x) = x^2 + 1$, and $h(x) = 5x - 2$.

15. $F(x) = \sqrt{5x - 2}$

16. $G(x) = x + 1$

17. $H(x) = 25x^2 - 20x + 5$

18. $K(x) = \sqrt{x^2 + 1}$

19. $U(x) = x^4 + 2x^2 + 2$

20. $V(x) = 5\sqrt{x} - 2$

In Exercises 21–26, find the inverse of f and determine its domain of definition.

21. $f(x) = \dfrac{1}{2 - x}$, $x \ne 2$

22. $f(x) = \dfrac{-2}{3x + 1}$, $x \ne -\dfrac{1}{3}$

23. $f(x) = x^2 - 4$, $x \ge 0$

24. $f(x) = \sqrt{2x - 3}$, $x \ge \dfrac{3}{2}$

25. $f(x) = \sqrt[3]{2x - 4}$

26. $f(x) = x^3 - 5$

Show that each of the following functions is its own inverse. What can you say about their graphs?

27. $f(x) = \dfrac{x}{x-1}$

28. $g(x, \quad \dfrac{2x+1}{x-2}$

In Exercises 29–34, show that the given functions are inverses of each other and sketch their graphs on the same coordinate plane.

29. $f(x) = 4x + 5;\ g(x) = \dfrac{x-5}{4}$

30. $f(x) = 2 - 3x;\ g(x) = \dfrac{2-x}{3}$

31. $f(x) = \dfrac{1}{x-1},\ x \ne 1;\ g(x) = \dfrac{x+1}{x},\ x \ne 0$

32. $f(x) = \dfrac{1}{2-x},\ x \ne 2;\ g(x) = \dfrac{2x-1}{x},\ x \ne 0$

33. $f(x) = \sqrt{3x-1},\ x \ge \dfrac{1}{3};\ g(x) = \dfrac{x^2+1}{3},\ x \ge 0$

34. $f(x) = x^2 + 3,\ x \le 0;\ g(x) = \sqrt{x-3},\ x \ge 3$

35. The demand D for a certain brand of toaster is given by $D = 300 - 0.002p^2$, where p is the price (in dollars) of each toaster. If the price p, as a function of the cost c (in dollars), is given by $p = 3c - 20$, find the demand in terms of the cost.

36. Suppose that the total number of units Q produced daily by a manufacturer is given by the following function of the number n of employees:

$$Q(n) = 6n - \frac{n^2}{10}.$$

Suppose also that the total revenue R received for selling Q units of the product is given by $R = 30Q$. Find $R \circ Q(n)$. What does this function represent?

● **37.** Let f and f^{-1} be inverses of each other. Show that (a, b) belongs to the graph of f if and only if (b, a) belongs to the graph of f^{-1}.

● **38.** Under what conditions on m and b does the linear function $f(x) = mx + b$ have an inverse?

● **39.** Let $f(x) = mx + b$ and $g(x) = nx + d$. Under what conditions on $m, b, n,$ and d do we have $f \circ g = g \circ f$?

● **40.** Find a formula for the inverse of the function

$$f(x) = \frac{ax + b}{cx + d}$$

where $a, b, c,$ and d are constants such that $ad - bc \ne 0$. Why is this restriction necessary?

CHAPTER SUMMARY

Polynomial functions are the simplest functions we can define using algebraic operations. In spite of their simplicity, certain polynomial functions may be very difficult to graph. However, the graphs of polynomials of degrees 1 and 2 are entirely understood.

A polynomial of degree 1 has a very simple graph: it is a *straight line*. The graph of a quadratic polynomial is a *parabola*, a curve that is simple to sketch. All the information about the parabola can be derived from the quadratic polynomial by performing simple algebraic manipulations. Let

$$f(x) = ax^2 + bx + c$$

be a quadratic function. If $a > 0$, the parabola *opens upward*; if $a < 0$, it *opens downward*. The sign of the *discriminant*

$$\Delta = b^2 - 4ac$$

tells us about the x-intercepts of the parabola. If $\Delta > 0$, the parabola has two x-intercepts; if $\Delta = 0$, it has only one x-intercept (in fact, the parabola just touches the x-axis); and if $\Delta < 0$, the parabola has no x-intercepts. By completing the square and writing the quadratic function as

$$f(x) = a(x - h)^2 + k,$$

we get the equation $x = h$ of the *axis of symmetry* and the coordinates (h, k) of the *vertex* of the parabola. The vertex is the highest point (*maximum*) if the parabola opens downward; it is the lowest point (*minimum*) if the parabola opens upward.

The task of graphing a polynomial of degree greater than 2 can be considerably simplified if you can express the polynomial as a product of linear factors. When a polynomial function $f(x)$ is written as a product of linear factors, you can immediately find the solutions of the equation $f(x) = 0$ and determine the x-intercepts of the graph. The x-intercepts divide the real line into a finite number of intervals in each of which the sign of the polynomial function remains constant. By analyzing the sign of the function in each interval you will have an idea of the location of the graph. By plotting points you will get a rough sketch of the graph. Polynomial functions have *high* and *low points*. These are *turning points* of the graph, where the function changes from increasing to decreasing or from decreasing to increasing. In most cases high and low points are difficult to find; they require techniques from calculus.

A *rational function* is a quotient of two polynomial functions. In most cases it is difficult to graph a rational function, especially when the methods of calculus are not available. For this reason, only simple rational functions are considered in the present chapter. To graph a rational function you should first find the domain of definition and then determine the *vertical asymptotes. Horizontal asymptotes* are found by comparing the degrees of the numerator and denominator and looking at the leading terms of the polynomials. Find the x- and y-intercepts of the graph. Study the sign variation of the function in the intervals where the function is defined. Look for points where the graph may cross a horizontal asymptote. Finally, make a table of function values and plot as many points as you think necessary to obtain a reasonable sketch of the graph.

Functions can be added, subtracted, multiplied, and divided to generate new functions. Another important operation between functions is the *composition of functions*. This operation can also be used to define inverse functions. Only *one-to-one functions* have inverses. In particular, *increasing* or *decreasing* functions have inverses. If you know the graph of a function, you may use the *horizontal line test* to determine whether or not the function has an inverse. In certain simple cases, you may be able to find the inverse of a function by just solving the equation associated with the function. This is called the *algebraic method for finding the inverse*. If f is a function and f^{-1} its inverse, then the graphs of these functions *are symmetric to each other about the line $y = x$.* So by knowing the graph of one of the functions, you may obtain the graph of the other by reflection about the line $y = x$.

REVIEW EXERCISES

Graph each of the following quadratic functions. Find x- and y-intercepts, and determine the axis of symmetry and the vertex.

1. $f(x) = 8 - 2x^2$

2. $f(x) = 27 - 3x^2$

3. $f(x) = 3x^2 + x - 4$

4. $f(x) = 4x^2 + 11x - 3$

5. $f(x) = -x^2 + x - 5$

6. $f(x) = -3x^2 + 2x - 4$

In Exercises 7–16, graph each polynomial function.

7. $f(x) = 3x^2(x + 2)$

8. $f(x) = 3x(x + 4)^2$

9. $f(x) = (x - 2)(x - 3)(x + 2)$

10. $f(x) = (x - 4)(x - 1)(x + 3)$

11. $f(x) = 2x^3 - x^2 - 8x + 4$

12. $f(x) = 3x^3 - 2x^2 - 3x + 2$

13. $f(x) = x(x + 1)^3$

14. $f(x) = x^3(x - 2)$

15. $f(x) = 16x^2 - x^4$

16. $f(x) = x^4 - 4x^2$

c *Sketch the graph of each polynomial function. Use a calculator to determine several points on each graph.*

17. $f(x) = x^3 + 2x + 1$

18. $f(x) = 2x^3 - x^2 + 2x - 1$

In Exercises 19–32, graph each rational function. Find horizontal and vertical asymptotes and x- and y-intercepts.

c *In order to improve the accuracy of your graph, use a calculator to plot more points of the graph.*

19. $f(x) = \dfrac{5}{x + 4}$

20. $f(x) = \dfrac{-2}{x + 4}$

21. $f(x) = \dfrac{x + 1}{x - 4}$

22. $f(x) = \dfrac{x - 3}{x + 5}$

23. $f(x) = \dfrac{2x + 3}{x - 2}$

24. $f(x) = \dfrac{x - 1}{3x + 4}$

25. $f(x) = \dfrac{x^2}{3x^2 + 1}$

26. $f(x) = \dfrac{4x^2}{x^2 + 1}$

27. $f(x) = \dfrac{2x}{x^2 - 9}$

28. $f(x) = \dfrac{3x}{x^2 - 16}$

29. $f(x) = \dfrac{x - 1}{x^2 - 16}$

30. $f(x) = \dfrac{x + 2}{x^2 - 9}$

31. $f(x) = \dfrac{x^2}{x^2 + 5x + 4}$

32. $f(x) = \dfrac{x^2}{x^2 - 5x + 4}$

Find horizontal asymptotes of the following functions:

33. $f(x) = \dfrac{3x - 4}{2x + 5}$

34. $f(x) = \dfrac{5x - 2}{3x + 4}$

35. $f(x) = \dfrac{4x^2 - 5}{3x^2 - 6x - 1}$

36. $f(x) = \dfrac{5x^2 - 3}{4x^2 - 3x + 1}$

37. $f(x) = \dfrac{3x + 7}{x^3 - 6x + 1}$

38. $f(x) = \dfrac{8x^2 - 4}{5x^3 - 4x - 2}$

In Exercises 39–42, find $(f \circ g)(x)$, $(g \circ f)(x)$, and $(f \circ f)(x)$.

39. $f(x) = \dfrac{1}{3x + 1}$, $g(x) = \sqrt{2x - 3}$

40. $f(x) = \dfrac{4x + 1}{x - 4}$, $g(x) = \dfrac{1}{x^2}$

41. $f(x) = \dfrac{x + 1}{x - 3}$, $g(x) = \dfrac{3x + 1}{x - 1}$

42. $f(x) = \dfrac{2x - 1}{x + 2}$, $g(x) = \dfrac{2x + 1}{2 - x}$

In Exercises 43–46, show that the given functions are inverses of each other. Graph the functions on the same coordinate plane.

43. $f(x) = \dfrac{2x + 1}{x}$, $g(x) = \dfrac{1}{x - 2}$

44. $u(x) = \dfrac{1}{1 - 3x}$, $v(x) = \dfrac{x - 1}{3x}$

45. $F(x) = \sqrt{3 - x}$, $x \le 3$, $G(x) = 3 - x^2$, $x \ge 0$

46. $f(x) = \sqrt{3x - 2}$, $x \ge \dfrac{2}{3}$, $g(x) = \dfrac{x^2 + 2}{3}$, $x \ge 0$

For each function in Exercises 47–52, find the inverse and determine its domain.

47. $f(x) = \dfrac{2}{3x - 5}$

48. $f(x) = \dfrac{2x + 1}{x - 3}$

49. $f(x) = \sqrt{3x - 4}$, $x \ge \dfrac{4}{3}$

50. $f(x) = 4 - x^2$, $0 \le x \le 2$

51. $f(x) = 9 - x^2$, $-3 \le x \le 0$

52. $f(x) = \sqrt[3]{2x - 1}$

● **53.** Show that if f and g are odd functions, then $f \circ g$ is an odd function.

● **54.** Show that if f is an even function and g is an odd function, then $f \circ g$ is an even function.

● **55.** Find the horizontal asymptote of the graph of

$$F(x) = \frac{x^n + 1}{1 - x^n} \quad (n \text{ a positive integer}).$$

● **56.** What is the horizontal asymptote of the graph of

$$G(x) = \frac{3x^n}{2x^{100} + 51}$$

when: a) $n < 100$;   b) $n = 100$? What happens when $n > 100$?

● **57.** The price P (in dollars) of a certain product is given by the equation

$$P = 100 - 5D^2,$$

where D is the demand (in hundreds of units). Express P as a function of the number t of years, if the demand has risen according to the equation

$$D = 3 + 2\sqrt{t}.$$

● **58.** Show that if the linear function $f(x) = mx + b$ has an inverse, then the inverse is also a linear function.

● **59.** Let $f(x) = mx + b$ and $g(x) = nx + d$. Under what conditions on m, n, b, and d are f and g inverses of each other?

● **60.** Let

$$f(x) = \frac{ax + b}{cx + d}$$

where a, b, c, and d are constants such that $ad - bc \ne 0$. For what values of these constants is $f(x)$ its own inverse?

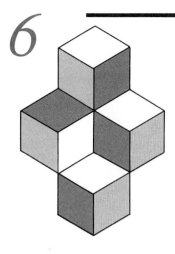

6

Exponential and Logarithmic Functions

$\mathbf{U}$p to now we have considered only algebraic functions, i.e., functions that are formed by a finite number of algebraic operations—sum, difference, product, quotient, and root extraction—on the constant and identity functions. A function that is not algebraic is called a transcendental function. In this chapter we study two of them: the exponential and logarithmic functions. Both play fundamental roles in mathematics and the applied sciences. Exponential functions describe certain growth and decay phenomena studied in biology, economics, and physics. Logarithmic functions are defined as inverses of exponential functions. Among them, the common logarithm (base 10 logarithm) was an important aid in numerical computations before the availability of calculators. The natural logarithm (base e logarithm) is of fundamental importance in both theoretical and applied mathematics.

6.1 EXPONENTIAL FUNCTIONS

It has been observed that, under favorable conditions, the number of bacteria in a culture doubles after a fixed period of time, independent of the size of the culture with which the experiment was started, and independent of the time when the first observation was made.

Suppose, for example, that the number of bacteria in a culture doubles every hour. If initially there are n_0 bacteria present, then the number of bacteria after one hour will be $n_0 2$. Since the number doubles again in another hour, two hours after starting our observation the number of bacteria will be $n_0 2^2$. It follows similarly that for each integer t, after t hours the number of bacteria in the culture will be given by the expression

$$n(t) = n_0 2^t.$$

Thus we obtain a function defined for *positive integral values* of t. But, according to the definition of rational exponents [Section 1.8], this function is also defined for *rational values* of t. For instance, we know that $2^{1/2} = \sqrt{2}$ and

$2^{5/2} = \sqrt{2^5} = 4\sqrt{2}$ so it is reasonable to expect that

$$n\left(\frac{1}{2}\right) = n_0\sqrt{2} \quad \text{and} \quad n\left(\frac{5}{2}\right) = n_0 4\sqrt{2}$$

represent the number of bacteria in the culture after 1/2 hour and 5/2 hours, respectively.

Can the function defined above for rational values of t be defined for *all* real values of t? In other words, can we give a meaning to expressions like $2^{\sqrt{2}}$ or 2^π? The answer, which is affirmative, depends on concepts and properties of the real numbers system that can be fully explained only in more advanced courses. Nevertheless, we describe in intuitive terms what is involved in defining the number $2^{\sqrt{2}}$.

First, recall that $\sqrt{2}$ is an irrational number and, as such, has a nonterminating decimal representation 1.4142136. . . . . Next, consider the rational numbers

$$1 < 1.4 < 1.41 < 1.414 < 1.4142 < 1.41421 < \cdots .$$

They are *decimal approximations* of $\sqrt{2}$ and form an *increasing sequence* of rational numbers. (Sequences are studied in Chapter 9.)

Given a rational number, such as 1.41, it follows from the definition of rational exponents that

$$2^{1.41} = 2^{141/100} = \sqrt[100]{2^{141}}.$$

An approximation to this power of 2 can be obtained with the help of calculator by performing the following keystrokes: 2 $\boxed{y^x}$ 1.41 $\boxed{=}$ 2.6573716. Table 6.1 was obtained by using a calculator and rounding off the results to five decimal places:

Table 6.1

t	2^t
1	2
1.4	2.63902
1.41	2.65737
1.414	2.66475
1.4142	2.66512
1.41421	2.66514
$\vdots$	$\vdots$

The table indicates that the sequence of powers of 2

$$2 < 2^{1.4} < 2^{1.41} < 2^{1.414} < 2^{1.4142} < 2^{1.41421} < \cdots$$

increases as the exponents get closer to $\sqrt{2}$. Moreover, it can be proved that this sequence of powers of 2 approximates to any desired degree of accuracy a real number that we denote by $2^{\sqrt{2}}$. As an exercise, you should compare the approximations in the right column of the table above with the approximation

$$2^{\sqrt{2}} \simeq 2.6651441$$

obtained using the square root and power keys of a calculator.

Similarly, it is possible to give a meaning to the number 2^π by using decimal approximations of π. Generally, any real number t can be approximated to any degree of accuracy by an increasing sequence $r_1 < r_2 < \cdots < r_n < \cdots$ of rational numbers. (This is a fundamental property of the real number system.) The number 2^t can then be approximated to any degree of accuracy by the sequence of powers

$$2^{r_1} < 2^{r_2} < \cdots < 2^{r_n} < \cdots.$$

The previous discussion shows that

$$f(x) = 2^x$$

is a function defined for *all* real numbers x not merely for rational numbers x. The function defined in this way is called *the exponential function with base 2*. The graph of this function is illustrated in Figure 6.1. In the table accompanying the graph, we show function values corresponding to some integral values of x. This has been done in order to simplify the plotting of points and because they suffice to give us an idea of the curve.

x	-3	-2	-1	0	1	2	3
2^x	$\dfrac{1}{8}$	$\dfrac{1}{4}$	$\dfrac{1}{2}$	1	2	4	8

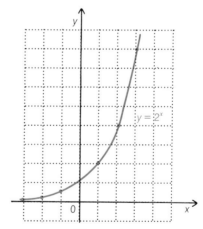

Figure 6.1

Since the function values of $f(x) = 2^x$ are always positive, the graph lies above the x-axis. Increasing values of x correspond to increasing values of 2^x; that is, if $x_1 < x_2$, then $2^{x_1} < 2^{x_2}$. Thus $f(x) = 2^x$ is an *increasing function* of x. Moreover, as x increases in the positive direction, the values of 2^x increase without bound. On the other hand, as x decreases in the negative direction the values of 2^x approach 0.

For future reference, we also sketch the graph of the function

$$f(x) = 2^{-x}$$

(Figure 6.2) which you should compare to the graph of $f(x) = 2^x$. Note that $2^{-x} = (1/2)^x$.

x	-3	-2	-1	0	1	2	3
2^{-x}	8	4	2	1	$\dfrac{1}{2}$	$\dfrac{1}{4}$	$\dfrac{1}{8}$

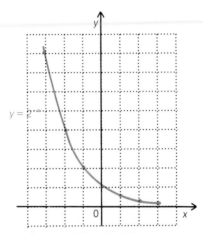

Figure 6.2

We see that the graph of $f(x) = 2^{-x}$ is the reflection about the y-axis of the graph of $f(x) = 2^x$. Also, observe that to increasing values of x correspond decreasing values of 2^{-x}. Thus, $f(x) = 2^{-x}$ is a *decreasing function*. As x increases in the positive direction, the values of 2^{-x} decrease and approach 0. As x moves leftward along the x-axis, the values of $f(x) = 2^{-x}$ increase without bound.

Exponential Functions

Let $a > 0$ and $a \neq 1$. The exponential function with base a is defined by

$$f(x) = a^x$$

where x is a real number.

If $x = n$ is a natural number, then

$$f(n) = a^n = a \cdot a \cdot \ldots \cdot a \quad (n \text{ times}).$$

If $x = p/q$ is a positive rational number, then

$$a^x = a^{p/q} = \sqrt[q]{a^p}$$

and

$$a^{-x} = a^{-p/q} = \frac{1}{\sqrt[q]{a^p}}.$$

If x is an irrational number approximated by a sequence $r_1, r_2, \ldots$ of rational numbers, then the powers $a^{r_1}, a^{r_2}, \ldots$ are defined and give approximations of a^x.

Remark

In the definition of exponential functions, we have assumed that the base a is greater than 0 and not equal to 1, for the following reasons.

If $a = 1$, then $1^x = 1$ for each value of x, and the exponential function becomes trivial: it coincides with the constant function $f(x) = 1$.

If $a < 0$, then the exponential function is not defined. For example, if $a = -3$ and $x = 1/2$, then $(-3)^{1/2} = \sqrt{-3}$ is a complex number, but we are considering only real valued functions.

Properties of the Exponential Functions

The exponential function $f(x) = a^x$, $a > 0$, $a \neq 1$, satisfy the following properties:

I. $a^x a^y = a^{x+y}$
II. $(a^x)^y = a^{xy}$
III. $a^{-x} = 1/a^x$
IV. If $a > 1$ and $x < y$, then $a^x < a^y$.
V. If $0 < a < 1$ and $x < y$, then $a^x > a^y$.

We already know that properties I, II, and III are true when the exponents x and y are rational numbers [Section 1.8]. Now, we are saying that these properties extend to the case of real exponents.

Property IV states that if $a > 1$, then the exponential function $f(x) = a^x$ is an *increasing function*. On the other hand, if $0 < a < 1$, property V states that $f(x) = a^x$ is a *decreasing function*.

The graphs of both functions are illustrated in Figure 6.3. Since a^0 is always equal to 1 for all $a \neq 0$, we see that both graphs cross the y-axis at the point $(0, 1)$.

When $a > 1$, as the values of x increase without bound the corresponding values a^x also increase without bound. As the values of x decrease without bound, the corresponding values of a^x decrease and approach 0 (Figure 6.3a).

Analogous remarks can be made in the case $0 < a < 1$ (Figure 6.3b). In both cases, the x-axis is a horizontal asymptote.

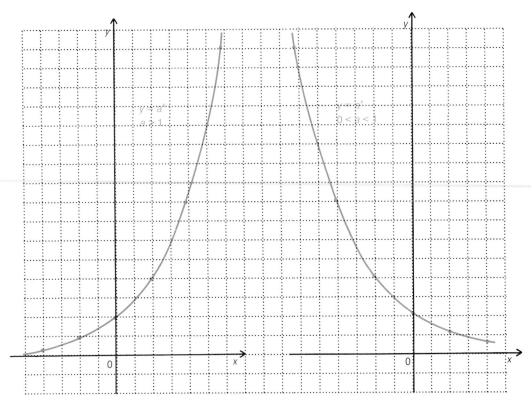

$y = a^x$
$a > 1$

$y = a^x$
$0 < a < 1$

Figure 6.3

◆ **Example 1.** Sketch the graphs of the exponential functions $f(x) = 2^x$ and $g(x) = (4/3)^x$ on the same coordinate plane and compare them.

Solution. Consider the following table of values for the two functions:

x	-2	-1	0	1	2
2^x	$\dfrac{1}{4}$	$\dfrac{1}{2}$	1	2	4
$\left(\dfrac{4}{3}\right)^x$	$\dfrac{9}{16}$	$\dfrac{3}{4}$	1	$\dfrac{4}{3}$	$\dfrac{16}{9}$

By plotting the corresponding points and joining them, we obtain the graphs illustrated in Figure 6.4.

Notice that the two graphs cross at the point $(0, 1)$. When $x > 0$, the graph of $f(x) = 2^x$ lies above the graph of $f(x) = (4/3)^x$. When $x < 0$, the graph of $f(x) = 2^x$ lies below the graph of $f(x) = (4/3)^x$.

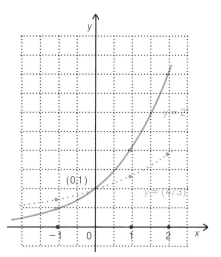

Figure 6.4

Example 1 illustrates the following properties of exponential functions.

> If $b > a > 1$, then $b^x > a^x$ holds for all $x > 0$, but $b^x < a^x$ holds for
> all $x < 0$. For $x = 0$, one has $b^0 = a^0 = 1$.

◇ **Practice Exercise 1.**   Sketch and compare the graphs of the functions $f(x) = 2^x$ and $g(x) = (3/2)^x$.

Answer.

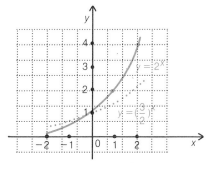

x	$y = 2^x$	$y = \left(\frac{3}{2}\right)^x$
-1	$\dfrac{1}{2}$	$\dfrac{2}{3}$
0	1	1
1	2	$\dfrac{3}{2}$
2	4	$\dfrac{9}{4}$

◆ **Example 2.**   Sketch and compare the graphs of the functions $f(x) = (1/2)^x$ and $g(x) = (1/3)^x$.

Solution.   Consider the following table of values:

x	-2	-1	0	1	2
$\left(\dfrac{1}{2}\right)^x$	4	2	1	$\dfrac{1}{2}$	$\dfrac{1}{4}$
$\left(\dfrac{1}{3}\right)^x$	9	3	1	$\dfrac{1}{3}$	$\dfrac{1}{9}$

The two graphs illustrated in Figure 6.5 cross at the point $(0, 1)$. For negative values of x, the graph of $g(x) = (1/3)^x$ is located above the graph of $f(x) = (1/2)^x$, while, for positive values of x, the graph of $g(x)$ is below the graph of $f(x)$.

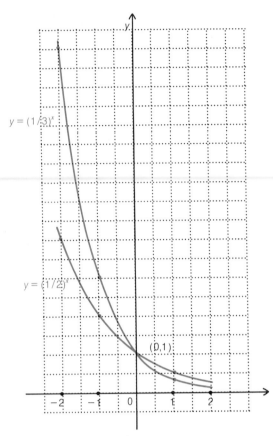

Figure 6.5

Example 2 illustrates the following properties of exponential functions.

If $0 < b < a < 1$, then $b^x < a^x$ holds for all $x > 0$, but $b^x > a^x$ holds for all $x < 0$. For $x = 0$, one has $b^0 = a^0 = 1$.

◇ **Practice Exercise 2.**   Sketch and compare the graphs of $f(x) = (2/3)^x$ and $g(x) = (1/2)^x$.

Answer.

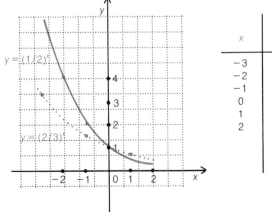

x	$(2/3)^x$	$(1/2)^x$
-3	$27/8$	8
-2	$9/4$	4
-1	$3/2$	2
0	1	1
1	$2/3$	$1/2$
2	$4/9$	$1/4$

The Number *e*

Among the examples of irrational numbers discussed in Section 1.2 of Chapter 1, we mentioned the number *e*, which is of fundamental importance in mathematics and the applied sciences. The following is an approximation of *e* to 20 decimal places:

$$e \simeq 2.71828\ 18284\ 59045\ 23536.$$

Since *e* is an irrational number, it can be approximated, in many ways, by increasing sequences of rational numbers. To obtain one of these sequences, we may proceed as follows. With each natural number *n*, we associate the rational number

$$\left(1 + \frac{1}{n}\right)^n = \left(\frac{n + 1}{n}\right)^n.$$

Evaluating this expression for $n = 1, 2, 3, \ldots$, we obtain a sequence of rational numbers that approximate *e*. Table 6.2 shows several approximations for *e* obtained with the help of a calculator. The numbers in the right column were rounded off to five decimal places.

Table 6.2

n	$\left(1 + \dfrac{1}{n}\right)^n$
1	2
2	2.25000
3	2.37037
4	2.44141
5	2.48832
10	2.59374
10^2	2.70481
10^3	2.71692
10^4	2.71815
10^5	2.71827
$\vdots$	$\vdots$

The Natural Exponential Function

The exponential function

$$f(x) = e^x,$$

whose base is the number *e*, is called the *natural exponential function*.

It is worth mentioning that since the number *e* can be approximated by $(1 + 1/n)^n$, the function e^x can also be approximated by

$$\left[\left(1 + \frac{1}{n}\right)^n\right]^x \quad \text{for } n = 1, 2, 3, \ldots.$$

Because of the importance of the exponential functions with base *e* and $1/e$, mathematicians have produced tables giving decimal approximations of the numerical values of e^x and e^{-x} for moderate-sized values of *x*. A short table of these values is given in the Appendix (Table 1). Nowadays, the use of such tables has been overshadowed by the advent of scientific calculators, which compute numerical values of these functions (correct to the number of decimal places the calculator can show) at the press of a key. Students who own scientific calculators are encouraged to use them to speed up their calculations as they work from this book. Students working without such calculators will find tables like the one in the Appendix indispensable.

Using Your Calculator

As we have already mentioned, the key $\boxed{y^x}$ on your calculator can be used to find functional values of exponential functions.

Certain scientific calculators have the key $\boxed{10^x}$ for the computation of powers of 10. For example, if you enter 3 and press the $\boxed{10^x}$ key, the display will show 1000.

Also, some calculators have the key $\boxed{e^x}$ or $\boxed{EXP}$ to compute the natural exponential function. If you own such a calculator, then keying in

$$1 \boxed{e^x}$$

gives you the number e to as many places as your calculator will display. Keying in

$$0.8 \boxed{e^x}$$

gives you an approximation for $e^{0.8}$.

However, there are calculators that do not have the key $\boxed{e^x}$, such as the TI-30-II. In order to display the number e, you have to enter 1 and press the keys $\boxed{INV}\ \boxed{\ln x}$. To obtain an approximation for $e^{0.8}$, enter 0.8 and press the keys $\boxed{INV}\ \boxed{\ln x}$. This will give you

$$e^{0.8} = 2.2255409.$$

The reason for this, explained in Section 6 of this chapter, is that the natural exponential function e^x is the inverse of the natural logarithmic function $\ln x$.

EXERCISES 6.1

In Exercises 1–16, sketch the graph of each function. Whenever applicable, use the techniques of shifting, stretching, and reflecting graphs.

1. $f(x) = 3^x$
2. $f(x) = \left(\frac{1}{4}\right)^x$
3. $f(x) = \left(\frac{3}{2}\right)^x$
4. $f(x) = \left(\frac{4}{3}\right)^x$
5. $f(x) = 2^x + 3$
6. $f(x) = 2^x - 1$
7. $f(x) = 3^x - 1$
8. $f(x) = 3^x + 2$
9. $f(x) = 2^{x-1}$
10. $f(x) = 2^{x+3}$
11. $f(x) = 3^{x+2}$
12. $f(x) = 3^{x-4}$
13. $f(x) = 3^{2-x}$
14. $f(x) = 2^{3-x}$
15. $f(x) = 1 - 2^x$
16. $f(x) = 3 - 2^{-x}$

Without using a table or a calculator, find the indicated function values.

17. $f(x) = 4^x$; $f(2), f(-2), f(1/2)$
18. $f(x) = 3(2^x)$; $f(3), f(-3), f(0)$
19. $f(x) = 5(9^x)$; $f(1/2), f(-1/2), f(1)$
20. $f(x) = 2(16^{-x})$; $f(1/4), f(-1/2), f(1/2)$
21. $f(x) = 3 + 2(1/8)^{x+1}$; $f(-2), f(0), f(-2/3)$
22. $f(x) = 2 - 3(4)^{x-2}$; $f(2), f(0), f(5/2)$

Using Table 1, approximate the following values.

23. e^2

24. e^{-2}

25. $e^{1/2}$

26. $e^{0.25}$

27. $e^{-1.2}$

28. $e^{-2.25}$

29. $e^{1/20}$

30. $e^{4/25}$

In Exercises 31–36, sketch the graph of the given function. Using Table 1 or a calculator, plot points corresponding to integral values of x and join them by a smooth curve.

31. $f(x) = e^x$

32. $f(x) = e^{-x}$

33. $f(x) = e^{x^2}$

34. $f(x) = e^{-x^2}$

35. $f(x) = \dfrac{e^x + e^{-x}}{2}$

36. $f(x) = \dfrac{e^x - e^{-x}}{2}$

● **c** *In Exercises 37–46, use a calculator to approximate the given numbers.*

37. $3^{\sqrt{2}}$

38. $5^{-\sqrt{3}}$

39. $4^{-\sqrt{5}}$

40. 3^{π}

41. $(\sqrt{2})^{\sqrt{2}}$

42. $\pi^2 - 2^{\pi}$

43. π^{π}

44. $3(2.015)^{3.271}$

45. $7\left(\dfrac{1}{2}\right)^{-0.15}$

46. $5(3)^{1.02}$

● **c** **47.** If $f(x) = x(2^x)$, find $f(-0.51)$, $f(0.01)$, and $f(0.51)$.

● **c** **48.** If $g(x) = 2xe^{-x}$, find $g(-0.5)$, $g(0.5)$ and $g(1.01)$.

● **c** **49.** Let $f(x) = [1 + (1/x)]^x$. Approximate the following functional values to five decimal places: $f(4)$, $f(10)$, $f(10^3)$, and $f(10^5)$. Compare your answers with Table 6.2.

● **c** **50.** Let $g(x) = (1 + x)^{1/x}$. Approximate the following functional values to five decimal places: $g(1/3)$, $g(1/5)$, $g(10^{-2})$, and $g(10^{-4})$. Compare your answers with Table 6.2. What conclusion can you draw from this?

6.2 APPLICATIONS OF EXPONENTIAL FUNCTIONS

Exponential functions appear in many problems in mathematics and the applied sciences.

Compound Interest

Assume that a sum of money p_0 (the *initial principal*) is invested at an interest rate of $100r\%$ per year *compounded annually*. Then, at the end of one year, the amount of principal $p(1)$ is equal to the initial principal p_0 plus the earned interest $p_0 r$. That is,

$$p(1) = p_0 + p_0 r$$
$$= p_0(1 + r).$$

At the end of two years, the new principal $p(2)$ will be equal to the sum of $p_0(1 + r)$, the principal at the end of the first year, plus the earned interest $p_0(1 + r)r$, that is,

$$p(2) = p_0(1 + r) + p_0(1 + r)r$$
$$= p_0(1 + r)^2.$$

By repeating this reasoning, we see that the principal $p(t)$ is given by

$$(6.1) \quad p(t) = p_0(1 + r)^t$$

which is an exponential function of t. If we set $t = 0$, then $p(0) = p_0$ is the initial principal.

When money is deposited in a savings account, the interest is compounded *quarterly* (four times a year). There are certain types of investment accounts that compound interest on a monthly or even on a daily basis. In such cases, formula (6.1) has to be changed.

Suppose that an interest rate of $100r\%$ per year is being *compounded N times a year*. Then, the interest rate will be $(100r/N)\%$ *per period* and, in t years, the *number of periods* will be Nt. Thus, the amount of principal $p(t)$ at the end of t years compounded N times a year will be given by

$$(6.2) \quad p(t) = p_0\left(1 + \frac{r}{N}\right)^{Nt}$$

which is also an exponential function of t.

◆ Example 1.   A sum of $1000 is invested at an annual rate of 8% compounded quarterly. Find the principal at the end of   **a)** one year;   **b)** four years.

Solution.   Since the interest is 8% per year, compounded quarterly, the variable r in formula (6.2) is equal to 0.08. The period is $N = 4$, and the rate per period is $r/N = 0.08/4 = 0.02$.
a) By using (6.2) with $t = 1$, we get

$$p(1) = 1000(1 + 0.02)^4$$
$$= 1000(1.02)^4$$
$$\simeq 1082.43$$

b) By using (6.2) with $t = 4$, we obtain

$$p(4) = 1000(1.02)^{16}$$
$$\simeq \$1372.80$$

In this example, the value of $(1.02)^{16}$ was obtained with the help of a calculator and rounded off to the nearest dime. If a calculator is unavailable, a compound interest table, found in most libraries, would have to be used.

◇ **Practice Exercise 1.** If $2000 is invested at an annual rate of 10% compounded semiannually, find the principal **a)** in two years, **b)** in five years.

Answer. **a)** $2431.00, **b)** $3257.79.

Returning to formula (6.2), we may ask what happens if the period N increases without bound. This corresponds to compounding the interest daily, hourly, every minute, and so on. First, write (6.2) as follows:

$$p(t) = p_0\left(1 + \frac{r}{N}\right)^{Nt}$$

$$= p_0\left(1 + \frac{1}{N/r}\right)^{Nt}.$$

Next, set $N/r = n$ or $N = nr$ and obtain

$$p(t) = p_0\left(1 + \frac{1}{n}\right)^{nrt}$$

$$= p_0\left[\left(1 + \frac{1}{n}\right)^{n}\right]^{rt}.$$

If N increases without bound, so does n. But it was remarked in the previous section that as n increases without bound, $(1 + 1/n)^n$ approaches the number e. Therefore, as the number of periods N increases without bound, formula (6.2) becomes

$$(6.3) \quad p(t) = p_0 e^{rt}$$

which represents the amount of principal after t years if the interest is *compounded continuously.*

Illustration Table

Principal after one year from an initial amount of $10,000 deposited at 12% annual interest for various frequencies of compounding.

	N	$p(1)$
Annually	1	11,200.00
Quarterly	4	11,255.09
Weekly	52	11,273.41
Hourly	8760	11,274.95
Continuously (formula (6.3))		11,274.97

Compare the last two numbers. Do you realize that you earn only 2¢ more by compounding continuously instead of hourly?

◆ **Example 2.**   A sum of \$1000 is invested at an annual rate of 8%. Find the principal at the end of 10 years, if the interest is   **a)** compounded quarterly,   **b)** compounded continuously.

Solution.   **a)** Applying (6.2) with $r = 0.08$, $n = 4$, and $t = 10$, we get

$$p(10) = 1000(1.02)^{40}$$
$$\simeq 1000 \times 2.2080$$
$$\simeq \$2208.00$$

The approximation of $(1.02)^{40}$ to four decimal places was obtained with the help of a calculator.

b) By using (6.3) with $r = 0.08$ and $t = 10$, we obtain

$$p(10) = 1000e^{0.8}$$
$$\simeq 1000 \times 2.2255409$$
$$\simeq \$2225.54,$$

where the approximation for $e^{0.8}$ was obtained using a calculator. Notice that the two answers differ by \$17.54.

◇ **Practice Exercise 2.**   If \$4000 is invested at an annual rate of 12%, find the principal at the end of 8 years if the interest is compounded   **a)** quarterly,   **b)** continuously.

Answer.   **a)** \$10,300.33,   **b)** \$10,446.79.

Population Growth

Under stable conditions, the population of a given species *increases* at a rate that is proportional to the population at each time. If $r > 0$ denotes the *rate of increase per individual per unit of time* and P_0 denotes the population at time $t = 0$, it can be shown that the formula

$$(6.4) \quad P(t) = P_0 e^{rt}$$

gives the population at time t.

Notice that the last formula is analogous to formula (6.3).

◆ **Example 3.**   In 1960, the Earth's human population was estimated at 3×10^9 and was increasing at a rate of 2% per year. What was the Earth's estimated population in 1970?

Solution.   If 1960 represents the time $t = 0$, then $P_0 = 3 \times 10^9$ in formula (6.4). Since the rate of increase is 2% per year, we have $r = 0.02$. Thus, we can write

$$P(t) = (3 \times 10^9)e^{0.02t}.$$

Setting $t = 10$, we obtain the estimated population in 1970:

$$P(10) = (3 \times 10^9)e^{0.02 \times 10}$$
$$= (3 \times 10^9)e^{0.2}$$
$$= (3 \times 10^9) \times 1.2214$$
$$= 3.6642 \times 10^9.$$

◇ **Practice Exercise 3.** Using the data of Example 3, find the Earth's estimated population in 1980.

Answer. 4.4755×10^9.

The graph in Figure 6.6 illustrates the Earth's estimated population based on the information of Example 3 and Practice Exercise 3.

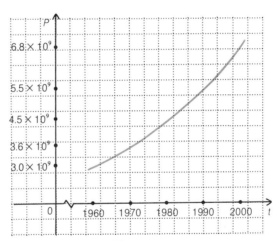

Figure 6.6

Doubling Time

Every exponential growth problem can be formulated using the concept of *doubling time*. This is the time interval required for an exponentially growing quantity to double in size.

Assume that T is the doubling time of a certain population and that P_0 is the initial population. Then formula (6.4) can be rewritten as an exponential with base 2 in the following way. If it takes a time interval T for the initial population P_0 to double, then according to formula (6.4), we have

$$2P_0 = P(T) = P_0 e^{rT}$$

or

$$e^{rT} = 2,$$

hence

$$e^r = 2^{1/T}.$$

Substituting this expression for e^r in formula (6.4), we obtain

$$(6.5) \quad P(t) = P_0 2^{t/T}$$

a formula giving the population at time t in terms of the doubling time T.

◆ **Example 4.**   A culture of bacteria growing exponentially doubles every 4 days. If the initial number of bacteria was 1500, find   **a)** the number of bacteria after 8 days, **b)** the number of bacteria after 10 days.

Solution.   **a)** Since the doubling time is 4 days and 8 is a multiple of 4, we can make the following table in order to solve the first part of the example.

t days	$P(t)$
0	1500 [initial number]
4	3000 [first doubling time]
8	6000 [second doubling time]

Thus, after 8 days the number of bacteria is 6000.

b) For the second part, we use formula (6.5) with $P_0 = 1500$ and $T = 4$:

$$P(t) = 1500 \cdot 2^{t/4}.$$

Thus,

$$\begin{aligned}
P(10) &= 1500 \cdot 2^{10/4} \\
&= 1500 \cdot 2^{5/2} \\
&= 1500 \cdot \sqrt{2^5} \\
&= 1500 \cdot 2^2 \cdot \sqrt{2} \\
&= 60000\sqrt{2} \\
&\simeq 8485.
\end{aligned}$$

◇ **Practice Exercise 4.**   In a certain culture, the number of bacteria is doubling every day. If the original number of bacteria was 1200, find the number of bacteria after 3 days. What is the number after $4\frac{1}{2}$ days?

Answer.   9600; 27,153.

Radioactive Decay

It has been observed that a radioactive material *decays exponentially* at a rate proportional to the amount of material present at a given time. If Q_0 is the initial amount of material and $r < 0$ is the *rate of decay per unit of time*, then

the formula

$$(6.6) \quad Q(t) = Q_0 e^{rt}$$

gives the amount of material remaining at time t.

HISTORICAL NOTE

Radioactivity

In 1896, A. H. Becquerel (1852–1908), a French physicist, discovered radioactivity in uranium. He noted that if salts of uranium were brought close to an unexposed photographic plate that was protected from light, the plate became exposed in spite of its protective wrapping. Becquerel's work on radioactivity was extended by Marie and Pierre Curie, who discovered the highly radioactive element radium. The Curies shared with Becquerel the 1903 Nobel Prize in Physics.

c ◆ Example 5. Carbon-14 decays at a rate of 0.012% per year. How much of a 60 gram sample will be left in 1000 years?

Solution. Since the rate of decay is 0.012% per year, then $r = -0.012/100 = -0.00012$. According to formula (6.6), we write

$$Q(1000) = 60 \cdot e^{-0.00012 \times 1000}$$
$$= 60 \cdot e^{-0.12}$$
$$= 60 \times 0.8869204$$
$$= 53.215226$$
$$\simeq 53 \text{ grams.}$$

c ◇ Practice Exercise 5. A laboratory worker places in a container 100 mg of thorium-234, a radioactive element whose rate of decay is 2.88% per day. Find the amount left after 30 days.

Answer. 42 mg.

Half-Life

The rate of decay of a radioactive element is often measured in terms of the *half-life* of the element. This is the time interval required for any initial amount of the element to be reduced by half.

If the half-life T of a radioactive element is known, then formula (6.6) can be rewritten as an exponential with base $1/2$, in the following way. If it takes a time interval T for an initial amount Q_0 of material to be reduced by half, then from formula (6.6) we have

$$\frac{Q_0}{2} = Q_0 e^{rT}$$

or

$$e^{rT} = \frac{1}{2},$$

so

$$e^r = \left(\frac{1}{2}\right)^{1/T}.$$

Substituting in (6.6), we obtain

$$(6.7)\quad Q(t) = Q_0\left(\frac{1}{2}\right)^{t/T},$$

a formula giving the amount at time t of a radioactive element whose half-life is T.

When the half-life of an element is given, the amount of the element can be easily computed, without the help of the above formulas, at periods of time that are multiples of the half-life.

◆ **Example 6.**   Radioactive iodine has a half-life of 25 minutes. If the initial amount of iodine was 100 mg, find   **a)** the amount of iodine left after 50 minutes,   **b)** the amount of iodine left after 62.5 minutes.

Solution.   **a)** Since half of the amount of iodine decays in 25 minutes, and half of the remaining amount decays in another 25 minutes, the following table suffices to solve the first part of this example.

t min	$Q(t)$ mg
0	100
25	50
50	25

Thus, the amount of iodine remaining after 50 minutes is 25 mg.
   b) Here, we use formula (6.7) with $T = 25$ and $Q_0 = 100$.

$$Q(t) = 100\left(\frac{1}{2}\right)^{t/25}$$

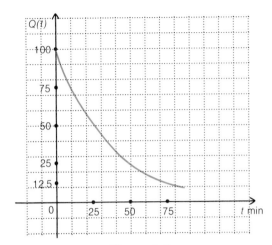

Figure 6.7

represents the amount of iodine at t minutes. Setting $t = 62.5$, we obtain

$$Q(t) = 100\left(\frac{1}{2}\right)^{62.5/25}$$

$$= 100\left(\frac{1}{2}\right)^{5/2}$$

$$= 100\left(\frac{1}{\sqrt{2^5}}\right)$$

$$= 100\left(\frac{1}{4\sqrt{2}}\right)$$

$$= \frac{25}{\sqrt{2}}$$

$$\simeq 17.67767$$

$$\simeq 18 \text{ mg.}$$

◇ **Practice Exercise 6.**   The half-life of polonium-128 is 3 minutes. What amount of a 120 gram sample of polonium-128 will be left after 9 minutes? After 4.5 minutes?

Answer.   15 grams, 42 grams.

Carbon Dating

Radioactive decay is used in archeology to determine the age of fossils. When cosmic rays strike nitrogen-14 in the upper atmosphere, a radioactive form of carbon is produced. This is carbon-14 (^{14}C), whose half-life is 5750 years. Carbon-14 has two more neutrons in its nucleus than ordinary carbon-12 (^{12}C); in radioactive decay, ^{14}C loses those two neutrons and becomes ^{12}C. All living creatures absorb and metabolize both ordinary and radioactive carbon at a constant ratio: roughly one atom of carbon-14 to 10^{12} atoms of carbon-12. As long as a plant or an animal is alive, the level of carbon-14 in its tissue remains constant. When death occurs, the absorption process stops and the level of carbon-14 diminishes with time. By measuring the amount of carbon-14 in an organic sample of a fossil, it is possible to determine the date of its death. This method is called *carbon-14 dating.*

◆ Example 7.   Tests show that an animal skeleton contains 1/8 of the original amount of ^{14}C. How old is the skeleton?

Solution.   Since the half-life of ^{14}C is 5750 years, formula (6.7) yields

$$Q(t) = Q_0\left(\frac{1}{2}\right)^{t/5750}.$$

Now, we must find out how long it takes for an initial amount Q_0 of carbon-14 to be reduced to $Q_0/8$. We have

$$\frac{Q_0}{8} = Q_0\left(\frac{1}{2}\right)^{t/5750}$$

or, after simplification,

$$\left(\frac{1}{2}\right)^3 = \left(\frac{1}{2}\right)^{t/5750}.$$

The last equation implies that

$$3 = \frac{t}{5750}$$

or

$$t = 5750 \times 3$$
$$= 17250 \text{ years.}$$

Could you have solved Example 7 without using formulas (6.6) or (6.7)? Explain.

◇ Practice Exercise 7.   An animal bone found in an archeological site contains 25% of the original amount of carbon-14. How long ago did the animal die?

Answer.   11,500 years.

Illustration Table

Half-life of certain radioactive elements

Element	*Half-life*
Bismuth-214	19.7 minutes
Polonium-218	3 minutes
Rubidium-87	6×10^{10} years
Thorium-234	24 days
Uranium-238	4.5×10^9 years

EXERCISES 6.2

In the exercises that follow, exponential behavior is assumed. Use a table or calculator when necessary.

1. The sum of $10,000 is invested at a rate of 7.5% per year compounded quarterly. Find the amount of principal at the end of a) one year, b) two years. If the interest is compounded continuously, find the principal at the end of the same periods of time.

2. If $1200 is invested at an annual rate of 6%, find a) the amount of money at the end of 5 years if the interest is compounded monthly; b) the amount of money at the end of 5 years if the interest is compounded continuously.

3. How much would have to be invested at 6% today, compounded yearly, to have $8000 in 4 years?

4. How much would have to be invested at a rate of 7.5% per year, compounded quarterly, to have $1500 in 3.5 years?

5. In 1790 the population of the United States was 4 million. From 1790 to 1960 the average rate of growth of the population was 2.24% per year. Find the population of the United States in 1960.

6. In 1960 the population of a certain country was 100 million. If the rate of growth is 3% per year, what will the country's population be in the year 2000?

7. It is estimated that the population of a certain city is given by

$$P(t) = 20,000(1.02)^t$$

where t is the number of years after 1980. How large will the population be in 1983?

8. The population of a city is 1 million and is increasing at an annual rate of 3%. Find the population after: a) 2 years; b) 5 years.

9. In a certain culture, the number of bacteria doubles every hour. If the initial number was 15,000, write a formula giving the number of bacteria after t hours. What is the number of bacteria after 3 hours?

10. In a culture the number of bacteria quadruples in 6 hours. Find the number of bacteria after 9 hours if the initial number of bacteria was 1500.

11. At sea level, the atmospheric pressure is $p_0 = 760$ millimeters of mercury. At an altitude of x kilometers above sea level, the atmospheric pressure in millimeters of mercury is given by the formula

$$p = p_0 e^{-0.11445x}.$$

Find the atmospheric pressure at an altitude of a) 2 km; b) 5 km.

12. At an altitude of h miles above sea level, the atmospheric pressure $p(h)$ in pounds per square inch is

$$p(h) = 13.7e^{-0.21h}.$$

Find the atmospheric pressure at a) 1 mile; b) 2.5 miles.

13. The half-life of radium is 1620 years. If the initial amount is 100 mg, write an equation giving the amount $Q(t)$ of radium after t years. Find the amount of radium after a) 630 years; b) 3240 years.

14. The half-life of krypton-91 is 10 seconds. If an initial amount of 500 mg is placed in a container, find the amount left after a) 15 sec; b) 20 sec.

15. A vertical beam of light with intensity I_0, when hitting the ocean surface, has intensity $I = I_0 e^{-1.4d}$ at a depth of d meters. If $I_0 = 180$ lumens, what is the intensity of the light at a depth of 4 m?

16. At a depth of 2.5 meters below the ocean surface, the intensity of a vertical beam of light is 8 lumens. What is the intensity of the beam at the ocean surface? (See Exercise 15.)

17. A sample of charcoal from an archeological site contains 6.25% of the original amount of ^{14}C. How old is the charcoal?

18. Carbon-14 decomposes at a rate of 0.012%. How much of a 120 gram sample decomposes in 2000 years?

19. Assume that it takes 15 seconds for half of a certain amount of sugar to dissolve in water and that the formula

$$A(t) = C\left(\frac{1}{2}\right)^{t/15}$$

gives the amount of sugar undissolved after t seconds. If 40 grams of sugar are placed in water, how long will it take for all but 10 grams of sugar to dissolve? How long will it take for all but 5 grams of sugar to dissolve?

20. Domestic wastewater is a mixture of biodegradable organic materials. In the presence of oxygen, such organic material decays as it is consumed by different kinds of bacteria. Under certain conditions, the decay is exponential and the amount L of organic material in wastewater

is given by

$$L = L_0 e^{-kt},$$

where the constant k, called the *reaction rate*, is about 0.2 per day for domestic water. What percentage of the initial amount of organic material will have decayed in 3.5 days?

6.3 LOGARITHMIC FUNCTIONS

Every exponential function $y = a^x$ is a one-to-one function [Section 5.4]; that is, if $x_1 \neq x_2$ then $a^{x_1} \neq a^{x_2}$. Or, in geometrical terms, each horizontal line intersects the graph of an exponential function in at most one point. Thus, we can define the inverse of an exponential function.

HISTORICAL NOTE

Logarithm
The word logarithm was introduced by John Napier (1550–1617), a Scottish mathematician. His work "Mirifici logarithmorum canonis descriptio" (A Description of the Marvelous Rule of Logarithms), published in 1641, contains the first logarithm table. Napier used logarithms to the base $(1 - 10^{-7})^{10^7}$, which is a number very close to $1/e$. For this reason, natural logarithms are also referred to as Napierian logarithms.

The Logarithmic Function

Let $a > 0$, $a \neq 1$, be a real number. The *logarithmic function with base a of x* is defined by

(6.8) $y = \log_a x$ if and only if $x = a^y$.

The domain of definition is the set of all positive real numbers, and the range is the set of all real numbers.

From (6.8) we derive the following identities:

$$a^{\log_a x} = x$$

and

$$y = \log_a(a^y),$$

which express the fact that the two functions $y = a^x$ and $y = \log_a x$ are inverses of each other.

Also, (6.8) tells us that logarithmic equations can be converted into exponential equations and vice-versa. For example:

Logarithmic form	*Exponential form*
$2 = \log_2 4$	$4 = 2^2$
$\log_{10} 1000 = 3$	$1000 = 10^3$
$\log_5\left(\dfrac{1}{25}\right) = -2$	$\dfrac{1}{25} = 5^{-2}$
$\log_e x = 4$	$x = e^4$

Example 1. Rewrite each expression in logarithmic form: **a)** $2^4 = 16$, **b)** $5^{-3} = 1/125$.

Solution. **a)** According to (6.8),

$$2^4 = 16 \quad \text{if and only if} \quad \log_2 16 = 4.$$

b) Analogously,

$$5^{-3} = \frac{1}{125} \quad \text{if and only if} \quad \log_5\left(\frac{1}{125}\right) = -3.$$

◇ **Practice Exercise 1.** Write as logarithms the following exponentials: **a)** $3^5 = 243$, **b)** $4^{-3} = 1/64$.

Answer. **a)** $\log_3 243 = 5$, **b)** $\log_4\left(\frac{1}{64}\right) = -3$.

Example 2. Find each of the following values: **a)** $\log_2 8$, **b)** $\log_{10}(0.001)$.

Solution. **a)** Set $y = \log_2 8$. According to (6.8), $y = \log_2 8$ is equivalent to $8 = 2^y$. Since $8 = 2^3$, we can rewrite $8 = 2^y$ as $2^3 = 2^y$, from which it follows that $y = 3$. Thus, $\log_2 8 = 3$.
  b) We have

$$\log_{10}(0.001) = y \quad \text{exactly when} \quad 0.001 = 10^y.$$

Since $0.001 = 10^{-3}$, the last expression can be rewritten as $10^{-3} = 10^y$. Thus, $y = -3$ and $\log_{10}(0.001) = -3$.

◇ **Practice Exercise 2.** Find the following numbers: **a)** $\log_4 2$, **b)** $\log_3(1/27)$.

Answer. **a)** $\frac{1}{2}$, **b)** -3.

Graphing Logarithmic Functions

As was explained in Section 5.4, the graph of $f^{-1}(x)$, the inverse of a function $f(x)$, is the reflection of the graph of f about the main diagonal, that is, the line $y = x$. Thus, the graph of $y = \log_a x$, is the reflection of the graph of $y = a^x$ about the line $y = x$. We have two cases to consider, depending on whether $a > 1$ or $0 < a < 1$.
  The graph of $y = \log_a x$, when $a > 1$, is illustrated in Figure 6.8. In the same coordinate plane, we have sketched the graph of a^x (light color) and the graph of $\log_a x$ (bright color).

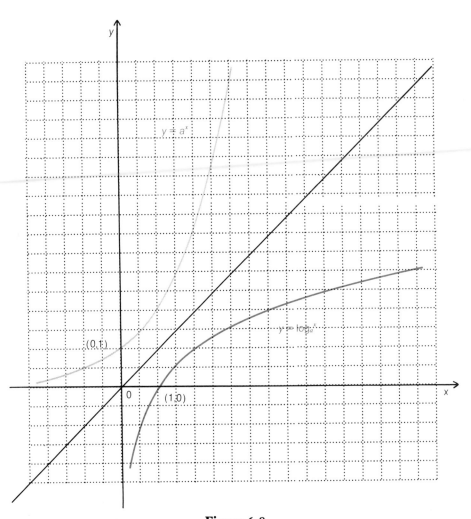

Figure 6.8

The graph shows the following important properties of the function $f(x) = \log_a x$, when $a > 1$.

1. The logarithmic function—defined only for positive values of x—is an increasing function of x; that is, if $x < y$, then $\log_a x < \log_a y$.
2. The x-intercept of the graph is 1. In other words, $\log_a 1 = 0$ because $a^0 = 1$.
3. If $0 < x < 1$, then $\log_a x < 0$; and, as x approaches 0, $\log_a x$ decreases without bound.
4. As x increases without bound, $\log_a x$ increases without bound.

We rarely use $\log_a x$ when $0 < a < 1$. But it is useful to know the graph of $f(x) = \log_a x$, when $0 < a < 1$. This is illustrated in Figure 6.9, where the graphs of $g(x) = a^x$, $0 < a < 1$, and its inverse (the logarithmic function) are shown.

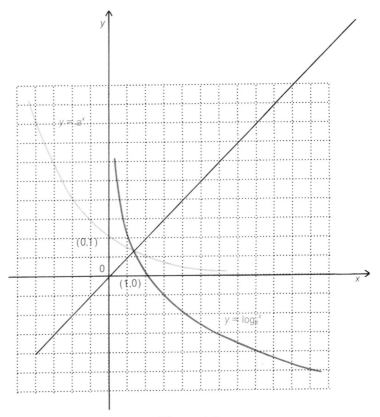

Figure 6.9

When $0 < a < 1$, the following properties of the function
$f(x) = \log_a x$ hold:
1. The logarithmic function—defined only for positive values of
   x—is a decreasing function of x; that is, if $x < y$, then
   $\log_a x > \log_a y$.
2. The x-intercept of the graph is 1, because $\log_a 1 = 0$.
3. If $0 < x < 1$, then $\log_a x > 0$. Moreover, as x approaches 0, the
   logarithm function increases without bound.
4. As x increases without bound, the function decreases without
   bound.

Notice that since exponential functions are one-to-one, logarithmic functions are also one-to-one.

Common Logarithms

When the base is the number 10, we write $\log x$ instead of $\log_{10} x$. Logarithms
to the base 10 are called *common logarithms*. Before the development of electronic calculators and computers, common logarithms were extensively em-
ployed in numerical computations, due to the fact that we use the decimal

system for writing numerals. Nowadays, common logarithms are rarely used in numerical computations. However, they are still found in many applications, such as measuring the acidity or basicity of chemical solutions, the loudness of sound, the magnitude of earthquakes, and the magnitude of stars.

Natural Logarithms

When the base is the number e, we write $\ln x$ (read "natural log of x") instead of $\log_e x$. Logarithms to the base e are called *natural* or *Napierian logarithms*, after John Napier (1550–1617), a Scottish mathematician who invented logarithms. Natural logarithms play an important role in theoretical work in mathematics and the applied sciences.

◆ **Example 3.** Evaluate x in each of the following:
a) $\log 1000 = x$ **b)** $\ln e^5 = x$
c) $\log_2 x = 3$ **d)** $\log_x 64 = 3$

Solution. By using (6.8), transform each logarithmic equation into an equivalent one involving exponents.
a) $\log 1000 = x$ if and only if $1000 = 10^x$, thus $x = 3$.
b) $\ln e^5 = x$ if and only if $e^5 = e^x$, so $x = 5$.
c) $\log_2 x = 3$ if and only if $x = 2^3 = 8$.
d) $\log_x 64 = 3$ if and only if $64 = x^3$, so $x = 4$.

◇ **Practice Exercise 3.** Find x such that
a) $\log_2 128 = x$ **b)** $\ln x = 2$
c) $\log x = 4$ **d)** $\log_x 125 = 3$

Answer. **a)** $x = 7$ **b)** $x = e^2$ **c)** $x = 10{,}000$ **d)** $x = 5$

Using Your Calculator

If your calculator has the keys $\boxed{\log}$ and $\boxed{\ln}$, it is very easy to obtain numerical values for the functions $\log x$ and $\ln x$. You simply enter x and press the desired key. For example, if $x = 5.92$, then entering 5.92 and pressing the $\boxed{\log}$ key will give you

$$\log 5.92 \simeq 0.7723217.$$

If you enter 5.92 and press the $\boxed{\ln}$ key, you obtain

$$\ln 5.92 \simeq 1.7783364.$$

Some calculators have keys for each of the functions e^x and $\ln x$. Others do not have a key for e^x. If this is the case, you must calculate e^x as the *inverse* of $\ln x$, using the keys $\boxed{\text{INV}}\,\boxed{\ln}$. For example, to find $e^{2.81}$, enter 2.81 and press the keys $\boxed{\text{INV}}\,\boxed{\ln}$, obtaining $e^{2.81} \simeq 16.609918$.

Properties of Logarithms

The properties of the exponential functions listed in Section 6.1 imply properties of the logarithmic functions, which are of fundamental importance in both practical and theoretical work with logarithms.

I. The logarithm of a product is the sum of the logarithms:

$$(6.9) \quad \log_a(mn) = \log_a m + \log_a n.$$

To prove this, set

$$u = \log_a m \quad \text{and} \quad v = \log_a n.$$

According to (6.8),

$$m = a^u \quad \text{and} \quad n = a^v.$$

Multiplying m and n, and using property I of exponential functions, we get

$$mn = a^u a^v = a^{u+v}.$$

From (6.8) it follows that

$$u + v = \log_a(mn).$$

Substituting $\log_a m$ for u and $\log_a n$ for v, we obtain (6.9).

II. The logarithm of the reciprocal of a number is the additive inverse of the logarithm of the number:

$$(6.10) \quad \log_a \frac{1}{n} = -\log_a n.$$

Indeed, if $u = \log_a n$, then $n = a^u$ and vice-versa. By property II of exponential functions,

$$a^{-u} = \frac{1}{a^u} = \frac{1}{n}.$$

It follows from (6.8) that

$$-u = \log_a \frac{1}{n}.$$

Substituting $\log_a n$ for u, we obtain

$$-\log_a n = \log_a \frac{1}{n}$$

which is relation (6.10).

Combining properties I and II, we obtain the following property.

III. The logarithm of a quotient is the difference of the logarithms:

$$(6.11) \quad \log_a \frac{m}{n} = \log_a m - \log_a n.$$

The proof is left as an exercise.

IV. The logarithm of a power of a number is the exponent of that power multiplied by the logarithm of the number:

$$(6.12) \quad \log_a(m^r) = r \log_a m.$$

This proof is also left as an exercise.

Since a radical is a fractional power, from property IV we derive the following property concerning the logarithm of a radical.

$$(6.13) \quad \log_a \sqrt[n]{m} = \frac{1}{n} \log_a m.$$

◆ **Example 4.**   If $\log 2 = 0.3010$ and $\log 3 = 0.4771$, find the numbers

a) $\log 6$ **b)** $\log \dfrac{2}{3}$

c) $\log 81$ **d)** $\log \sqrt[5]{3}$.

Solution.   **a)** Since $6 = 2 \cdot 3$, we write

$$\begin{aligned}
\log 6 &= \log(2 \cdot 3) \\
&= \log 2 + \log 3 \qquad \text{[Property I]} \\
&= 0.3010 + 0.4771 \\
&= 0.7781.
\end{aligned}$$

b) By property III

$$\begin{aligned}
\log \frac{2}{3} &= \log 2 - \log 3 \\
&= 0.3010 - 0.4771 \\
&= -0.1761.
\end{aligned}$$

c) Since $81 = 3^4$, then according to property IV,

$$\begin{aligned}
\log 81 &= \log 3^4 \\
&= 4 \log 3 \\
&= 4 \times 0.4771 \\
&= 1.9084.
\end{aligned}$$

d) Using (6.13), we obtain

$$\log \sqrt[5]{3} = \frac{1}{5} \log 3$$

$$= \frac{1}{5} \, 0.4771$$

$$= 0.0954.$$

A word of caution. Notice that

$$\log(2 \cdot 3) \neq (\log 2)(\log 3).$$

From part **a)** of Example 4, $\log(2 \cdot 3) = 0.7781$. However, $(\log 2)(\log 3) = (0.3010)(0.4771) = 0.1436071$.

Also, $\log(2/3) \neq \log 2/\log 3$, because $\log(2/3) = -0.1761$ and $\log 2/\log 3 = 0.3010/0.4771 = 0.6308949$.

◇ **Practice Exercise 4.**   Given $\log 4 = 0.6021$ and $\log 5 = 0.6990$, find the numbers

a) $\log 20$ **b)** $\log \dfrac{5}{4}$

c) $\log 4^5$ **d)** $\log \sqrt[5]{4}$

Answer.   **a)** 1.3011,   **b)** 0.0969,   **c)** 3.0105,   **d)** 0.12042.

◆ **Example 5.**   **a)** Express $\log_a(x^5 y^2 / \sqrt[3]{z})$ in terms of $\log_a x$, $\log_a y$, and $\log_a z$.
  b) Write $\log \sqrt[4]{x^2(x-1)/(x-2)^3}$ in terms of $\log x$, $\log(x-1)$, and $\log(x-2)$.

Solution.   **a)** Using the properties of logarithms described above, we get

$$\log_a(x^5 y^2 / \sqrt[3]{z}) = \log_a(x^5 y^2) - \log_a \sqrt[3]{z}$$

$$= \log_a x^5 + \log_a y^2 - \log_a \sqrt[3]{z}$$

$$= 5 \log_a x + 2 \log_a y - \frac{1}{3} \log_a z.$$

b) Analogously, we have

$$\log \sqrt[4]{x^2(x-1)/(x-2)^3} = \frac{1}{4} \log x^2[(x-1)/(x-2)^3]$$

$$= \frac{1}{4} \{ \log[x^2(x-1)] - \log(x-2)^3 \}$$

$$= \frac{1}{4} \{ \log x^2 + \log(x-1) - \log(x-2)^3 \}$$

$$= \frac{1}{4} \{ 2 \log x + \log(x-1) - 3 \log(x-2) \}$$

$$= \frac{1}{2} \log x + \frac{1}{4} \log(x-1) - \frac{3}{4} \log(x-2).$$

◇ **Practice Exercise 5.**   Use the properties of logarithms to write $\log m^3 \sqrt{(x-1)/(y+2)^4}$ in terms of $\log m$, $\log(x-1)$, and $\log(y+2)$.

Answer. $3 \log m + \dfrac{1}{2} \log(x-1) - 2 \log(y+2)$

◆ **Example 6.**   Write each of the following expressions as a single logarithm:

a) $2 \log 5 - \log 7 + \log 4 - \log 3$,   **b)** $\ln(x+1) - 3 \ln x + \dfrac{1}{2} \ln(x+2)$.

Solution.   **a)** We have

$$
\begin{aligned}
2 \log 5 - \log 7 + \log 4 - \log 3 &= \log 5^2 + \log 4 - (\log 7 + \log 3) \\
&= \log(5^2 \cdot 4) - \log(7 \cdot 3) \\
&= \log 100 - \log 21 \\
&= \log \frac{100}{21}.
\end{aligned}
$$

b) Also

$$
\begin{aligned}
\ln(x+1) - 3 \ln x + \frac{1}{2} \ln(x+2) &= \ln(x+1) - \ln x^3 + \ln \sqrt{x+2} \\
&= \ln[(x+1)\sqrt{x+2}] - \ln x^3 \\
&= \ln \frac{(x+1)\sqrt{x+2}}{x^3}.
\end{aligned}
$$

◆ **Practice Exercise 6.**   **Write as a single logarithm the expression**

$$
\ln 15 + 5 \ln x - 2 \ln(x+1) + \frac{1}{2} \ln(x-1).
$$

Answer. $\ln \dfrac{15x^5 \sqrt{x-1}}{(x+1)^2}.$

EXERCISES 6.3

Use the definition of logarithms to find each of the following values.

1. $\log_2 16$

2. $\log_3 27$

3. $\log_2 \dfrac{1}{8}$

4. $\log_5 125$

5. $\log 10^4$

6. $\ln e^{1/2}$

7. $\log(0.1)^2$

8. $\log \dfrac{1}{1000}$

9. $\ln\left(\dfrac{1}{e}\right)^3$

10. $\log_4 \dfrac{1}{64}$

In Exercises 11–18, rewrite each expression in logarithmic form.

11. $2^5 = 32$ **12.** $3^4 = 81$ **13.** $4^{-2} = \dfrac{1}{16}$ **14.** $5^{-4} = \dfrac{1}{625}$

15. $\left(\dfrac{1}{3}\right)^4 = \dfrac{1}{81}$ **16.** $\left(\dfrac{1}{2}\right)^{-6} = 64$ **17.** $a^b = c$ **18.** $u^{-v} = w$

In Exercises 19–26, rewrite each expression in exponential form.

19. $\log_2 128 = 7$ **20.** $\log_3 243 = 5$ **21.** $\log_2 \dfrac{1}{1024} = -10$ **22.** $\log_5 \dfrac{1}{3125} = -5$

23. $\log 100 = 2$ **24.** $\log 1000 = 3$ **25.** $\log q = r$ **26.** $\log_n m = -p$

In Exercises 27–40, given $\log 2 \simeq 0.30$, $\log 3 \simeq 0.48$, $\log 4 \simeq 0.60$, *and* $\log 5 \simeq$
0.70, find approximations of the following numbers.

27. $\log 10$ **28.** $\log 12$ **29.** $\log 18$ **30.** $\log 15$

31. $\log 20$ **32.** $\log 60$ **33.** $\log \dfrac{3}{5}$ **34.** $\log \dfrac{5}{2}$

35. $\log \dfrac{10}{3}$ **36.** $\log \dfrac{4}{15}$ **37.** $\log 8$ **38.** $\log 16$

39. $\log \sqrt{10}$ **40.** $\log \sqrt[3]{15}$

Find x in Exercises 41–50.

41. $\log_6 x = 3$ **42.** $\log_4 x = 0$ **43.** $\log_{3/2} x = 4$ **44.** $\log_{2/3} x = 3$
45. $\log_x 4 = 5$ **46.** $\log_x 216 = 3$ **47.** $\log_{2x} 8 = 3$ **48.** $\log_{3x} 225 = 2$
49. $\log_7 49 = x$ **50.** $\log_5 125 = x$

In Exercises 51–60, write each expression as a single logarithm.

51. $\log 8 + \log 6 - \log 12$ **52.** $\ln 12 - \ln 4 + \ln 5$

53. $3 \log_5 2 + 2 \log_5 3 - \dfrac{1}{2} \log_5 4$ **54.** $2 \log 8 - 5 \log 2 - \dfrac{1}{3} \log 27$

55. $2 \log x + \dfrac{1}{3} \log y - 4 \log x$ **56.** $\dfrac{1}{4} \log x - \dfrac{1}{2} \log y + \dfrac{1}{4} \log z$

57. $\log(x + 1) - \log(x - 2) + 2 \log x$ **58.** $3 \log x - \dfrac{1}{2} \log(x + 1) - \log(2x - 1)$

59. $3 \log(x + 2) + 2 \log x - 4 \log(x + 1)$ **60.** $4 \log(x + 1) + 2 \log(x - 1) - \dfrac{1}{2} \log(x + 3)$

In Exercises 61–66, write each expression in terms of $\log x$ *and* $\log(x - 2)$.

61. $\log \dfrac{x^2}{(x - 2)^3}$ **62.** $\log \dfrac{\sqrt{x}}{(x - 2)^2}$ **63.** $\log[x^2(x - 2)^4]$

64. $\log \sqrt{\dfrac{x^3}{x - 2}}$ **65.** $\log \sqrt{(x - 2)^3 x^4}$ **66.** $\log \left(\dfrac{x}{x - 2}\right)^4$

In Exercises 67–70, write each expression in terms of $\log x$, $\log(x + 1)$ *and* $\log(x - 2)$.

67. $\log \dfrac{x^2(x + 1)^3}{(x - 2)^5}$

68. $\log \dfrac{x(x - 2)^3}{\sqrt{x + 1}}$

69. $\log\left(\dfrac{1}{x}\right)\sqrt[3]{\dfrac{x + 1}{(x - 2)^2}}$

70. $\log \sqrt{\dfrac{x^2(x + 1)^3}{(x - 2)^5}}$

6.4 COMPUTING COMMON LOGARITHMS; APPLICATIONS OF COMMON LOGARITHMS

In the past, common logarithms constituted an important computational tool. They were widely used in astronomy, engineering, and statistics to reduce the computational work of certain numerical expressions. For this purpose, the use of tables, such as Table 2 in the Appendix, was absolutely necessary.

Today the importance of tables has considerably declined. Inexpensive calculators can perform power and root extraction operations, and compute exponential, logarithmic, and trigonometric functions with great accuracy and speed.

Despite this fact, it is useful for students to learn how to use tables. You may be in a situation where no calculator is available, or the teacher may feel that, at this stage, calculators should not be used.

Figure 6.10 reproduces a portion of Table 2 found in the Appendix. The table gives approximations of common logarithms to four decimal places. Suppose that we want to find the common logarithm of 592. First, we write the number in scientific notation

$$592 = 5.92 \times 10^2.$$

Next, use properties of logarithms to obtain

$$
\begin{aligned}
\log 592 &= \log(5.92 \times 10^2) \\
&= \log 5.92 + \log 10^2 &&\text{[Property I of logarithms]} \\
&= \log 5.92 + 2 &&\text{[}\log 10^2 = 2\text{]}
\end{aligned}
$$

n	0	1	2	3	4	5	6	7	8	9
5.5	.7404	.7412	.7419	.7427	.7435	.7443	.7451	.7459	.7466	.7474
5.6	.7482	.7490	.7497	.7505	.7513	.7520	.7528	.7536	.7543	.7551
5.7	.7559	.7566	.7574	.7582	.7589	.7597	.7604	.7612	.7619	.7627
5.8	.7634	.7642	.7649	.7657	.7664	.7672	.7679	.7686	.7694	.7701
5.9	.7709	.7716	.7723	.7731	.7738	.7745	.7752	.7760	.7767	.7774
6.0	.7782	.7789	.7796	.7803	.7810	.7818	.7825	.7832	.7839	.7846
6.1	.7853	.7860	.7868	.7875	.7882	.7889	.7896	.7903	.7910	.7917
6.2	.7924	.7931	.7938	.7945	.7952	.7959	.7966	.7973	.7980	.7987
6.3	.7993	.8000	.8007	.8014	.8021	.8028	.8035	.8041	.8048	.8055
6.4	.8062	.8069	.8075	.8082	.8089	.8096	.8102	.8019	.8116	.8122

Figure 6.10

Now, we have to find the value of log 5.92. Referring to the table, first locate the number 5.9 in the left column and then move across until you reach the column headed by 2, which is the last digit of 5.92, to obtain

$$\log 5.92 \simeq 0.7723.$$

Therefore,

$$\log 592 = \log 5.92 + 2$$
$$\simeq 0.7723 + 2$$
$$= 2.7723.$$

Observe that in our calculations, we have obtained an *integral part*, 2, called the *characteristic* of the logarithm, and a *decimal number between 0 and 1*, 0.7723, called the *mantissa* of the logarithm.

In general, if a number n is written in scientific notation as

$$n = m \times 10^k,$$

where m is a number such that $1 \leq m < 10$ and k is an integer, then the common logarithm of n is

$$\log n = \log(m \times 10^k)$$
$$= \log m + \log 10^k$$
$$= \log m + k.$$

The integer k is the *characteristic* and the number $\log m$ is the *mantissa* of $\log n$. Since $\log 1 = 0$, $\log 10 = 1$, and the common logarithm is an increasing function, we have

$$0 \leq \log m < 1;$$

that is, the mantissa of a logarithm is always a nonnegative number less than 1.

◆ **Example 1.** Find each of the following: **a)** log 57,200, **b)** log 0.0583. Determine the characteristic and mantissa of each logarithm.

Solution. **a)** We have

$$\log 57{,}200 = \log(5.72 \times 10^4) \qquad \text{[Property I of logarithms]}$$
$$= \log 5.72 + \log 10^4 \qquad \text{[Using the table]}$$
$$\simeq 0.7574 + 4$$
$$= 4.7574.$$

The characteristic of log 57,200 is 4 and the mantissa is 0.7574.

b)
$$\log 0.0583 = \log(5.83 \times 10^{-2})$$
$$= \log 5.83 + \log 10^{-2}$$
$$\simeq 0.7657 - 2$$
$$= -1.2343.$$

The characteristic of log 0.0583 is -2 and the mantissa is 0.7657.

If you are using a calculator, it is not necessary to convert the given numbers into scientific notation. The sequence of keystrokes

$$57,200 \; \boxed{\log} \; \boxed{=}$$

will give you log $57,200 \simeq 4.757396$. Also, using your calculator you can check that

$$\log 0.0583 \simeq -1.2343314.$$

◇ **Practice Exercise 1.**   Approximate the following numbers: **a)** log 6240,   **b)** log 0.00635.

Answer.   **a)** log $6240 \simeq 3.7952$,   **b)** log $0.00635 \simeq -2.1972$.

Antilogarithms

Now, consider the opposite problem: suppose we are given the logarithm of a number and want to find the number or an approximation of it.

◆ **Example 2.**   Find x such that $\log x = 5.7664$. The number x is called the *antilogarithm* of 5.7664.

Solution.   First, write

$$\log x = 5.7664$$
$$= 0.7664 + 5$$

in order to determine the characteristic, 5, and the mantissa, 0.7664. Next, according to the table (Figure 6.10), log $5.84 \simeq 0.7664$, so

$$\log x = 0.7664 + 5$$
$$\simeq \log 5.84 + \log 10^5 \qquad [5 = \log 10^5]$$
$$= \log(5.84 \times 10^5) \qquad \text{[Property I of logarithms]}.$$

Hence

$$x \simeq 5.84 \times 10^5.$$

◇ **Practice Exercise 2.**   If $\log x = 2.7832$, then find an approximation for x.

Answer. $x \simeq 607$.

Before using the table, make sure that the mantissa of the logarithm is a number between 0 and 1, and that the characteristic is an integer.

◆ **Example 3.**   Find x if $\log x = -1.2396$.

Solution.   As before, we write

$$\log x = -1.2396$$
$$= -0.2396 - 1.$$

Since -0.2396 is negative, it *cannot* be the mantissa of $\log x$. On the other hand, the mantissa of a logarithm is always a number ≥ 0 and <1. Thus we have to express the right-hand side of the last equality as a sum of a number ≥ 0 and <1 and an integer. This can be achieved by adding and subtracting 2 to the right-hand side of that equality:

$$\log x = (2 - 1.2396) - 2$$
$$= 0.7604 - 2.$$

The number 0.7604 is the mantissa of $\log x$ and -2 is its characteristic. Referring to the table (Figure 6.10), we see that $\log 5.76 \simeq 0.7604$. Thus,

$$\log x = 0.7604 - 2$$
$$\simeq \log 5.76 + \log 10^{-2}$$
$$= \log(5.76 \times 10^{-2});$$

hence

$$x \simeq 5.76 \times 10^{-2}.$$

◇ **Practice Exercise 3.**   Determine x if $\log x = -2.1952$.

Answer. $x \simeq 6.38 \times 10^{-3}$.

If you are using a calculator, it is very simple to compute the antilogarithm of a number. For example, to find x in Example 2, enter 5.7664 and press the keys $\boxed{\text{INV}}$ $\boxed{\text{log}}$ to obtain

$$x \simeq 583982.72$$

or, in scientific notation,

$$x \simeq 5.8398 \times 10^5.$$

To solve Example 3 with a calculator, enter 1.2396, and press the keys $\boxed{+/-}$, $\boxed{\text{INV}}$, and $\boxed{\text{log}}$. You will get

$$x \simeq 0.057597,$$

or

$$x \simeq 5.7597 \times 10^{-2}.$$

All the numbers considered in Examples 1, 2, and 3 were found in the table of logarithms. If a number is not found in the table, then the method of *linear interpolation*, discussed in the Appendix, can be used to approximate the logarithm or antilogarithm of a number. When speed and more accuracy are desired, a calculator should be used.

Applications of Common Logarithms

Logarithmic functions are used in chemistry, physics, psychology, and other applied sciences.

For example, the acidity or basicity of a chemical solution, denoted by pH, is defined by the formula

$$\text{pH} = -\log[\text{H}^+]$$

where $[H^+]$ is the hydrogen ion concentration in moles per liter. Acid solutions have pH < 7 and basic solutions have pH > 7.

◆ **Example 4.**   **a)** Find the pH of eggs, knowing that $[H^+] = 1.6 \times 10^{-8}$ moles per liter.
 c   **b)** Find the hydrogen ion concentration, in moles per liter, of tomatoes whose pH is 4.2.

Solution.   **a)** We have

$$
\begin{aligned}
\text{pH} &= -\log(1.6 \times 10^{-8}) \\
&= -\log 1.6 - \log 10^{-8} \\
&= -\log 1.6 + 8 \\
&\simeq -0.2 + 8 \\
&= 7.8.
\end{aligned}
$$

b) If pH $= 4.2$, then $\log [H^+] = -4.2$. Hence, using a calculator,

$$[H^+] \simeq 6.31 \times 10^{-5}.$$

◇ **Practice Exercise 4.**   **a)** Apricots have a pH of 3.8. Find the hydrogen ion concentration in moles per liter.   **b)** Find the pH of apples if $[H^+] \simeq 8 \times 10^{-4}$ moles per liter.

Answer.   **a)** 1.6×10^{-4},   **b)** 3.1.

The human ear is sensitive to sound over an enormous range of intensity, from 10^{-16} to 10^{-4} watt/cm^2. For this reason, the loudness of sound is measured on a logarithmic scale. The sound level L, in decibels (db), of a sound intensity I is given by

$$L = 10 \log \frac{I}{I_0},$$

where $I_0 = 10^{-16}$ watt/cm^2 is the minimum detectable intensity.

◆ **Example 5.**   **a)** Find the sound level, in decibels, of ordinary conversation, where $I = 10^{-10}$ watt/cm^2.
 c   **b)** The clatter of a chain saw equals 105 db. What is the corresponding sound intensity?

Solution.   **a)** Substitute 10^{-10} for I and 10^{-16} for I_0 into the above equation, obtaining

$$
\begin{aligned}
L &= 10 \log \frac{10^{-10}}{10^{-16}} \\
&= 10 \log 10^6 \\
&= 60 \text{ db.}
\end{aligned}
$$

b) We have

$$105 = 10 \log \frac{I}{10^{-16}}$$

or
$$\log(10^{16}I) = 10.5.$$

Hence
$$10^{16}I = 10^{10.5} \simeq 3.1623 \times 10^{10},$$

or
$$I \simeq 3.1623 \times 10^{-6} \text{ watt/cm}^2.$$

◇ **Practice Exercise 5.**   **a)** The sound level inside a car moving at 55 miles per hour is 72 decibels. Find the sound intensity.   **b)** What is the sound level of whisper whose sound intensity is 10^{-14} watt/cm^2?

Answer.   **a)** 1.6×10^{-9} watt/cm^2,   **b)** 20 db.

 EXERCISES 6.4

In Exercises 1–10, use Table 2 or a calculator to approximate each of the following to four decimal places.

1. log 3.72	**2.** log 5.38	**3.** log 3250	**4.** log 1230
5. log 71300	**6.** log 10500	**7.** log 0.0225	**8.** log 0.00518
9. log 51.2	**10.** log 5120		

In Exercises 11–20, use Table 2 or a calculator to find an approximation of x.

11. log x = 3.5391	**12.** log x = 5.6513	**13.** log x = 4.9253	**14.** log x = 2.9713
15. log x = -2.2636	**16.** log x = -0.4609	**17.** log x = 1.3181	**18.** log x = -2.0141
19. log x = 5.3243 $-$ 8	**20.** log x = 2.1004 $-$ 5		

c 21. Given [H$^+$], find the pH of each substance: a) Carrots: [H$^+$] = 1.1×10^{-5}; b) Soil: [H$^+$] = 3.1×10^{-7}.

c 22. Compare the acidity of the following substances: a) Milk: [H$^+$] = 4×10^{-7}; b) Vinegar: [H$^+$] = 6.3×10^{-3}.

c 23. Find the hydrogen ion concentration [H$^+$] in each of the following substances: a) Sea water: pH $\simeq$ 7.8; b) Soil: pH $\simeq$ 5.

c 24. Gastric juice has pH $\simeq$ 1.7 and pancreatic juice has pH $\simeq$ 7.8. Determine their hydrogen ion concentrations.

c 25. Sound levels above 90 db are considered dangerous to human ears. The sound intensity of a jet airliner during take-off is approximately 10^{-4} watt/cm^2. Is the decibel level dangerous?

c 26. Determine the sound intensity corresponding to: a) 90 db; b) 110 db.

● 27. On the Richter scale, the magnitude R of an earthquake of intensity I is given by the formula

$$R = \log \frac{I}{I_0}$$

where I_0 is a minimum intensity used for comparison. Find the magnitude on the Richter scale of an earthquake of intensity $10^{6.5}I_0$.

● c **28.** If on the Richter scale the magnitude of an earthquake is 8.2, compare its intensity to the minimum intensity I_0.

● c **29.** The magnitude M of a star is defined by

$$M = -2.5 \log kI$$

where k is a constant and I is the intensity of the light from the star. Sirius, the brightest star in the sky, has magnitude -1.42, and Canopus, the second brightest star, has magnitude -0.72. What is the ratio of the intensities of light from Sirius and Canopus?

● **30.** Find the magnitude of a star whose light intensity is $1/100$ that of Sirius.

6.5   EXPONENTIAL AND LOGARITHMIC EQUATIONS; CHANGE OF BASE

In certain equations the unknown may appear as an exponent or a logarithm. To solve such equations, we use the properties of the exponential and logarithmic functions.

◆ **Example 1.**   Solve the equation $3^{2x-1} = 11$.

Solution.   Taking common logarithms of both sides, we have

$$\log 3^{2x-1} = \log 11,$$
$$(2x - 1) \log 3 = \log 11,$$
$$2x - 1 = \frac{\log 11}{\log 3}.$$

So

$$x = \frac{1}{2}\left(1 + \frac{\log 11}{\log 3}\right)$$

is the exact solution of the given equation. If an approximation is desired, then using Table 2 or a calculator, we obtain

$$x \simeq \frac{1}{2}\left(1 + \frac{1.0414}{0.4771}\right)$$
$$\simeq 1.5913.$$

◇ **Practice Exercise 1.**   **a)** Find the exact solution of the equation $4^{3x+1} = 30$.
      c   **b)** Approximate the solution to four decimal places.

Answer.   **a)** $x = \frac{1}{3}\left[\left(\frac{\log 30}{\log 4}\right) - 1\right]$ **b)** $\simeq 0.4845$.

◆ **Example 2.**   Solve the equation

$$\log(3x - 1) = 1 + \log(1 - x).$$

Solution.   The given equation can be written as follows:

$$\log(3x - 1) - \log(1 - x) = 1$$

or, using property IV of logarithms,

$$\log \frac{3x - 1}{1 - x} = 1,$$

which, according to the definition of logarithm, is equivalent to

$$\frac{3x - 1}{1 - x} = 10.$$

Solving for x, we obtain

$$3x - 1 = 10(1 - x),$$
$$13x = 11,$$
$$x = \frac{11}{13}.$$

You can check that this is the desired solution, by substituting $11/13$ for x into the original equation.

◇ **Practice Exercise 2.** Find the solution of the logarithmic equation $\log(2x - 1) - 1 = \log(x - 1)$.

Answer. $x = \dfrac{9}{8}$.

◆ **Example 3.** Solve $\log(2x - 1) + \log(x - 1) = 1$.

Solution. Use property I of logarithms, to write the given equation in its equivalent form

$$\log(2x - 1)(x - 1) = 1.$$

From definition (6.8), obtain

$$(2x - 1)(x - 1) = 10,$$
$$2x^2 - 3x + 1 = 10,$$
$$2x^2 - 3x - 9 = 0,$$
$$(2x + 3)(x - 3) = 0.$$

Hence

$$x = 3 \quad \text{or} \quad x = -\frac{3}{2}.$$

If $x = -3/2$, then $x - 1$ is negative and so $\log(x - 1)$ is not defined. The number $-3/2$ is called an *extraneous* solution, and it must be discarded. The only solution of the given equation is then $x = 3$.

◇ **Practice Exercise 3.** Solve the equation

$$\log(x + 1) = 1 - \log(2x + 1).$$

Answer. $x = \dfrac{3}{2}$.

◆ **Example 4.**   Solve the equation $(e^x - e^{-x})/2 = 1$.

Solution.   Write the given equation as

$$e^x - e^{-x} = 2.$$

Since $e^{-x} = 1/e^x$, it follows that if we denote e^x by u, then the last equation becomes

$$u - \frac{1}{u} = 2.$$

Multiplying both sides by u, we obtain the quadratic equation

$$u^2 - 1 = 2u$$

or

$$u^2 - 2u - 1 = 0.$$

Solving it for u, we obtain the solutions

$$u = 1 \pm \sqrt{2}.$$

Since $u = e^x$ is always positive, the negative solution $1 - \sqrt{2}$ must be discarded. The positive solution $1 + \sqrt{2}$ yields the exponential equation

$$e^x = 1 + \sqrt{2},$$

which can be solved by taking natural logarithms of both sides:

$$x \ln e = \ln(1 + \sqrt{2}),$$

or

$$x = \ln(1 + \sqrt{2}).$$

This is the exact solution of the given equation. If an approximation is desired, we obtain

$$x \simeq \ln 2.4142$$

$$\simeq 0.8814.$$

◇ **Practice Exercise 4.**   What is the exact solution of the exponential equation $(10^x - 10^{-x})/2 = 1$? Approximate the solution to four decimal places.

Answer. $x = \log(1 + \sqrt{2}) \simeq 0.3828$.

Doubling Time and Rate of Increase of a Population

In Section 6.2, we dealt with two formulas that describe population growth:

$$P(t) = P_0 e^{rt} \quad \text{and} \quad P(t) = P_0 2^{t/T}.$$

In the first formula, the positive number r denotes the *rate of increase per individual per unit of time*; in the second formula, T denotes the *doubling time*.

Sometimes, it is useful to derive a relation between these two quantities. Assuming that both formulas represent the growth of the same population, it follows that

$$P_0 e^{rt} = P_0 2^{t/T},$$

or

$$e^{rt} = 2^{t/T},$$
$$[e^r]^t = [2^{1/T}]^t.$$

If we set $t = 1$, then we obtain

$$e^r = 2^{1/T}.$$

By taking natural logarithms of both sides, an expression for the rate of increase in terms of the doubling time is obtained:

$$r = \frac{\ln 2}{T}$$

This shows that the rate of increase is *inversely proportional* to the doubling time, and the *constant of proportionality* is $\ln 2$.

◆ **Example 5.**   In a culture of bacteria growing exponentially, the doubling time is 4 days. What is the daily rate of increase?

Solution.   According to the above formula,

$$r = \frac{\ln 2}{4}$$
$$= \frac{0.6931471}{4}$$
$$= 0.1732868$$
$$\simeq 0.18.$$

Thus, the rate of increase is approximately 18% per day.

◇ **Practice Exercise 5.**   What is the doubling time for a population increasing at a rate of 2% per year?

Answer.   34.657 years $\simeq$ 35 years.

Half-Life and Rate of Decrease of a Radioactive Element

Two formulas were used in Section 6.2 to describe the decay of a radioactive element:

$$Q(t) = Q_0 e^{rt} \quad \text{and} \quad Q(t) = Q_0 \left(\frac{1}{2}\right)^{t/T},$$

where r (a negative number) represents the *rate of decay per unit of time* and T is the *half-life* of the radioactive element.

Proceeding in exactly the same manner as for the rate of increase of a population and the doubling time, you can check that the following relation holds:

$$r = -\frac{\ln 2}{T}$$

That is, the rate of decay of a radioactive element varies *inversely* as the half-life of the element, and the *constant of proportionality* is $-\ln 2$.

◆ **Example 6.**   What is the half-life of radioactive iodine if 2.77% of the iodine decays every minute?

Solution.   The rate of decay is $r = -2.77/100 = -0.0277$. Since

$$r = -\frac{\ln 2}{T},$$

then

$$T = -\frac{\ln 2}{r}.$$

Substituting -0.0277 for r, get

$$T = \frac{\ln 2}{0.0277}$$
$$= 25.023364$$
$$\simeq 25 \text{ minutes.}$$

◇ **Practice Exercise 6.**   The half-life of polonium-128 is 3 minutes. Determine its rate of decrease.

Answer.   23.1% per minute.

Change of Base

When dealing with logarithms it is sometimes necessary to change the base. For example, suppose that we want to determine the value of $\log_2 7$. We would need a table of logarithms to the base 2, but such a table is unavailable. On the other hand, we cannot use a calculator directly, since the only keys available are for common and natural logarithms. However, we can change the base and express $\log_2 7$ as the quotient of the common logarithms of 7 and 2. This is done as follows. If we set $x = \log_2 7$, then $2^x = 7$. Taking common logarithms of both sides gives us

$$x \log 2 = \log 7,$$

or

$$x = \frac{\log 7}{\log 2}.$$

We conclude that

$$\log_2 7 = \frac{\log 7}{\log 2}.$$

The method described in this example applies to logarithms to any base. Suppose that we want to express $\log_a x$ in terms of $\log_b x$. If we set

$$y = \log_a x,$$

then

$$a^y = x.$$

Taking logarithms to the base b of both sides gives us

$$y \log_b a = \log_b x,$$

or

$$y = \frac{\log_b x}{\log_b a}.$$

Substituting for y, we obtain the *change of base formula*

$$\log_a x = \frac{\log_b x}{\log_b a}.$$

If we set $x = b$ in the last formula and use the fact that $\log_b b = 1$, we obtain

$$\log_a b = \frac{1}{\log_b a}$$

◆ **Example 7.** Find an approximation for $\log_6 8$ using natural logarithms.

Solution. Apply the change of base formula with $x = 8$, $a = 6$, and $b = e$, to get

$$\log_6 8 = \frac{\log_e 8}{\log_e 6} = \frac{\ln 8}{\ln 6}.$$

Using Table 3 in the Appendix or a calculator, you can obtain the approximation

$$\log_6 8 \simeq \frac{2.0794}{1.7918}$$
$$= 1.1605.$$

◇ **Practice Exercise 7.** Approximate the value $\log_2 7$.

Answer. $\log_2 7 \simeq 2.807$.

EXERCISES 6.5

In Exercises 1–24, solve each of the given equations.

1. $7^x = 2$

2. $10^x = 5$

3. $6^x = 8$

4. $2^x = 16$

5. $3^{-x} = 8$

6. $2^{-x} = 10$

7. $10^{5x+1} = 21$

8. $3^{x-4} = 9$

9. $2^{3x-1} = 3^{1-2x}$

10. $5^{2-x} = 6^{x+1}$

11. $\log x - \log 2 = 1$

12. $\log x + \log 7 = 2$

13. $\log(2x + 4) = 1 + \log(x - 2)$

14. $\log x - \log(x + 1) = 2$

15. $\log(x + 1) = 1 - \log(x + 2)$

16. $\log(x^2 + 1) - \log(x - 1) = 1 + \log(x + 1)$

17. $(\log x)^2 + \log x^2 = 0$

18. $\log x^3 = (\log x)^2$

19. $\log(\log x) = 1$

20. $\log(\log x^2) = 1$

21. $\dfrac{e^x + e^{-x}}{2} = 1$

22. $\dfrac{4^x - 4^{-x}}{2} = 1$

23. $\dfrac{e^x - e^{-x}}{e^x + e^{-x}} = \dfrac{1}{2}$

24. $\dfrac{e^x + e^{-x}}{e^x - e^{-x}} = 4$

Given $\ln 2 \simeq 0.7$, $\ln 3 \simeq 1.1$, *and* $\ln 5 \simeq 1.6$, *evaluate the logarithms in Exercises 25–32 without the use of tables or calculators.*

25. $\log_5 4$

26. $\log_3 25$

27. $\log 18$

28. $\log 15$

29. $\log_6 30$

30. $\log_4 12$

31. $\log_{15} 24$

32. $\log_9 45$

Given $\log 2 \simeq 0.30$, $\log 3 \simeq 0.50$ *and* $\log 5 \simeq 0.70$, *evaluate the logarithms in Exercises 33–40 without using tables or calculators.*

33. $\log_2 9$

34. $\log_3 8$

35. $\log_3 12$

36. $\log_4 20$

37. $\log_6 24$

38. $\log_{12} 10$

39. $\log_4 \dfrac{1}{6}$

40. $\log_5 \dfrac{4}{9}$

In Exercises 41–48, using the change of base formula with Table 2, Table 3, or a calculator, find each of the following numbers to four decimal places.

41. $\log_7 8$

42. $\log_4 15$

43. $\log_2 5.4$

44. $\log_5 3.1$

45. $\log_3 1.23$

46. $\log_4 2.01$

47. $\log_4(2.3 \times 10^{-2})$

48. $\log_3(7 \times 10^{-3})$

49. If $\log_4 11 \simeq 1.7$ and $\log_4 2 \simeq 0.5$, find $\log_2 11$.

50. If $\log_3 12 \simeq 2.2619$, find $\log_{12} 3$.

51. Assume that the formula $n(t) = n_0 2^{2t}$ gives the number of bacteria in a certain culture at time t. If $n_0 = 10{,}000$, how long will it take for the number of bacteria to equal a) 320,000, b) 2,560,000?

c 52. The population of a city of 20,000 inhabitants is growing at a rate of 3% per year. After t years the population will be $P(t) = 20{,}000 e^{0.03t}$. How long will it take for the population to double?

c 53. Iodine-128 decays according to the formula

$$Q(t) = Q_0 \left(\frac{1}{2}\right)^{t/25}$$

where Q_0 is the initial amount and $Q(t)$ is the amount of iodine after t minutes. How long will it take for the remaining amount to be 1/10 of the initial amount?

c **54.** The equation $Q(t) = Q_0 e^{-0.00012t}$ represents the amount remaining, after t years, of an original amount Q_0 of carbon-14. Find the half-life of carbon-14.

c **55.** The half-life of rubidium-87 is 6×10^{11} years. Find its rate of decay.

c **56.** The intensity I of a vertical beam of light at a depth of x feet below the surface of Crystal Lake in Wisconsin is given by the formula

$$I = I_0 e^{-0.0485x}.$$

At what depth is the light intensity equal to 1/10 of its value at the surface of the lake?

● c **57.** The formula

$$PV = Ae^{-rt}$$

gives the dollar amount (called *present value*) that should be invested now at an annual interest rate of $100r\%$ compounded continuously to achieve A dollars in t years.
a) What is the present value at 8% interest of a gift of $5000 to be made 5 years from now?
b) What is the annual interest rate of a savings account if the present value of $8000 in 10 years is $4282?

● c **58.** According to Newton's law of cooling, the temperature T of a body after time t is given by

$$T = T_0 + Ce^{-kt}$$

where T_0 is the temperature of the surrounding medium, and C and k are physical constants. A cup of coffee at a temperature of 150°F is placed in a room whose temperature is 75°F. After 3/4 hr it has cooled to 100°F. Find its temperature after 1/2 hr.

● c **59.** The formula $p(h) = 14.7e^{-0.12h}$ gives the atmospheric pressure in pounds per square inch at an altitude of h miles above sea level. At what altitude will the atmospheric pressure be one half of the sea level pressure? One third of the sea level pressure?

● c **60.** A sum of $1200 is invested at an annual rate of 7.5% compounded quarterly. How long will it take for it to double? When will the principal exceed $5000?

CHAPTER SUMMARY

If a is any real number such that $a > 0$ and $a \neq 1$, the *exponential function with base a* is defined by

$$f(x) = a^x$$

for all real numbers x. Exponential functions play a fundamental role in mathematics and the applied sciences. Exponential growth is described by $f(x) = a^x$, with $a > 1$. If $0 < a < 1$, the function describes exponential decay. Of special interest are the *natural exponential* functions e^x and e^{-x}, with bases e and $1/e$.

Logarithmic functions are defined as inverses of exponential functions. Among other functions, the *common logarithm* (base 10) and the *natural logarithm* (base e) are widely used. Common logarithms are still used in chemistry, geology, and physics. Natural logarithms are used extensively in theoretical and applied mathematics.

Exponential and logarithmic functions are used in a wide variety of problems. Formulas to compute compound interest are of exponential type. Certain mathematical models describing population growth or radioactive decay lead to exponential functions. On the other hand, logarithmic scales are used to measure the pH of chemical solutions, the loudness of sound, or the intensity of earthquakes.

When dealing with exponential growth or decay, the concepts of *doubling time* and *half-life* are of particular importance. For example, given the doubling time of an exponentially growing population, we can determine the rate of increase of the population per individual per unit of time, and vice-versa. Both concepts are tied to the idea of change of base for exponential or logarithmic functions.

In view of the practical importance of logarithmic and exponential functions, tables were constructed as an aid to numerical computations. Tables are now becoming obsolete because of the availability of inexpensive calculators that can compute logarithms and exponents with great accuracy and speed.

In the last section of this chapter, techniques from algebra, together with properties of exponential and logarithmic functions, are combined to solve exponential and logarithmic equations.

REVIEW EXERCISES

1. Find each of the following values:
   a) $5^{\log_5 8}$ b) $3^{\log_3 7}$
   c) $e^{\ln 25}$ d) $10^{\log 2}$

2. Determine the numbers:
   a) $\log_{1/2} 4$ b) $\log_{0.1} 100$
   c) $\log_{1/4} 64$ d) $\log_{1/5} 625$

3. Find each of the functional values:

   a) $f(x) = \left(\dfrac{1}{4}\right)^x$; $f(3)$, $f(2)$, $f(-2)$

   c   b) $g(x) = x^2(2^{-x})$; $g(-0.5)$, $g(0.01)$, $g(0.5)$

4. Find each of the functional values:

   a) $F(x) = 5\left(\dfrac{1}{2}\right)^x$; $F(-16)$, $F(-4)$, $F(2)$

   c   b) $G(x) = \sqrt{x}(3^{x-2})$; $G(0.2)$, $G(0.4)$, $G(2.5)$.

5. Let $q(t) = ab^t$, where a and b are constants. If $q(0) = 10$ and $q(1) = 20$, find a and b. Next, find $q(3)$.

6. Let $p(t) = p_0 r^t$, where p_0 and r positive are constants. Determine p_0 and r, knowing that $p(0) = 128$ and that $p(2) = 32$. Find $p(5)$.

7. The function $S(t)$ is defined by $S(t) = S_0\left(1 + \dfrac{r}{2}\right)^t$,

   where S_0 and r are constants.
   a) If $S(0) = 16$ and $S(3) = 250$, find S_0 and r.
   b) Find $S(2)$.

8. Let $A(t) = A_0\left(1 + \dfrac{r}{2}\right)^t$. If $A(0) = 16$ and $A(3) = 54$, find the constants A_0 and r. Also, find $A(2)$.

9. Graph the following functions:

   a) $f(x) = 2^x - 2^{-x}$ b) $f(x) = \dfrac{2}{e^x - e^{-x}}$.

10. Sketch the graph of each function:

   a) $f(x) = 3^x + \left(\dfrac{1}{3}\right)^x$ b) $f(x) = \dfrac{2}{e^x + e^{-x}}$.

In Exercises 11–26, solve the equations.

11. $\log_4(x + 2) = 2$
12. $\log_2(3x + 1) = 4$
13. $\log_8(2x + 1) = \dfrac{2}{3}$
14. $\log_{16}(3x + 2) = \dfrac{3}{4}$
15. $\log(2x + 4) = \log 10 - \log 6$
16. $\log(3 - 2x) = \log 4 - \log 9$
17. $2 \log_3 x = 3 \log_3 4$
18. $2 \log(x + 1) = 4 \log 5$
19. $\log x - \log(x - 1) = 2 \log 4$
20. $\log_6(2x - 1) - \log_6(x + 1) = \log_6 5 - \log_6 4$
21. $4^{2x-1} = 16$

22. $3^{x+1} = 9$

23. $3^{2x+1} = 4^{3x-2}$

24. $5^{3-x} = 2^{3x+2}$

25. $\dfrac{3^x + 3^{-x}}{2} = 1$

26. $\dfrac{2^x - 2^{-x}}{2} = 1$

Write each of the following expressions as a single logarithm.

27. $\ln 7 + \ln 12 - \ln 5$

28. $\log 9 + \log 15 - \log 21$

29. $3 \ln 4 - 2 \ln 5 + 3 \ln 8$

30. $2 \log 8 - 3 \log 4 + \dfrac{1}{5} \log 12$

31. $7 \log a - \dfrac{1}{2} \log b + \dfrac{3}{4} \log c$

32. $3 \ln u - \dfrac{2}{3} \ln v + 4 \ln w$

33. $2 \log(x + 1) - 3 \log(x + 2) + 4 \log x$

34. $3 \ln(x + 2) - \dfrac{1}{3} \ln(x + 1) + 2 \ln(x + 3)$

Write each expression in terms of $\ln x$, $\ln(x - 1)$, and $\ln(2x + 1)$.

35. $\ln \dfrac{x(x - 1)^2}{(2x + 1)^3}$

36. $\ln \dfrac{x^3 \sqrt{x - 1}}{\sqrt[3]{2x + 1}}$

37. $\ln \dfrac{\sqrt{x(x - 1)^3}}{(2x + 1)^4}$

38. $\ln \dfrac{x^2 \sqrt[3]{(x - 1)^2}}{(2x + 1)^3}$

c 39. Approximate the following logarithms:
a) $\log (32.14 \times 10^{-3})$ b) $\log (257.1 \times 10^{-5})$
c) $\log 0.006254$ d) $\log 4516000$

c 40. Approximate x in each of the following.
a) $\log x = 5.3313$ b) $\log x = 1.4449$
c) $\log x = -4.4409$ d) $\log x = -2.3079$

c 41. Find the pH, of each substance. Round off your answer to two significant digits.
a) Lemon juice: $[H^+] = 5.1 \times 10^{-3}$
b) Beer: $[H^+] = 5 \times 10^{-5}$.

c 42. Find the hydrogen ion concentration $[H^+]$ in moles per liter of each substance. Round off your answers to two significant digits.
a) Blood: pH = 7.4;
b) Intestinal juice: pH = 7.7.

c 43. Find the sound level, in decibels, of heavy traffic, where the sound intensity is $I = 10^{-8}$ watt/cm^2.

c 44. The sound level of a pneumatic drill is 110 db. Find the corresponding sound intensity.

c 45. The threshold intensity for human hearing (also called the threshold of pain) is 10^{-4} watt/cm^2. Determine the corresponding sound level in decibels.

c 46. Compare the intensity of an earthquake whose magnitude is 6.5, on the Richter scale, with the minimum intensity I_0.

c 47. Compare the intensities of two earthquakes of magnitudes 5.5 and 7 measured on the Richter scale.

c 48. How many times more intense is the light emanating from Canopus, a star of magnitude -0.72, than the light emanating from Alpha Centauri, a star of magnitude -0.27?

c 49. Vega, a star in the constellation Lyra, has magnitude 0.04. Find the magnitude of a star whose light intensity is 1/50 that of Vega.

c 50. The half-life of thorium-230 is 80,000 years. Find its yearly rate of decay.

c 51. Potassium-40 has a half-life of 1.3×10^9 years. Given an initial amount of 1.25×10^{-3} gm of potassium-40, find how much is left after 5×10^9 years.

c 52. According to a recently completed census, the population of China reached 1 billion in 1982 and is growing at an annual rate of 1.45%. Estimate how large the population of China will be at the turn of the century.

c 53. How much would have to be invested today at an annual rate of 7.5% compounded continuously to have $12,500 in 8 years?

c 54. A sum of $25,120 is invested at an annual rate of 8.25% compounded continuously. How long will it take for the sum to double?

● **c 55.** The 200% *declining balance* method is one of the depreciation methods allowed by the IRS. According to this method, the original cost C of an item depreciates over N years at a rate of $200/N$% per year. Thus the value V of the item after t years is

$$V = C\left(1 - \frac{2}{N}\right)^t.$$

If a typewriter costing $1200 is depreciated by this method over 6 years, what is its value after four years? Six years?

● c 56. The value of a truck costing $15,000 after 3 years is $556. Using the formula of Exercise 55, find the number of years over which the truck was depreciated.

● c 57. An apple pie baked at 325°F is removed from the oven and placed in a room at a constant temperature of 70°F. One minute later the temperature of the pie is 250°F. What is its temperature after 5 minutes?

● c 58. A thermometer registering 50°C is placed in a freezer kept at a constant temperature of −10°C. What temperature will the thermometer register after 5 minutes if it registers 5°C after 2 minutes?

● 59. Show that $\log_a(m/n) = \log_a m - \log_a n$.

● 60. Show that $\log_a \sqrt[n]{m} = (1/n)\log_a m$.

7

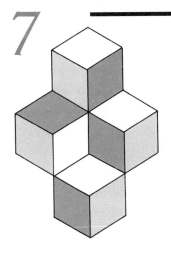

Systems of Equations

In most of this chapter we discuss various methods of solving systems of linear equations. We begin with a review of the methods of substitution and elimination for solving two-by-two linear systems. Next, we explain the powerful and simple Gaussian elimination method of solving general systems of linear equations. Matrices and their properties are discussed in Section 7.3. Every system of linear equations can be written in matrix form, and if the matrix of the coefficients is invertible, then the system is solvable by matrix algebra. In Section 7.4 we define determinants, describe their properties, and discuss Cramer's rule for solving linear systems. Section 7.5 contains a brief discussion on nonlinear systems of equations. Finally, in the last section, we study systems of linear inequalities and discuss linear programming.

7.1 SYSTEMS OF TWO LINEAR EQUATIONS

Suppose that a manufacturer produces two types of toys, cars and trucks, and that each requires wheels and axles according to the following table:

	Cars	Trucks
Wheels	4	10
Axles	2	3

Such a table is called a *production matrix*. How many cars and trucks can be produced if the manufacturer has a stock of 300 wheels and 110 axles and uses his entire stock? To solve this problem, let x be the number of cars and y the number of trucks to be produced. According to the table above, 4 wheels are needed for each car and 10 wheels for each truck, so the equation describing the use of the wheels is $4x + 10y = 300$. Also, 2 axles are needed for each car and 3 axles for each truck, so we get the equation $2x + 3y = 110$. Thus, we obtain the system of two linear equations in two *unknowns* x and y:

$$\begin{cases} 4x + 10y = 300 \\ 2x + 3y = 110. \end{cases}$$

Such a system is also referred to as a *two-by-two linear system*. To *solve the system* means to find two numbers which, when substituted for x and y respectively, satisfy both equations. You can check, by direct substitution, that the solution of the system is $x = 25$ and $y = 20$. Thus, if the entire stock of wheels and axles is used, the manufacturer will produce 25 cars and 20 trucks.

We frequently encounter systems of linear equations in the course of solving problems in mathematics or the applied sciences. Depending on the particular problem, the system whose solution we are supposed to find may have two, three, or more linear equations in several variables.

Geometric Interpretation

Consider the problem of finding all real numbers x and y satisfying the above system of equations. Each equation represents a line in the Cartesian plane, as illustrated in Figure 7.1. Since the two lines are neither parallel nor identical, they have a unique point of intersection whose coordinates are obtained by solving the linear system of equations. Since the solution of the system is $x = 25$ and $y = 20$, the coordinates of the point of intersection are (25, 20). Conversely, if we can find the coordinates of the point of intersection by geometric means, then we also have the solution of the two-by-two system.

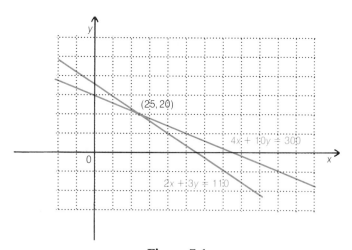

Figure 7.1

Consistent, Inconsistent, and Dependent Systems

There are systems of equations that do not have solutions. For example, the two equations in the system

$$\begin{cases} 2x - y = 3 \\ 4x - 2y = 1 \end{cases}$$

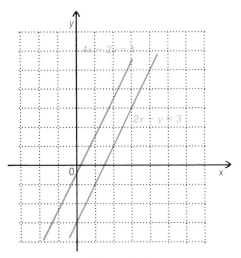

Figure 7.2

represent parallel lines (Figure 7.2). Since these lines are distinct, they do not intersect. Thus, the system does not have solutions. We can also reason by contradiction. Multiplying both sides of the first equation by 2, we get

$$\begin{cases} 4x - 2y = 6 \\ 4x - 2y = 1. \end{cases}$$

But this implies that $6 = 1$, which is absurd. Thus, there cannot be any pair of values for x and y that make both equations true simultaneously.

There are systems of equations that have infinitely many solutions. For example, consider the system

$$\begin{cases} 2x + 3y = 1 \\ 4x + 6y = 2. \end{cases}$$

Although the two equations are formally different, they represent the same line (Figure 7.3). In fact, the second equation is a multiple of the first. Every

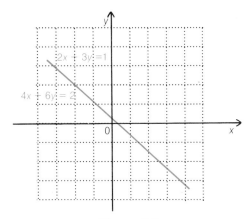

Figure 7.3

point on the line—such as $(0, 1/3)$, $(1/2, 0)$, $(2, -1)$, and so on—is a solution of the system. Since the line contains infinitely many points, the system has infinitely many solutions.

These examples show that a two-by-two linear system may have a *unique solution, no solutions,* or an *infinite number of solutions.* A system that has a unique solution is called *consistent;* a system with no solutions is called *inconsistent;* and a system with infinitely many solutions is called *dependent.*

Now we review the methods of substitution and elimination for solving two-by-two linear systems, which are already familiar to you.

Solving by Substitution

In this method we solve one of the equations for one of the variables. Next we substitute the expression just obtained into the other equation and solve for the remaining variable.

◆ **Example 1.** Solve the system

$$\begin{cases} 2x + 4y = 10 \\ x - 2y = -7. \end{cases}$$

Solution. Solving the second equation for x, we get

(7.1) $\quad x = 2y - 7.$

Next we substitute $2y - 7$ for x in the first equation, obtaining an equation in the variable y that we solve for y:

$$2(2y - 7) + 4y = 10$$
$$4y - 14 + 4y = 10$$
$$8y = 24$$
$$y = 3.$$

Substituting 3 for y in (7.1), we obtain x:

$$x = 2(3) - 7 = -1.$$

We could instead have substituted 3 for y in either of the two equations of the given system to obtain the same $x = -1$.

Now to check that $x = -1$ and $y = 3$ are solutions of the given system, we substitute -1 for x and 3 for y into the two equations and verify that the result is an identity in each case:

$$2(-1) + 4(3) = 10$$
$$(-1) - 2(3) = -7.$$

◇ **Practice Exercise 1.** Find the solution of the system

$$\begin{cases} 3x - 2y = 7 \\ 4x - \ \ y = 6 \end{cases}$$

Answer. $x = 1$, $y = -2$.

Solving by Elimination

In this method we eliminate one of the variables between the two equations, obtaining a single equation in the remaining variable. This equation is then solved for the remaining variable.

◆ **Example 2.** Solve the system

$$\begin{cases} x - y = 1 \\ 2x + y = 5. \end{cases}$$

Solution. By adding the two equations, we eliminate the variable y:

$$+ \ \frac{\begin{aligned} x - y &= 1 \\ 2x + y &= 5 \end{aligned}}{3x \quad \ \ = 6}$$

so

$$x = 2.$$

Substituting 2 for x in the first equation, say, and solving for y, we get

$$2 - y = 1,$$
$$y = 1.$$

You can check that $(2, 1)$ is the solution of the system.

◇ **Practice Exercise 2.** Solve the two by two system

$$\begin{cases} 4x - 3y = \ \ 9 \\ \ \ x + 3y = 21 \end{cases}$$

Answer. $x = 6$, $y = 5$.

Sometimes it is necessary to multiply one equation, or both, by constants before eliminating one of the variables.

◆ **Example 3.** Solve the system

$$\begin{cases} 4x + 10y = 300 \\ 2x + \ \ 3y = 110. \end{cases}$$

Solution. If a linear system has a solution, then multiplying one equation, or both, by nonzero numbers does not affect the solution of the system. Multiplying the second equation by -2 and leaving the first equation unchanged, we obtain the new system

$$\begin{cases} 4x + 10y = 300 \\ -4x - 6y = -220. \end{cases}$$

Adding the two equations, we eliminate x and get

$$4y = 80,$$
$$y = 20.$$

Substituting 20 for y in the first equation gives us

$$4x + 10 \cdot 20 = 300,$$
$$4x = 100,$$
$$x = 25.$$

The solution of the system is then $x = 25$ and $y = 20$.

◇ **Practice Exercise 3.** Solve the system

$$\begin{cases} 3x - 5y = 11 \\ 4x + 3y = 5 \end{cases}$$

Answer. $x = 2, y = -1$.

Note that in Example 2, we *added* the two equations in order to *eliminate* one of the variables, while in Example 3, we *multiplied* the second equation by a *nonzero constant* before adding the two equations. The method described in both examples is a particular case of a general method for solving systems of linear equations called *Gaussian elimination*, which will be discussed in detail in the next section.

Many word problems can be solved by using two-by-two linear systems.

◆ **Example 4.** A silversmith has two bars of silver alloy in stock. One alloy is 50% silver and the other is 75% silver. How many grams of each alloy should he use to obtain 40 grams of a 60% silver alloy?

Solution. If x denotes the number of grams of the 50% silver alloy and y the number of grams of the 75% silver alloy to be used, then

$$x + y = 40.$$

The amount of silver in x grams of the 50% alloy is $0.5x$, the amount of silver in y grams of the 75% alloy is $0.75y$, and the amount of silver in 40 grams of the 60% alloy to be obtained is $0.6 \times 40 = 24$ grams. Thus

$$0.5x + 0.75y = 24.$$

We must now solve the system

$$\begin{cases} x + y = 40 \\ 0.5x + 0.75y = 24. \end{cases}$$

HISTORICAL NOTE

Linear Systems
Linear systems have a long history in East Asia. The consideration of systems of linear equations that are defined by the square array of their coefficients and are solved by means of what are now called matrix transformations is found in the Chinese classic *Nine Chapters on the Mathematical Art*, which dates from the period of the Han dynasty (206 B.C.–220 A.D.).

Proceeding by substitution, we get

$$y = 40 - x.$$

Next,

$$0.5x + 0.75(40 - x) = 24$$
$$0.5x + 30 - 0.75x = 24$$
$$-0.25x + 30 = 24$$
$$-0.25x = -6$$
$$x = \frac{-6.00}{-0.25}$$
$$x = 24 \text{ grams.}$$

Then

$$y = 40 - 24 = 16 \text{ grams.}$$

Thus he should use 24 grams of the 50% silver alloy and 16 grams of the 75% silver alloy.

◇ **Practice Exercise 4.** Find the dimensions of a rectangle whose perimeter measures 54 ft, if its length is 3 ft longer than its width.

Answer.   15 ft and 12 ft.

EXERCISES 7.1

Solve each linear system.

1. $\begin{cases} 3x - 2y = 7 \\ x + 4y = 21 \end{cases}$

2. $\begin{cases} 2x - 4y = 10 \\ 5x + 3y = 12 \end{cases}$

3. $\begin{cases} 4x - 3y = 6 \\ 2x + 5y = 16 \end{cases}$

4. $\begin{cases} x + 2y = 5 \\ 5x + y = 2 \end{cases}$

5. $\begin{cases} 5x + 7y = 9 \\ 2x + 6y = -6 \end{cases}$

6. $\begin{cases} 2x - 4y = 8 \\ x + 5y = -17 \end{cases}$

7. $\begin{cases} 8x + 6y = -1 \\ 4x + 10y = -4 \end{cases}$

8. $\begin{cases} 9x - 5y = 1 \\ 6x + 15y = 8 \end{cases}$

9. $\begin{cases} 5x - 6y = 1 \\ 3x - 2y = 7 \end{cases}$

10. $\begin{cases} 2x + 5y = 7 \\ -4x + 3y = 25 \end{cases}$

11. $\begin{cases} 6x - 4y = 2 \\ 9x - 2y = 5 \end{cases}$

12. $\begin{cases} 3x - 10y = -1 \\ 6x + 5y = 3 \end{cases}$

13. $\begin{cases} 0.5x - 0.25y = 3.5 \\ 1.2x - 1.5y = 3 \end{cases}$

14. $\begin{cases} 0.8x + 4y = 10 \\ 2x - 10y = -9 \end{cases}$

15. $\begin{cases} \dfrac{2}{3}x - \dfrac{1}{2}y = 2 \\ \dfrac{3}{4}x - \dfrac{2}{3}y = \dfrac{11}{6} \end{cases}$

16. $\begin{cases} \dfrac{5}{6}x - \dfrac{3}{8}y = 2 \\ \dfrac{1}{2}x + \dfrac{3}{4}y = 9 \end{cases}$

17. $\begin{cases} \dfrac{x}{5} + \dfrac{y}{3} = \dfrac{1}{15} \\ -\dfrac{x}{2} + 2y = \dfrac{11}{2} \end{cases}$

18. $\begin{cases} 2x - \dfrac{1}{2}y = \dfrac{13}{2} \\ \dfrac{x}{2} + \dfrac{y}{3} = -\dfrac{2}{3} \end{cases}$

In Exercises 19–22, solve each system. (Hint: Set $u = 1/x$ and $v = 1/y$ and first solve the corresponding linear system.)

19.
$$\begin{cases} \dfrac{5}{x} - \dfrac{1}{y} = 7 \\ \dfrac{3}{x} + \dfrac{2}{y} = -1 \end{cases}$$

20.
$$\begin{cases} \dfrac{1}{x} - \dfrac{3}{y} = 9 \\ -\dfrac{2}{x} - \dfrac{4}{y} = 2 \end{cases}$$

21.
$$\begin{cases} \dfrac{4}{x} - \dfrac{3}{y} = 5 \\ -\dfrac{3}{x} + \dfrac{7}{y} = 1 \end{cases}$$

22.
$$\begin{cases} \dfrac{5}{x} + \dfrac{2}{y} = 4 \\ -\dfrac{3}{x} - \dfrac{5}{y} = 9 \end{cases}$$

In Exercises 23–30, write the two by two linear system representing each problem and solve the system.

23. Two zinc alloys contain 20% and 50% of zinc. How many kilograms of each alloy should be mixed to obtain 30 kg of a 45% zinc alloy?

24. A chemical manufacturer wishes to prepare 600 gallons of a 50 percent acid solution by mixing a 60 percent acid solution with a 36 percent acid solution. How many gallons of each solution should he mix?

25. On a river, a boat took 3 h to travel 24 mi upstream and $1\frac{1}{2}$ h to return. Find the speed of the boat in still water and the speed of the current.

26. With the aid of a tail wind, an airplane travels 1050 miles in 3 hours and 30 minutes. Flying against the same wind, the airplane takes 4 hours and 12 minutes to travel the same distance. Find the speed of the airplane in still air and the speed of the wind.

27. A girl has 30 coins consisting of nickels and dimes. If the total amount is $2.10, how many nickels and dimes does the girl have?

28. Peter spends $6.32 buying 11-cent and 18-cent stamps. If he buys 10 more 11-cent stamps than 18-cent stamps, how many stamps of each denomination does he buy?

29. If a rectangle has perimeter 140 cm and its width is equal to 3/4 of its length, what are its dimensions?

30. The table below shows the percentage of zinc and copper in two alloys A and B.

	Zinc	Copper
Alloy A	30	50
Alloy B	20	60

How many kilograms of each alloy should be mixed to obtain 100 kg of a new alloy containing 26% of zinc?

7.2 SYSTEMS OF LINEAR EQUATIONS IN MORE THAN TWO VARIABLES; GAUSSIAN ELIMINATION

A *linear equation in n unknowns* $x_1, x_2, \ldots, x_n$ is an equation of the form

$$a_1 x_1 + a_2 x_2 + \cdots + a_n x_n = b,$$

where $a_1, a_2, \ldots, a_n$ and b are real numbers, and $x_1, x_2, \ldots, x_n$ (also called *variables*) are quantities to be determined. The number a_i is the *coefficient* of

x_i in the equation. A *solution* of the equation is an *n*-tuple of real numbers $(\alpha_1, \alpha_2, \ldots, \alpha_n)$ such that

$$a_1\alpha_1 + a_2\alpha_2 + \cdots + a_n\alpha_n = b.$$

For example, $3x_1 - x_2 + 2x_3 = 5$ is a linear equation in three unknowns. The numbers 3, -1, and 2 are respectively the coefficients of x_1, x_2, and x_3. The triples $(1, 0, 1)$, $(2, 1, 0)$, and $(0, -1, 2)$ are examples of solutions of this equation.

A collection of linear equations is called a *system of linear equations*. For example,

$$\begin{cases} 2x_1 - 3x_2 + x_3 = 1 \\ x_1 + 2x_2 - x_3 = 3 \end{cases}$$

is a system of two linear equations in three unknowns.

A *solution* of a system in *n* unknowns is an *n*-tuple of real numbers which satisfies each equation of the system. For example, $(1, 2, 3)$ is a solution of the system

$$\begin{aligned} 2x - y + z &= 3 \\ x + 2y - z &= 2 \\ 3x - 2y + 2z &= 5. \end{aligned}$$

Here, for simplicity of notation, the variables are being denoted by x, y, and z instead of x_1, x_2, and x_3.

General Form of Three by Three Equations

The general form of a system of three linear equations in three unknowns is

$$(7.2) \quad \begin{cases} a_1 x + b_1 y + c_1 z = d_1 \\ a_2 x + b_2 y + c_2 z = d_2 \\ a_3 x + b_3 y + c_3 z = d_3. \end{cases}$$

Consistent, Inconsistent, and Dependent Systems

A system has either a *unique solution, no solution,* or *infinitely many solutions*. If the solution is unique, the system is said to be *consistent*; if no solutions exist the system is said to be *inconsistent*; and if there is an infinite number of solutions, the system is said to be *dependent*. The same definitions apply to the case of a system of *n* linear equations in *n* unknowns.

Geometric Interpretation of Solutions

We already know that a Cartesian system of coordinates can be defined on a plane. Points are represented by ordered pairs of real numbers and each ordered pair of real numbers determines a point. In a similar way, a Cartesian system of coordinates can be defined in a three dimensional Euclidean space. Points are now represented by *ordered triples* (x, y, z) of real numbers, and each ordered triple of real numbers determines a unique point in the Euclidean space.

If a Cartesian coordinate system is defined on a three-dimensional space, then it can be shown that a linear equation

$$ax + by + cz = d$$

represents a *plane*. Thus finding the solution of the system (7.2) corresponds to locating the points of intersection of three planes.

If the system has a unique solution, then the three planes intersect at a single point P (Figure 7.4).

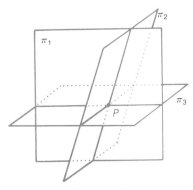

Figure 7.4

If the system has no solution, then the three planes have no point in common. In this case, the three planes are parallel (Figure 7.5a); or two planes are parallel and the third plane either coincides with one of the two or crosses them (Figure 7.5b); or the three planes intersect along three parallel lines (Figure 7.5c).

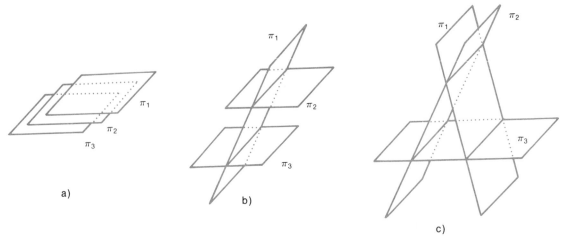

Figure 7.5

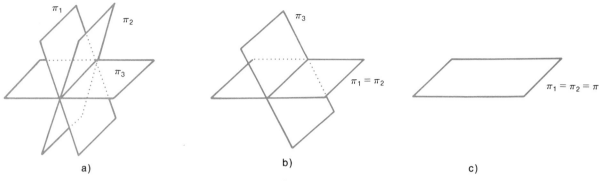

Figure 7.6

If the system has an infinite number of solutions, then either the three planes have a common line of intersection (Figure 7.6a), or two planes coincide and the third plane is different (Figure 7.6b), or the three planes coincide (Figure 7.6c).

Solving by Substitution

Systems of linear equations such as (7.2) can be solved by the method of substitution: solve any one of the equations for one of the variables; replace it in the other two equations to obtain a two by two linear system; solve the new system using one of the methods described in the previous section.

◆ Example 1. Solve the system

$$(7.3) \quad \begin{cases} 2x - y + z = 3 \\ x + 2y - z = 2 \\ 3x - 2y + 2z = 5 \end{cases}$$

Solution. Solving the second equation for z, we get

$$z = -2 + x + 2y$$

Substituting for z in the first and third equations, we obtain

$$\begin{cases} 2x - y + (-2 + x + 2y) = 3 \\ 3x - 2y + 2(-2 + x + 2y) = 5 \end{cases}$$

which simplifies to the two by two linear system

$$\begin{cases} 3x + y = 5 \\ 5x + 2y = 9. \end{cases}$$

Solving the first equation for y gives us

$$y = 5 - 3x.$$

Substituting for y in the second equation and simplifying, we get

$$5x + 2(5 - 3x) = 9,$$

or

$$-x = -1$$

hence $x = 1$. To find y we substitute 1 for x in the equation $y = 5 - 3x$, obtaining

$$y = 5 - 3(1)$$
$$y = 2.$$

Finally, to find z, we substitute for x and y in the equation $z = -2 + x + 2y$, and obtain

$$z = -2 + 1 + 2(2)$$
$$z = 3.$$

The ordered triple (1, 2, 3) is then the unique solution of the system. We leave it to you to check that the triple (1, 2, 3) satisfies each equation of the system.

Remember that to check the solution you must substitute 1 for x, 2 for y, and 3 for z into all three equations of the system (7.3) and verify that the result is an identity in each case.

◇ **Practice Exercise 1.**   Solve the three by three linear system

$$\begin{cases} x - 2y + z = -2 \\ 3x + y - z = -3 \\ -x + 2y + 3z = 6 \end{cases}$$

Answer. $x = -1, y = 1, z = 1.$

The method of substitution extends without difficulty to any system of linear equations. However, this method has several drawbacks: it can be too lengthy, time consuming, and tedious. It is preferable to use instead the *Gaussian elimination method*, which allows us to transform the given system into a simpler *equivalent* system.

Equivalent Systems

Two systems of linear equations in n unknowns are said to be *equivalent* if they have the same solutions.

For example, the systems

$$\begin{cases} x - y = 1 \\ 2x + y = 5 \end{cases} \quad \text{and} \quad \begin{cases} 2x - 2y = 2 \\ 4x - y = 7 \end{cases}$$

are equivalent since they have the same solution (2, 1). Also, the system (7.3) is equivalent to the system

(7.4) $$\begin{cases} x + 2y - z = 2 \\ -5y + 3z = -1 \\ z = 3. \end{cases}$$

The linear system (7.4) is said to be in *upper triangular form*. To solve it, notice that the last equation is already solved for z. Substituting 3 for z in the second equation and solving for y, we obtain $y = 2$. Finally, setting $y = 2$ and $z = 3$ in the first equation and solving for x, we get $x = 1$.

Gaussian Elimination

Any operation that transforms a system of linear equations into an equivalent one is called an *admissible operation*. In the Gaussian elimination method, a given system is transformed into an equivalent one in upper (or lower) triangular form, by means of admissible operations. The solution of the triangular system which is easily obtained is the solution of the original system.

The following are three admissible operations that may be applied to a system to get an equivalent one:

I. Changing the order in which the equations are listed.

II. Multiplying one equation by a nonzero number, and leaving the other equations unchanged.

For example, the systems

$$\begin{cases} x - y = 1 \\ 2x + y = 5 \end{cases} \quad \text{and} \quad \begin{cases} 2x - 2y = 2 \\ 2x + y = 5 \end{cases}$$

are equivalent. The first equation in the second system is twice the first equation in the first system.

III. Adding one equation to another equation, while leaving the other equations unchanged.

For example, the systems

$$\begin{cases} 2x - 2y = 2 \\ 2x + y = 5 \end{cases} \quad \text{and} \quad \begin{cases} 2x - 2y = 2 \\ 4x - y = 7 \end{cases}$$

are equivalent. The second equation in the second system is obtained by adding the first and second equations in the first system.

When transforming a system into an equivalent one, we often combine the admissible operations II and III. For example, the two systems

$$\begin{cases} x - y = 1 \\ 2x + y = 5 \end{cases} \quad \text{and} \quad \begin{cases} x - y = 1 \\ 4x - y = 7 \end{cases}$$

are equivalent since they have the same solution (2, 1). The second system is obtained from the first one by multiplying the first equation by 2 and adding it to the second.

In the following example we show how the three by three linear system (7.3) of Example 1 can be transformed into the upper triangular system (7.4).

◆ **Example 2.** Solve the system

$$\begin{cases} 2x - y + z = 3 \\ x + 2y - z = 2 \\ 3x - 2y + 2z = 5. \end{cases}$$

Solution. First, we interchange the first and second equations (admissible operation I) and write the system as follows:

$$\begin{cases} x + 2y - z = 2 \\ 2x - y + z = 3 \\ 3x - 2y + 2z = 5. \end{cases}$$

By doing this, we obtain a system in which the coefficient of x in the first equation is 1. Next, using admissible operations II and III, we can eliminate the terms containing x in the second and third equations. Multiplying the first equation by -3 and adding it to the third, we obtain the equivalent system

$$\begin{cases} x + 2y - z = 2 \\ 2x - y + z = 3 \\ -8y + 5z = -1. \end{cases}$$

Multiplying the first equation by -2 and adding it to the second, we get the system

$$\begin{cases} x + 2y - z = 2 \\ -5y + 3z = -1 \\ -8y + 5z = -1. \end{cases}$$

We now work with the second and third equations, seeking to eliminate the term containing y in the third equation. Multiplying the second equation by $-8/5$ and adding it to the third, we get the upper triangular system

$$\begin{cases} x + 2y - z = 2 \\ -5y + 3z = -1 \\ \dfrac{1}{5}z = \dfrac{3}{5}. \end{cases}$$

Finally, if we multiply the last equation by 5, we obtain the system (7.4),

$$\begin{cases} x + 2y - z = 2 \\ -5y + 3z = -1 \\ z = 3 \end{cases}$$

whose solution is $x = 1$, $y = 2$, and $z = 3$.

◇ **Practice Exercise 2.**   Using the Gaussian elimination method, transform the linear system

$$\begin{cases} x - 2y + z = -2 \\ 3x + y - z = -3 \\ -x + 2y + 3z = 6 \end{cases}$$

into an equivalent system in upper triangular form and solve it.

Answer. $x = -1$, $y = 1$, $z = 1$.

The method of elimination for two by two systems described in Section 7.1 is a variant of the Gaussian elimination method. To see this, consider the two by two linear system

$$\begin{cases} 4x + 10y = 300 \\ 2x + 3y = 110 \end{cases}$$

discussed in Example 3 of Section 7.1. Multiplying the second equation by -2 (admissible operation II) we get the equivalent system

$$\begin{cases} 4x + 10y = 300 \\ -4x - 6y = -220. \end{cases}$$

Adding the first equation to the second, while leaving the first equation unchanged (admissible operation III), we obtain the upper triangular system

$$\begin{cases} 4x + 10y = 300 \\ 4y = 80. \end{cases}$$

Solving the last equation for y, we get $y = 20$. Substituting 20 for y in the first equation yields $x = 25$.

The Matrix of Coefficients and the Augmented Matrix

It is useful to observe that when we solve a system of linear equations by Gaussian elimination, the computations involve only the coefficients and constant terms of the equations. Since the variables play no role in the calculations, the process can be speeded up by omitting them. Referring to the system considered in Example 2, we see that the array of numbers

$$\begin{bmatrix} 2 & -1 & 1 \\ 1 & 2 & -1 \\ 3 & -2 & 2 \end{bmatrix}$$

consisting of the coefficients of x, y, and z is called the *matrix of coefficients* of the system. If we add a column consisting of the constant terms 3, 2, and 5, we obtain the *augmented matrix* of the system,

$$\begin{bmatrix} 2 & -1 & 1 & \vdots & 3 \\ 1 & 2 & -1 & \vdots & 2 \\ 3 & -2 & 2 & \vdots & 5 \end{bmatrix}$$

Matrices will be studied in detail in the next section.

Two augmented matrices are said to be *row-equivalent* if they are augmented matrices of two equivalent systems of equations. For instance, the matrices

$$\begin{bmatrix} 2 & -1 & 1 & | & 3 \\ 1 & 2 & -1 & | & 2 \\ 3 & -2 & 2 & | & 5 \end{bmatrix} \quad \text{and} \quad \begin{bmatrix} 1 & 2 & -1 & | & 2 \\ 0 & -5 & 3 & | & -1 \\ 0 & 0 & 1 & | & 3 \end{bmatrix}$$

are row-equivalent since they are augmented matrices of the two equivalent systems considered in Example 2.

Admissible Row Operations

An augmented matrix can be transformed into a row-equivalent matrix by performing *admissible row operations*. Such operations are the following:

I. Interchanging two rows.
II. Multiplying a row by a nonzero number, and leaving the other rows unchanged.
III. Adding a row to another one, while leaving the other rows unchanged.

Observe that these admissible row operations correspond to the admissible operations for systems of equations listed above.

In the example that follows, a system of linear equations is solved by the augmented matrix method.

◆ **Example 3.** Solve the linear system

$$\begin{cases} 2x - y + 3z = 4 \\ x \quad\quad - 4z = 6 \\ 5x - y + 2z = 11 \end{cases}$$

Solution. We first write the augmented matrix of the system:

$$\begin{bmatrix} 2 & -1 & 3 & | & 4 \\ 1 & 0 & -4 & | & 6 \\ 5 & -1 & 2 & | & 11 \end{bmatrix}$$

Interchanging the first and second rows according to I, we obtain the row-equivalent matrix

$$\begin{bmatrix} 1 & 0 & -4 & | & 6 \\ 2 & -1 & 3 & | & 4 \\ 5 & -1 & 2 & | & 11 \end{bmatrix}$$

This matrix corresponds to interchanging the first and second equations of the given system of equations. Next we want to obtain zeros in the first column of the second and third rows. This is the same as eliminating the terms containing x in the second and third equations of the corresponding system. Multiplying the first row by -5 and adding it to the third one, we get

$$\begin{bmatrix} 1 & 0 & -4 & | & 6 \\ 2 & -1 & 3 & | & 3 \\ 0 & -1 & 22 & | & -19 \end{bmatrix}$$

Multiplying the first row by -2 and adding it to the second row, we have

$$\begin{bmatrix} 1 & 0 & -4 & | & 6 \\ 0 & -1 & 11 & | & -8 \\ 0 & -1 & 22 & | & -19 \end{bmatrix}$$

Next, without changing the first column, we want to obtain a zero in the third row and second column. To do this, we multiply the second row by -1 and add it to the third:

$$\begin{bmatrix} 1 & 0 & -4 & | & 6 \\ 0 & -1 & 11 & | & -8 \\ 0 & 0 & 11 & | & -11 \end{bmatrix}$$

This is the augmented matrix of the upper triangular system

$$\begin{cases} x & -4z = & 6 \\ & -y + 11z = & -8 \\ & 11z = & -11 \end{cases}$$

which is equivalent to the given system. Solving the last system, we obtain the solution $x = 2$, $y = -3$ and $z = -1$.

◇ **Practice Exercise 3.**   Solve the following system by the augmented matrix method.

$$\begin{cases} 3x - 2y + 2z = 0 \\ -x - y + 3z = 9 \\ -2x - z = 2 \end{cases}$$

Answer. $x = -2$, $y = -1$, $z = 2$.

Observe that our objective, in Example 3, was to transform the original augmented matrix into a matrix of the form

$$\begin{bmatrix} a & b & c & | & l \\ 0 & d & e & | & m \\ 0 & 0 & f & | & n \end{bmatrix}$$

where the matrix of coefficients is in *upper triangular form*, that is, it contains zeros below the *main diagonal*. Once this is achieved we write the corresponding system of linear equations and solve it.

Similarly, by performing row-admissible operations we could have obtained a matrix such as

$$\begin{bmatrix} a' & 0 & 0 & | & l' \\ b' & d' & 0 & | & m' \\ c' & e' & f' & | & n' \end{bmatrix}$$

where the matrix of coefficients is now in *lower triangular* form, that is, it contains zeros above the main diagonal. The corresponding system is readily solvable and the solution, of course, would be the same.

A System with Infinitely Many Solutions

The method of Gaussian elimination applies in general to any system of linear equations. If the system has solutions, they can be obtained by this method. If no solutions exist, then we obtain a contradiction.

◆ **Example 4.** Solve the system

$$\begin{cases} x - 3y + z = 2 \\ 2x - 5y + z = 5. \end{cases}$$

Solution. This is a system of two equations in three unknowns, and we cannot expect it to have a unique solution. As we already said, each equation represents a plane; two planes may intersect or coincide—in which cases there will be infinitely many solutions—or they may be parallel, in which case there will be no solution. Multiplying the first equation by -2 and adding it to the second, we eliminate x from the second equation and obtain the equivalent system

$$\begin{cases} x - 3y + z = 2 \\ y - z = 1. \end{cases}$$

It is not possible to eliminate y or z from the second equation. You should check that any attempt to do so would reintroduce x in the second equation. Now we solve the second equation $y - z = 1$ for one of the variables, say y, obtaining

$$y = 1 + z.$$

(We could have solved it for z. The conclusion we are about to reach would be the same.) Substituting for y in the first equation and solving for x, we get

$$x - 3(1 + z) + z = 2,$$

or

$$x = 5 + 2z.$$

The given system has an infinite number of solutions: to each value assigned to z there corresponds a value of y given by $y = 1 + z$ and a value of x given by $x = 5 + 2z$. Thus, the solutions of the system consist of all ordered triples

$(5 + 2z, 1 + z, z)$ where z is an arbitrary number. For example, if $z = -1$ we obtain the solution $(3, 0, -1)$; if $z = 0$, we get $(5, 1, 0)$; if $z = 1/2$, we have $(6, 3/2, 1/2)$; and so on.

◇ **Practice Exercise 4.**   Use the augmented matrix method to solve the system

$$\begin{cases} 3x - y + 2z = 1 \\ x + y - z = 0 \end{cases}$$

Answer.   Infinitely many solutions: $(x, 1 - 5x, 1 - 4x)$ with x any number.

A System with No Solution

◆ **Example 5.**   Solve the system

$$\begin{cases} x - 2y - z = 1 \\ 3x - 5y - 6z = 5 \\ x - y - 4z = 6. \end{cases}$$

Solution.   Multiplying the first equation by -3 and adding it to the second, and then multiplying the first equation by -1 and adding it to the third, leads us to the equivalent system

$$\begin{cases} x - 2y - z = 1 \\ y - 3z = 2 \\ y - 3z = 5. \end{cases}$$

Since the last two equations cannot be solved for y and z, the given system does not have solutions. This can also be seen by continuing with the Gaussian elimination and trying to eliminate y from the third equation. Multiplying the second equation by -1 and adding it to the third gives us

$$\begin{cases} x - 2y - z = 1 \\ y - 3z = 2 \\ 0 = 3. \end{cases}$$

We have reached a contradiction: $0 = 3$! Therefore, the given system has no solution. We have an inconsistent system of equations.

◇ **Practice Exercise 5.**   Use Gaussian elimination or the augmented matrix method to solve the system

$$\begin{cases} x - y + 3z = 2 \\ 3x - z = 1 \\ 2x + y - 4z = -1 \end{cases}$$

Answer.   No solution.

Homogeneous Systems

If all the numbers appearing on the right-hand side of a system of linear equations are zero, the system is said to be *homogeneous*. The general form of a homogeneous system of three linear equations in three unknowns is

$$(7.5) \quad \begin{cases} a_1 x + b_1 y + c_1 z = 0 \\ a_2 x + b_2 y + c_2 z = 0 \\ a_3 x + b_3 y + c_3 z = 0. \end{cases}$$

Every homogeneous system has at least one solution, namely, the triple $(0, 0, 0)$. This is called the *trivial solution*. A homogeneous system may have nontrivial solutions. They can be found by the same methods used for non-homogeneous systems.

◆ **Example 6.**   Find nontrivial solutions of the system

$$\begin{cases} x + y + z = 0 \\ 3x + y - z = 0 \\ x + 2y + 3z = 0 \end{cases}$$

Solution.   If we multiply the first equation by -3 and add it to the second, and then multiply the first equation by -1 and add it to the third, we obtain the equivalent system

$$\begin{cases} x + y + z = 0 \\ -2y - 4z = 0 \\ y + 2z = 0. \end{cases}$$

In the new system, the second equation $-2y - 4z = 0$ is equal to the third equation $y + 2z = 0$ multiplied by -2. Thus the given system is equivalent to

$$\begin{cases} x + y + z = 0 \\ y + 2z = 0. \end{cases}$$

Solving the last equation for y, we get $y = -2z$. Substituting for y into the first equation and solving for x gives us $x = z$. Thus the variables x and y depend on the variable z. To each value assigned to z there corresponds a value of x and a value of y. The ordered triple obtained in this way is a solution of the original system. For example, if $z = 1$, then $x = 1$ and $y = -2$, and we obtain the triple $(1, -2, 1)$, which is a solution of the given system. Similarly, $(-2, 4, -2)$, $(1/2, -1, 1/2)$ are also solutions. We conclude that the system has infinitely many solutions given by the ordered triples $(z, -2z, z)$ where z is an arbitrary number.

Also, notice that the solutions can be represented in a different manner. Since $x = z$, we can write $y = -2z$ as $y = -2x$. By doing so, the solutions are now given by the ordered triples $(x, -2x, x)$ where x is an arbitrary number.

As an exercise, you can check that the solutions can also be given by the ordered triples $(-y/2, y, -y/2)$ where y is an arbitrary number.

◇ Practice Exercise 6.   Find all solutions of the homogeneous system.

$$\begin{cases} 2x \quad\quad + 4z = 0 \\ \quad\quad y + 2z = 0 \\ x - y \quad\quad = 0 \end{cases}$$

Answer. $(x, x, -x/2)$ with x any number.

An Application

We end this section by discussing a word problem that leads us to a system of linear equations.

◆ Example 7.   A manufacturer produces three kinds of equipment, A, B, and C, that have to be transported by truck. He has three types of trucks, F, G, and H, that can carry various units of equipment according to the following table:

Equipment	Trucks		
	F	G	H
A	2	1	1
B	1	3	3
C	3	2	1

If the manufacturer has to deliver 15 units of equipment A, 20 units of equipment B, and 22 units of equipment C, how many trucks of each type should he use so that all trucks operate at full capacity?

Solution.   Let x be the number of trucks of type F, y the number of trucks of type G, and z the number of trucks of type H. According to the table, truck F carries 2 units of equipment A, truck G carries 1 unit of equipment A, and truck H carries 1 unit of equipment A. Together the three trucks must carry 15 units of equipment A, so we get the equation $2x + y + z = 15$. With similar reasoning for equipments B and C, we obtain two other equations: $x + 3y + 3z = 20$ and $3x + 2y + z = 22$. We thus obtain the three by three system

$$\begin{cases} 2x + y + z = 15 \\ x + 3y + 3z = 20 \\ 3x + 2y + z = 22. \end{cases}$$

Using Gaussian elimination or the augmented matrix method, you can check that the solution of this system is $x = 5$, $y = 2$, and $z = 3$. Thus the manufacturer should use 5 trucks of type F, 2 trucks of type G, and 3 trucks of type H.

◇ Practice Exercise 7.   Judy spent $6.66 buying 11-, 20-, and 40-cent stamps. She bought twice as many 20-cent stamps as 11-cent stamps and half as many 40-cent stamps as the sum of 11- and 20-cent stamps. How many stamps of each denomination did she buy?

Answer. 6 stamps of 11¢, 12 of 20¢, and 9 of 40¢.

 EXERCISES 7.2

In Exercises 1–6, solve each system by substitution.

1. $\begin{cases} 3x - y - z = 2 \\ 2y - z = 5 \\ 2z = 6 \end{cases}$

2. $\begin{cases} 2y = 4 \\ y - 3z = 5 \\ x - 2y - 3z = 2 \end{cases}$

3. $\begin{cases} y + z = 5 \\ x + z = 4 \\ x + y = 3 \end{cases}$

4. $\begin{cases} x - 2y + z = 4 \\ 2x - y + 2z = 2 \\ x - 2y = 5 \end{cases}$

5. $\begin{cases} 2x - y = 3 \\ y - 3z = -8 \\ x + 2z = 8 \end{cases}$

6. $\begin{cases} x - 2y - 3z = 0 \\ y - 2z = 4 \\ x + 2y + 3z = 2 \end{cases}$

Solve each of the systems in Exercises 7–14 by Gaussian elimination.

7. $\begin{cases} y - 2z = 5 \\ x + z = 0 \\ 2x - y = 3 \end{cases}$

8. $\begin{cases} x - y = -1 \\ x + z = 4 \\ y - 3z = -3 \end{cases}$

9. $\begin{cases} 2x + y - 3z = 3 \\ 4x - 2y - 3z = 1 \\ 2x - y + 3z = -1 \end{cases}$

10. $\begin{cases} 3x + 4y - z = 2 \\ 3x - 2y - 2z = -3 \\ 6x - 4y - 3z = -4 \end{cases}$

11. $\begin{cases} x + 2y + 3z = 6 \\ 2x + z = -1 \\ 3x + y + z = 0 \end{cases}$

12. $\begin{cases} 3x + 2y + z = 3 \\ x + 2z = 4 \\ 2y + z = -3 \end{cases}$

13. $\begin{cases} x + y - z + w = 1 \\ 2x - y + 2z - w = 2 \\ x + 2y + z + 3w = 2 \\ 2x - y - z - 2w = 1 \end{cases}$

14. $\begin{cases} x - 3y + 4w = 0 \\ 3x - 2z = 1 \\ y + 2z - w = 2 \\ 2x - 3w = 5 \end{cases}$

In Exercises 15–22, use row-admissible operations with the augmented matrix to solve each system.

15. $\begin{cases} 3x + 2y = 2 \\ 9x - 4y = 1 \end{cases}$

16. $\begin{cases} 3x - 4y = -1 \\ 6x + 8y = 10 \end{cases}$

17. $\begin{cases} x + 2y = 2 \\ 3x - 4y = 1 \end{cases}$

18. $\begin{cases} x + 4y = 1 \\ 2x + 2y = -1 \end{cases}$

19. $\begin{cases} x - 2y - 3z = 4 \\ 2x + y - 2z = 4 \\ 3x + 2y + z = 2 \end{cases}$

20. $\begin{cases} 2x - y - z = 4 \\ x + 2y - z = 0 \\ x + y + 3z = -2 \end{cases}$

21. $\begin{cases} x + 2y = -1 \\ 2x + z = 8 \\ y + 2z = -7 \end{cases}$

22. $\begin{cases} x + y + z = 4 \\ 2x - 2z = -4 \\ 3x - 2y = 17 \end{cases}$

In Exercises 23–28, solve each system in terms of an arbitrary variable.

23. $\begin{cases} 2x - y + 4z = 1 \\ x \quad\;\; + 3z = 0 \end{cases}$

24. $\begin{cases} x + 2y \quad\;\; = 3 \\ 2x - 3y + z = 1 \end{cases}$

25. $\begin{cases} 2x + 3y - \;z = 4 \\ x - 2y + 3z = 2 \end{cases}$

26. $\begin{cases} x - \;y + 3z = 1 \\ 3x + 2y - \;z = 8 \end{cases}$

27. $\begin{cases} x - y + \;z = 0 \\ 2x + y - 3z = 0 \\ 4x - y - \;z = 0 \end{cases}$

28. $\begin{cases} 3x - 4y + z = 0 \\ x + \;y \quad\;\; = 0 \\ 2x - 5y + z = 0 \end{cases}$

In Exercises 29–34, use any of the methods described in this section to find the solution (if any) of each system.

29. $\begin{cases} 2x - 3y = 5 \\ -3x + \;y = 3 \\ x - \;y = 1 \end{cases}$

30. $\begin{cases} 2x - 3y = 1 \\ 3x + 2y = 8 \\ x + \;y = 2 \end{cases}$

31. $\begin{cases} x + 2y = -1 \\ 5x - \;y = \;\;6 \\ 2x - \;y = \;\;0 \end{cases}$

32. $\begin{cases} x + \;y = 1 \\ 2x + 4y = 3 \\ 6x - 2y = 2 \end{cases}$

33. $\begin{cases} x + y + 2z = 0 \\ y + \;z = 0 \\ 3x \quad\;\; + \;z = 0 \end{cases}$

34. $\begin{cases} 2x - \;y + \;z = 0 \\ x + 2y - \;z = 0 \\ 3x - \;y + 2z = 0 \end{cases}$

35. Mary spends $8.86 on 11-, 18-, and 40-cent stamps. The total number of 11- and 18-cent stamps she buys is the same as the number of 40-cent stamps, and there are four more 18-cent stamps than 11-cent stamps. How many stamps of each type does she buy?

36. A boy has $5.00 in nickels, dimes and quarters. The total number of nickels and quarters is twice the number of dimes, and the number of nickels is twice the difference between the numbers of dimes and quarters. Find how many nickels, dimes, and quarters the boy has.

37. The table below shows the percentages of copper, zinc, and tin that are used to produce alloys *A*, *B*, and *C*.

	A	*B*	*C*
Copper	50	60	40
Zinc	30	5	30
Tin	20	25	30

How many kilograms of each alloy must be used to produce 100 kg of a new alloy containing 52% of copper and 24% of tin?

38. The longest side of a triangle is 20 cm longer than one of the sides and 7 cm longer than the other side. Find the length of each side if the perimeter of the triangle measures 138 cm.

39. A grocer blends three types of coffee that sell for $2.60, $2.80, and $3.20 per pound so as to obtain 150 pounds of coffee worth $3.00 per pound. If he uses equal amounts of the two lower-priced coffees, how many pounds of each type of coffee should he blend?

40. In a given year an investor received $515 in dividends corresponding to an investment in three types of bonds yielding 7.5%, 8.5%, and 9% per year. If the amount invested at 8.5% is twice the amount invested at 7.5% and the amount invested at 9% equals the sum of the amounts invested at 7.5% and 8.5%, find how much money is invested in each type of bond.

7.3 MATRICES

In the previous section we introduced the notion of matrices when we were solving a system of linear equations. A matrix is an important concept that has many applications in mathematics and the applied sciences.

Matrices

A rectangular array of numbers, written within brackets, is called a *matrix*.

$$(7.6) \quad \begin{bmatrix} a_{11} & a_{12} & \cdots & a_{1n} \\ a_{21} & a_{22} & \cdots & a_{2n} \\ \cdots & \cdots & \cdots & \cdots \\ a_{m1} & a_{m2} & \cdots & a_{mn} \end{bmatrix}$$

The numbers are called the *elements* or *entries* of the matrix.

For example,

$$\begin{bmatrix} 2 & 1 \\ 3 & 2 \end{bmatrix} \qquad [1, 3, 5] \qquad \begin{bmatrix} 2 \\ 4 \\ 6 \end{bmatrix}$$

$$\begin{bmatrix} 3 & 0 & -3 \\ 1 & 4 & -1 \end{bmatrix} \qquad \begin{bmatrix} 2 & 3 & 1 \\ 0 & -1 & 4 \\ -2 & 5 & 2 \end{bmatrix}$$

are matrices.

   It is common practice to denote each entry a_{ij} of a matrix with two subscripts i and j. The first subscript i identifies the *row*, while the second j identifies the *column* where the element is located. For example, in the matrix

$$\begin{bmatrix} 2 & 3 & 1 \\ 0 & -1 & 4 \\ -2 & 5 & 2 \end{bmatrix}$$

the element a_{32}, located in the third row and second column, is 5. Also, $a_{13} = 1$, $a_{21} = 0$, and so on.

   We often use the short notation $[a_{ij}]_{\substack{1 \le i \le m \\ 1 \le j \le n}}$ or $[a_{ij}]$, when no confusion is possible, to denote the matrix (7.6).

   A matrix with m rows and n columns is said to be an $m \times n$ *matrix* or an m by n *matrix*. For example, $[1, 3, 5]$ is a 1 by 3 matrix, and $\begin{bmatrix} 3 & 0 & -3 \\ 1 & 4 & -1 \end{bmatrix}$ is a 2×3 matrix.

When $m = n$, we say that the matrix is a *square matrix*. An $n \times n$ matrix is also called a matrix of *order n*. For example,

$$\begin{bmatrix} 2 & 1 \\ 3 & 2 \end{bmatrix} \quad \text{and} \quad \begin{bmatrix} 2 & 3 & 1 \\ 0 & -1 & 4 \\ -2 & 5 & 2 \end{bmatrix}$$

are square matrices of order 2 and 3, respectively. A matrix of order 1 consists of a single element.

Equality

Two matrices are *equal* if and only if they have the same number of rows and columns, and the same entry in each position (i, j).

In other words $[a_{ij}]_{\substack{1 \le i \le m \\ 1 \le j \le n}} = [b_{ij}]_{\substack{1 \le i \le p \\ 1 \le j \le q}}$ if and only if $m = p$, $n = q$, and $a_{ij} = b_{ij}$ for all indices i and j.

Sum

If $A = [a_{ij}]$ and $B = [b_{ij}]$ are two $m \times n$ matrices, then their sum $A + B$ is defined to be the matrix $[a_{ij} + b_{ij}]$.

◆ **Example 1.**   Find the sum of the matrices

$$A = \begin{bmatrix} 3 & 0 & -3 \\ 1 & 4 & -1 \end{bmatrix} \quad \text{and} \quad B = \begin{bmatrix} -1 & 2 & 3 \\ -2 & -4 & 5 \end{bmatrix}$$

Solution.   We have

$$A + B = \begin{bmatrix} 3 & 0 & -3 \\ 1 & 4 & -1 \end{bmatrix} + \begin{bmatrix} -1 & 2 & 3 \\ -2 & -4 & 5 \end{bmatrix}$$

$$= \begin{bmatrix} 2 & 2 & 0 \\ -1 & 0 & 4 \end{bmatrix}.$$

◇ **Practice Exercise 1.**   Find the sum of the matrices

$$M = \begin{bmatrix} 1 & -1 \\ -3 & 4 \\ 2 & -5 \end{bmatrix} \quad \text{and} \quad N = \begin{bmatrix} 3 & 2 \\ 0 & -2 \\ -4 & 4 \end{bmatrix}.$$

Answer. $M + N = \begin{bmatrix} 4 & 1 \\ -3 & 2 \\ -2 & -1 \end{bmatrix}.$

The sum of matrices satisfies the commutative and associative properties. If $A = [a_{ij}]$, $B = [b_{ij}]$, and $C = [c_{ij}]$ are $m \times n$ matrices, then:

$$A + B = B + A \qquad \text{(commutativity)}$$

and

$$(A + B) + C = A + (B + C) \qquad \text{(associativity)}.$$

We can also write

$$[a_{ij}] + [b_{ij}] = [b_{ij}] + [a_{ij}]$$

and

$$([a_{ij}] + [b_{ij}]) + [c_{ij}] = [a_{ij}] + ([b_{ij}] + [c_{ij}]).$$

The Zero Matrix

A matrix whose entries are all zeros is called a *zero matrix*.

If $A = [a_{ij}]$ is an $m \times n$ matrix and 0 is the $m \times n$ zero matrix, then $A + 0 = 0 + A = A$.

The Additive Inverse of a Matrix

If $A = [a_{ij}]$ is an $m \times n$ matrix, there is always an $m \times n$ matrix $B = [b_{ij}]$ such that $A + B = 0$. The matrix B is called the *additive inverse* of A.

If B is the additive inverse of A, then its entries b_{ij} are such that $b_{ij} = -a_{ij}$, and we write $B = -A$.

Difference

If $A = [a_{ij}]$ and $B = [b_{ij}]$ are two $m \times n$ matrices, then their difference $A - B$ is defined by

$$A - B = A + (-B).$$

◆ **Example 2.** Find $A - B$ if

$$A = \begin{bmatrix} 2 & -3 & 6 \\ -4 & 1 & -2 \end{bmatrix} \quad \text{and} \quad B = \begin{bmatrix} 4 & 2 & 3 \\ -1 & 3 & -5 \end{bmatrix}$$

Solution. We have

$$A - B = \begin{bmatrix} 2 & -3 & 6 \\ -4 & 1 & -2 \end{bmatrix} - \begin{bmatrix} 4 & 2 & 3 \\ -1 & 3 & -5 \end{bmatrix}$$

$$= \begin{bmatrix} 2 & -3 & 6 \\ -4 & 1 & -2 \end{bmatrix} + \begin{bmatrix} -4 & -2 & -3 \\ 1 & -3 & 5 \end{bmatrix}$$

$$= \begin{bmatrix} -2 & -5 & 3 \\ -3 & -2 & 3 \end{bmatrix}.$$

◇ **Practice Exercise 2.** For the matrices of Example 2, find $B - A$.

Answer. $B - A = \begin{bmatrix} 2 & 5 & -3 \\ 3 & 2 & -3 \end{bmatrix}.$

Scalar Multiplication

Let $A = [a_{ij}]$ be an $m \times n$ matrix and let α be a number. The product of α by A is the matrix αA defined by $[\alpha a_{ij}]$.

We say that αA is the *scalar product* of α by A.

◆ **Example 3.** Multiply 3 by the matrix $A = \begin{bmatrix} 5 & 0 & 5 \\ -2 & 1 & -4 \end{bmatrix}.$

Solution. We have

$$3A = 3 \cdot \begin{bmatrix} 5 & 0 & 5 \\ -2 & 1 & -4 \end{bmatrix} = \begin{bmatrix} 15 & 0 & 15 \\ -6 & 3 & -12 \end{bmatrix}.$$

◇ **Practice Exercise 3.** Let $\alpha = \dfrac{1}{2}$ and $A = \begin{bmatrix} 4 & -2 \\ 3 & 8 \end{bmatrix}$. Find αA.

Answer. $\alpha A = \begin{bmatrix} 2 & -1 \\ \dfrac{3}{2} & 4 \end{bmatrix}.$

From the definitions of additive inverse and scalar product it follows that $-A = (-1)A$; that is, the additive inverse of a matrix is equal to the scalar product of the matrix by -1.

Properties of the Scalar Product

If α and β are numbers, and A and B are $m \times n$ matrices, then the following properties of scalar multiplication are easy to check:

$$\alpha(A + B) = \alpha A + \alpha B$$
$$(\alpha + \beta)A = \alpha A + \beta A$$
$$\alpha(\beta A) = (\alpha\beta)A$$

Matrix Multiplication

Let $A = [a_{ij}]$ be an $m \times p$ matrix and $B = [b_{ij}]$ be a $p \times n$ matrix. The product (row by column) of A and B is the $m \times n$ matrix C whose entries are defined by

$$(7.7) \quad c_{ij} = a_{i1}b_{1j} + a_{i2}b_{2j} + \cdots + a_{ip}b_{pj}.$$

We write $C = AB$.

Observe that to find the entry c_{ij} of the product AB, we multiply each element in the ith row of A by the corresponding element in the jth column of B, and sum the products.

$$
i\begin{bmatrix} a_{i1} & a_{i2} & \cdots & a_{ip} \\ & & \vdots & \end{bmatrix}
\begin{bmatrix} & & b_{1j} & & \\ \cdots & & b_{2j} & & \cdots \\ & & \vdots & & \\ & & b_{pj} & & \end{bmatrix}
= \begin{bmatrix} \cdots & c_{ij} & \cdots \end{bmatrix} i
$$

with c_{ij} given by (7.7).

It is important to observe that the product (row by column) of A and B is defined only if *the number of columns of A is the same as the number of rows of B*. Then the product AB has *the same number of rows as A and the same number of columns as B*.

◆ **Example 4.** Find AB if

$$A = \begin{bmatrix} 1 & -1 & 2 \\ 3 & 0 & 1 \end{bmatrix} \quad \text{and} \quad B = \begin{bmatrix} 2 & 0 \\ 1 & -1 \\ -2 & 3 \end{bmatrix}.$$

Solution. According to (7.7), we have

$$AB = \begin{bmatrix} 1 & -1 & 2 \\ 3 & 0 & 1 \end{bmatrix} \begin{bmatrix} 2 & 0 \\ 1 & -1 \\ -2 & 3 \end{bmatrix}$$

$$= \begin{bmatrix} 1 \cdot 2 + (-1) \cdot 1 + 2 \cdot (-2) & 1 \cdot 0 + (-1)(-1) + 2 \cdot 3 \\ 3 \cdot 2 + 0 \cdot 1 + 1 \cdot (-2) & 3 \cdot 0 + 0 \cdot (-1) + 1 \cdot 3 \end{bmatrix}$$

$$= \begin{bmatrix} -3 & 7 \\ 4 & 3 \end{bmatrix}.$$

◇ **Practice Exercise 4.** If A and B are the matrices of Example 4, find BA.

Answer. $BA = \begin{bmatrix} 2 & -2 & 4 \\ -2 & -1 & 1 \\ 7 & 2 & -1 \end{bmatrix}.$

Properties of the Product of Matrices

Whenever it is defined, the product of matrices satisfies the following properties:

$$A(BC) = (AB)C \qquad \text{(associativity)}$$
$$A(B + C) = AB + AC$$
$$(B + C)A = BA + CA \qquad \text{(distributivity)}$$

In general, the multiplication of matrices is a *noncommutative* operation; that is, $AB \neq BA$ for most matrices. Indeed, if A and B are not square matrices, one of the two products may be undefined.

◆ **Example 5.** If $A = \begin{bmatrix} 1 & 2 \\ 0 & 1 \end{bmatrix}$ and $B = \begin{bmatrix} 0 & -1 \\ -2 & 2 \end{bmatrix}$, verify that $AB \neq BA$.

Solution. We have

$$AB = \begin{bmatrix} 1 & 2 \\ 0 & 1 \end{bmatrix} \begin{bmatrix} 0 & -1 \\ -2 & 2 \end{bmatrix} = \begin{bmatrix} -4 & 3 \\ -2 & 2 \end{bmatrix}$$

and

$$BA = \begin{bmatrix} 0 & -1 \\ -2 & 2 \end{bmatrix} \begin{bmatrix} 1 & 2 \\ 0 & 1 \end{bmatrix} = \begin{bmatrix} 0 & -1 \\ -2 & -2 \end{bmatrix}$$

As we can see, $AB \neq BA$.

◇ **Practice Exercise 5.** Let $A = \begin{bmatrix} 1 & 1 \\ 0 & 1 \end{bmatrix}$ and $B = \begin{bmatrix} 0 & 0 \\ 2 & 0 \end{bmatrix}$. Find AB and BA.

Answer. $AB = \begin{bmatrix} 2 & 0 \\ 2 & 0 \end{bmatrix}$, $BA = \begin{bmatrix} 0 & 0 \\ 2 & 2 \end{bmatrix}$.

The Product of Square Matrices

Of special interest is the multiplication of square matrices. If A and B are square matrices of order n, then the products AA, AB, BA, and BB are defined, and are all matrices of order n.

Identity Matrices

The $n \times n$ matrix $I_n = [\delta_{ij}]$ whose entries are

$$\delta_{ij} = 1 \quad \text{if} \quad i = j$$
$$\delta_{ij} = 0 \quad \text{if} \quad i \neq j$$

is called the *identity matrix of order n.*

Such a matrix has *ones* along the main diagonal and *zeros* elsewhere. For example,

$$I_2 = \begin{bmatrix} 1 & 0 \\ 0 & 1 \end{bmatrix} \quad \text{and} \quad I_3 = \begin{bmatrix} 1 & 0 & 0 \\ 0 & 1 & 0 \\ 0 & 0 & 1 \end{bmatrix}$$

are the identity matrices of order 2 and 3, respectively.

It can be shown that if A is an $n \times n$ matrix, then $AI_n = I_nA = A$. For example, you can easily apply the definition of matrix multiplication to see that:

$$\begin{bmatrix} a & b \\ c & d \end{bmatrix}\begin{bmatrix} 1 & 0 \\ 0 & 1 \end{bmatrix} = \begin{bmatrix} 1 & 0 \\ 0 & 1 \end{bmatrix}\begin{bmatrix} a & b \\ c & d \end{bmatrix} = \begin{bmatrix} a & b \\ c & d \end{bmatrix}.$$

The Multiplicative Inverse of a Matrix

If A is an $n \times n$ matrix, and if there is a matrix B of order n such that

$$AB = BA = I_n,$$

then A is said to be *invertible* and B is the *multiplicative inverse* or, simply, *inverse* of A.

For example, the matrix $B = \begin{bmatrix} 3 & -4 \\ -2 & 3 \end{bmatrix}$ is the inverse of $A = \begin{bmatrix} 3 & 4 \\ 2 & 3 \end{bmatrix}$. To verify this, compute the products

$$AB = \begin{bmatrix} 3 & 4 \\ 2 & 3 \end{bmatrix}\begin{bmatrix} 3 & -4 \\ -2 & 3 \end{bmatrix}$$

$$= \begin{bmatrix} 3 \cdot 3 + 4 \cdot (-2) & 3 \cdot (-4) + 4 \cdot 3 \\ 2 \cdot 3 + 3 \cdot (-2) & 2 \cdot (-4) + 3 \cdot 3 \end{bmatrix}$$

$$= \begin{bmatrix} 1 & 0 \\ 0 & 1 \end{bmatrix} = I_2$$

and

$$BA = \begin{bmatrix} 3 & -4 \\ -2 & 3 \end{bmatrix}\begin{bmatrix} 3 & 4 \\ 2 & 3 \end{bmatrix}$$

$$= \begin{bmatrix} 3 \cdot 3 + (-4) \cdot 2 & 3 \cdot 4 + (-4) \cdot 3 \\ (-2) \cdot 3 + 3 \cdot 2 & (-2) \cdot 4 + 3 \cdot 3 \end{bmatrix}$$

$$= \begin{bmatrix} 1 & 0 \\ 0 & 1 \end{bmatrix} = I_2.$$

Not every matrix has an inverse. However, if a matrix has an inverse, it can be proved that the inverse is *unique*. In Section 7.4 on *determinants* we shall state a necessary and sufficient condition for a matrix to have an inverse. If A is an invertible matrix, its inverse is denoted by A^{-1}.

◆ **Example 6.** Find the inverse of the matrix $A = \begin{bmatrix} 3 & 4 \\ 2 & 3 \end{bmatrix}$.

Solution. We wish to find a 2×2 matrix B such that $AB = BA = I_2$. If we set

$$B = \begin{bmatrix} x & u \\ y & v \end{bmatrix},$$

then we must have

$$\begin{bmatrix} 3 & 4 \\ 2 & 3 \end{bmatrix}\begin{bmatrix} x & u \\ y & v \end{bmatrix} = \begin{bmatrix} 1 & 0 \\ 0 & 1 \end{bmatrix}.$$

Computing the product, we get

$$\begin{bmatrix} 3x + 4y & 3u + 4v \\ 2x + 3y & 2u + 3v \end{bmatrix} = \begin{bmatrix} 1 & 0 \\ 0 & 1 \end{bmatrix}$$

from which we obtain the two systems of linear equations

$$\begin{cases} 3x + 4y = 1 \\ 2x + 3y = 0 \end{cases} \quad \text{and} \quad \begin{cases} 3u + 4v = 0 \\ 2u + 3v = 1. \end{cases}$$

Solving these two systems, we obtain

$$x = 3, y = -2 \quad \text{and} \quad u = -4, v = 3.$$

Thus

$$B = \begin{bmatrix} 3 & -4 \\ -2 & 3 \end{bmatrix}.$$

We already checked, before this example, that $AB = BA = I_2$; therefore B is the multiplicative inverse of matrix A.

◇ **Practice Exercise 6.** What is the inverse of the matrix $A = \begin{bmatrix} 5 & 3 \\ 3 & 2 \end{bmatrix}$?

Answer. $A^{-1} = \begin{bmatrix} 2 & -3 \\ -3 & 5 \end{bmatrix}.$

Systems of Linear Equations in Matrix Form

Every $n \times n$ system of linear equations

$$(7.8) \begin{cases} a_{11}x_1 + a_{12}x_2 + \cdots + a_{1n}x_n = b_1 \\ a_{21}x_1 + a_{22}x_2 + \cdots + a_{2n}x_n = b_2 \\ \cdot \quad \cdot \quad \cdot \quad \cdot \quad \cdot \quad \cdot \quad \cdot \quad \cdot \quad \cdot \quad \cdot \\ a_{n1}x_1 + a_{n2}x_2 + \cdots + a_{nn}x_n = b_n \end{cases}$$

can be expressed in compact form using matrix notation. In fact, if we set $A = [a_{ij}]_{\substack{1 \le i \le n \\ 1 \le j \le n}}$, $X = [x_j]_{1 \le j \le n}$, and $B = [b_j]_{1 \le j \le n}$, then we can write the system (7.8) as

$$(7.9) \quad AX = B.$$

The matrix A is called the *matrix of coefficients.*

For example, the two by two system

$$\begin{cases} 3x + 4y = 1 \\ 2x + 3y = -2 \end{cases}$$

can be written in matrix notation as

$$\begin{bmatrix} 3 & 4 \\ 2 & 3 \end{bmatrix} \begin{bmatrix} x \\ y \end{bmatrix} = \begin{bmatrix} 1 \\ -2 \end{bmatrix}.$$

Using the Inverse Matrix to Solve a System

If A^{-1} is the inverse of A and $B \neq 0$, then the system (7.8) has a unique solution given by

$$(7.10) \quad X = A^{-1}B.$$

Indeed, multiplying both sides of (7.9) by A^{-1}, we have

$$A^{-1}(AX) = A^{-1}B.$$

Using the associativity of the product of matrices, we get

$$(A^{-1}A)X = A^{-1}B.$$

Since $AA^{-1} = A^{-1}A = I_n$, it follows that

$$I_nX = A^{-1}B,$$

from which we obtain (7.10).

◆ **Example 7.** Solve the given system by using the inverse matrix.

$$\begin{cases} 3x + 4y = 1 \\ 2x + 3y = -2 \end{cases}$$

Solution. We know from Example 6 that the matrix of coefficients $A = \begin{bmatrix} 3 & 4 \\ 2 & 3 \end{bmatrix}$ has inverse $A^{-1} = \begin{bmatrix} 3 & -4 \\ -2 & 3 \end{bmatrix}$. Thus, according to (7.10), we have

$$\begin{bmatrix} x \\ y \end{bmatrix} = \begin{bmatrix} 3 & -4 \\ -2 & 3 \end{bmatrix}\begin{bmatrix} 1 \\ -2 \end{bmatrix}$$
$$= \begin{bmatrix} 3 \cdot 1 + (-4) \cdot (-2) \\ (-2) \cdot 1 + 3 \cdot (-2) \end{bmatrix}$$
$$= \begin{bmatrix} 11 \\ -8 \end{bmatrix}.$$

The solution of the given system is then (11, 8).

Efficient computer programs have been written to find the inverse of a matrix and to solve a system of n equations in n unknowns by using exactly the method of Example 7.

◇ **Practice Exercise 7.** Use the inverse of the matrix of coefficients to solve the system

$$\begin{cases} 5x + 3y = 1 \\ 3x + 2y = 0. \end{cases}$$

Answer. $x = 2, y = -3$.

Formula (7.10) is important from a theoretical viewpoint. In computations it may require a lot of work, especially for higher order systems. However, if the matrix of coefficients A remains the same and $b_1, b_2, \ldots, b_n$ are given many different values, the use of formula (7.10) will save much work in solving the resulting systems.

Finding the Inverse of a Matrix by Row Operations

As we already remarked, if the matrix of coefficients of an $n \times n$ nonhomogeneous system of linear equations has a multiplicative inverse, then we can solve the system. Thus, it is important to have methods to find the inverse of a matrix. In Example 6, we described a way of finding the inverse of a 2×2 matrix. The method extends to higher order matrices, but it has a serious drawback: the number of systems of linear equations to be solved increases with the order of the matrix.

A more effective method, which we now discuss in the particular case of 3×3 matrices, uses the admissible row operations of Section 7.2. Suppose that we want to find the inverse of the matrix

$$A = \begin{bmatrix} a & b & c \\ d & e & f \\ g & h & i \end{bmatrix}.$$

Form the augmented matrix

$$\left[\begin{array}{ccc|ccc} a & b & c & 1 & 0 & 0 \\ d & e & f & 0 & 1 & 0 \\ g & h & i & 0 & 0 & 1 \end{array} \right]$$

by placing the identity matrix I_3 to the right of the matrix A. If the inverse A^{-1} exists, then by performing as many admissible row operations as necessary, it is possible to transform the augmented matrix into a matrix of the form

$$\left[\begin{array}{ccc|ccc} 1 & 0 & 0 & a' & b' & c' \\ 0 & 1 & 0 & d' & e' & f' \\ 0 & 0 & 1 & g' & h' & i' \end{array} \right]$$

where I_3 now appears to the left of a 3×3 matrix. It can be proved that the matrix

$$B = \begin{bmatrix} a' & b' & c' \\ d' & e' & f' \\ g' & h' & i' \end{bmatrix}$$

is the desired multiplicative inverse A^{-1}. If it is not possible to transform the augmented matrix $[A \mid I_3]$ into a matrix of the form $[I_3 \mid B]$, then the multiplicative inverse of A does not exist.

The same method also extends to the case of $n \times n$ matrices.

◆ **Example 8.** Find the inverse of the matrix

$$A = \begin{bmatrix} 2 & 0 & 3 \\ 1 & -1 & 0 \\ 0 & 1 & -1 \end{bmatrix}$$

by transforming the augmented matrix $[A \mid I_3]$ to the form $[I \mid B]$.

Solution. First form the matrix $[A \mid I_3]$:

$$\left[\begin{array}{ccc|ccc} 2 & 0 & 3 & 1 & 0 & 0 \\ 1 & -1 & 0 & 0 & 1 & 0 \\ 0 & 1 & -1 & 0 & 0 & 1 \end{array}\right]$$

Next, perform admissible row operations until the left matrix is transformed into the identity matrix

$$\left[\begin{array}{ccc|ccc} 2 & 0 & 3 & 1 & 0 & 0 \\ 1 & -1 & 0 & 0 & 1 & 0 \\ 0 & 1 & -1 & 0 & 0 & 1 \end{array}\right]$$

$\downarrow$ Interchange the first and second rows

$$\left[\begin{array}{ccc|ccc} 1 & -1 & 0 & 0 & 1 & 0 \\ 2 & 0 & 3 & 1 & 0 & 0 \\ 0 & 1 & -1 & 0 & 0 & 1 \end{array}\right]$$

$\downarrow$ Multiply the first row by -2 and add it to the second

$$\left[\begin{array}{ccc|ccc} 1 & -1 & 0 & 0 & 1 & 0 \\ 0 & 2 & 3 & 1 & -2 & 0 \\ 0 & 1 & -1 & 0 & 0 & 1 \end{array}\right]$$

$\downarrow$ Multiply the second row by $1/2$

$$\left[\begin{array}{ccc|ccc} 1 & -1 & 0 & 0 & 1 & 0 \\ 0 & 1 & \frac{3}{2} & \frac{1}{2} & -1 & 0 \\ 0 & 1 & -1 & 0 & 0 & 1 \end{array}\right]$$

$\downarrow$ Multiply the second row by -1 and add it to the third

$$\left[\begin{array}{ccc|ccc} 1 & -1 & 0 & 0 & 1 & 0 \\ 0 & 1 & \frac{3}{2} & \frac{1}{2} & -1 & 0 \\ 0 & 0 & -\frac{5}{2} & -\frac{1}{2} & 1 & 1 \end{array}\right]$$

$\downarrow$ Multiply the third row by $-2/5$

$$\left[\begin{array}{ccc|ccc} 1 & -1 & 0 & 0 & 1 & 0 \\ 0 & 1 & \frac{3}{2} & \frac{1}{2} & -1 & 0 \\ 0 & 0 & 1 & \frac{1}{5} & -\frac{2}{5} & -\frac{2}{5} \end{array}\right]$$

$\downarrow$ Multiply the third row by $-3/2$ and add it to the second

$$\left[\begin{array}{ccc|ccc} 1 & -1 & 0 & 0 & 1 & 0 \\ 0 & 1 & 0 & \frac{1}{5} & -\frac{2}{5} & \frac{3}{5} \\ 0 & 0 & 1 & \frac{1}{5} & -\frac{2}{5} & -\frac{2}{5} \end{array}\right]$$

$$\begin{bmatrix} 1 & 0 & 0 & \vdots & \dfrac{1}{5} & \dfrac{3}{5} & \dfrac{3}{5} \\ 0 & 1 & 0 & \vdots & \dfrac{1}{5} & -\dfrac{2}{5} & \dfrac{3}{5} \\ 0 & 0 & 1 & \vdots & \dfrac{1}{5} & -\dfrac{2}{5} & -\dfrac{2}{5} \end{bmatrix} \begin{array}{l} \downarrow \quad \text{Add the second row to the first} \end{array}$$

Thus the inverse of the given matrix is

$$A^{-1} = \begin{bmatrix} \dfrac{1}{5} & \dfrac{3}{5} & \dfrac{3}{5} \\ \dfrac{1}{5} & -\dfrac{2}{5} & \dfrac{3}{5} \\ \dfrac{1}{5} & -\dfrac{2}{5} & -\dfrac{2}{5} \end{bmatrix} = \frac{1}{5} \begin{bmatrix} 1 & 3 & 3 \\ 1 & -2 & 3 \\ 1 & -2 & -2 \end{bmatrix}.$$

◇ **Practice Exercise 8.**   Find the inverse of the matrix

$$A = \begin{bmatrix} 0 & 1 & 1 \\ 1 & 1 & 0 \\ 1 & 0 & -2 \end{bmatrix}$$

by row operations.

Answer. $A^{-1} = \begin{bmatrix} -2 & 2 & -1 \\ 2 & -1 & 1 \\ -1 & 1 & -1 \end{bmatrix}.$

◆ **Example 9.**   Solve the three by three system by using the inverse of the matrix of coefficients.

$$\begin{cases} 2x & + 3z = 2 \\ x - y & = 5 \\ y - z = 1 \end{cases}$$

Solution.   In matrix form the given system can be written as

$$AX = B,$$

where

$$A = \begin{bmatrix} 2 & 0 & 3 \\ 1 & -1 & 0 \\ 0 & 1 & -1 \end{bmatrix}, \quad X = \begin{bmatrix} x \\ y \\ z \end{bmatrix}, \quad \text{and} \quad B = \begin{bmatrix} 2 \\ 5 \\ 1 \end{bmatrix}.$$

Now the solution can be found by formula (7.10),

$$X = A^{-1}B$$

where we know from Example 8 that

$$A^{-1} = \frac{1}{5}\begin{bmatrix} 1 & 3 & 3 \\ 1 & -2 & 3 \\ 1 & -2 & -2 \end{bmatrix}$$

is the inverse of the matrix A. Note that our calculations will be easier if we leave the scalar multiple $1/5$ outside the matrix. We have

$$\begin{bmatrix} x \\ y \\ z \end{bmatrix} = \frac{1}{5}\begin{bmatrix} 1 & 3 & 3 \\ 1 & -2 & 2 \\ 1 & -2 & -2 \end{bmatrix}\begin{bmatrix} 2 \\ 5 \\ 1 \end{bmatrix}$$

$$= \frac{1}{5}\begin{bmatrix} 2 + 15 + 3 \\ 2 - 10 + 3 \\ 2 - 10 - 2 \end{bmatrix}$$

$$= \frac{1}{5}\begin{bmatrix} 20 \\ -5 \\ -10 \end{bmatrix} = \begin{bmatrix} 4 \\ -1 \\ -2 \end{bmatrix}$$

The solution is then $x = 4$, $y = -1$, and $z = -2$.

◇ **Practice Exercise 9.** Use the inverse of the matrix of coefficients to solve the system

$$\begin{cases} y + z = -1 \\ x + y = 6 \\ x - 2z = 10. \end{cases}$$

Answer. $x = 4$, $y = 2$, $z = -3$.

EXERCISES 7.3

Perform the indicated operations.

1. $\begin{bmatrix} 2 & 1 \\ 3 & -2 \end{bmatrix} + \begin{bmatrix} 4 & 0 \\ -5 & 3 \end{bmatrix}$

2. $\begin{bmatrix} 1 & -5 \\ -2 & 4 \end{bmatrix} - \begin{bmatrix} 3 & -2 \\ -2 & -4 \end{bmatrix}$

3. $\begin{bmatrix} 2 \\ -4 \\ 5 \end{bmatrix} + \begin{bmatrix} -3 \\ 2 \\ -1 \end{bmatrix}$

4. $[3 \quad -2 \quad 4] - [1 \quad -3 \quad 2]$

5. $\begin{bmatrix} 3 & 1 & -2 \\ 0 & 1 & 4 \end{bmatrix} - \begin{bmatrix} 1 & -3 & 4 \\ -3 & 1 & -2 \end{bmatrix}$

6. $\begin{bmatrix} 4 & -2 & 6 \\ -3 & 5 & -4 \end{bmatrix} + \begin{bmatrix} -3 & 3 & -3 \\ 3 & -2 & 6 \end{bmatrix}$

7. $\begin{bmatrix} 1 & -2 \\ 2 & -3 \\ 0 & -1 \\ 2 & 1 \end{bmatrix} - \begin{bmatrix} 2 & 3 \\ -1 & 0 \\ -2 & 2 \\ 0 & 2 \end{bmatrix}$

8. $\begin{bmatrix} 1 & -1 & 0 & 1 \\ -2 & 0 & 2 & -1 \end{bmatrix} + \begin{bmatrix} -2 & 1 & -1 & 2 \\ 1 & -2 & 3 & 0 \end{bmatrix}$

In Exercises 9–16, find A + B, A − B, 3A, and 2A − 3B.

9. $A = \begin{bmatrix} 2 & 1 \\ 3 & -2 \end{bmatrix}$, $B = \begin{bmatrix} 1 & -2 \\ 2 & -3 \end{bmatrix}$

10. $A = \begin{bmatrix} 3 & 0 \\ 0 & -3 \end{bmatrix}$, $B = \begin{bmatrix} 2 & -2 \\ -2 & 2 \end{bmatrix}$

11. $A = \begin{bmatrix} 2 & -1 & 5 \\ -1 & 3 & -2 \end{bmatrix}$, $B = \begin{bmatrix} -2 & 1 & -2 \\ 1 & -3 & 4 \end{bmatrix}$

12. $A = \begin{bmatrix} 3 & 2 \\ -1 & -2 \\ -3 & 0 \end{bmatrix}$, $B = \begin{bmatrix} -3 & 5 \\ 1 & 4 \\ 5 & -2 \end{bmatrix}$

13. $A = \begin{bmatrix} 1 \\ -2 \\ 1 \end{bmatrix}$, $B = \begin{bmatrix} -2 \\ 3 \\ 4 \end{bmatrix}$

14. $A = \begin{bmatrix} 1 & -1 \\ 2 & -2 \\ 3 & -3 \end{bmatrix}$, $B = \begin{bmatrix} -2 & 1 \\ 0 & 0 \\ 1 & -2 \end{bmatrix}$

15. $A = \begin{bmatrix} 1 \\ 2 \\ 0 \\ 2 \end{bmatrix}$, $B = \begin{bmatrix} 3 \\ 0 \\ 2 \\ 2 \end{bmatrix}$

16. $A = \begin{bmatrix} 1 & -1 & 0 & 1 \end{bmatrix}$, $B = \begin{bmatrix} -2 & 1 & -1 & 2 \end{bmatrix}$

In Exercises 17–26, find AB and BA.

17. $A = \begin{bmatrix} 1 & -1 \\ 3 & 0 \end{bmatrix}$, $B = \begin{bmatrix} 2 & 0 \\ 1 & -1 \end{bmatrix}$

18. $A = \begin{bmatrix} -1 & 2 \\ 0 & 1 \end{bmatrix}$, $B = \begin{bmatrix} 1 & -1 \\ -2 & 3 \end{bmatrix}$

19. $A = \begin{bmatrix} 3 \\ -2 \\ 4 \end{bmatrix}$, $B = \begin{bmatrix} 1 & -3 & 2 \end{bmatrix}$

20. $A = \begin{bmatrix} 4 & 0 & -1 \end{bmatrix}$, $B = \begin{bmatrix} 0 \\ -2 \\ 3 \end{bmatrix}$

21. $A = \begin{bmatrix} 2 & -2 & 1 \\ 3 & 0 & 1 \end{bmatrix}$, $B = \begin{bmatrix} 0 & -2 \\ -1 & -1 \\ 3 & 2 \end{bmatrix}$

22. $A = \begin{bmatrix} 1 & -3 & 1 \\ 2 & -1 & 2 \end{bmatrix}$, $B = \begin{bmatrix} 2 & 1 \\ 1 & 0 \\ -4 & -2 \end{bmatrix}$

23. $A = \begin{bmatrix} 2 & 0 & -2 \\ 0 & 2 & 2 \\ -2 & 0 & 2 \end{bmatrix}$, $B = \begin{bmatrix} 0 & 1 & 0 \\ 1 & 0 & 1 \\ 0 & 1 & 0 \end{bmatrix}$

24. $A = \begin{bmatrix} 1 & 2 & 3 \\ 0 & 1 & 2 \\ 0 & 0 & 1 \end{bmatrix}$, $B = \begin{bmatrix} 1 & 2 & 1 \\ 0 & 1 & 2 \\ 0 & 0 & 1 \end{bmatrix}$

25. $A = \begin{bmatrix} 4 \\ 1 \\ 0 \\ 2 \end{bmatrix}$, $B = \begin{bmatrix} 2 & 0 & -1 & 3 \end{bmatrix}$

26. $A = \begin{bmatrix} 1 & -2 & 0 & 0 \end{bmatrix}$, $B = \begin{bmatrix} 0 \\ -1 \\ 3 \\ 2 \end{bmatrix}$

In Exercises 27–40, find the inverse of each matrix if the inverse exists.

27. $\begin{bmatrix} 1 & 1 \\ 0 & 1 \end{bmatrix}$

28. $\begin{bmatrix} 1 & 0 \\ 1 & 1 \end{bmatrix}$

29. $\begin{bmatrix} 2 & -1 \\ 1 & 1 \end{bmatrix}$

30. $\begin{bmatrix} 1 & -1 \\ 1 & 1 \end{bmatrix}$

31. $\begin{bmatrix} 2 & 2 \\ 0 & 2 \end{bmatrix}$

32. $\begin{bmatrix} 2 & 0 \\ -1 & -3 \end{bmatrix}$

33. $\begin{bmatrix} 1 & 0 & 1 \\ 0 & 1 & 0 \\ 0 & 0 & 1 \end{bmatrix}$ 34. $\begin{bmatrix} 1 & 0 & 1 \\ 0 & 1 & 0 \\ 1 & 0 & 0 \end{bmatrix}$ 35. $\begin{bmatrix} 1 & 2 & 3 \\ 0 & 1 & 2 \\ 0 & 0 & 1 \end{bmatrix}$

36. $\begin{bmatrix} 2 & 0 & 0 \\ 1 & 2 & 0 \\ 1 & 1 & 2 \end{bmatrix}$ 37. $\begin{bmatrix} 2 & 1 & 1 \\ 0 & 0 & 1 \\ 0 & 0 & 2 \end{bmatrix}$ 38. $\begin{bmatrix} 1 & 0 & 0 \\ 2 & 1 & 0 \\ 3 & 2 & 0 \end{bmatrix}$

39. $\begin{bmatrix} 1 & 0 & 0 \\ 0 & 2 & 0 \\ 0 & 0 & 3 \end{bmatrix}$ 40. $\begin{bmatrix} 0 & 0 & 1 \\ 0 & 2 & 0 \\ 3 & 0 & 0 \end{bmatrix}$

In Exercises 41–50, solve the linear system using formula (7.10).

41. $\begin{cases} 2x - y = 2 \\ 4x - y = 1 \end{cases}$ 42. $\begin{cases} 3x - 2y = 4 \\ 2x + 6y = -1 \end{cases}$ 43. $\begin{cases} x - y = 5 \\ 2x - 3y = 12 \end{cases}$ 44. $\begin{cases} -x + 3y = 1 \\ x - 2y = -2 \end{cases}$

45. $\begin{cases} 2y = 4 \\ y - 3z = 5 \\ x - 2y - 3z = 2 \end{cases}$ 46. $\begin{cases} x - y = -1 \\ x + z = 4 \\ y - 3z = -3 \end{cases}$ 47. $\begin{cases} x - z = -2 \\ 2y + z = 13 \\ y + z = 9 \end{cases}$

48. $\begin{cases} 2x - z = 3 \\ y - 3z = 8 \\ x + z = 0 \end{cases}$ 49. $\begin{cases} 2x - y = 7 \\ 2y + z = -10 \\ x + 3y + 2z = 15 \end{cases}$ 50. $\begin{cases} x + y = 0 \\ x + 2y + z = 2 \\ y + 3z = 8 \end{cases}$

7.4 DETERMINANTS AND CRAMER'S RULE

With every *square* matrix A we can associate a number, called the *determinant* of A and denoted by det A. The notion of determinants is important in both applied and theoretical mathematics. In this section, we define determinants and discuss several of their properties. Also, we study *Cramer's rule*, which uses determinants to solve systems of linear equations.

Determinants will be defined in an *inductive way*. First determinants of 1×1 matrices are defined. (This is the trivial case.) Next we define determinants of 2×2 matrices. Using this definition, we define determinants of 3×3 matrices, and so on. Assuming that the definition of determinants of $(n - 1) \times (n - 1)$ matrices is known, we define determinants of $n \times n$ matrices. This way of defining determinants uses the *principle of mathematical induction*, a method of proof discussed in Chapter 9.

The Definition of Determinants

If $A = [a]$ is a 1×1 matrix, then by definition we set

$$\det A = a.$$

If $A = \begin{bmatrix} a_{11} & a_{12} \\ a_{21} & a_{22} \end{bmatrix}$ is a 2×2 matrix, then its determinant is defined by

(7.11) $\det A = a_{11}a_{22} - a_{21}a_{12}.$

The determinant of A is also denoted by

$$\det A = \begin{vmatrix} a_{11} & a_{12} \\ a_{21} & a_{22} \end{vmatrix}.$$

◆ **Example 1.** If $A = \begin{bmatrix} 2 & 1 \\ 4 & -5 \end{bmatrix}$, find det A.

Solution.   According to (7.11), we have

$$\det A = \begin{vmatrix} 2 & 1 \\ 4 & -5 \end{vmatrix} = 2 \cdot (-5) - (1 \cdot 4) = -10 - 4 = -14.$$

◇ **Practice Exercise 1.** Find the determinant of the matrix

$$B = \begin{bmatrix} -1 & 2 \\ -3 & -4 \end{bmatrix}.$$

Answer.   det $B = 10$.

In order to extend the definition of determinant to $n \times n$ matrices, with $n \geq 3$, we have to introduce the notions of *minors* and *cofactors*.

Minors and Cofactors

If $A = [a_{ij}]$ is an $n \times n$ matrix, we denote by M_{ij} the $(n - 1) \times (n - 1)$ matrix obtained by deleting the *i*th row and *j*th column from the matrix A. The matrix M_{ij} is called the *complementary matrix* of the element a_{ij}.

Next, assuming that determinants of $(n - 1) \times (n - 1)$ matrices are known, we define the *minor* of the element a_{ij} as the determinant of the matrix M_{ij}. Also, the number $A_{ij} = (-1)^{i+j} \det M_{ij}$ is said to be the *cofactor* of the element a_{ij}. Thus the cofactor of a_{ij} is the determinant of the complementary matrix M_{ij} multiplied by 1 if the sum $i + j$ is *even*, or by -1 if the sum $i + j$ is *odd*.

◆ **Example 2.** Find the minor and cofactor of the element a_{23} in the matrix

$$A = \begin{bmatrix} 3 & 0 & -2 \\ 0 & 1 & 4 \\ 1 & -2 & 1 \end{bmatrix}$$

Solution.   By deleting the second row and third column of the given matrix as shown,

$$\begin{bmatrix} 3 & 0 & -2 \\ 0 & 1 & 4 \\ 1 & -2 & 1 \end{bmatrix}.$$

we obtain the 2×2 complementary matrix

$$M_{23} = \begin{bmatrix} 3 & 0 \\ 1 & -2 \end{bmatrix}.$$

The minor of the element $a_{23} = 4$ of A is then

$$\det M_{23} = \begin{vmatrix} 3 & 0 \\ 1 & -2 \end{vmatrix} = 3(-2) - 1 \cdot 0 = -6.$$

The cofactor of a_{23} is then

$$\begin{aligned} A_{23} &= (-1)^{2+3} \det M_{23} \\ &= (-1)^{2+3} \begin{vmatrix} 3 & 0 \\ 1 & -2 \end{vmatrix} \\ &= (-1)[(3)(-2) - (1)(0)] \\ &= (-1)(-6) = 6. \end{aligned}$$

◇ **Practice Exercise 2.** For the matrix in Example 2, find the minor and cofactor of the element a_{32}.

Answer. $M_{32} = \begin{bmatrix} 3 & -2 \\ 0 & 4 \end{bmatrix}$, $A_{32} = -12$.

By using the definition of cofactors, we may redefine the determinant of a 2×2 matrix as follows:

(7.12) $\det A = a_{11}A_{11} + a_{12}A_{12}.$

Indeed, the cofactor of a_{11} is

$$A_{11} = (-1)^{1+1} \det[a_{22}] = a_{22},$$

while the cofactor of a_{12} is

$$A_{12} = (-1)^{1+2} \det[a_{21}] = -a_{21}.$$

Thus, formula (7.12) follows from formula (7.11). We conclude that, to find the determinant of a 2×2 matrix, we multiply each element in the first row by the corresponding cofactor, and add the products.

The Determinant of a Three by Three Matrix

We can now extend the definition of determinants to 3×3 matrices. If

$$A = \begin{bmatrix} a_{11} & a_{12} & a_{13} \\ a_{21} & a_{22} & a_{23} \\ a_{31} & a_{32} & a_{33} \end{bmatrix}$$

is a 3×3 matrix, its determinant is defined by

(7.13) $\det A = a_{11}A_{11} + a_{12}A_{12} + a_{13}A_{13}.$

That is, *the determinant of A is the sum of all the products of the elements in the first row by their corresponding cofactors.*

◆ **Example 3.** If

$$A = \begin{bmatrix} 3 & 0 & -2 \\ 0 & 1 & 4 \\ 1 & -2 & 1 \end{bmatrix},$$

find det A.

Solution.   We have

$$M_{11} = \begin{bmatrix} 1 & 4 \\ -2 & 1 \end{bmatrix}, \quad M_{12} = \begin{bmatrix} 0 & 4 \\ 1 & 1 \end{bmatrix}, \quad M_{13} = \begin{bmatrix} 0 & 1 \\ 1 & -2 \end{bmatrix}$$

Hence

$$A_{11} = (-1)^{1+1} \begin{vmatrix} 1 & 4 \\ -2 & 1 \end{vmatrix}, \quad A_{12} = (-1)^{1+2} \begin{vmatrix} 0 & 4 \\ 1 & 1 \end{vmatrix}$$

and

$$A_{13} = (-1)^{1+3} \begin{vmatrix} 0 & 1 \\ 1 & -2 \end{vmatrix}.$$

Thus according to (7.13), we have

$$\det A = (3)(-1)^{1+1} \begin{vmatrix} 1 & 4 \\ -2 & 1 \end{vmatrix} + (0)(-1)^{1+2} \begin{vmatrix} 0 & 4 \\ 1 & 1 \end{vmatrix}$$
$$+ (-2)(-1)^{1+3} \begin{vmatrix} 0 & 1 \\ 1 & -2 \end{vmatrix}$$
$$= (3) \begin{vmatrix} 1 & 4 \\ -2 & 1 \end{vmatrix} + (0) \begin{vmatrix} 0 & 4 \\ 1 & 1 \end{vmatrix} + (-2) \begin{vmatrix} 0 & 1 \\ 1 & -2 \end{vmatrix}$$
$$= (3)[(1)(1) - (-2)(4)] + (-2)[(0)(-2) - (1)(1)]$$
$$= (3)(9) + (-2)(-1)$$
$$= 27 + 2$$
$$= 29.$$

◇ **Practice Exercise 3.**   Find the determinant of the matrix

$$B = \begin{bmatrix} 1 & -1 & 0 \\ 1 & 0 & 1 \\ 0 & 1 & -3 \end{bmatrix}$$

Answer.   det $B = -4$.

The Determinant of an $n \times n$ Matrix

In general, if

$$A = \begin{bmatrix} a_{11} & a_{12} & \cdots & a_{1n} \\ a_{21} & a_{22} & \cdots & a_{2n} \\ \cdots & \cdots & \cdots & \cdots \\ a_{n1} & a_{n2} & \cdots & a_{nn} \end{bmatrix}$$

is an $n \times n$ matrix and we assume that determinants of $(n-1) \times (n-1)$ matrices are defined, then

(7.14) $\det A = a_{11}A_{11} + a_{12}A_{12} + \cdots + a_{1n}A_{1n}.$

This formula is similar to formulas (7.12) and (7.13). We shall refer to it as the *cofactor expansion along the first row* of the determinant of the matrix A.

The determinant of A is also denoted by

$$\det A = \begin{vmatrix} a_{11} & a_{12} & \cdots & a_{1n} \\ a_{21} & a_{22} & \cdots & a_{2n} \\ \cdots & \cdots & \cdots & \cdots \\ a_{n1} & a_{n2} & \cdots & a_{nn} \end{vmatrix}$$

In what follows we shall refer to *elements, rows,* and *columns of a determinant,* instead of elements, rows, and columns of the matrix corresponding to the determinant.

An important theorem that we quote without proof states that the determinant of an $n \times n$ matrix can be obtained as the *cofactor expansion along any of its rows or columns.* Thus, definition (7.14) is equivalent to

(7.15) $\det A = a_{i1}A_{i1} + a_{i2}A_{i2} + \cdots + a_{in}A_{in},$

which is the cofactor expansion along the ith row of the determinant of the matrix A.

According to this result, it may be possible to shorten the computation of a determinant by choosing a row or a column with the largest number of zeros. Notice that if all entries of a row or a column are zeros, then the determinant is zero.

◆ **Example 4.**   Find $\det A$ if

$$A = \begin{bmatrix} 3 & -1 & 2 & 0 \\ 2 & 1 & -3 & 1 \\ 1 & 0 & -1 & 0 \\ 4 & 1 & -2 & 0 \end{bmatrix}$$

Solution.   Since the last column has three zeros, we use the cofactor expansion along the fourth column, to get

$$\det A = (-1)^{2+4}\begin{vmatrix} 3 & -1 & 2 \\ 1 & 0 & -1 \\ 4 & 1 & -2 \end{vmatrix} = \begin{vmatrix} 3 & -1 & 2 \\ 1 & 0 & -1 \\ 4 & 1 & -2 \end{vmatrix}$$

We now expand the 3×3 determinant by the elements of the second row or column. Choosing the second row, we have

$$\det A = \begin{vmatrix} 3 & -1 & 2 \\ 1 & 0 & -1 \\ 4 & 1 & -2 \end{vmatrix} = (-1)^{2+3} \begin{vmatrix} -1 & 2 \\ 1 & -2 \end{vmatrix} + (-1)^{3+3} \begin{vmatrix} 3 & -1 \\ 4 & 1 \end{vmatrix}$$

$$= -[(-1) \cdot (-2) - 1 \cdot 2] + [3 \cdot 1 - 4 \cdot (-1)] = 7.$$

◇ **Practice Exercise 4.**   Let

$$A = \begin{bmatrix} 2 & 0 & 0 & 0 \\ 1 & 2 & 0 & -1 \\ -1 & 3 & 0 & 2 \\ 3 & -1 & 1 & 2 \end{bmatrix}.$$

Find det A.

Answer.   det $A = -14$.

Computing Third Order Determinants

The determinant of a three by three matrix can be computed by either of the two following schemes:

$$= a_{11}a_{22}a_{33} + a_{12}a_{23}a_{31} + a_{13}a_{21}a_{32}$$
$$- a_{11}a_{32}a_{23} - a_{12}a_{21}a_{33} - a_{13}a_{22}a_{31}.$$

Equivalently, think of reproducing the first two columns to the right of the matrix:

$$\begin{vmatrix} a_{11} & a_{12} & a_{13} \\ a_{21} & a_{22} & a_{23} \\ a_{31} & a_{32} & a_{33} \end{vmatrix} = a_{11}a_{22}a_{33} + a_{12}a_{23}a_{31} + a_{13}a_{21}a_{33}$$
$$- a_{13}a_{22}a_{31} - a_{11}a_{23}a_{32} - a_{12}a_{21}a_{33}$$

This can be checked by computing the right-hand side of (7.13). Note that we have six terms, three with a plus sign and three with a minus sign, as the above diagram indicates.

◆ Example 5. Use the above scheme to evaluate the determinant

$$\begin{vmatrix} 3 & 0 & -2 \\ 0 & 1 & 4 \\ 1 & -2 & 1 \end{vmatrix}$$

Solution. This determinant was computed in Example 3 by the cofactor expansion along the first row. Now, we compute it by using the scheme described above.

$$\begin{vmatrix} 3 & 0 & -2 \\ 0 & 1 & 4 \\ 1 & -2 & 1 \end{vmatrix} = 3 \cdot 1 \cdot 1 + 0 \cdot 4 \cdot 1 + (-2) \cdot (-2) \cdot 0$$

$$- 3 \cdot (-2) \cdot 4 - 0 \cdot 0 \cdot 1 - (-2) \cdot 1 \cdot 1$$
$$= 3 + 24 + 2$$
$$= 29.$$

The student should be very much aware that the above scheme applies *only* to 3×3 determinants. The method *does not apply to higher order determinants.*

◇ Practice Exercise 5. Use the scheme explained above to compute the determinant of the matrix

$$B = \begin{bmatrix} 1 & -1 & 0 \\ 1 & 0 & 1 \\ 0 & 1 & -3 \end{bmatrix}.$$

Answer. $\det B = -4.$

Properties of Determinants

The properties that follow greatly simplify the task of evaluating determinants of matrices of order 3 or greater.

> I. Interchanging two rows (or columns) of a determinant changes the sign of the determinant.

For example,

$$\begin{vmatrix} a_{21} & a_{22} \\ a_{11} & a_{12} \end{vmatrix} = a_{21}a_{12} - a_{11}a_{22}$$

$$= -(a_{11}a_{22} - a_{21}a_{12})$$

$$= -\begin{vmatrix} a_{11} & a_{12} \\ a_{21} & a_{22} \end{vmatrix}.$$

> II. If we multiply a row (or column) of a determinant by a constant, the new determinant is equal to the original determinant times the constant.

For example,

$$\begin{vmatrix} ca_{11} & ca_{12} \\ a_{21} & a_{22} \end{vmatrix} = (ca_{11})a_{22} - a_{21}(ca_{12})$$

$$= c(a_{11}a_{22} - a_{21}a_{12})$$

$$= c\begin{vmatrix} a_{11} & a_{12} \\ a_{21} & a_{22} \end{vmatrix}.$$

> III. If some multiple of a row (or column) of a determinant is added to another row (or column), the value of the determinant is unchanged.

For example,

$$\begin{vmatrix} a_{11} & a_{12} \\ a_{21} + ca_{11} & a_{22} + ca_{12} \end{vmatrix} = a_{11}(a_{22} + ca_{12}) - a_{12}(a_{21} + ca_{11})$$

$$= a_{11}a_{22} + ca_{11}a_{12} - a_{12}a_{21} - ca_{12}a_{11}$$

$$= a_{11}a_{22} - a_{12}a_{21}$$

$$= \begin{vmatrix} a_{11} & a_{12} \\ a_{21} & a_{22} \end{vmatrix}.$$

Properties I, II, and III that we proved for 2×2 determinants extend to the general case of $n \times n$ determinants. The proofs, which are beyond the scope of this book, will not be given here.

Property I implies that if the corresponding elements in two rows (or columns) of a matrix are equal, then its determinant is zero. To see this, note that if a matrix A has two equal rows, then interchanging the rows does not change the matrix. But, by property I, we have

$$\det A = -\det A,$$

which implies that $\det A = 0$.

◆ **Example 6.**   Without computing the determinant, verify that

$$\begin{vmatrix} 1 & 2 & -1 \\ 2 & 1 & 2 \\ 5 & 7 & -1 \end{vmatrix} = 0.$$

Solution.   By using properties I, II, and III above, we have

$$\begin{vmatrix} 1 & 2 & -1 \\ 2 & 1 & 2 \\ 5 & 7 & -1 \end{vmatrix}$$

$\downarrow$ Multiply the second row by -1 and add it to the third

$$= \begin{vmatrix} 1 & 2 & -1 \\ 2 & 1 & 2 \\ 3 & 6 & -3 \end{vmatrix}$$

$\downarrow$ Property II

$$= 3 \begin{vmatrix} 1 & 2 & -1 \\ 2 & 1 & 2 \\ 1 & 2 & -1 \end{vmatrix} = 0$$

since the first and third rows are equal.

◇ **Practice Exercise 6.**   Using Properties I, II, and III of determinants, show that

$$\begin{vmatrix} 1 & 0 & -2 \\ 3 & 1 & -4 \\ 5 & 1 & -8 \end{vmatrix} = 0.$$

without actually expanding the determinant.

◆ **Example 7.**   Show that

$$\begin{vmatrix} x & y & 1 \\ x_1 & y_1 & 1 \\ x_2 & y_2 & 1 \end{vmatrix} = 0$$

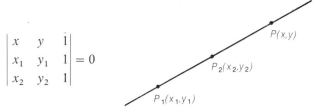

represents an equation of the line through the points $P_1(x_1, y_1)$ and $P_2(x_2, y_2)$, where $P_1 \neq P_2$.

Solution.   Expanding the determinant by the cofactors of the elements of the first row and equating to the result zero, we get

$$\begin{vmatrix} y_1 & 1 \\ y_2 & 1 \end{vmatrix} x - \begin{vmatrix} x_1 & 1 \\ x_2 & 1 \end{vmatrix} y + \begin{vmatrix} x_1 & y_1 \\ x_2 & y_2 \end{vmatrix} = 0,$$

or

$$(y_1 - y_2)x + (x_2 - x_1)y + (x_1 y_2 - x_2 y_1) = 0.$$

The latter is an equation of the form $Ax + By + C = 0$, where $A = y_1 - y_2$, $B = x_2 - x_1$, and $C = x_1 y_2 - x_2 y_1$. The coefficients A and B are not both zero (because $P_1 \neq P_2$), so the equation represents a line (Section 3.2). Substituting (x_1, y_1) for (x, y) in the determinant makes the first two rows equal, so the determinant is zero. Thus the point $P_1(x_1, y_1)$ belongs to the line. Similarly, $P_2(x_2, y_2)$ also belongs to the line.

◇ **Practice Exercise 7.** Use the method described in Example 7 to find an equation of a line through the points $P_1(2, -3)$ and $P_2(-1, 4)$.

Answer. $\begin{vmatrix} x & y & 1 \\ 2 & -3 & 1 \\ -1 & 4 & 1 \end{vmatrix} = 0$ or $7x + 3y - 5 = 0$.

◆ **Example 8.** Evaluate det A if

$$A = \begin{bmatrix} 1 & 0 & -2 & 1 \\ 2 & -1 & 3 & 0 \\ -2 & 0 & 2 & 0 \\ -3 & 2 & -1 & 1 \end{bmatrix}$$

Solution. Noticing that the last column of A already has two zeros, we try to produce another zero in it. This can be achieved by multiplying the fourth row by -1 and adding it to the first.

$$\det A = \begin{vmatrix} 4 & -2 & -1 & 0 \\ 2 & -1 & 3 & 0 \\ -2 & 0 & 2 & 0 \\ -3 & 2 & -1 & 1 \end{vmatrix}$$

Now, we use the cofactor expansion along the fourth column:

$$\det A = \begin{vmatrix} 4 & -2 & -1 & 0 \\ 2 & -1 & 3 & 0 \\ -2 & 0 & 2 & 0 \\ -3 & 2 & -1 & 1 \end{vmatrix}$$

$$= (-1)^{4+4} \begin{vmatrix} 4 & -2 & -1 \\ 2 & -1 & 3 \\ -2 & 0 & 2 \end{vmatrix} = \begin{vmatrix} 4 & -2 & -1 \\ 2 & -1 & 3 \\ -2 & 0 & 2 \end{vmatrix}$$

There is a zero in the third row of the last determinant. We can produce another zero by adding the first column to the third, obtaining

$$\det A = \begin{vmatrix} 4 & -2 & 3 \\ 2 & -1 & 5 \\ -2 & 0 & 0 \end{vmatrix}$$

By the cofactor expansion along the last row, we get

$$\det A = (-2)(-1)^{1+3} \begin{vmatrix} -2 & 3 \\ -1 & 5 \end{vmatrix}$$

$$= (-2)[(-2) \cdot 5 - (-1) \cdot 3]$$

$$= (-2)(-10 + 3) = (-2)(-7) = 14.$$

◇ **Practice Exercise 8.** Find

$$
\begin{vmatrix}
1 & 0 & 2 & 0 \\
-2 & 1 & -4 & 3 \\
3 & 4 & 6 & 1 \\
-1 & 2 & -1 & 2
\end{vmatrix}
$$

by first producing as many zeros as possible along a row or column and then expanding the determinant along that row or column.

Answer. 11.

Cramer's Rule

HISTORICAL NOTE

Determinants
The notion of determinant was known to the Japanese mathematician Seki Takakusu before 1683. In Western mathematics, determinants occur in an unpublished 1693 letter of G. W. Leibniz, the German-coinventor (simultaneously with Isaac Newton) of calculus. The Swiss mathematician Gabriel Cramer (1704–1752) rediscovered them; the method for solving linear systems, called *Cramer's rule*, appeared in 1750. The use of determinants became widespread after the appearance, in 1841, of a paper *On the Formation and Properties of Determinants* by the Konigsberg mathematician Carl G. Jacobi (1804–1851).

Determinants provide a general method of solving systems of linear equations. Consider, for example, the two by two system

$$
(7.16) \quad \begin{cases} a_{11}x_1 + a_{12}x_2 = b_1 \\ a_{21}x_1 + a_{22}x_2 = b_2 \end{cases}
$$

and assume that the determinant of the matrix of coefficients is nonzero, that is,

$$
D = \begin{vmatrix} a_{11} & a_{12} \\ a_{21} & a_{22} \end{vmatrix} = a_{11}a_{22} - a_{21}a_{12} \neq 0.
$$

Next, we solve the system (7.16) by elimination. Multiplying the first equation by a_{22} and the second by a_{12}, we obtain

$$
\begin{cases} a_{11}a_{22}x_1 + a_{12}a_{22}x_2 = b_1a_{22} \\ a_{21}a_{12}x_1 + a_{22}a_{12}x_2 = b_2a_{12} \end{cases}
$$

Subtracting the second equation from the first one, we eliminate the terms containing x_2, and get

$$
(a_{11}a_{22} - a_{21}a_{12})x_1 = b_1a_{22} - b_2a_{12};
$$

hence

$$
x_1 = \frac{b_1a_{22} - b_2a_{12}}{a_{11}a_{22} - a_{21}a_{12}}.
$$

Proceeding in a similar manner with system (7.16) and eliminating the terms containing x_1, we obtain

$$
x_2 = \frac{b_2a_{11} - b_1a_{21}}{a_{11}a_{22} - a_{21}a_{22}}.
$$

The solution (x_1, x_2) of system (7.16) can be expressed in terms of determinants as follows:

$$
(7.17) \quad x_1 = \frac{\begin{vmatrix} b_1 & a_{12} \\ b_2 & a_{22} \end{vmatrix}}{\begin{vmatrix} a_{11} & a_{12} \\ a_{21} & a_{22} \end{vmatrix}}
$$

and

$$(7.18) \quad x_2 = \frac{\begin{vmatrix} a_{11} & b_1 \\ a_{21} & b_2 \end{vmatrix}}{\begin{vmatrix} a_{11} & a_{12} \\ a_{21} & a_{22} \end{vmatrix}}$$

Formulas (7.17) and (7.18) are *Cramer's rule* for a 2 × 2 linear system.

◆ **Example 9.** Using Cramer's rule, solve the system

$$\begin{cases} 2x - y = 8 \\ 3x + 2y = -2 \end{cases}$$

Solution. First, make sure that the determinant of the matrix of coefficients is different from zero:

$$\begin{vmatrix} 2 & -1 \\ 3 & 2 \end{vmatrix} = 2 \cdot 2 - (-1) \cdot 3 = 4 + 3 = 7 \neq 0.$$

Next, according to (7.17) and (7.18), obtain the solution of the system.

$$x = \frac{\begin{vmatrix} 8 & -1 \\ -2 & 2 \end{vmatrix}}{7} = \frac{16 - 2}{7} = \frac{14}{7} = 2$$

and

$$y = \frac{\begin{vmatrix} 2 & 8 \\ 3 & -2 \end{vmatrix}}{7} = \frac{-4 - 24}{7} = \frac{-28}{7} = -4.$$

◇ **Practice Exercise 9.** Solve by Cramer's rule the system

$$\begin{cases} 2x - y = 1 \\ x + 2y = -12. \end{cases}$$

Answer. $x = -2, y = -5.$

The General Case

Formulas (7.17) and (7.18) generalize to $n \times n$ linear systems. Let

$$(7.19) \quad \begin{cases} a_{11}x_1 + a_{12}x_2 + \cdots + a_{1n}x_n = b_1 \\ a_{21}x_1 + a_{22}x_2 + \cdots + a_{2n}x_n = b_2 \\ \cdot \quad \cdot \quad \cdot \quad \cdot \quad \cdot \quad \cdot \quad \cdot \quad \cdot \quad \cdot \\ a_{n1}x_1 + a_{n2}x_2 + \cdots + a_{nn}x_n = b_n \end{cases}$$

be an $n \times n$ linear system and assume that the determinant D of the matrix of coefficients is nonzero, that is

$$D = \begin{vmatrix} a_{11} & a_{12} & \cdots & a_{1n} \\ a_{21} & a_{22} & \cdots & a_{2n} \\ \cdots & \cdots & \cdots & \cdots \\ a_{n1} & a_{n2} & \cdots & a_{nn} \end{vmatrix} \neq 0.$$

Then, the jth component of the solution $(x_1, x_2, \ldots, x_n)$ of the system (7.19) is given by the formula

$$
(7.20) \quad x_j = \frac{\begin{vmatrix} a_{11} & \cdots & b_1 & \cdots & a_{1n} \\ a_{21} & \cdots & b_2 & \cdots & a_{2n} \\ \cdots & \cdots & \cdots & \cdots & \cdots \\ a_{n1} & \cdots & b_n & \cdots & a_{nn} \end{vmatrix}}{D}, \quad 1 \leq j \leq n
$$

In other words, the jth component x_j of the solution is the quotient of two determinants. The denominator is the determinant of the matrix of coefficients. The numerator is the determinant of the matrix obtained from the matrix of coefficients by replacing the jth column with the constants $b_1, b_2, \ldots, b_n$.

Except for 2×2 and 3×3 systems, Cramer's rule is not very useful in practice. In most cases the Gaussian elimination method is more efficient. However, Cramer's rule is interesting for historical reasons, besides being a valuable theoretical tool. In fact, Cramer's rule may become important again: recent research seems to indicate that parallel-processing computer techniques could implement determinant calculations and Cramer's rule relatively quickly and efficiently. Present-day computer methods for solving linear systems are for the most part versions of Gaussian elimination.

◆ *Example 10.* Use Cramer's rule to solve the system

$$
\begin{cases}
x & + 2z = 1 \\
3x - y & = -3 \\
2y + z = 1.
\end{cases}
$$

Solution. We have

$$
D = \begin{vmatrix} 1 & 0 & 2 \\ 3 & -1 & 0 \\ 0 & 2 & 1 \end{vmatrix} = 1 \begin{vmatrix} -1 & 0 \\ 2 & 1 \end{vmatrix} - 0 \begin{vmatrix} 3 & 0 \\ 0 & 1 \end{vmatrix} + 2 \begin{vmatrix} 3 & -1 \\ 0 & 2 \end{vmatrix}
$$

$$
= 1[(-1)(1) - 2(0)] + 2[3(2) - 0(-1)]
$$

$$
= 1(-1) + 2(6)
$$

$$
= 11 \neq 0
$$

According to formula (7.20), we get

$$
x = \frac{\begin{vmatrix} 1 & 0 & 2 \\ -3 & -1 & 0 \\ 1 & 2 & 1 \end{vmatrix}}{11} = \frac{-11}{11} = -1,
$$

$$
y = \frac{\begin{vmatrix} 1 & 1 & 2 \\ 3 & -3 & 0 \\ 0 & 1 & 1 \end{vmatrix}}{11} = \frac{0}{11} = 0,
$$

and

$$z = \frac{\begin{vmatrix} 1 & 0 & 1 \\ 3 & -1 & -3 \\ 0 & 2 & 1 \end{vmatrix}}{11} = \frac{11}{11} = 1$$

where the computation of each determinant is left as an exercise to the reader.

◇ **Practice Exercise 10.** Solve by Cramer's rule the system

$$\begin{cases} x + y & = 0 \\ x + 2y + z = 2 \\ y + 3z = 8 \end{cases}$$

Answer. $x = 1$, $y = -1$, $z = 3$.

⬛ EXERCISES 7.4

For each matrix in Exercises 1–6, write in determinant form the minor and cofactor of each of the given entries.

1. $\begin{bmatrix} 2 & -1 \\ 3 & -2 \end{bmatrix}$; a_{11}, a_{21}

2. $\begin{bmatrix} -3 & 1 \\ -2 & 2 \end{bmatrix}$; a_{12}, a_{22}

3. $\begin{bmatrix} 1 & -2 & 0 \\ 3 & -1 & 2 \\ 0 & -2 & 1 \end{bmatrix}$; a_{11}, a_{32}, a_{13}

4. $\begin{bmatrix} 0 & 1 & 2 \\ 1 & 0 & 1 \\ 2 & 1 & 0 \end{bmatrix}$; a_{13}, a_{22}, a_{31}

5. $\begin{bmatrix} 1 & 0 & 0 & 2 \\ 2 & 1 & 0 & 3 \\ -2 & 0 & 0 & -1 \\ -3 & 0 & 1 & 2 \end{bmatrix}$; a_{11}, a_{13}, a_{43}

6. $\begin{bmatrix} -2 & 0 & 1 & 2 \\ -1 & 3 & 0 & 1 \\ 2 & 2 & -1 & 1 \\ 1 & -1 & 3 & -2 \end{bmatrix}$; a_{21}, a_{33}, a_{42}

In Exercises 7–22, evaluate each determinant.

7. $\begin{vmatrix} 2 & -1 \\ 3 & -2 \end{vmatrix}$

8. $\begin{vmatrix} -3 & 1 \\ -2 & 2 \end{vmatrix}$

9. $\begin{vmatrix} 2 & -3 \\ -1 & -2 \end{vmatrix}$

10. $\begin{vmatrix} -2 & -1 \\ 3 & -1 \end{vmatrix}$

11. $\begin{vmatrix} 0 & 1 \\ 2 & 0 \end{vmatrix}$

12. $\begin{vmatrix} 2 & 0 \\ 0 & 3 \end{vmatrix}$

13. $\begin{vmatrix} 1 & -2 & 3 \\ 3 & -1 & 2 \\ 0 & -2 & 1 \end{vmatrix}$

14. $\begin{vmatrix} 0 & 1 & 2 \\ 1 & 0 & 1 \\ 2 & 1 & 0 \end{vmatrix}$

15. $\begin{vmatrix} 1 & 2 & 3 \\ 0 & -1 & -2 \\ -3 & -2 & 1 \end{vmatrix}$

16. $\begin{vmatrix} 2 & 0 & 0 \\ 1 & 3 & 4 \\ -2 & -1 & -2 \end{vmatrix}$

17. $\begin{vmatrix} 2 & 1 & 2 \\ 0 & 3 & 1 \\ 0 & 0 & 4 \end{vmatrix}$

18. $\begin{vmatrix} 1 & 2 & -2 \\ -2 & 3 & 0 \\ 4 & 0 & 0 \end{vmatrix}$

19. $\begin{vmatrix} 2 & -1 & 3 \\ 1 & 2 & 1 \\ 4 & -2 & 6 \end{vmatrix}$

20. $\begin{vmatrix} 1 & 0 & -2 \\ 2 & 1 & 3 \\ 4 & 1 & -1 \end{vmatrix}$

21. $\begin{vmatrix} 1 & 0 & 0 & 2 \\ 2 & 1 & 0 & 3 \\ -2 & 0 & 0 & -1 \\ -3 & 0 & 1 & 2 \end{vmatrix}$

22. $\begin{vmatrix} -2 & 0 & 1 & 2 \\ -1 & 3 & 0 & 1 \\ 2 & 2 & -1 & 1 \\ 1 & -1 & 3 & -2 \end{vmatrix}$

23. Write in determinant form an equation of the line through the points $(2, 3)$ and $(-1, -2)$.

24. Write in determinant form an equation of the line through the points $(-3, 1)$ and $(2, -4)$.

In Exercises 25–28, compute the 3 × 3 determinants as in Example 5.

25. $\begin{vmatrix} 1 & 2 & -1 \\ 0 & -2 & 1 \\ -1 & 2 & 3 \end{vmatrix}$

26. $\begin{vmatrix} 2 & 0 & -2 \\ 1 & -1 & 1 \\ 3 & 0 & -3 \end{vmatrix}$

27. $\begin{vmatrix} 1 & 2 & 3 \\ 0 & -1 & -2 \\ -3 & -2 & 1 \end{vmatrix}$

28. $\begin{vmatrix} 2 & 0 & 0 \\ 1 & 3 & 4 \\ -2 & -1 & -2 \end{vmatrix}$

In Exercises 29–34, verify each statement without expanding the determinant.
Use only Properties I, II, and III of determinants.

29. $\begin{vmatrix} 1 & 2 & 3 \\ 0 & -1 & -2 \\ 3 & 6 & 9 \end{vmatrix} = 0$

30. $\begin{vmatrix} -2 & -1 & -2 \\ 3 & 2 & 1 \\ 4 & 2 & 4 \end{vmatrix} = 0$

31. $\begin{vmatrix} 1 & 2 & -1 \\ 0 & -2 & 1 \\ 2 & 2 & -1 \end{vmatrix} = 0$

32. $\begin{vmatrix} 2 & 0 & -2 \\ 1 & -1 & 1 \\ -1 & 3 & -5 \end{vmatrix} = 0$

● **33.** $\begin{vmatrix} 2 & 1 & 1 & 3 \\ -1 & -2 & 2 & 1 \\ 1 & 0 & 0 & -1 \\ 5 & 2 & 2 & 5 \end{vmatrix} = 0$

● **34.** $\begin{vmatrix} 0 & 0 & 0 & 3 \\ 1 & 2 & -2 & 1 \\ -2 & 5 & -5 & 4 \\ -1 & 1 & -1 & 1 \end{vmatrix} = 0$

Use Cramer's rule to solve the systems.

35. $\begin{cases} 2x - y = 4 \\ 3x - 2y = 3 \end{cases}$

36. $\begin{cases} 2x + 2y = -1 \\ 2x + 3y = 6 \end{cases}$

37. $\begin{cases} 10x - 5y = -1 \\ 9x + 6y = 3 \end{cases}$

38. $\begin{cases} x - y = 1 \\ x - 4y = 2 \end{cases}$

39. $\begin{cases} y - 2z = 5 \\ x + z = 0 \\ 2x - y = 3 \end{cases}$

40. $\begin{cases} x - y = -1 \\ x + z = 4 \\ y - 3z = -3 \end{cases}$

7.5 NONLINEAR SYSTEMS OF EQUATIONS

In the previous sections we described several methods of solving systems of linear equations, such as substitution, Gaussian elimination, the inverse matrix method, and Cramer's rule. These are general methods that can be applied to solve any given linear system. In this section we shall study certain *nonlinear systems of equations*. A system is said to be *nonlinear* if it contains at least one equation that is not a linear equation. Nonlinear systems are in most cases more difficult to solve than linear systems. Moreover, there is no general method that could be applied to solve any given nonlinear system. In what follows we shall discuss only examples of nonlinear systems in two unknowns which can be solved either by substitution or elimination.

◆ Example 1. Find all solutions of the system

$$\begin{cases} 2x + y = -1 \\ x^2 - y = 4. \end{cases}$$

Solution. Solve the second equation for y and get

$$y = x^2 - 4.$$

Substituting for y in the first equation, obtain

$$2x + x^2 - 4 = -1$$
$$x^2 + 2x - 3 = 0$$
$$(x + 3)(x - 1) = 0.$$

Hence, $x = -3$ or $x = 1$ are solutions. Now substitute these two values into either of the two equations. Setting $x = -3$ into $y = x^2 - 4$, obtain

$$y = (-3)^2 - 4 = 9 - 4 = 5.$$

Similarly, substituting 1 for x into the same equation, get

$$y = 1^2 - 4 = 1 - 4 = -3.$$

As an exercise, you can check that the same values for y can be obtained by substituting $x = -3$ and $x = 1$ into the equation $2x + y = -1$. Thus the two pairs $(-3, 5)$ and $(1, -3)$ are the solutions of the system.

We can now give a geometrical interpretation of this result. The equation $2x + y = -1$ represents a line, while the equation $x^2 - y = 4$ represents a parabola (Figure 7.8). The two pairs $(-3, 5)$ and $(1, -3)$ are the coordinates of the points of intersection of the line and the parabola.

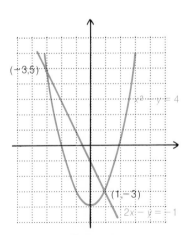

Figure 7.8

◇ **Practice Exercise 1.** Find the intersections of the line $2x - y = -1$ and the parabola $x^2 + y = 1$.

Answer. $(-2, -3), (0, 1)$.

◆ **Example 2.** Find the solution of the system

$$\begin{cases} -x + 6y = 1 \\ 3xy = 1. \end{cases}$$

Solution. Solve the first equation for y and obtain

$$y = \frac{1}{6}x + \frac{1}{6}.$$

Substitute in the second equation and get

$$3x\left[\frac{1}{6}x + \frac{1}{6}\right] = 1$$
$$\frac{x^2}{2} + \frac{x}{2} = 1$$
$$x^2 + x - 2 = 0$$
$$(x - 1)(x + 2) = 0.$$

Thus $x = 1$ or $x = -2$. You may now substitute these two values into either of the two equations of the system. Working with the second equation, obtain

$$3 \cdot 1 \cdot y = 1,$$
$$y = \frac{1}{3}.$$

Similarly,

$$3(-2)y = 1,$$
$$y = -\frac{1}{6}.$$

Thus, $(1, 1/3)$ and $(-2, -1/6)$ are the two solutions of the given system. Figure

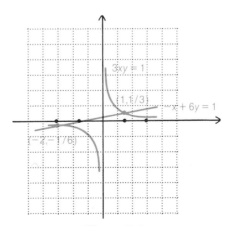

Figure 7.9

7.9 shows the graphs of the equations $-x + 6y = 1$ and $3xy = 1$ and the two points of intersection of the graphs.

◇ **Practice Exercise 2.**   Find all solutions of the system

$$\begin{cases} xy + 1 = 0 \\ 2x + y = -1. \end{cases}$$

Answer. $(1/2, -2), (-1, 1)$.

◆ **Example 3.**   Find the intersections of the parabola $x = y^2$ and the circle $x^2 + y^2 = 6$.

Solution.   The two curves are sketched in Figure 7.10, which indicates that there are two points of intersection. The coordinates of these points are the solutions of the nonlinear system

$$\begin{cases} x = y^2 \\ x^2 + y^2 = 6. \end{cases}$$

Substituting x for the term y^2 in the second equation, we obtain a quadratic equation in x:

$$x^2 + x = 6,$$
$$x^2 + x - 6 = 0,$$
$$(x - 2)(x + 3) = 0,$$

whose solutions are $x = 2$ and $x = -3$. Now, the first equation of the system $x = y^2$ tells us that x is always nonnegative. Thus the solution $x = -3$ of the quadratic equation $x^2 + x - 6 = 0$ must be discarded. Next, substituting $x = 2$ into the first equation gives us

$$2 = y^2,$$
$$y = \pm\sqrt{2}.$$

Thus the points of intersection of the given curves are $(2, \sqrt{2})$ and $(2, -\sqrt{2})$.

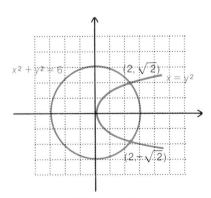

Figure 7.10

◇ **Practice Exercise 3.**   Find all solutions of the system

$$\begin{cases} 4x + y^2 = 0 \\ x^2 + y^2 = 5. \end{cases}$$

What is the geometrical interpretation of these solutions?

Answer. $(-1, 2)$ and $(-1, -2)$ are the points of intersection of the parabola $4x + y^2 = 0$ and the circle $x^2 + y^2 = 5$.

◆ **Example 4.**   Solve the system

$$\begin{cases} x^2 + y^2 = 1 \\ \dfrac{x^2}{4} + 4y^2 = 1. \end{cases}$$

Solution.   The first equation represents a circle and the second an ellipse (Figure 7.11). The two curves may intersect in at most four points, which correspond to the solutions of the system. These can be found by elimination. Multiply the first equation by -4 and add the result to the second.

$$-4x^2 - 4y^2 = -4$$
$$\frac{x^2}{4} + 4y^2 = 1$$
$$\overline{\qquad\qquad\qquad}$$
$$-4x^2 + \frac{x^2}{4} = -3 \;\cdot$$
$$-16x^2 + x^2 = -12$$
$$-15x^2 = -12$$
$$x^2 = \frac{12}{15} = \frac{4}{5}$$
$$x = \pm\frac{2}{\sqrt{5}} = \pm\frac{2\sqrt{5}}{5}.$$

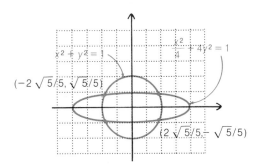

Figure 7.11

Now substitute $2/\sqrt{5}$ for x into either one of the original equations. Working with the first equation, we have

$$x^2 + y^2 = 1$$

$$\left(\frac{2}{\sqrt{5}}\right)^2 + y^2 = 1$$

$$\frac{4}{5} + y^2 = 1$$

$$y^2 = \frac{1}{5}$$

$$y = \pm\frac{1}{\sqrt{5}} = \pm\frac{\sqrt{5}}{5}.$$

Similarly, substituting $-2\sqrt{5}/5$ for x into the first equation yields $y = \pm\sqrt{5}/5$. Thus we have four solutions: $(2\sqrt{5}/5, \sqrt{5}/5), (2\sqrt{5}/5, -\sqrt{5}/5), (-2\sqrt{5}/5, \sqrt{5}/5)$, and $(-2\sqrt{5}/5, -\sqrt{5}/5)$. You should check that these are actually solutions of the given system.

◇ **Practice Exercise 4.**   The two ellipses $2x^2 + 3y^2 = 1$ and $3x^2 + 2y^2 = 1$ have four points of intersection. Find the coordinates of these points.

Answer. $(\sqrt{5}/5, \sqrt{5}/5), (\sqrt{5}/5, -\sqrt{5}/5), (-\sqrt{5}/5, \sqrt{5}/5), (-\sqrt{5}/5, -\sqrt{5}/5).$

 EXERCISES 7.5

In Exercises 1–24, find all solutions of each of the given systems.

1. $\begin{cases} x = y^2 \\ x - 3y = 0 \end{cases}$

2. $\begin{cases} x^2 + y = 0 \\ 2x - y = 0 \end{cases}$

3. $\begin{cases} y = x - 2 \\ x^2 + y = 0 \end{cases}$

4. $\begin{cases} y + x = 6 \\ y^2 - x = 0 \end{cases}$

5. $\begin{cases} x^2 - 4y = 0 \\ 3x - 4y = -4 \end{cases}$

6. $\begin{cases} 4x^2 - y = 0 \\ 4x + y = 0 \end{cases}$

7. $\begin{cases} 3x - y = 0 \\ xy = 3 \end{cases}$

8. $\begin{cases} 4xy = 3 \\ y = \dfrac{3}{4}x \end{cases}$

9. $\begin{cases} 2xy = 1 \\ 2x - 4y = 3 \end{cases}$

10. $\begin{cases} x - 2y = 6 \\ xy + 4 = 0 \end{cases}$

11. $\begin{cases} xy = -1 \\ y + x = 1 \end{cases}$

12. $\begin{cases} xy = 2 \\ y - x = 4 \end{cases}$

13. $\begin{cases} x^2 - 2y = 0 \\ x^2 + y^2 = 8 \end{cases}$

14. $\begin{cases} x^2 + 3y = 0 \\ x^2 + y^2 = 18 \end{cases}$

15. $\begin{cases} y = x^2 - 1 \\ x^2 + y^2 = 13 \end{cases}$

16. $\begin{cases} x - y^2 = 1 \\ x^2 + y^2 = 5 \end{cases}$

17. $\begin{cases} x^2 + y^2 = 7 \\ x - y^2 = -1 \end{cases}$

18. $\begin{cases} x^2 + y = 1 \\ x^2 + y^2 = 3 \end{cases}$

19. $\begin{cases} 2x^2 + 2y^2 = 1 \\ x^2 + 4y^2 = 1 \end{cases}$

20. $\begin{cases} x^2 + y^2 = 1 \\ 5x^2 + y^2 = 1 \end{cases}$

21. $\begin{cases} x^2 + 4y^2 = 1 \\ 2x^2 + y^2 = 1 \end{cases}$

22. $\begin{cases} 3x^2 + y^2 = 1 \\ 4x^2 + y^2 = 1 \end{cases}$

23. $\begin{cases} x^2 - 2x + y^2 = 0 \\ x^2 + y^2 = 1 \end{cases}$

24. $\begin{cases} x^2 + y^2 - 4y = 0 \\ x^2 + y^2 = 1 \end{cases}$

25. The difference of two positive numbers is 8 and their product is 240. What are the numbers?

26. Find two numbers whose sum is 4 and whose product is -96.
27. What are the dimensions of a rectangle whose perimeter is 36 m and whose area is 80 m^2?
28. Find the dimensions of a rectangle with an area of 75 ft^2 and perimeter of 40 feet.
29. The area of a right triangle is 54 m^2 and the hypotenuse is 25 m. What are the dimensions of the two legs?
30. The sum of squares of the sides of a rectangle is 74 cm^2 and the area of the rectangle is 35 cm^2. Find the dimensions of the rectangle.

7.6 SYSTEMS OF INEQUALITIES. LINEAR PROGRAMMING

A line lying in a plane divides the plane into two parts each of which is called a *half-plane*. Each half-plane may or may not contain the line. If a Cartesian coordinate system is defined in the plane and the line is represented by a linear equation, then we may ask what can be said about the coordinates of points on each of the half-planes. To answer this question consider, as an example, the line whose equation is $2x - y + 1 = 0$ (Figure 7.12). As we already know, points such as $(1, 3)$, $(0, 1)$, and $(-1/2, 0)$ that satisfy the equation $2x - y + 1 = 0$ lie on the line. If a point $P(x, y)$ does not lie on the line, then $2x - y + 1 \neq 0$. Consequently, *either* $2x - y + 1 > 0$ *or* $2x - y + 1 < 0$. For example, the point $(1, 4)$ does not lie on the line because $2 \cdot 1 - 4 + 1 = -1 < 0$. Similarly, the point $(1, 2)$ does not belong to the line because $2 \cdot 1 - 2 + 1 = 1 > 0$. Moreover, it can be shown that all points (x, y) on the half-plane that contains $(1, 4)$ satisfy the inequality $2x - y + 1 < 0$, and all points (x, y) on the same half-plane as the point $(1, 2)$ satisfy the inequality $2x - y + 1 > 0$.

Thus each half-plane is characterized as the set of points whose coordinates satisfy one of the strict inequalities

$$2x - y + 1 > 0 \quad or \quad 2x - y + 1 < 0.$$

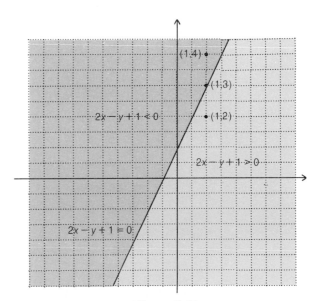

Figure 7.12

If the half-plane contains the line itself, then it is the set of points whose coordinates satisfy

$$2x - y + 1 \geq 0 \quad or \quad 2x - y + 1 \leq 0.$$

The following is a general result about linear inequalities in two variables x and y.

Linear Inequalities

A line $Ax + Bx + C = 0$, with A and B not both equal to zero, divides the coordinate plane into two half-planes. If a point $P(x, y)$, not on the line, belongs to one of the half-planes, then the coordinates of P satisfy one of the *strict linear inequalities*

$$Ax + By + c > 0 \quad or \quad Ax + By + C < 0.$$

Each inequality

$$Ax + By + C \geq 0 \quad or \quad Ax + By + C \leq 0$$

represents one of the half-planes and the line itself.

◆ **Example 1.**   Graph the inequality $4x - 2y + 6 \leq 0$.

Solution.   **First method.** The graph consists of the line $4x - 2y + 6 = 0$ and one of the half-planes determined by this line. To find the half-plane, solve the given inequality for y:

$$4x - 2y + 6 \leq 0$$
$$-2y \leq -4x - 6 \qquad \text{[Multiply by } -1/2. \text{ Note that the sense of the inequality is reversed.]}$$
$$y \geq 2x + 3.$$

Any point with first coordinate x lies *on* the line if its second coordinate is *equal to* $2x + 3$. Any point with first coordinate x and second coordinate y *greater than* $2x + 3$ lies *above* the line. The graph, shown in Figure 7.13a, consists of the line and the region above the line. A solid line is used to indicate that the line is included in the graph.

Second method. Here we choose a *test point* and check to see if the coordinates of the point satisfy the given inequality. For example, the point $(0, 5)$ is such that

$$4 \cdot 0 - 2 \cdot 5 + 6 = -10 + 6 = -4 < 0.$$

Thus it satisfies the inequality $4x - 2y + 6 \leq 0$. It follows that *all* points on the side of the line where the point $(0, 5)$ is located also satisfy the same inequality (Figure 7.13a).

Notice that the origin $(0, 0)$ is such that

$$4 \cdot 0 - 2 \cdot 0 + 6 = 6 > 0.$$

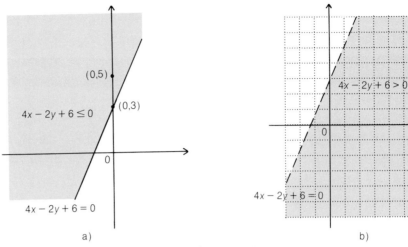

Figure 7.13

Thus $(0, 0)$ *does not* satisfy the inequality $4x - 2y + 6 \leq 0$. It follows that the origin belongs to the region defined by the inequality $4x - 2y + 6 > 0$. The graph of this inequality is illustrated in Figure 7.13b. The dashed line indicates that the line $4x - 2y + 6 = 0$ is not included in the graph.

◇ **Practice Exercise 1.**   Graph the inequality $2x + 3y - 6 > 0$.

Answer.

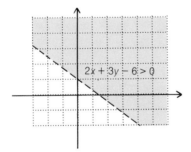

The graph of a function may also divide the plane into two regions.

Nonlinear Inequalities

Let $y = f(x)$ be a function with domain D. All points (x, y) with $x \in D$ and such that $y > f(x)$ are said to be *above* the graph of $y = f(x)$. All points (x, y) with $x \in D$ and such that $y < f(x)$ are said to be *below* the graph of $y = f(x)$.

The *graph of the inequality* $y > f(x)$ $(y < f(x))$ consists of all points above (below) the graph of $y = f(x)$.

Consider, for example, the function $y = x^2$. Figure 7.14a illustrates the graph of the inequality $y \geq x^2$. The solid line indicates that the parabola $y = x^2$

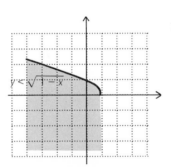

Figure 7.14

is included in the graph. Figure 7.14b shows the graph of the inequality $y < x^2$. The dotted line indicates that the curve $y = x^2$ is not included in the graph.

◆ **Example 2.**   Graph the inequality $y < \sqrt{1 - x}$.

Solution.   The domain of the function defined by $y = \sqrt{1 - x}$ consists of all points such that $1 - x \geq 0$, that is, $x \leq 1$. The graph of this function as well as the graph of the given inequality are shown in Figure 7.15.

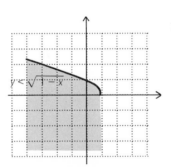

Figure 7.15

◇ **Practice Exercise 2.**   Graph the inequality $y \geq \sqrt{x + 4}$.

Answer.

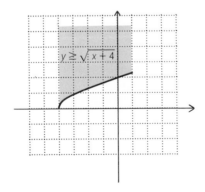

Systems of Inequalities in Two Variables

Instead of a single inequality as discussed in Examples 1 and 2, we may have to deal simultaneously with several linear and/or nonlinear inequalities. Two or more inequalities form a *system of inequalities*. The *solution set* of a system consists of all ordered pairs (x, y) that satisfy all the inequalities of the system. To graph the solution set, just graph each inequality and find the region common to all graphs.

◆ **Example 3.**   Graph the system of linear inequalities

$$\begin{cases} x + y < 4 \\ 2x - y + 3 > 0. \end{cases}$$

Solution.   Solving each inequality for y, obtain the equivalent system

$$\begin{cases} y < -x + 4 \\ y < 2x + 3. \end{cases}$$

Now graph each inequality as shown in Figure 7.16. The region common to the two graphs is the graph of the given system.

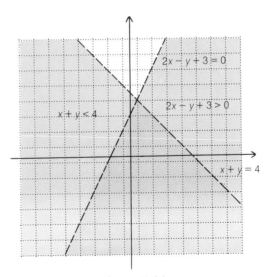

Figure 7.16

◇ **Practice Exercise 3.**   Graph the system of linear inequalities

$$\begin{cases} -x + y < 4 \\ 2x + y - 2 > 0. \end{cases}$$

Answer.

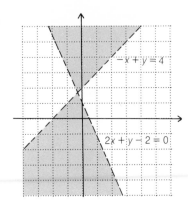

In the next example we consider a nonlinear system of inequalities.

◆ **Example 4.**   Graph the system of inequalities

$$\begin{cases} y > x^2 - 1 \\ y \leq x + 1. \end{cases}$$

Solution.   The graphs of these inequalities are the lightly shaded areas shown in Figure 7.17. Notice that the graph of the inequality $y \leq x + 1$ contains the line $y = x + 1$. The region common to both graphs (heavily shaded area) is the graph of the given system of inequalities.

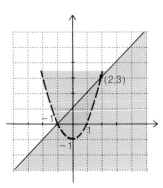

Figure 7.17

◇ **Practice Exercise 4.**   Graph the system of inequalities

$$\begin{cases} y \leq 3 - x^2 \\ y < -x + 1. \end{cases}$$

Answer.

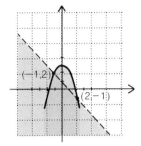

Linear Programming

An *optimization problem* is a problem of the following form: given a function, and given some subset of its domain, find the points in the subset where the function takes its largest (smallest) value and find that largest (smallest) value. We have already solved some simple optimization problems. For example, to find the vertex of a parabola we had to find a point at which a quadratic function attained its maximum (minimum) on $\mathbb{R}$.

The theory of *linear programming* studies optimization problems in which the given function is a *linear function of two variables* x, y (or, in more advanced settings, of more than two variables) and the given subset of the domain is *the set of solutions of a system of linear inequalities*. The given function is called the *objective function*; the set of solutions of the system of linear inequalities is called the *feasible set*; and the inequalities are called the *constraints* of the problem. The solution of the optimization problem may be called the *maximum (minimum) feasible* solution. (Linear programming has innumerable applications in management and economics, and the language of the theory—words like *feasible, objective*—reflects its business orientation.)

We shall use the technique of graphing systems of linear inequalities to solve linear programming problems. As an example, consider the following linear programming problem, which arises in attempting to manage a business efficiently.

◆ **Example 5.**   A manufacturer of furniture can produce at most 30 desks and at most 40 chairs per week. Each desk requires 4 hours of labor and each chair requires 2 hours of labor. The manufacturer has a maximum of 160 hours of labor available per week. If the profit on each desk is $40 and the profit on each chair is $30, find the number of desks and chairs the manufacturer should produce per week for maximum profit.

Solution.   Let x represent the number of desks and y the number of chairs produced per week. Since the number of desks produced per week is at most 30, it follows that the inequality $0 \le x \le 30$ is a constraint on the number of desks produced in a week. Similarly, the inequality $0 \le y \le 40$ is a constraint on the number of chairs produced weekly. In addition, another constraint on desks and chairs is derived from the maximum number of hours of labor available per week. If each desk requires 4 hours of labor and each chair requires 2 hours of labor, then x desks and y chairs require $4x + 2y$ hours of labor per week. Since this quantity cannot exceed 160 hours, the inequality $4x + 2y \leq 160$ gives us a third constraint to our problem. Thus the set of constraints on the number of desks and chairs is given by the following system of linear inequalities:

$$\begin{cases} 0 \leq x \leq \ 30 \\ 0 \leq y \leq \ 40 \\ 4x + 2y \leq 160. \end{cases}$$

The graph of this system (set of all feasible solutions) is the region of Figure 7.18.

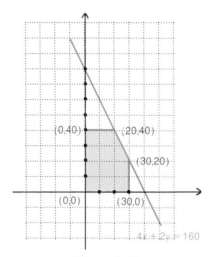

Figure 7.18

Since the profit on each desk is $40 and the profit on each chair is $30, it follows that the profit on x desks and y chairs is $P = 40x + 30y$. This is the objective function that we have to maximize on the set of all feasible solutions.

We now analyze the values of P on the set of constraints. For convenience, this set is again illustrated in Figure 7.19. Suppose that no desks or chairs were produced. Then, the profit would be zero and the equation $40x + 30y = 0$ would represent the line of zero profit.

Now, if 30 desks and no chairs were produced, then the profit would be $P = 40 \cdot 30 + 30 \cdot 0 = 1200$, and the equation $40x + 30y = 1200$ would represent the line along which profit is $1,200 (Figure 7.19). Notice that the point $(0, 40)$ corresponding to a production of no desks and 40 chairs also lies along this line. Thus, the same profit of $1,200 would be achieved by producing a total of 40 chairs and no desks. Consider, now, a production of 30 desks and 20 chairs. This yields a profit $P = 40 \cdot 30 + 30 \cdot 20 = 1200 + 600 = 1800$. The equation $40x + 30y = 1800$ represents the line of profit $1,800. All lines of constant profit are parallel to each other (they all have slope $-4/3$), and as they move away from the origin they represent higher profit. We can see (Figure 7.19) that the maximum profit is attained at the point $(20, 40)$. When 20 desks and 40 chairs are produced, the profit is $P = 40 \cdot 20 + 30 \cdot 40 = 800 + 1200 = 2000$, and this is the maximum profit on our set of constraints. We con-

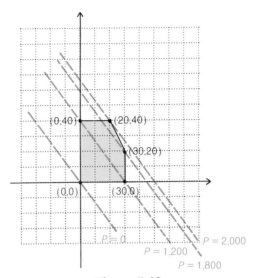

Figure 7.19

clude that the manufacturer should produce 20 desks and 40 chairs per week for maximum profit.

This example shows that the optimal solution of our problem has occurred at a *vertex* of the set of constraints. This is justified by a theorem which asserts that if the optimal solution of a linear programming problem exists, then it occurs at a vertex of the set of feasible solutions.

Linear Programming Theorem

Consider a linear expression $L = Ax + By + C$ where x and y are subject to constraints given by a linear system of inequalities. If L has a maximum (or a minimum) on the solution set of the system, then this occurs at a vertex of the set.

In view of this theorem we may now go back to Example 5 and make the following table to find the optimal solution of the problem:

Vertex	$P = 40x + 30y$
(0, 0)	$40 \cdot 0 + 30 \cdot 0 = 0$
(30, 0)	$40 \cdot 30 + 30 \cdot 0 = 1200$
(0, 40)	$40 \cdot 0 + 30 \cdot 20 = 1200$
(30, 20)	$40 \cdot 30 + 30 \cdot 20 = 1800$
(20, 40)	$40 \cdot 20 + 30 \cdot 40 = 2000$

In this table we are just computing the values of $P = 40x + 30y$ on the vertices of the set of all feasible solutions. As we already know, the largest value of P is \$2,000 and the corresponding optimal solution is (20, 40).

The importance of the Linear Programming Theorem is that it reduces the problem of finding the maximum or minimum of the objective function over an *infinite* set (the feasible set) to a finite calculation: we only have to check the values of the objective function at *finitely many points* (the vertices of the feasible set).

◇ **Practice Exercise 5.**   Under the same assumptions of Example 5, suppose that the profit on each desk is $50 and the profit on each chair is $20. What is the number of desks and chairs the manufacturer should produce for maximum profit? How much is his profit?

Answer.   30 desks and 20 chairs; $1,900.

◆ **Example 6.**   An advertising agency is comparing ad costs for various products in two monthly magazines. A half-page ad costs $240 in magazine *X* and $320 in magazine *Y*. According to a survey, during a given month and for each issue published, 7200 readers per ad of magazine *X* and 6000 readers per ad of magazine *Y* will notice the advertisement. Also, 300 readers per ad of magazine *X* and 500 readers per ad of magazine *Y* are likely to complete an attached questionnaire card asking for additional information. The agency also estimates that to profit from the advertising campaign at least 72,000 readers should be reached and at least 4500 should answer the questionnaire. How many monthly ads should be placed with each magazine in order to minimize the cost? What is the minimum cost?

Solution.   If x is the number of ads with magazine *X* and y the number with magazine *Y*, then the monthly cost to run the advertising campaign is $C = 240x + 320y$. We have to minimize this function on the set of constraints of our problem. These are the following:

$$x \geq 0 \qquad y \geq 0$$

Readers noticing the ads $7200x + 6000y \geq 72,000$
Readers returning the cards $300x + 500y \geq 4,500$

After simplification we obtain the system of linear inequalities

$$x \geq 0 \qquad y \geq 0$$
$$6x + 5y \geq 60$$
$$3x + 5y \geq 45.$$

The set of feasible solutions (i.e., the solution set of this system) is the shaded area in Figure 7.20. The vertices of this set are (0, 12), (5, 6), and (15, 0).

According to the Linear Programming Theorem, the optimal solution (i.e., the minimum cost) occurs at one of these vertices. We may now arrange our work in tabular form as follows:

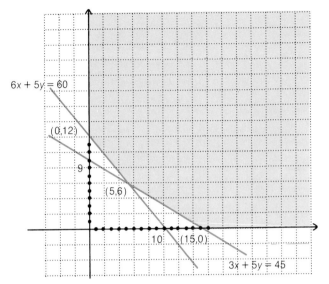

Figure 7.20

Vertex	$C = 240x + 320y$
(0, 12)	$240 \cdot 0 + 320 \cdot 12 + 0 + 3840 = 3840$
(5, 6)	$240 \cdot 5 + 320 \cdot 6 = 1200 + 1920 = 3120$
(15, 0)	$240 \cdot 15 + 320 \cdot 0 = 3600 + 0 = 3600$

Thus to minimize cost, five ads should be placed with magazine X and six ads with magazine Y. The minimum cost is \$3120.

◇ **Practice Exercise 6.** Under the same assumptions of Example 6, suppose that the advertising agency estimates that at least 108,000 readers should be reached and at least 6000 should answer the cards. How many monthly ads should now be placed with each magazine to minimize the cost? Find the minimum cost.

Answer. 10 ads with magazine X, 6 with magazine Y; \$4,320.

 EXERCISES 7.6

In Exercises 1–4, decide whether or not each of the pairs satisfies the given inequality.

1. $3x - 2y < 1$
$(1, -3), (-2, 3)$

2. $x + 5y > 2$
$(1, -4), (3, 1)$

3. $3x - 4y + 1 \geq 0$
$\left(\dfrac{1}{3}, \dfrac{1}{2}\right), (-2, 3)$

4. $-2x + 5y - 3 \leq 0$
$(-2, 3), \left(\dfrac{1}{2}, \dfrac{4}{5}\right)$

Graph each of the following inequalities in a Cartesian plane.

5. $2x + 1 > 0$

6. $-3x + 1 < 0$

7. $x + 5y \geq 2$

8. $3x - 2y \leq 1$

9. $-2x + 4y - 3 < 0$ **10.** $3x - 6y + 9 > 0$ **11.** $y \leq \sqrt{x - 4}$ **12.** $y \geq \sqrt{9 + x}$

13. $y > x^2 - 1$ **14.** $y \leq x^2 - 4$

In Exercises 15–24, graph each nonlinear system of inequalities.

15. $\begin{cases} y > x^2 \\ y < x + 2 \end{cases}$ **16.** $\begin{cases} y < 1 - x^2 \\ y > -6(x + 1) \end{cases}$ **17.** $\begin{cases} x^2 + y^2 < 1 \\ y \geq x \end{cases}$ **18.** $\begin{cases} x^2 + y^2 < 5 \\ y \leq 2x \end{cases}$

19. $\begin{cases} 4x^2 + y^2 < 4 \\ y > 4x \end{cases}$ **20.** $\begin{cases} x^2 + 9y^2 < 1 \\ 2y > x \end{cases}$ **21.** $\begin{cases} y \geq x^2 - 1 \\ x^2 + y^2 < 3 \end{cases}$ **22.** $\begin{cases} y < 4 - x^2 \\ x^2 + y^2 \geq 6 \end{cases}$

23. $\begin{cases} x^2 + y^2 < 2 \\ \dfrac{x^2}{9} + y^2 < 1 \end{cases}$ **24.** $\begin{cases} x^2 + y^2 \geq 2 \\ \dfrac{x^2}{4} + y^2 < 1 \end{cases}$

In Exercises 25–34, graph each system of linear inequalities. In each case specify the vertices.

25. $\begin{cases} y \geq 3x \\ y < -x + 4 \end{cases}$ **26.** $\begin{cases} 4y > -x \\ y \geq 2x + 6 \end{cases}$ **27.** $\begin{cases} x - y \geq 2 \\ 3x + y < 0 \end{cases}$ **28.** $\begin{cases} 2x - 4y < 6 \\ x - 3y \geq 0 \end{cases}$

29. $\begin{cases} 3x - 4y - 1 > 0 \\ 5x + 8y + 2 \leq 0 \end{cases}$ **30.** $\begin{cases} 6x - 3y + 2 \geq 0 \\ 2x + 8y - 3 < 0 \end{cases}$ **31.** $\begin{cases} 3x + 2y \geq 6 \\ 3x + 2y \leq 12 \\ x \geq 0 \\ y \leq 0 \end{cases}$ **32.** $\begin{cases} x - 4y < 6 \\ 2x - 8y > 16 \\ x \leq 0 \\ y \leq 0 \end{cases}$

33. $\begin{cases} 3x + 4y \leq 24 \\ x - 4y \geq -8 \\ x \leq 6 \\ y \geq 0 \end{cases}$ **34.** $\begin{cases} 2x - y + 4 < 0 \\ 3x + 2y - 15 < 0 \\ x \leq 3 \\ y \geq 0 \end{cases}$

In Exercises 35–40, an objective function L and a set of contraints are given. Find the maximum and minimum values of the objective function on the set of constraints. Specify the corresponding optimal solutions in each case.

35. $L = 2x + 3y - 5$
$\begin{cases} x + y \leq 4 \\ 2x - y \geq -3 \\ 0 \leq x \leq 3 \\ 0 \leq y \leq 3 \end{cases}$

36. $L = x - 2y$
$\begin{cases} x - y \leq 4 \\ 2x + y \geq 2 \\ -3 \leq x \leq 0 \\ 0 \leq y \leq 2 \end{cases}$

37. $L = 4x - 2y + 1$
$\begin{cases} 3x - y \geq 0 \\ x + y \geq 0 \\ 1 \leq x \leq 3 \\ y \geq 0 \end{cases}$

38. $L = 3x + 4y - 2$
$\begin{cases} x + y \geq -6 \\ y - 2x \geq 0 \\ -4 \leq x \leq 0 \\ y \leq 0 \end{cases}$

39. $L = 5x + 4y$
$\begin{cases} 2x + 3y \geq 12 \\ 2x + y \geq 8 \\ x \geq 0 \\ y \geq 0 \end{cases}$

40. $L = 7x + 4y - 2$
$\begin{cases} 3x + y \geq 6 \\ 3x + 4y \geq 15 \\ x \geq 0 \\ y \geq 0 \end{cases}$

41. A refinery with the capacity of refining at most 2500 barrels of oil per day produces gasoline and heating oil. The profit on each barrel of gasoline is $4, and the profit on each barrel of heating oil is $3. If at least 300 barrels of gasoline and 400 barrels of heating oil must be produced each day, find how many barrels of gasoline and heating oil should be produced for maximum profit.

42. A merchant has a roasting plant that can produce 1800 lb of roasted coffee per day. A shipment of Brazilian and Colombian coffee has just arrived, and she has to roast at least 200 lb of Brazilian coffee and 350 lb of Colombian coffee per day. The profit on each pound of Brazilian coffee is 50 cents, and that on each pound of Colombian coffee is 40 cents. How many pounds of each coffee should she produce for maximum profit?

43. The assembly line of a manufacturer produces two models of golf carts, Birdie Custom and Bogey De Luxe. It takes 10 hours of labor to produce a Birdie Custom and 15 hours of labor to produce a Bogey De Luxe, and the manufacturer has up to 90 hours of labor available per day. The assembly line can produce as many as 6 Birdie Customs and as many as 4 Bogey De Luxe per day. The profit on each Birdie Custom is $150, and that on each Bogey De Luxe is $240. How many of each model should the manufacturer assemble per day to have maximum profit? What is the maximum profit?

44. A factory assembles two types of motors, small and large. One hour of labor is required to assemble a small motor and four hours are needed for a large motor, and there are 80 hours of labor available each day. The maximum number of small motors assembled in a day is 60 and the maximum number of large motors is 15. The owner can make a profit of $120 on each small motor and $250 on each large motor. What is the maximum daily profit that can be realized?

45. The table below gives the number of units of protein and fiber in one gram of each of two ingredients, A and B, used in the preparation of a certain brand of cat food.

	A	B
Protein	3	2
Fiber	3	4

Each gram of ingredient A costs 8¢ and each gram of ingredient B costs 6¢. If each can of food has to contain at least 180 units of protein and 240 units of fiber, find the number of grams of each ingredient that should be used to minimize the cost.

46. The owner of a dog kennel prepares her own brand of dog food in such a way that her dogs receive at least 6 ounces of protein and 4 ounces of fat per day. For this she mixes two brand-name dog foods: Dog's Delight and Pet's Treat. Each brand contain protein and fat as follows:

	Protein (percentage)	Fat (percentage)
x Dog's Delight	40	20
y Pet's Treat	25	25

Dog's Delight costs 5¢ an ounce and Pet's Treat 4¢ an ounce. What is the minimum cost mix of the two brand-name dog foods?

47. A farmer has 360 acres available for planting barley and corn. It takes 1/2 hour of labor per acre to plant barley and 3/4 hour of labor per acre to plant corn, and the farmer has a maximum of 210 hours available for this job. He expects a profit of $35 per acre from barley and $45 per acre from corn. How many acres of each crop should he plant to maximize his profit? What is this maximum profit?

48. Find the maximum daily profit that can be realized on manufacturing *x* tables and *y* bookcases under the following assumptions:
    a) each table requires 4 hours of labor and each bookcase requires 2 hours;
    b) there are 90 hours of labor available each day;
    c) at most 15 tables and 25 bookcases can be produced each day;
    d) the profit on each table is $35 and on each bookcase is $20.

49. A town operates two centers, *A* and *B*, which collect paper, glass, and aluminum for recycling. Center *A* receives an average of 360 pounds of paper, 180 pounds of glass, and 120 pounds of aluminum per day. Center *B* receives an average of 120 pounds of paper, 540 pounds of glass, and 120 pounds of aluminum per day. The operating costs for centers *A* and *B* are, respectively, $30 and $20 per day. The town has agreed to supply at least 1800 pounds of paper, 2700 pounds of glass, and 1080 pounds of aluminum per week to a recycling company. To minimize the operating cost, how many days a week should each center remain open?

50. A farmer wishes to prepare a supply of cattle food by mixing two commercially available foods, *A* and *B*. Food *A* costs 40 cents per pound and *B* costs 50 cents per pound. The farmer mixes the two products so that the mixture contains at least 25 units of protein, at least 25 units of carbohydrates, and at least 20 units of fat per pound. The specifications for the two commercially available foods are listed in the following table:

Food	Protein (units per pound)	Carbohydrates (units per pound)	Fat (units per pound)
A	3	2	2
B	2	3	2

How many pounds of each food should be mixed to minimize the cost?

CHAPTER SUMMARY

There are many problems in mathematics and applied sciences whose solution depends on the solution of a system of linear equations. A linear system is said to be *consistent* if it has a *unique solution*, *inconsistent* if it has *no solutions*, and *dependent* if it has *infinitely many solutions*. A linear system can always be solved by the method of *substitution*. This method, which is easy to apply in the case of a two by two linear system, may become too lengthy and tedious when applied to linear systems of more than two equations and unknowns. It is preferable to use *Gaussian elimination*. This is a method of relative simplicity which can be applied to any linear system. The method consists of transforming the original system into an equivalent system in *upper triangular form*. Such a system is then easy to solve. To transform the original system, a sequence of *admissible operations* is performed. At each step the linear system is transformed into an equivalent system. Two linear systems are said to be *equivalent* if they have the same solutions. An operation is said to be admissible when it transforms a given system into an equivalent system.

When using Gaussian elimination, we can work with the complete system of equations or just with the *augmented matrix* of the system. This is the matrix formed by adding to the *matrix of coefficients* of the system a column consisting of the *constant terms*. To each admissible operation performed on the system there corresponds an *admissible row operation* for the augmented matrix, and vice-versa. The final product is an upper triangular matrix whose corresponding system is in upper triangular form.

Another method of solving a linear system uses the *algebra of matrices*. An *m by n matrix* is an array of $m \cdot n$ numbers written in *m rows* and *n columns*. Matrices can be added and subtracted. We can also multiply a matrix by a number (*scalar multiplication*). In certain cases the *product* of two matrices can be defined. If A is an n by n matrix (*square matrix*) and there is an n by n matrix B such that $AB = BA = I$, *the identity matrix*, then A is said to be *invertible* and the matrix B is the *inverse* of the matrix A. A linear system can be written in *matrix form* as explained in Section 7.3. If the *matrix of coefficients* is invertible, then left multiplication by the inverse of that matrix gives us the solution of the system. The inverse of a matrix, if it exists, can be found by performing *admissible row operations*.

A third method of solving linear systems uses *Cramer's rule*, which depends on the notion of the *determinant* of a square matrix. The definition and computation of a determinant, which is a simple matter in the case of 2 by 2 or 3 by 3 matrices, becomes more difficult task for matrices of higher order. In theory, a determinant can be computed by the *cofactor expansion along any row or column*. But, in practice, the actual computation may be quite involved. Cramer's rule, as explained in Section 7.4, provides a general method for solving linear systems. In most cases, the method is not as easy to apply as the Gaussian elimination method. However, Cramer's rule is interesting for historical reasons, besides being a valuable theoretical tool.

A *nonlinear system* is a system that contains at least one nonlinear equation. Nonlinear systems are in most cases more difficult to solve than linear systems. Moreover, there is no general method of solution for a nonlinear system. In Section 7.5, we considered special cases that can be solved either by substitution or by elimination.

In a *linear programming problem* we try to maximize or minimize a linear function, known as the *objective function* on the *feasible set*, that is, the solution set of a *system of linear inequalities*. The *constraints* are the inequalities that define the set. Hence, it is important to learn techniques of graphing systems of linear inequalities. The Linear Programming Theorem states that the maximum or minimum of the objective function occurs at a vertex of the feasible set. Linear programming has innumerable applications in management sciences and economics.

REVIEW EXERCISES

In Exercises 1–4, find 2A + B, A − 3B, and 5B.

1. $A = \begin{bmatrix} 2 & 1 \\ 2 & -3 \end{bmatrix}$, $B = \begin{bmatrix} 1 & -2 \\ 3 & 2 \end{bmatrix}$

2. $A = \begin{bmatrix} 1 & -1 \\ 2 & 3 \\ -1 & 1 \end{bmatrix}$, $B = \begin{bmatrix} 4 & -2 \\ 3 & -1 \\ 2 & 0 \end{bmatrix}$

3. $A = \begin{bmatrix} 5 & 0 \\ 0 & -5 \end{bmatrix}$, $B = \begin{bmatrix} 0 & 2 \\ -2 & 0 \end{bmatrix}$

4. $A = \begin{bmatrix} 1 & 3 & -1 \\ -3 & 1 & 3 \end{bmatrix}$, $B = \begin{bmatrix} 2 & 0 & -2 \\ 0 & 3 & 0 \end{bmatrix}$

In Exercises 5–8, find AB and BA.

5. $A = \begin{bmatrix} 0 & 2 & -1 & -2 \end{bmatrix}$, $B = \begin{bmatrix} 2 \\ 1 \\ -3 \\ 2 \end{bmatrix}$

6. $A = \begin{bmatrix} -2 & 4 \\ 0 & 2 \end{bmatrix}$, $B = \begin{bmatrix} 2 & -1 \\ -2 & 4 \end{bmatrix}$

7. $A = \begin{bmatrix} 1 & -1 & 2 \\ 4 & 0 & 3 \end{bmatrix}$, $B = \begin{bmatrix} 3 & 2 \\ -2 & 0 \\ 1 & -2 \end{bmatrix}$

8. $A = \begin{bmatrix} 1 & 1 & 1 \\ 0 & 1 & 1 \\ 0 & 0 & 1 \end{bmatrix}$, $B = \begin{bmatrix} 1 & 0 & 0 \\ 1 & 1 & 0 \\ 1 & 1 & 1 \end{bmatrix}$

In Exercises 9–14, find the inverse of each matrix.

9. $\begin{bmatrix} 2 & -1 \\ 3 & 2 \end{bmatrix}$

10. $\begin{bmatrix} 3 & 4 \\ 1 & 2 \end{bmatrix}$

11. $\begin{bmatrix} 1 & 2 & 3 \\ 0 & 2 & 3 \\ 0 & 0 & 3 \end{bmatrix}$

12. $\begin{bmatrix} 1 & 1 & 2 \\ 1 & 2 & 0 \\ 2 & 0 & 0 \end{bmatrix}$

● **13.** $\begin{bmatrix} 1 & 0 & 0 & 0 \\ 2 & 1 & 0 & 0 \\ 3 & 3 & 1 & 0 \\ 4 & 4 & 4 & 1 \end{bmatrix}$

● **14.** $\begin{bmatrix} 0 & -1 & -1 & -1 \\ 1 & 2 & 1 & 1 \\ 1 & 2 & 2 & 1 \\ 1 & 2 & 2 & 2 \end{bmatrix}$

In Exercises 15–20, evaluate the determinants.

15. $\begin{vmatrix} 0 & 2 & 2 \\ 3 & 0 & 2 \\ 3 & 3 & 0 \end{vmatrix}$

16. $\begin{vmatrix} 3 & 1 & 0 \\ 1 & 0 & 1 \\ 0 & 1 & 3 \end{vmatrix}$

17. $\begin{vmatrix} 3 & 2 & 0 \\ 1 & 0 & 2 \\ 0 & 1 & 2 \end{vmatrix}$

18. $\begin{vmatrix} 2 & 0 & 1 \\ 0 & -2 & 0 \\ -1 & 0 & 2 \end{vmatrix}$

19. $\begin{vmatrix} 1 & 2 & 0 & 3 \\ -1 & -2 & 0 & 1 \\ -1 & 2 & 1 & 1 \\ -3 & 2 & 1 & 2 \end{vmatrix}$

20. $\begin{vmatrix} 1 & 0 & 1 & 0 \\ 0 & 1 & 0 & 1 \\ 1 & 0 & 1 & 1 \\ 0 & 1 & 1 & 1 \end{vmatrix}$

● **21.** If $ab \neq 0$, find the inverse of the matrix $\begin{bmatrix} a & 0 \\ 0 & b \end{bmatrix}$.

● **22.** If $abc \neq 0$, find the inverse of the matrix

$$\begin{bmatrix} a & 0 & 0 \\ 0 & b & 0 \\ 0 & 0 & c \end{bmatrix}.$$

In Exercises 23–28, solve each system.

23. $\begin{cases} x - 3y = 7 \\ 2x - y = -4 \end{cases}$

24. $\begin{cases} 6x + 5y = 4 \\ 3x - 15y = -5 \end{cases}$

25. $\begin{cases} x + y - z = 0 \\ 2x - 3y - 4z = 6 \\ 2x - y - z = 8 \end{cases}$

26. $\begin{cases} x + y - 2z = -1 \\ 3x - 2y - z = 2 \\ 2x - 3y - 3z = -5 \end{cases}$

27. $\begin{cases} x + y = -1 \\ 2y + z = 0 \\ 4x - 3z = 0 \end{cases}$

28. $\begin{cases} y - z = 2 \\ x + 2z = 11 \\ x - y = 0 \end{cases}$

Solve for x.

29. $\begin{vmatrix} x & 2 \\ -3 & -2 \end{vmatrix} = 0$

30. $\begin{vmatrix} 1 & x \\ -2 & 4 \end{vmatrix} = 6$

31. $\begin{vmatrix} 1 & x & -1 \\ 3 & 0 & 2 \\ 5 & 0 & 4 \end{vmatrix} = -8$

32. $\begin{vmatrix} 1 & 2 & -1 \\ x & 0 & 1 \\ 3 & 1 & -2 \end{vmatrix} = 14$

33. Find a and b so that the line of equation $ax + by + 1 = 0$ passes through the points $(1, 2)$ and $(-2, -3)$.

34. Find a and b so that the line of equation $ax + by - 1 = 0$ contains the points $(6, -1)$ and $(1/2, 5/6)$.

35. Find a and b so that the lines $ax + by + 1 = 0$ and $bx - ay - 8 = 0$ intersect at the point $(3, -2)$.

36. Find a and b so that the lines $ax + by + 1 = 0$ and $bx - ay - 5 = 0$ intersect at the point $(1, -1)$.

37. John has \$500 more invested at 8% per year than he has invested at 7.5% per year. Together the two investments provide him an annual income of \$425. How much does John have invested at each rate?

38. A man has \$6000 invested in two types of bonds paying 9% and 8% per year. If the annual income from his investments is \$515, how much does he have invested in each type of bond?

39. A boy has four times more nickels than dimes, for a total of \$4.80. How many nickels and dimes does he have?

40. Ann has \$2.40 in nickels and dimes. If she had half as many nickels and twice as many dimes as she does, the new total would be \$3.90. How many nickels and dimes does Ann have?

41. A company pays its salespeople on a basis of a percentage of the first \$15,000 in sales, plus another percentage of any amount over \$15,000. If a salesperson earns \$1900 on sales of \$27,000 and \$2600 on sales of \$36,500, find the two percentages.

42. In a biological experiment, two species of fish A_1 and A_2 are kept in the same environment and are fed two types of food F_1 and F_2. The table below gives the number of units of each type of food that each species consumes daily.

	A_1	A_2
F_1	2	2
F_2	3	1

If 74 units of food F_1 and 67 units of food F_2 are available per day, find the population size of each species that would exactly consume all the available food.

43. A manufacturer uses two machines M and N to produce two products P and Q. The amount of time, in hours, that each machine is used to produce one unit of each product is given by the table:

	P	Q
M	1	$\dfrac{1}{2}$
N	$\dfrac{3}{4}$	$\dfrac{1}{4}$

Machine M is available 16 hours a day, and machine N is available 11 hours a day. Find how many units of P and Q should be produced daily so that M and N are fully in operation.

44. Determine the values for a, b, and c so that the graph of the parabola $y = ax^2 + bx + c$ contains the points $(-1, -3)$, $(1, 3)$, and $(-2, -3)$.

45. Find a, b, and c so that the graph of $y = ax^2 + bx + c$ contains the points $(0, 1)$, $(1, 0)$, and $(2, 3)$.

46. The perimeter of a triangle is 110 meters. Side c is 30 meters longer than the sum of sides a and b, and side a is $1/3$ the difference between sides a and b. Find the length of each side.

47. A merchant wants to mix peanuts costing $2.10 per lb, almonds costing $4.20 per lb, and cashews costing $4.00 per lb to obtain 100 lb of a mixture costing $2.90 per lb. He also wants the amount of peanuts to be three times the amount of almonds. How many pounds of each variety should he mix?

48. A family consists of a father, a mother, and a son. The mother is three times as old as the son. Two years from now the father will be three times as old as the son. Eight years ago, the age of the father was the sum of the ages of the mother and the son. Find the present age of each member of the family.

49. A man has $425 in one-dollar, five-dollar, and ten-dollar bills. He has 75 bills in all, and 5 more ten-dollar bills than one-dollar bills. How many bills of each kind does the man have?

50. A lab technician mixes three different sulfuric acid solutions whose concentrations are 20%, 45%, and 50%, to obtain 180 liters of a 42% solution. If he uses twice as much of the 50% solution as the 20% solution, how many liters of each solution does he use?

51. A company employs 120 workers and pays $6, $8, and $10 an hour. The total of the hourly wages is $840. Three times as many workers are paid $8 an hour as are paid $10 an hour. How many workers are paid $6? $8? $10?

52. Copper, zinc, and lead are mixed in different amounts according to the table below to produce three types of alloys A, B, and C.

	A	B	C
Copper	45%	55%	60%
Zinc	30%	20%	25%
Lead	25%	25%	15%

How many pounds of each alloy should be mixed to obtain 200 lb of a new alloy that is 25.25% zinc and 22.50% lead?

● **53.** Show that

$$\begin{vmatrix} 1 & 1 & 1 \\ x & y & z \\ x^2 & y^2 & z^2 \end{vmatrix} = (x - y)(y - z)(z - x).$$

● **54.** Show that

$$\begin{vmatrix} a_{11} & a_{12} & a_{13} \\ 0 & a_{22} & a_{23} \\ 0 & 0 & a_{33} \end{vmatrix} = a_{11}a_{22}a_{33}.$$

● **55.** Without expanding the determinant, show that

$$\begin{vmatrix} a & b+c & 1 \\ b & c+a & 1 \\ c & a+b & 1 \end{vmatrix} = 0.$$

● **56.** Show that

$$\begin{vmatrix} a_{11} & a_{12} & a_{13} & a_{14} \\ 0 & a_{21} & a_{23} & a_{24} \\ 0 & 0 & a_{33} & a_{34} \\ 0 & 0 & 0 & a_{44} \end{vmatrix} = a_{11}a_{22}a_{33}a_{44}.$$

57. Show that

$$\begin{vmatrix} a+1 & a-1 \\ a-1 & a+1 \end{vmatrix} = 4a.$$

58. Find all the roots of the following equation:

$$\begin{vmatrix} 1 & x_1 & x_2 \\ 1 & x & x_2 \\ 1 & x_1 & x \end{vmatrix} = 0.$$

59. Find all the roots of the equation

$$\begin{vmatrix} 1 & x_1 & x_2 & x_3 \\ 1 & x & x_2 & x_3 \\ 1 & x_1 & x & x_3 \\ 1 & x_1 & x_2 & x \end{vmatrix} = 0.$$

60. Show that

$$\begin{vmatrix} a & b & 0 & 0 \\ c & d & 0 & 0 \\ 0 & 0 & e & f \\ 0 & 0 & g & h \end{vmatrix} = \begin{vmatrix} a & b \\ c & d \end{vmatrix} \cdot \begin{vmatrix} e & f \\ g & h \end{vmatrix}.$$

Solve each nonlinear system of equations by algebraic methods.

61. $\begin{cases} y = x^2 + 3 \\ x^2 + y^2 = 9 \end{cases}$
62. $\begin{cases} xy = 3 \\ 2y + x = -5 \end{cases}$
63. $\begin{cases} 3x^2 - 2y = 0 \\ 5x - 2y = -2 \end{cases}$
64. $\begin{cases} x^2 + y = 2 \\ x^2 + y^2 = 8 \end{cases}$

Graph each nonlinear system of inequalities.

65. $\begin{cases} 3y > x \\ x - y^2 > 0 \end{cases}$

66. $\begin{cases} y - 2x > 0 \\ x^2 + y^2 < 5 \end{cases}$

67. $\begin{cases} x^2 + y^2 < 8 \\ x^2 + y < 2 \end{cases}$

68. $\begin{cases} x^2 - 2x + y^2 < 0 \\ x^2 + y^2 < 1 \end{cases}$

Graph each of the following systems of linear inequalities.

69. $\begin{cases} 2x - 5y < 9 \\ 3x + y > 5 \end{cases}$

70. $\begin{cases} 6x - 2y - 8 > 0 \\ 3x + y - 10 < 0 \end{cases}$

71. $\begin{cases} 3x + 4y \geq 10 \\ x + 2y \geq 4 \\ x \geq 0 \\ y \geq 0 \end{cases}$

72. $\begin{cases} 2x + 3y \leq -8 \\ x + y \leq -3 \\ x \leq 0 \\ y \leq 0 \end{cases}$

In Exercises 73–76, find the maximum and the minimum values of the objective function L on the given set of constraints. Specify the optimal solution in each case.

73. $L = 3x + 4y$
$\begin{cases} x + y \geq 4 \\ 2x - y \geq -3 \\ 0 \leq x \leq 3 \\ 0 \leq y \leq 3 \end{cases}$

74. $L = 2x - 4y + 1$
$\begin{cases} x - y \leq 4 \\ 2x + y \leq 2 \\ -3 \leq x \leq 1 \\ 0 \leq y \leq 2 \end{cases}$

75. $L = x + 3y - 4$
$\begin{cases} x + y \geq -6 \\ y - 2y \geq 0 \\ -4 \leq x \leq 0 \\ y \leq 0 \end{cases}$

76. $L = 2x + 5y$
$\begin{cases} 3x + y \geq 6 \\ 3x + 4y \geq 15 \\ x \geq 0 \\ y \geq 0 \end{cases}$

77. The difference of two positive numbers is 7 and the product is 450. Find the two numbers.

78. The perimeter of a rectangle is 46 m and its area is 120 m². What are the dimensions of the rectangle?

79. A manufacturer assembles two models of bicycles: Standard and De Luxe. It takes 2 hours of labor to assemble a Standard model and 4 hours to assemble a De Luxe, and the manufacturer has up to 40 hours of labor available per day. The assembly line can produce at most 10 Standard and 8 De Luxe per day. The profit on each Standard model is $60, and that on each De Luxe is $80. Find how many bicycles of each model the manufacturer should assemble per day to realize maximum profit. What is the maximum profit?

80. The table below gives the number of units of fat and protein contained in one pound of each of two ingredients *A* and *B* used in the preparation of a brand of dog food.

	A	B
Fat	3	2
Protein	3	4

Each pound of ingredient *A* costs 50¢ and each pound of *B* costs 40¢. If each bag of dog food has to contain at least 12 units of fat and 18 units of protein, find the number of pounds of each ingredient that should be used to minimize the cost.

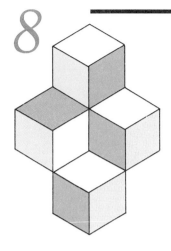

8

Zeros of Polynomials

The present chapter begins with a brief review of the definition of polynomials and the operations of sum and product of polynomials. Next, we describe the division algorithm for polynomials, the division of a polynomial by $x - a$, and the Remainder and Factor Theorems. The notion of multiplicity of a zero is defined, and the theorem about the number of zeros of a polynomial is proved. If a polynomial has integer coefficients, then its rational zeros, if they exist, can be found with the help of a simple test explained in Section 8.4. In the last section of this chapter, we discuss the method of partial fraction decomposition.

8.1 PRELIMINARIES

Polynomials were defined in Section 1.4 of Chapter 1. Recall that if n is a positive integer, a polynomial of degree n is an expression of the form

$$a_n x^n + a_{n-1} x^{n-1} + \cdots + a_1 x + a_0$$

where $a_0, a_1, \ldots, a_n$ are real numbers and $a_n \neq 0$. The term $a_k x^k$ in a given polynomial is called the *term of degree k* or the *kth-degree term*. If $a_n \neq 0$, then $a_n x^n$ is the *leading term* and a_n is the *leading coefficient*. The term a_0 is the *constant term* of the polynomial.

As we already know, the polynomial function defined by

$$A(x) = a_n x^n + \cdots + a_1 x + a_0$$

has for its domain of definition the set $\mathbb{R}$ of all real numbers. If $b \in \mathbb{R}$, the function value of $A(x)$ at $x = b$ is defined by

$$A(b) = a_n b^n + \cdots + a_1 b + a_0.$$

◆ **Example 1.**   If $A(x) = 3x^2 + 5x + 3$, find $A(-1)$, $A(2)$, and $A(1)/A(2)$.

Solution.   We have

$$A(-1) = 3(-1)^2 + 5(-1) + 3 = 3 - 5 + 3 = 1$$
$$A(2) = 3(2)^2 + 5(2) + 3 = 12 + 10 + 3 = 25$$
$$\frac{A(1)}{A(2)} = \frac{1}{25}.$$

◇ **Practice Exercise 1.**   Let $P(x) = 4x^3 - 5x + 6$. Find $P(-2)$, $P(1)$, and $3P(1) + P(-2)$.

Answer. $P(-2) = -16$, $P(1) = 5$, $3P(1) + P(-2) = -1$.

◆ **Example 2.**   If $P(x) = x^3 + ax^2 - 4x + 1$ and $P(-2) = -11$, find a.

Solution.   We have

$$-11 = P(-2) = (-2)^3 + a(-2)^2 - 4(-2) + 1$$
$$-11 = -8 + 4a + 8 + 1$$
$$-12 = 4a$$
$$\frac{-12}{4} = a.$$

Thus $a = -3$.

◇ **Practice Exercise 2.**   If $A(x) = x^3 + bx - 4$ and $A(-1) = -3$, find b.

Answer. $b = -2$.

Constant Polynomials

A constant function $A(x) = a_0$, with $a_0 \neq 0$, can be regarded as a polynomial function of degree 0. The constant function $A(x) = 0$ is sometimes called the *zero polynomial* and its degree is not defined.

Throughout this chapter, we denote polynomial functions with upper case Romans such as A, B, P, Q, R.

Equality

Two polynomials $A(x)$ and $B(x)$ are *identically equal* if they are equal as functions, that is, if $A(x) = B(x)$ *for all values of x.*

When no confusion is possible, we simply say of two polynomials that they are *equal* instead of saying that they are *identically equal*. The following theorem, stated without proof, tells us exactly when two given polynomials are equal.

Polynomials Identically Equal

Two polynomials $A(x)$ and $B(x)$ are equal if and only if they have the same degree and the coefficients of terms of same degree are equal.

For example, let

$$A(x) = 2x^3 - 4x^2 + 7$$

and

$$B(x) = b_4 x^4 + b_3 x^3 + b_2 x^2 + b_1 x + b_0.$$

If $A(x)$ is identically equal to $B(x)$, then the theorem implies that

$$b_4 = 0, \quad b_3 = 2, \quad b_2 = -4, \quad b_1 = 0, \quad b_0 = 7.$$

At this point we suggest that you review the examples and exercises about sum and product of polynomials that were discussed in Section 1.4 of Chapter 1. They may help you understand the following formal definitions of sum and product.

Let

$$B(x) = b_m x^m + b_{m-1} x^{m-1} + \cdots + b_1 x + b_0$$

be a polynomial of degree m.

Sum

The sum $A(x) + B(x)$ is defined by

$$A(x) + B(x) = (a_0 + b_0) + (a_1 + b_1)x + (a_2 + b_2)x^2 + \cdots$$

That is, to add polynomials we add the coefficients of terms of the same degree.

Product

The product $A(x) \cdot B(x)$ is defined by

$$\begin{aligned}
A(x) \cdot B(x) &= (a_n x^n + \cdots + a_1 x + a_0)(b_m x^m + \cdots + b_1 x + b_0) \\
&= a_0 b_0 + (a_1 b_0 + a_0 b_1)x \\
&\quad + (a_2 b_0 + a_1 b_1 + a_0 b_2)x^2 \cdots + a_n b_m x^{n+m}.
\end{aligned}$$

The definitions of sum and product of two polynomials extend to the case of any *finite* number of polynomials. The degree of a product of polynomials is equal to the *sum* of the degrees of the polynomials. Moreover, the zero polynomial 0 is the identity for the sum, while the constant polynomial 1 is the identity for the product.

The Division Algorithm

If a and b are two nonnegative integers, with $b \neq 0$, then there is a unique pair q and r of nonnegative integers, such that

$$a = bq + r, \quad \text{with } 0 \leq r < b.$$

The number a is called the *dividend*, b is the *divisor*, q is the *quotient*, and r is the *remainder* in the division of a by b. When $r = 0$, we say that a is *divisible* by b, and b is a *factor* of a.

The following theorem, known as the division algorithm for polynomials, is analogous to the preceding result for integers.

The Division Algorithm

Let $A(x)$ be a polynomial of degree n and let $B(x) \neq 0$ be a polynomial of degree m, such that $n \geq m$. Then there are two polynomials $Q(x)$ and $R(x)$ such that

$$(8.1) \quad A(x) = B(x) \cdot Q(x) + R(x),$$

where the degree of $Q(x)$ is $n - m$, and either $R(x) = 0$ or the degree of $R(x)$ is less than m.

The polynomial $A(x)$ is called the *dividend*, $B(x)$ is the *divisor*, $Q(x)$ is the *quotient*, and $R(x)$ is the *remainder* in the division of $A(x)$ by $B(x)$. It is important to observe that the degree of $Q(x)$ is the *difference* of the degree of $A(x)$ and $B(x)$, and the degree of $R(x)$ is always *smaller* than the degree of $B(x)$, unless $R(x) = 0$.

◆ **Example 3.** Find the quotient and remainder in the division of $4x^3 - 8x^2 + 5x + 1$ by $2x - 1$.

Solution. Use the method of long division as follows:

$$
\begin{array}{r}
2x^2 - 3x + 1 \leftarrow \text{Quotient} \\
2x - 1 \overline{\smash{)}\, 4x^3 - 8x^2 + 5x + 1} \\
\text{Subtract} \longrightarrow \quad 4x^3 - 2x^2 \\
\hline
-6x^2 + 5x \\
\text{Subtract} \longrightarrow \quad -6x^2 + 3x \\
\hline
2x + 1 \\
\text{Subtract} \longrightarrow \quad 2x - 1 \\
\hline
2 \leftarrow \text{Remainder}
\end{array}
$$

Thus the *quotient* is $2x^2 - 3x + 1$ and the *remainder* is 2. By performing the algebraic operations, you can check that the following relation holds:

$$4x^3 - 8x^2 + 5x + 1 = (2x - 1)(2x^2 - 3x + 1) + 2.$$

This is exactly relation (8.1).

◇ **Practice Exercise 3.** Divide $6x^3 + 19x^2 + x + 3$ by $3x + 2$.

Answer. Quotient: $2x^2 + 5x - 3$, remainder: -3
$(6x^3 + 19x^2 + x + 3) = (3x + 2)(2x^2 + 5x - 3) - 3$.

◆ **Example 4.** Use the method of long division to find the quotient and remainder in the division of $x^4 - x^2 + 5$ by $x^2 + 2x + 1$.

Solution. The terms of degree 3 and 1 are missing from the dividend $x^4 - x^2 + 5$. It is helpful to allow spaces for the two terms and work as follows.

$$
\begin{array}{r}
x^2 - 2x + 2 \\
x^2 + 2x + 1 \enclose{longdiv}{x^4 \qquad - x^2 \qquad + 5} \\
\underline{x^4 + 2x^3 + x^2} \\
-2x^3 - 2x^2 \\
\underline{-2x^3 - 4x^2 - 2x} \\
2x^2 + 2x + 5 \\
\underline{2x^2 + 4x + 2} \\
-2x + 3
\end{array}
$$

Thus the quotient is $x^2 - 2x + 2$ and the remainder is $-2x + 3$. We may now write

$$x^4 - x^2 + 5 = (x^2 + 2x + 1)(x^2 - 2x + 2) + (-2x + 3).$$

◇ **Practice Exercise 4.** Find the quotient and remainder in the division of $x^5 - 3x^4 + 5x^2$ by $x^2 - 3x + 2$.

Answer. Quotient: $x^3 - 2x - 1$, remainder: $x + 2$
$x^5 - 3x^4 + 5x^2 = (x^2 - 3x + 2)(x^3 - 2x - 1) + (x + 2)$.

Exact Division

If $R(x) = 0$, the expression (8.1) becomes

(8.2) $A(x) = B(x)Q(x).$

In this case, the division is *exact*, and $A(x)$ is said to be *divisible by* $B(x)$. Also $B(x)$ is called a *factor* of $A(x)$.

The reader should notice that this is exactly what happens with integers: if one integer is a factor of another then the division is exact.

◆ **Example 5.** Divide $x^5 + x + 1$ by $x^2 + x + 1$.

Solution. Again we proceed by long division

$$
\begin{array}{r}
x^3 - x^2 + 1 \\
\hline
x^2 + x + 1 \,\big|\, x^5 \cdot \qquad\qquad x + 1 \\
\underline{x^5 + x^4 + x^3} \\
-x^4 - x^3 \\
\underline{-x^4 - x^3 - x^2} \\
x^2 + x + 1 \\
\underline{x^2 + x + 1} \\
0
\end{array}
$$

Since the remainder is 0, it follows that $x^5 + x + 1$ is divisible by $x^2 + x + 1$ and we have

$$x^5 + x + 1 = (x^2 + x + 1)(x^3 - x^2 + 1).$$

◇ **Practice Exercise 5.** Find the quotient and remainder in the division of $x^6 + x^4 - 5x^2 + 3$ by $x^2 + 3$.

Answer. Quotient: $x^4 - 2x^2 + 1$, remainder: 0
$$x^6 + x^4 - 5x^2 + 3 = (x^2 + 3)(x^4 - 2x^2 + 1).$$

___

▢ EXERCISES 8.1

1. If $P(x) = 2x^3 - x^2 + 2x - 1$, find:
   a) $P(1)$ b) $P(2)$ c) $P(1)/P(2)$
   d) $P(1/2)$ e) $2P(1) + 3P(2)$
2. If $Q(x) = x^4 - 16$, find:
   a) $Q(2)$ b) $Q(0)$ c) $Q(1/2)$
   d) $1/Q(0)$ e) $Q(2) \cdot Q(1/2)$
3. If $A(x) = x^3 - 2x^2 + ax + 1$ and $A(1) = 4$, find a.
4. If $B(x) = ax^4 + 2x - 3$ and $Q(2) = 33$, find a.

In Exercises 5–10, find constants A, B, C and D so that P(x) = Q(x).

5. $P(x) = 4x^2 - 3x + 2$, $Q(x) = (A + 1)x^2 + (B + 1)x + (C - 3)$
6. $P(x) = x^2 - 4x + 8$, $Q(x) = (2A)x^2 + (B - 2)x + 4C$
7. $P(x) = 3x^2 - 4x + 6$, $Q(x) = Ax^2 + (B - 2A)x + (C + 1)$
8. $P(x) = 4x^2 - x - 3$, $Q(x) = (A + B)x^2 + (B + C)x + C$
9. $P(x) = x^3 - 2x + 1$, $Q(x) = (A - B)x^3 + (B + C)x^2 + (C - 2D)x + D$
10. $P(x) = 2x^3 - 1$, $Q(x) = Ax^3 + (B - 2A)x^2 + (A - 2C)x + (3A - D)$

In Exercises 11–20, use long division to find the quotient and the remainder.

11. $(6x^2 + 11x + 4) \div (2x + 5)$
12. $(8x^2 - 8x + 6) \div (2x - 3)$
13. $(x^3 - 2x^2 + 4x - 1) \div (3x - 4)$
14. $(2x^3 - 7x^2 - 7x + 12) \div (2x + 3)$
15. $(x^4 + 2x^3 - 2x^2 + 7x + 2) \div (x^2 - x + 2)$
16. $(x^5 - 3x^4 + x^3 + 4x^2 - x + 2) \div (x^2 - 3x - 1)$

17. $(8x^6 - 3x^2 + 1) \div (4x^3 - x - 2)$

18. $(2x^5 - x) \div (2x^2 - 1)$

19. $(x^6 - x^2 - 1) \div (2x^3 - 4)$

20. $(x^4 - 3x^3 + 2x^2 + x + 1) \div (2x^2 - 1)$

Formula (8.1) can be written as

$$\frac{A(x)}{B(x)} = Q(x) + \frac{R(x)}{B(x)}$$

where the polynomial $Q(x)$ is the quotient of $A(x)$ by $B(x)$ and $R(x)/B(x)$ is a proper rational expression, that is, a rational expression where the degree of the numerator is smaller than the degree of the denominator. In Exercises 21–30, write each rational expression as a sum of a polynomial and a proper rational expression.*

21. $\dfrac{x^2 - 4x + 5}{x - 1}$

22. $\dfrac{2x^2 + x - 3}{x + 2}$

23. $\dfrac{2x^3 - 4x - 2}{2x + 4}$

24. $\dfrac{3x^3 - 7x^2 + 11x + 4}{3x - 1}$

25. $\dfrac{x^4 - 5x^2 + 3}{2x - 1}$

26. $\dfrac{6x^4 + 2x + 1}{2x + 1}$

27. $\dfrac{10x^6 + 5x^3 - x}{2x^4 - x^2 - 1}$

28. $\dfrac{x^6 - 1}{x^4 + 1}$

29. $\dfrac{3x^4 - 2x^3 + x^2 + 2x + 3}{x^2 - x - 1}$

30. $\dfrac{2x^5 - 4x^3 + x - 1}{3x^2 - x - 1}$

8.2 DIVISION BY $x - a$

The division algorithm yields interesting consequences when the divisor is a first degree polynomial of the form

$$x - a,$$

where the leading coefficient is 1 and a is an arbitrary number.

If the dividend $A(x)$ has degree n, then the quotient $Q(x)$ is a polynomial of degree $n - 1$. The remainder is either 0 or a polynomial of degree 0, that is, a constant. We then have

$$(8.3) \quad A(x) = (x - a)Q(x) + R,$$

where R is a number.

If in the last formula a is substituted for x, we get

$$A(a) = (a - a)Q(a) + R$$
$$= 0 \cdot Q(a) + R$$
$$= R.$$

Thus,

$$(8.4) \quad R = A(a)$$

and we have proved the following theorem.

The Remainder Theorem

The remainder in the division of a polynomial $A(x)$ by $x - a$ is $A(a)$.

This result is extremely useful in applications. Among other things, it allows us to find the remainder when $A(x)$ is divided by $(x - a)$, without performing the division.

◆ **Example 1.**   Find the remainder in the division of $A(x) = 2x^4 - 4x^3 - 4x^2 + 13x - 7$ by $B(x) = x + 2$.

Solution.   First, write

$$B(x) = x + 2 = x - (-2).$$

Thus, in this example, $a = -2$. To find the remainder, compute the function value of the polynomial $A(x)$ at $x = -2$:

$$A(-2) = 2 \cdot (-2)^4 - 4 \cdot (-2)^3 - 4 \cdot (-2)^2 + 13(-2) - 7$$
$$= 32 + 32 - 16 - 26 - 7$$
$$= 15.$$

By the Remainder Theorem, 15 is the remainder in the division of the given polynomial by $x + 2$. As an exercise, you can check this result by carrying out the division of $2x^4 - 4x^3 - 4x^2 + 13x - 7$ by $x + 2$.

◇ **Practice Exercise 1.**   Find the remainder in the division of $A(x) = 2x^4 - 4x^3 - 4x^2 + 13x - 7$ by $B(x) = x - 2$.

Answer. $R = 3$.

When $R = A(a) = 0$, formula (8.3) becomes

$$(8.5) \quad A(x) = (x - a)Q(x).$$

Thus, $A(x)$ is *divisible* by $x - a$, and $x - a$ is a factor of $A(x)$. Conversely, if $A(x)$ is divisible by $x - a$, then the remainder R is 0. Since $R = A(a)$ according to (8.4), we get $A(a) = 0$. Thus, we have proved the following theorem.

Factor Theorem

A linear polynomial $x - a$ is a factor of $A(x)$ if and only if $A(a) = 0$.

◆ **Example 2.**   Verify that $P(x) = x^3 - 3x^2 + 6x - 8$ is divisible by $x - 2$.

Solution.   According to the Remainder Theorem, it suffices to compute the numerical value of $P(x)$ at $x = 2$:

$$R = P(2)$$
$$= 2^3 - 3 \cdot (2)^2 + 6 \cdot 2 - 8$$
$$= 8 - 12 + 12 - 8$$
$$= 0.$$

As an exercise, you can check this result by long division.

◇ **Practice Exercise 2.**   Is $x + 3$ divisible by $x^4 - 9x^2 + x$?

Answer.   No. Remainder of division is -3.

Horner's Method

The numerical evaluation of a polynomial can be greatly simplified by proceeding as follows. Suppose that we want to compute the functional value of $P(x) = x^3 - 3x^2 + 6x - 8$ at $x = 2$. First, write the polynomial in nested form as follows

$$P(x) = x^3 - 3x^2 + 6x - 8$$
$$= [x^2 - 3x + 6]x - 8 \quad \text{[Factoring } x \text{ from the first three terms]}$$
$$= [(x - 3)x + 6]x - 8. \quad \text{[Factoring } x \text{ from the first two terms inside the square brackets]}$$

Next,

$$P(2) = [(2 - 3) \cdot 2 + 6] \cdot 2 - 8$$
$$= [-2 + 6] \cdot 2 - 8 \quad \text{[Eliminating parentheses]}$$
$$= 4 \cdot 2 - 8 \quad \text{[Eliminating brackets]}$$
$$= 0.$$

This procedure, known as *Horner's method*, applies to any polynomial

$$P(x) = a_n x^n + a_{n-1} x^{n-1} + \cdots + a_1 x + a_0$$

by writing it in nested form as

$$P(x) = \{\cdots [(a_n x + a_{n-1})x + a_{n-2}]x + \cdots + a_1\}x + a_0.$$

This method is extremely useful if you are computing function values of a polynomial with the help of a calculator, particularly if your calculator has a memory. Storing the numerical value of x at the beginning of the computation, and recalling it whenever necessary, will speed up your calculations.

◆ **Example 3.**   Find the remainder in the division of $P(x) = 0.5x^3 - 0.2x^2 + 0.3x - 0.4$ by $x - 1.5$.

Solution.   First, we write

$$P(x) = [(0.5x - 0.2)x + 0.3]x - 0.4.$$

Since, by the Remainder Theorem, $R = P(1.5)$, we compute the functional value of $P(x)$ at $x = 1.5$:

$$P(1.5) = [(0.5 \times 1.5 - 0.2) \times 1.5 + 0.3] \times 1.5 - 0.4$$
$$= [0.55 \times 1.5 + 0.3] \times 1.5 - 0.4$$
$$= 1.125 \times 1.5 - 0.4$$
$$= 1.6875 - 0.4$$
$$= 1.2875.$$

The following keystrokes were performed on a TI-30-II in order to obtain the function value of $P(x)$ at $x = 1.5$:

$$1.5\boxed{\text{STO}}\,0.5\,\boxed{\times}\,\boxed{\text{RCL}}\,\boxed{-}\,0.2\,\boxed{=}\,\boxed{\times}\,\boxed{\text{RCL}}\,\boxed{+}\,0.3\,\boxed{=}\,\boxed{\times}\,\boxed{\text{RCL}}\,\boxed{-}\,0.4\,\boxed{=}$$

c◇ **Practice Exercise 3.**   Find the functional value of $A(x) = 0.3x^4 - 1.5x^3 - 0.4x^2 + 2.1x - 1.3$ at $x = 0.16$.

Answer. $A(0.16) = -0.9801874$.

Synthetic Division

When dividing a polynomial by $x - a$, it is possible to avoid long division by using a shorthand method called *synthetic division*, which we now explain.

◆ **Example 4.**   Divide $A(x) = 4x^3 - x^2 - 8x + 5$ by $x - 2$.

Solution.   First, let us perform the long division for comparison.

$$
\begin{array}{r}
4x^2 + 7x + 6 \\
x - 2 \overline{)\,4x^3 - x^2 - 8x + 5} \\
\underline{4x^3 - 8x^2} \\
7x^2 - 8x \\
\underline{7x^2 - 14x} \\
6x + 5 \\
\underline{6x - 12} \\
17
\end{array}
$$

We obtain $Q(x) = 4x^2 + 3x + 1$ as quotient and $R = 17$ as remainder.

Next, we describe the method of *synthetic division*.

a) Write the coefficients of the dividend $A(x)$, and write the number a ($= 2$ in this example) in the corner to the left of the coefficients:

$$
\begin{array}{c|cccc}
2 & 4 & -1 & -8 & 5 \\
\hline
& & & & \\
& & & & \\
\end{array}
$$

b) Bring down the first coefficient of the dividend:

$$
\begin{array}{c|cccc}
2 & 4 & -1 & -8 & 5 \\
& \downarrow & & & \\
\hline
& 4 & & & \\
\end{array}
$$

This is the first coefficient of the quotient.

c) Multiply the first coefficient of the quotient by 2 (the number in the left corner) and add it to −1, the second coefficient of the dividend.

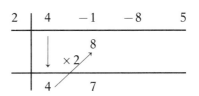

This is the *second coefficient of the quotient.*

d) Proceed in a similar manner until all coefficients of the dividend have been used.

$$
\begin{array}{r|rrrr}
2 & 4 & -1 & -8 & 5 \\
  &   & 8 & 14 & 12 \\
  & \downarrow & \times 2 & \times 2 & \times 2 \\
\hline
  & 4 & 7 & 6 & ⑰
\end{array}
$$

The circled number is the remainder. The other numbers are the coefficients of the quotient. As before, we get

$$Q(x) = 4x^2 + 7x + 6 \quad \text{and} \quad R = 17.$$

◇ **Practice Exercise 4.**   Use the method of synthetic division to find the quotient and remainder in the division of $-4x^3 + 3x^2 - x + 1$ by $x - 1$.

Answer. $Q(x) = -4x^2 - x - 2, \ R = -1.$

◆ **Example 5.**   Use the method of synthetic division to divide $x^4 + 1$ by $x + 1$.

Solution.   First, notice that $x + 1 = x - (-1)$, so $a = -1$. Next, the missing terms of degrees 3, 2, and 1 in the dividend correspond to coefficients equal to 0. We have

$$
\begin{array}{r|rrrrr}
-1 & 1 & 0 & 0 & 0 & 1 \\
   &   & -1 & 1 & -1 & 1 \\
   & \downarrow & & & & \\
\hline
   & 1 & -1 & 1 & -1 & ②
\end{array}
$$

Thus,

$$Q(x) = x^3 - x^2 + x - 1 \quad \text{and} \quad R = 2.$$

◇ **Practice Exercise 5.**   Divide by synthetic division $x^5 - 2x^3 - 7x$ by $x + 2$.

Answer. $Q(x) = x^4 - 2x^3 + 2x^2 - 4x + 1, \ R = -2.$

The method of synthetic division can also be used to evaluate polynomial functions.

◆ **Example 6.**   If $P(x) = 2x^3 - 5x + 6$, find $P(3)$.

Solution.   We have

$$
\begin{array}{r|rrrr}
3 & 2 & 0 & -5 & 6 \\
  &   & 6 & 18 & 39 \\
\hline
  & 2 & 6 & 13 & 45
\end{array}
$$

Thus, according to (8.4), $R = P(3) = 45$.
Of course the same result could have been obtained directly:

$$
\begin{aligned}
P(3) &= 2(3)^3 - 5(3) + 6 \\
&= 54 - 15 + 6 \\
&= 45.
\end{aligned}
$$

◇ **Practice Exercise 6.**   Let $P(x) = 4x^3 - x^2 - 2x + 12$. Use synthetic division to find $P(-2)$.

Answer. $R = P(-2) = -20$.

◼ **EXERCISES 8.2**

Without performing the division, find the remainder in the following divisions.

1. $(2x^2 - 6x - 4) \div (x + 2)$
2. $(-x^2 - x - 1) \div (x - 2)$
3. $(-x^3 - 2x^2 + x + 1) \div \left(x + \dfrac{1}{2}\right)$
4. $(5x^3 - x - 6) \div \left(x - \dfrac{2}{3}\right)$
5. $(x^5 - x^3 - 2x + 2) \div (x - 2)$
6. $(x^6 - 3x^4 + 4x^2 - 2) \div (x - 1)$
7. $(x^8 - 3x^6 - 4x^5 - 2x^4 + x^3 - 2x^2 + 3x - 4) \div x$
8. $(x^{10} - 5x^5 - 8x + 1) \div x$
c 9. $(0.2x^3 - 0.03x^2 + 0.21x - 0.18) \div (x + 0.5)$
c 10. $(0.5x^4 - 0.12x^2 + 0.03) \div (x - 0.02)$

In Exercises 11–22, find the quotient and remainder by synthetic division.

11. $(2x^2 - 6x - 4) \div (x - 2)$
12. $(-x^2 - x - 1) \div (x - 2)$
13. $(-x^3 - 2x^2 + x + 1) \div \left(x + \dfrac{1}{2}\right)$
14. $(5x^3 - x - 6) \div \left(x - \dfrac{2}{3}\right)$
15. $(4x^3 - 3x + 4) \div (x - 4)$
16. $(x^4 + x^2 + 1) \div (x + 2)$
17. $(x^5 - x^3 - 2x + 2) \div (x + 2)$
18. $(x^6 - 3x^4 + 4x^2 - 2) \div (x + 1)$
19. $(x^4 - 1) \div (x + 1)$
20. $(x^5 + 1) \div (x + 1)$
c 21. $(0.2x^3 - 0.03x^2 + 0.21x - 0.18) \div (x + 0.5)$
c 22. $(0.5x^4 - 0.12x^2 + 0.03) \div (x - 0.02)$

23. Show that $x - 3$ is a factor of the polynomial $x^3 - 6x^2 + 11x - 6$. Factor this polynomial completely.

24. Show that $x + 4$ is a factor of the polynomial $x^3 + 4x^2 - x - 4$. Determine the other factors.

25. Factor completely the polynomial $2x^3 - x^2 - 5x - 2$.

26. Factor completely the polynomial $x^4 + 3x^3 + x^2 - 3x - 2$.

In Exercises 27–32, apply Horner's method to find the indicated function values. Use your calculator if you wish.

27. $P(x) = 4x^3 - 3x^2 + 5x - 1$, $P(-2)$.

28. $Q(t) = t^4 - 3t^2 + 5$, $Q(-3)$.

29. $A(x) = x^3 - 2x^2 + 5x - 2$, $A(0.18)$.

30. $B(x) = 2x^3 - 5x + 3$, $B(1.25)$.

c 31. $A(t) = 0.1t^4 - 0.4t^3 + 0.3t^2 - 0.5$, $A(1.2)$.

c 32. $B(s) = 1.3s^4 + 2.5s^2 - 0.31$, $B(0.02)$.

33. Find a value of a such that $x + 4$ is a factor of $x^3 + x^2 - 17x + a$.

34. For what value of a is the remainder of $x^4 - 3x^2 + 2a$ divided by $x - 2$ equal to -3?

35. For what value of a is $x - 3$ a factor of $x^3 - 7x^2 + 3ax - 18$?

36. Find a value of a such that $2x^3 - 3x^2 + 2ax + 4$ is divisible by $x + 1/2$.

37. Find a value of a such that the remainder of $x^3 - ax^2 + 5$ divided by $x - 2$ is -3.

38. For what value of a is the remainder of $x^4 + ax + 10$ divided by $x + 3$ equal to -2?

● **39.** Use synthetic division to obtain the quotient and the remainder of the division of the polynomial $P(x) = 3x^2 - 2x + 1$ by $Q(x) = x - a$.

● **40.** Determine by synthetic division the quotient of $ax^2 + bx + c$ by $x - 1$.

8.3   ZEROS OF A POLYNOMIAL

> **Zero or Root of a Polynomial**
>
> A number r is said to be a *zero* or *root* of a polynomial $P(x)$, if the functional value of the polynomial at $x = r$ is equal to zero, that is, if $P(r) = 0$. In other words, r is a *solution* of the equation $P(x) = 0$.

If r is a zero of $P(x)$, then, by the Factor Theorem, the polynomial $P(x)$ is divisible by $x - r$ and vice-versa. For example, 1 is a root of $x^2 + x - 2$, since $1^2 + 1 - 2 = 0$. The polynomial $x^2 + x - 2$ is then divisible by $x - 1$ and we have the factorization

$$x^2 + x - 2 = (x - 1)(x + 2).$$

From the last expression, we see that -2 is the other zero of the equation $x^2 + x - 2 = 0$.

◆ **Example 1.**   Is 3 a root of $P(x) = x^3 - 2x^2 - 2x - 3$?

Solution.   To answer this question, evaluate $P(x)$ at $x = 3$:

$$P(3) = (3)^3 - 2(3)^2 - 2(3) + 3$$
$$= 27 - 18 - 6 - 3$$
$$= 0.$$

Thus 3 is a zero of $P(x)$.

◇ **Practice Exercise 1.**   Is -2 a zero of the polynomial $A(x) = 2x^4 - 3x^2 - x - 6$?

Answer.   No. $A(-2) = 16 \neq 0$.

◆ **Example 2.**   Show that $x + 2$ is a factor of the polynomial $2x^3 + x^2 - 5x + 2$.

Solution.   We can proceed in two different ways. One way is by evaluating the given polynomial at $x = -2$, as we did in the previous example. The other way is by using synthetic division:

$$
\begin{array}{r|rrrr}
-2 & 2 & 1 & -5 & 2 \\
   &   & -4 & 6 & -2 \\
\hline
   & 2 & -3 & 1 & \circled{0}
\end{array}
$$

The remainder is 0, so -2 is a solution of the equation $2x^3 + x^2 - 5x + 2 = 0$. It then follows that $x + 2$ is a factor of the polynomial $2x^3 + x^2 - 5x + 2$. The method of synthetic division also gives the other factor:

$$2x^3 + x^2 - 5x + 2 = (x + 2)(2x^2 - 3x + 1).$$

Notice that the quadratic polynomial $2x^2 - 3x + 1$ can be factored into $(2x - 1)(x - 1)$. Thus, we obtain the complete factorization of the given polynomial:

$$2x^3 + x^2 - 5x + 2 = (x + 2)(2x - 1)(x - 1).$$

◇ **Practice Exercise 2.**   Is $x - 3$ a factor of the polynomial $x^3 - 4x^2 + 4x - 3$?

Answer.   Yes: $x^3 - 4x^2 + 4x - 3 = (x - 3)(x^2 - x + 1)$. (Can you factor $x^2 - x + 1$ any further?)

Complex Zeros

Zeros of polynomials can be real or complex numbers. Even polynomials with real coefficients may have complex zeros. For example, the quadratic polynomial $x^2 + 1$, that led us to the definition of complex numbers, has zeros i and $-i$. Notice that the Remainder and Factor Theorems are also true in the case of complex zeros. For example, $x^2 + 1 = (x + i)(x - i)$. Similarly, $x^3 - 2x^2 + x - 2 = (x - 2)(x + i)(x - i)$ since 2, i, and $-i$ are the three roots of the polynomial $x^3 - 2x^2 + x - 2$. Both theorems also remain true in the case of polynomials with complex coefficients. Throughout this book we deal mostly with polynomials with real coefficients. However, in some instances we will have to consider polynomials with complex coefficients. The computations with such polynomials are no more difficult than the computations with real coefficient polynomials

The Fundamental Theorem of Algebra

When dealing with polynomials it is important to know how to find their zeros. It is very simple to find the zero of a linear polynomial. Quadratic equations can be solved by using the quadratic formula discussed in Section 2.5. As the degree of the polynomial increases, so does the difficulty in solving the corresponding polynomial equation.

It appears that the Babylonians (about 1950 B.C.) had already considered and solved quite general quadratic equations, seeking only positive roots. They must have known an equivalent version of the quadratic formula at least in certain special cases. The Hindu mathematician Bhaskara (about 1150 A.D.) investigated negative roots, possibly under Chinese influence, and observed that positive numbers had two real roots while negative numbers had none.

Efforts to find formulas to solve equations of higher degrees continued for many centuries without success. In 1545, Hieronimo Cardano (1501–1576), an Italian mathematician, published in his "Ars Magna" formulas to solve equations of degrees three and four. These formulas are known as *Cardano's formulas* although they were not discovered by him. Scipio del Ferro (1465–1526), professor at the University of Bologna, classified and solved all the cases of the general third degree equation. Del Ferro did not publish, but his methods were rediscovered about 1535 by Niccolo Tartaglia (1500–1577), an Italian engineer and mathematician. It was Tartaglia who revealed the methods to Cardano. Also, the solution of the general fourth degree equation was found by Cardano's young student Ludovico Ferrari (1527–1565).

Cardano's formulas involving rational combinations of the coefficients and root extraction are too complicated and almost never used in practice. For more than two centuries mathematicians tried unsuccessfully to find formulas involving rational combinations of the coefficients and root extractions to solve equations of degree $n > 4$. However, the Norwegian mathematician Niels Henrik Abel (1802–1829) proved this to be impossible. In other words, equations of higher degree than four cannot be solved by root extractions except for special values of the coefficients.

Even if a polynomial equation of degree higher than four cannot be solved by root extractions, it is conceivable that the equation has solutions. In 1799, Carl Friedrich Gauss (1777–1855), a German mathematician, proved the following important theorem.

HISTORICAL NOTE

Solving a Cubic Equation
In his "Ars Magna" (The Great Art), published in 1545, Cardano described the solution of the cubic equation

$$x^3 + px + q = 0,$$

with the following words:

"Cube one third of the coefficient of x; add to it the square of one-half of the constant of the equation; and take the square root of the whole. You will duplicate this and to one of the two you add one-half of the number you have already squared and from the other you subtract one-half of the same. You will have a binomium and its apotome. Then subtracting the cube root of the apotome from the cube root of the binomium, the remainder of that which is left is the value of x."*

In modern notation, here is the formula for the solution:

$$x = \sqrt[3]{-\frac{q}{2} + \sqrt{\frac{q^2}{4} + \frac{p^3}{27}}}$$
$$- \sqrt[3]{\frac{q}{2} + \sqrt{\frac{q^2}{4} + \frac{p^3}{27}}}$$

* From Lars Garding, "Encounters with Mathematics," Springer-Verlag (1977).

> **The Fundamental Theorem of Algebra**
>
> Every polynomial of degree n with real (or complex) coefficients has at least one zero.

The proof of this theorem involves techniques beyond the scope of this book. We point out that the Fundamental Theorem of Algebra is an *existential* theorem: it only asserts the *existence* of a zero without giving any indication of *how* to find it. Also, the zero may be a real or complex number.

Complex zeros of polynomials with *real* coefficients always occur in *conjugate pairs*. Let us state this more precisely.

Complex Zeros of Polynomials with Real Coefficients

If r is a complex zero of a polynomial $P(x) = a_n x^n + \cdots + a_1 x + a_0$ with real coefficients, then $\bar{r}$, the complex conjugate of r, is also a zero of P.

Indeed, if r is a zero of $P(x)$, then

$$P(r) = a_n r^n + \cdots + a_1 r + a_0 = 0.$$

Taking complex conjugates, we have

$$\overline{a_n r^n + \cdots + a_1 r + a_0} = 0.$$

Since the complex conjugate of a sum or product is the sum or product of complex conjugates [Section 1.9], we obtain

$$\bar{a}_n \bar{r}^n + \cdots + \bar{a}_1 \bar{r} + \bar{a}_0 = 0.$$

But the complex conjugate of a real number is the real number itself, so

$$a_n \bar{r}^n + \cdots + a_1 \bar{r} + a_0 = 0$$

which proves that $\bar{r}$ is also a zero of $P(x)$.

The Multiplicity of a Zero

Let r be a zero of a polynomial of degree $n \geq 1$. Then, by the Factor Theorem, the linear polynomial $x - r$ is a factor of $P(x)$ and we have

$$P(x) = (x - r)Q_1(x),$$

where $Q_1(x)$ is a polynomial of degree $n - 1$. Either r is also a root of $Q_1(x)$ or it is not. If it is, then we can apply the Factor Theorem and thus obtain

$$Q_1(x) = (x - r)Q_2(x).$$

This gives us the factorization

$$P(x) = (x - r)Q_1(x)$$
$$= (x - r)^2 Q_2(x)$$

where the degree of $Q_2(x)$ is $n - 2$. This process can be continued until we reach a stage where

$$P(x) = (x - r)^k Q_k(x)$$

and r is not a root of $Q_k(x)$ (we must always reach that stage, because no such factorization can exist with $k > n$). It is possible to show that the number k and the polynomial $Q_k(x)$, which has degree $n - k$, are uniquely determined by r.

The number k is called the *multiplicity* of the root r of $P(x)$. This means that if r is a root of $P(x)$ of multiplicity k, then $(x - r)^k$ is *a factor of* $P(x)$, *while* $(x - r)^{k+1}$ *is not.* In other words, k is the largest number m for which $(x - r)^m$ is a factor of $P(x)$.

For example, 1 is a zero of multiplicity 2 of $P(x) = x^3 + x^2 - 5x + 3$, because $P(x) = (x - 1)^2(x + 3)$. The number -3 is a zero of multiplicity 1 of $P(x)$.

A zero of multiplicity 2 is also called a *double zero*, and a zero of multiplicity 3 is called a *triple zero*.

When counting the number of zeros of a polynomial, *a zero of multiplicity k is counted as k zeros.*

The Number of Zeros of a Polynomial

By the Fundamental Theorem of Algebra, if $P(x)$ is a polynomial of degree $n \geq 1$, then it has at least one zero r_1. By the Factor Theorem, $P(x)$ is divisible by $x - r_1$, and we can write

$$P(x) = (x - r_1)Q_1(x),$$

where $Q_1(x)$ is a polynomial of degree $n - 1$. If $n = 1$, then the degree of $Q_1(x)$ is zero, so $Q_1(x) = a_1$ is a constant. If $n > 1$, then we reason with $Q_1(x)$ in the same way as we did with $P(x)$, to obtain

$$Q_1(x) = (x - r_2)Q_2(x),$$

where r_2 is a zero of $Q_1(x)$ and where $Q_2(x)$ is a polynomial of degree $n - 2$. Note that r_2 may be equal to r_1. Even if this is the case we continue to use different notations for the same root. Replacing $Q_1(x)$ in the expression for $P(x)$ yields

$$\begin{aligned}P(x) &= (x - r_1)[(x - r_2)Q_2(x)] \\ &= (x - r_1)(x - r_2)Q_2(x).\end{aligned}$$

Either $n = 2$, in which case $Q_2(x) = a_2$ is a constant, or $n > 2$, in which case we can again apply the Fundamental Theorem of Algebra. After a finite number of steps, we obtain the factorization

$$(8.6) \quad P(x) = (x - r_1)(x - r_2) \cdots (x - r_n)a_n,$$

where $a_n \neq 0$ is the leading coefficient of P, and $r_1, r_2, \ldots, r_n$ are the zeros (not necessarily distinct) of $P(x)$. Thus, every polynomial of degree n has *at least* n zeros.

Now, let r be a number different from the zeros above. We have

$$P(r) = a_n(r - r_1)(r - r_2) \cdots (r - r_n) \neq 0$$

since all factors are different from zero. We conclude that $r_1, r_2, \ldots, r_n$ are the only zeros of $P(x)$ and, therefore, that $P(x)$ has *at most* n zeros. Thus, we have proved the following theorem.

The Number of Zeros of a Polynomial

Every polynomial of degree $n \geq 1$ has exactly n zeros, counting multiplicities.

◆ **Example 3.** Verify that -2 is a double zero of $P(x) = x^4 + 2x^3 + x^2 + 12x + 20$, and find all roots of the polynomial.

Solution.   We divide $P(x)$ by $x + 2$:

$$
\begin{array}{r|rrrrr}
-2 & 1 & 2 & 1 & 12 & 20 \\
   &   & -2 & 0 & -2 & -20 \\
\hline
   & 1 & 0 & 1 & 10 & ⓪
\end{array}
$$

The remainder is zero, so -2 is a root of $P(x)$. The quotient in the division of $P(x)$ by $x + 2$ is $Q_1(x) = x^3 + x + 10$, and we have

$$P(x) = (x + 2)(x^3 + x + 10).$$

Next, we divide $Q_1(x)$ by $x + 2$:

$$
\begin{array}{r|rrrr}
-2 & 1 & 0 & 1 & 10 \\
   &   & -2 & 4 & -10 \\
\hline
   & 1 & -2 & 5 & ⓪
\end{array}
$$

The remainder is zero, so -2 is a root of $Q_1(x)$, and we have

$$Q_1(x) = (x + 2)(x^2 - 2x + 15).$$

Replacing $Q_1(x)$ in the expression for $P(x)$, we obtain

$$P(x) = (x + 2)^2(x^2 - 2x + 5).$$

Setting $Q_2(x) = x^2 - 2x + 5$, we can check that $Q_2(-2) = 13 \neq 0$. Thus, -2 is indeed a double root of $P(x)$. To find the other roots, we solve the quadratic equation

$$x^2 - 2x + 5 = 0.$$

Using the quadratic formula, we obtain two complex zeros

$$x = \frac{2 \pm \sqrt{4 - 20}}{2} = \frac{2 \pm 4i}{2} = 1 \pm 2i.$$

The four zeros of the given polynomial are: -2 with *multiplicity two*, $1 + 2i$, and $1 - 2i$.

◇ **Practice Exercise 3.** Verify that 2 is a double root of the polynomial $x^4 - 4x^3 + 5x^2 - 4x + 4$. Find all roots of the polynomial.

Answer.   2 (double root), i, $-i$.

◆ **Example 4.** Factor the polynomial $P(x) = x^6 - 5x^5 + 6x^4$ and determine its zeros.

Solution.   We have

$$P(x) = x^4(x^2 - 5x + 6)$$
$$= x^4(x - 2)(x - 3).$$

This shows that the given polynomial has six zeros: 0 with multiplicity four, 2, and 3.

◇ **Practice Exercise 4.**   Determine all zeros of the polynomial $x^5 + 3x^4 - 4x^3$. Factor this polynomial completely.

Answer.   Five zeros: 0 with multiplicity three, 1, -4. Thus, $x^5 + 3x^4 - 4x^3 = x^3(x^2 + 3x - 4) = x^3(x - 1)(x + 4)$.

◆ **Example 5.**   **a)** Find a polynomial of degree 3 with zeros -1, 1, and 2.
   b) Find a polynomial of degree 3 having the same zeros and taking the value 6 at $x = 0$.

Solution.   **a)** If a polynomial $P(x)$ has zeros -1, 1, and 2, then $x + 1$, $x - 1$, and $x - 2$ are factors of $P(x)$. Since we want $P(x)$ to have degree 3, it suffices to write

$$P(x) = (x - 1)(x + 1)(x - 2)$$
$$= (x^2 - 1)(x - 2)$$
$$= x^3 - 2x^2 - x + 2.$$

b) According to (8.6), we write

$$P(x) = a_3(x - 1)(x + 1)(x - 2).$$

When $x = 0$, we want $P(0) = 6$. Thus

$$P(0) = a_3(0 - 1)(0 + 1)(0 - 2) = 6,$$
$$2a_3 = 6$$
$$a_3 = 3.$$

Substituting into the above expression, we obtain

$$P(x) = 3(x - 1)(x + 1)(x - 2)$$
$$= 3(x^2 - 1)(x - 2)$$
$$= 3(x^3 - 2x^2 - x + 2)$$
$$= 3x^3 - 6x^2 - 3x + 6,$$

which is the desired polynomial.

◇ **Practice Exercise 5.**   Find a polynomial $P(x)$ of lowest degree with zeros -1, 0, and -2 and such that $P(1) = 12$.

Answer. $P(x) = 2x(x + 1)(x + 2)$.

Example 6.   Find a polynomial of third degree with real coefficients, and having zeros 1 and $2 + 3i$.

Solution.   If a polynomial has real coefficients and $2 + 3i$ is a zero, then the complex conjugate $2 - 3i$ is also a zero. The three zeros of the third degree polynomial we are looking for are 1, $2 + 3i$, and $2 - 3i$. Therefore,

$$
\begin{aligned}
P(x) &= (x - 1)[x - (2 + 3i)][x - (2 - 3i)] \\
&= (x - 1)[(x - 2) - 3i][(x - 2) + 3i] \\
&= (x - 1)[(x - 2)^2 + 9] \\
&= (x - 1)(x^2 - 4x + 13) \\
&= x^3 - 5x^2 + 17x - 13.
\end{aligned}
$$

HISTORICAL NOTE

Carl Friedrich Gauss (1777–1855),

a German mathematician and astronomer, made notable contributions in several branches of mathematics and physics. He investigated the geometry of curved surfaces, was an authority in geodesy, and applied mathematical theory to electricity and magnetism. The Fundamental Theorem of Algebra, proved in 1799, was part of Gauss's doctoral dissertation, "A New Proof that Every Rational Integral Function of One Variable Can Be Resolved into Real Factors of the First and Second Degree." The mathematical work of Gauss is so vast and important that he is often referred to as the Prince of Mathematicians. Gauss himself is placed in the same class as Archimedes and Newton.

◇ **Practice Exercise 6.**   Write a third degree polynomial with real coefficients, having leading coefficient 2 and zeros $1 - i$ and -3.

Answer. $2x^3 + 2x^2 - 8x + 12$.

 EXERCISES 8.3

In Exercises 1–10, find whether the given number is a zero of the given polynomial.

1. $x^2 - 3x + 3$, 1

2. $2x^2 - 4x - 6$, -1

3. $3x^3 - 7x + 10$, -2

4. $x^3 - 4x^2 + 9$, 3

5. $2x^2 - 5x - 3$, $-\dfrac{1}{2}$

6. $3x^2 - 7x - 20$, $\dfrac{5}{3}$

7. $x^4 - 2x^3 + 3x^2 - 3$, -2

8. $x^5 - 3x^2 - 6$, 2

9. $x^2 - 2x + 5$, $1 - 2i$

10. $x^2 - 6x + 13$, $3 + 2i$

In Exercises 11–14, find whether the linear polynomial is a factor of the given polynomial. Whenever possible, find the other factors.

11. $2x^3 - 4x^2 - 3x + 6$, $x - 2$

12. $3x^3 + 14x^2 + 8x$, $x + 4$

13. $2x^4 + 6x^3 + 5x^2 + 14x - 3$, $x + 3$

14. $x^4 + 4x^3 + 4x^2 - 4x + 8$, $x - 2$

In Exercises 15–18, a zero of P(x) is given. Find another zero.

15. $P(x) = x^2 - 4x + 5$, $2 - i$

16. $P(x) = x^2 - 2x + 10$, $1 + 3i$

17. $P(x) = 2x^3 - 7x^2 + 22x + 13$, $2 + 3i$

18. $P(x) = x^3 + 8$, $1 - i\sqrt{3}$

In Exercises 19–22, a polynomial and one of its roots are given. Find the other roots.

19. $x^3 - x^2 - 9x - 12$, 4

20. $2x^3 - x^2 - 15x + 18$, 2

21. $x^4 + 3x^2 - 4$, $2i$

22. $x^3 + 2x^2 + 5x - 26$, $-2 + 3i$

23. Show that 3 is a double root of $x^4 - 5x^3 + 4x^2 + 3x + 9$ and find the other roots.

24. Show that -1 is a double root of $2x^4 + 4x^3 + 3x^2 + 2x + 1$ and find the other roots.

In Exercises 25–32, find the zeros of the given polynomials and determine the multiplicity of each zero.

25. $P(x) = (x - 5)(3x + 2)$

26. $P(x) = (x + 2)^3(4x - 1)^2$

27. $P(x) = (2x + 3)^2(x - 1)^3(3x - 4)$

28. $P(x) = (4x^2 - 3)^2$

29. $P(x) = x^4 + 2x^3 + 2x^2$

30. $P(x) = (x^2 - 4x + 3)^2$

31. $P(x) = (x^2 - 2x + 4)^2$

32. $P(x) = x^3(x^2 - x + 1)$

In Exercises 33–40, find a polynomial P(x) of lowest degree having the indicated zeros and satisfying the given condition.

33. $-2, 2, 3$; $P(1) = 6$

34. $-2, -1, 4$; $P(2) = 8$

35. $-1, 2, 4$; $P(0) = 8$

36. $-2, 0, 3$; $P(1) = -6$

37. -1 (multiplicity 2), 3 (multiplicity 2); leading coefficient 1.

38. 2 (multiplicity 3), 1 (multiplicity 2); leading coefficient 2.

39. $1 + 2i$, $1 - 2i$, 3; $P(0) = -15$

40. $2 - 2i$, $2 + 2i$, 1 (multiplicity 2); leading coefficient 1.

8.4   TEST FOR RATIONAL ZEROS

If a polynomial has *integral coefficients*, there is a very useful test that allows us to determine whether the polynomial has *rational zeros*.

Let

$$P(x) = a_nx^n + a_{n-1}x^{n-1} + \cdots + a_1x + a_0$$

be a polynomial of degree n ($a_n \neq 0$) with integral coefficients and such that $a_0 \neq 0$. Assume that the rational number p/q, in its lowest terms, is a zero of $P(x)$. Then,

$$P\left(\frac{p}{q}\right) = a_n\left(\frac{p}{q}\right)^n + a_{n-1}\left(\frac{p}{q}\right)^{n-1} + \cdots + a_1\left(\frac{p}{q}\right) + a_0 = 0.$$

Multiplying by q^n, we obtain

$$(8.7) \quad a_n p^n + a_{n-1} p^{n-1} q + \cdots + a_1 p q^{n-1} + a_0 q^n = 0.$$

Solving for $a_0 q^n$, we get

$$a_0 q^n = -a_n p^n - a_{n-1} p^{n-1} q - \cdots - a_1 p q^{n-1}$$

which can be written as

$$a_0 q^n = p(-a_n p^{n-1} - a_{n-1} p^{n-2} q - \cdots - a_1 q^{n-1}).$$

This shows that p is a factor of $a_0 q^n$. Since p does not divide q, because the fraction p/q is in its lowest terms, we see that p must divide a_0. Therefore, p is a factor of a_0.

Similarly, solving (8.7) for $a_n p^n$, we get

$$a_n p^n = q(-a_{n-1} p^{n-1} - \cdots - a_1 p q^{n-2} - a_0 q^{n-1}).$$

Arguing as before, we conclude that q is a factor of a_n.

Thus, we have proved the following theorem.

Test for Rational Zeros

Let

$$P(x) = a_n x^n + a_{n-1} x^{n-1} + \cdots + a_1 x + a_0$$

be a polynomial with integral coefficients such that $a_n \neq 0$ and $a_0 \neq 0$. If a rational number p/q is a zero of $P(x)$, then p must divide the constant term a_0, and q must divide the leading coefficient a_n.

◆ **Example 1.**   Find the rational roots of $P(x) = 2x^3 + 9x^2 + 3x - 4$ and factor the polynomial.

Solution.   According to the above theorem, a fraction p/q, reduced to its lowest terms, is a zero of the given polynomial if p is a factor of -4 and q is a factor of 2. The choices for p are then

$$\pm 1, \quad \pm 2, \quad \text{and} \quad \pm 4,$$

and those for q are

$$\pm 1 \quad \text{and} \quad \pm 2.$$

Thus, the possible rational roots are

$$\pm 1, \quad \pm \frac{1}{2}, \quad \pm 2, \quad \text{and} \quad \pm 4.$$

Every rational zero of $P(x) = 2x^3 + 9x^2 + 3x - 4$ must belong to this set of rational numbers. Next, compute the function value of P for each of these numbers. Since

$$P(-1) = -2 + 9 - 3 - 4 = 0,$$

we see that -1 is a root and $x + 1$ is a factor of $P(x)$. Dividing $P(x)$ by $x + 1$, we obtain

$$2x^3 + 9x^2 + 3x - 4 = (x + 1)(2x^2 + 7x - 4).$$

The remaining zeros of $P(x)$ are the zeros of the quadratic polynomial $2x^2 + 7x - 4$. At this point, you could either use this polynomial to check for the other zeros of $P(x)$, or solve the quadratic equation $2x^2 + 7x - 4 = 0$. In any case, you can verify that $1/2$ and -4 are roots of $P(x)$ and obtain the factorization

$$2x^3 + 9x^2 + 3x - 4 = 2(x + 1)\left(x - \frac{1}{2}\right)(x + 4).$$

◇ **Practice Exercise 1.**   Find all the roots of $A(x) = 2x^3 - x^2 - 7x + 6$ and factor this polynomial completely.

Answer.   Roots: $-2, 1, \dfrac{3}{2}$

$$2x^3 - x^2 - 7x + 6 = 2\left(x - \frac{3}{2}\right)(x - 1)(x + 2).$$

It is important to observe that the test for rational roots does not say that every polynomial with integer coefficients has rational roots. It only states that if a rational number (in its lowest terms) is a root, then it must satisfy the conditions of the theorem. What the theorem does is enable us to list a finite set of rational numbers that includes every rational root of the polynomial. This reduces the problem of finding all the rational roots of a polynomial to the mechanical task of checking the possibilities on the list. There are no rational roots that are not on that list.

In particular, if the leading coefficient of a polynomial with integer coefficients is 1, and the polynomial has rational roots, then these roots are integers.

◆ **Example 2.**   Find the rational zeros of $P(x) = x^4 - x^3 - 4x^2 + 4x$ and factor the polynomial.

Solution.   Since the constant term of $P(x)$ is zero, we see that $x = 0$ is a root of $P(x)$, and we can write

$$P(x) = x(x^3 - x^2 - 4x + 4).$$

The remaining zeros are the roots of the third degree polynomial $Q(x) = x^3 - x^2 - 4x + 4$. Since the leading coefficient is 1, the only possible rational zeros are integers. According to the above theorem, they must be among the integral factors of 4, that is, $\pm 1, \pm 2, \pm 4$. Since

$$Q(-1) = (-1)^3 - (-1)^2 - 4(-1) + 4$$
$$= -1 - 1 + 4 + 4$$
$$= 6,$$

-1 is not a zero. On the other hand,

$$Q(1) = 1^3 - 1^2 - 4(1) + 4$$
$$= 1 - 1 - 4 + 4$$
$$= 0,$$

so 1 is a root. Dividing $Q(x)$ by $x - 1$, we have

$$x^3 - x^2 - 4x + 4 = (x - 1)(x^2 - 4).$$

Since -2 and 2 are the roots of the quadratic polynomial $x^2 - 4$, we can now write $P(x)$ as a product of linear factors:

$$x^4 - x^3 - 4x^2 + 4x = x(x - 1)(x - 2)(x + 2).$$

◇ **Practice Exercise 2.** Find all the zeros of $x^4 - 2x^3 - 5x^2 + 6x$ and factor the polynomial.

Answer. Zeros: $-2, 0, 1, 3$
$x^4 - 2x^3 - 5x^2 + 6x = x(x + 2)(x - 1)(x - 3).$

The test for rational zeros also applies to polynomials with *rational* coefficients. Indeed, if a polynomial equation has rational coefficients, then multiplying both sides by the least common denominator of the coefficients transforms the equation into an equivalent one with integral coefficients.

◆ **Example 3.** Find the zeros of $P(x) = \dfrac{1}{6}x^3 - x^2 + \dfrac{3}{2}x - \dfrac{1}{3}$.

Solution. We have to solve the equation

$$\frac{1}{6}x^3 - x^2 + \frac{3}{2}x - \frac{1}{3} = 0.$$

Multiplying both sides by 6, we obtain an equivalent equation with integral coefficients:

$$x^3 - 6x^2 + 9x - 2 = 0.$$

The leading coefficient is 1, so the only possible rational roots are the integral factors of -2, namely, ± 1 and ± 2. We have

$$
\begin{array}{r|rrrr}
2 & 1 & -6 & 9 & -2 \\
  &   & 2 & -8 & 2 \\
\hline
  & 1 & -4 & 1 & 0
\end{array}
$$

so 2 is a root. On the other hand,

$$
\begin{array}{r|rrrr}
-2 & 1 & -6 & 9 & -2 \\
   &   & -2 & 16 & -50 \\
\hline
   & 1 & -8 & 25 & -52
\end{array}
$$

so -2 is not a root. As an exercise, you can check that -1 and 1 are not zeros of $P(x)$. Since 2 is a root, $x - 2$ is a factor of $P(x)$ and we have

$$x^3 - 6x^2 + 9x - 2 = (x - 2)(x^2 - 4x + 1).$$

The other roots of $P(x)$ are the solutions of the equation $x^2 - 4x + 1 = 0$. Using the quadratic formula, we obtain the solutions $2 + \sqrt{3}$ and $2 - \sqrt{3}$. Therefore, the three zeros of $P(x)$ are 2, $2 + \sqrt{3}$, and $2 - \sqrt{3}$.

◇ **Practice Exercise 3.** Find all zeros, real or complex, of the polynomial $\dfrac{x^3}{12} - \dfrac{x^2}{3} + \dfrac{x}{12} - \dfrac{1}{3} = 0$.

Answer. Zeros: 4, $\pm i$.

EXERCISES 8.4

Find all roots of the polynomials in Exercises 1–14.

1. $x^3 + x^2 - 4x - 4$ **2.** $x^3 - 3x^2 - x + 3$ **3.** $x^3 - 3x^2 - 4x + 12$
4. $2x^3 - 6x^2 + 6x - 2$ **5.** $x^3 - x$ **6.** $x^3 + x^2 - 6x$
7. $x^3 + x^2 + x + 2$ **8.** $x^4 - 4x^2$ **9.** $2x^3 + x^2 - 2x - 1$
10. $3x^3 + 5x^2 - 11x + 3$ **11.** $4x^3 - x$ **12.** $25x^4 - x^2$
13. $x^3 - 4x^2 + 4x - 4$ **14.** $2x^3 + 2x^2 + 3x + 3$

Factor each polynomial in Exercises 15–32.

15. $x^3 - x$ **16.** $x^4 - 4x^2$ **17.** $\dfrac{1}{2}x^3 - 2x$

18. $3x^4 - 12x^2$ **19.** $x^3 + x^2 - 4x - 4$ **20.** $3x^3 - 9x^2 - 3x + 9$
21. $x^3 + x^2 - 6x$ **22.** $4x^3 - 13x - 6$ **23.** $x^4 - 16$
24. $3x^3 + 3x$ **25.** $x^3 - 4x^2 + x - 4$ **26.** $x^3 - x^2 + 2x - 2$

27. $2x^3 - 3x^2 + 4x - 6$ **28.** $8x^3 + 4x^2 - 2x - 1$ **29.** $\dfrac{3}{4}x^4 - \dfrac{3}{2}x^3 - \dfrac{3}{2}x^2 + 6x - 6$

30. $\dfrac{1}{3}x^5 - \dfrac{1}{2}x^4 - \dfrac{1}{3}x + \dfrac{1}{2}$ **31.** $\dfrac{1}{4}x^3 - \dfrac{5}{4}x^2 + \dfrac{3}{2}x - \dfrac{1}{2}$ **32.** $x^3 + \dfrac{2}{3}x^2 - \dfrac{1}{4}x - \dfrac{1}{6}$

Show that the equations in Exercises 33–40 have no rational roots.

33. $x^2 - 2 = 0$ **34.** $x^2 - 3 = 0$
35. $2x^2 - x + 3 = 0$ **36.** $4x^2 - 12x + 7 = 0$
37. $x^3 - 3x^2 + 4x - 6 = 0$ **38.** $3x^3 - 4x^2 + 7x + 5 = 0$
39. $x^4 + x^3 - x^2 - 2x - 2 = 0$ **40.** $2x^4 - x^3 + 4x^2 - 2x + 3 = 0$

8.5 PARTIAL FRACTION DECOMPOSITION

In Section 1.6, we learned how to combine two or more rational functions into a single one by addition or subtraction. For example,

$$\frac{5}{x + 1} + \frac{x}{x^2 + 2} = \frac{5(x^2 + 2) + x(x + 1)}{(x + 1)(x^2 + 2)}$$

$$= \frac{6x^2 + x + 10}{x^3 + x^2 + 2x + 2}.$$

In many situations (such as in integral calculus) the reverse process is required: we have to express a quotient of two polynomials as the sum of two or more simpler quotients called *partial fractions*. Before we describe the method of *partial fraction decomposition* of a rational function, a few remarks are in order.

Irreducible Polynomials

A polynomial with real coefficients is said to be *irreducible over* $\mathbb{R}$ if it cannot be expressed as a product of two polynomials of positive degree with real coefficients. Otherwise, the polynomial is said to be *reducible over* $\mathbb{R}$.

For example, $x + 1$, $2x + 3$, $x^2 + 1$, and $x^2 + x + 1$ are irreducible over $\mathbb{R}$. On the other hand, the polynomials $x^2 - 1$, $2x^2 - 5x - 3$, and $x^3 - 6x^2 + 11x - 6$ are reducible over $\mathbb{R}$, because

$$x^2 - 1 = (x - 1)(x + 1)$$
$$2x^2 - 5x - 3 = (2x + 1)(x - 3)$$
$$x^3 - 6x^2 + 11x - 6 = (x - 1)(x - 2)(x - 3).$$

Note that if all the roots of a polynomial are real, then the polynomial is reducible over $\mathbb{R}$, and it can be written as a product of linear factors.

A quadratic polynomial $ax^2 + bx + c$ with real coefficients is *irreducible over* $\mathbb{R}$ if and only if it has *complex roots*; that is, if and only if the *discriminant* $b^2 - 4ac$ is *negative*. For example, $x^2 + 1$, $x^2 + x + 1$, $3x^2 - 2x + 1$ are irreducible over $\mathbb{R}$. Note that if a quadratic polynomial with real coefficients is irreducible over $\mathbb{R}$, then it has exactly two complex roots which form a pair of complex conjugate numbers.

We now state without proof the following important theorem.

The Factorization Theorem

Every polynomial with real coefficients can always be expressed as a product of *linear polynomials* and/or *quadratic polynomials* *which have real coefficients* and are *irreducible over* $\mathbb{R}$.

For example,

$$2x^2 - 5x - 3 = (2x + 1)(x - 3)$$
$$x^3 - 6x + 11x - 6 = (x - 1)(x - 2)(x - 3)$$
$$x^3 - 1 = (x - 1)(x^2 + x + 1)$$
$$x^4 - 2x^2 + 1 = (x^2 + 2x + 1)(x^2 - 2x + 1).$$

It follows from the Factorization Theorem that every polynomial with real coefficients has two types of factors: $(ax + b)^k$ and/or $(ax^2 + bx + c)^m$ where $ax^2 + bx + c$ is irreducible over $\mathbb{R}$. The first factor corresponds to the real zero $-b/a$ with multiplicity k, while the second factor corresponds to a pair of complex conjugate zeros each of which has multiplicity m.

Partial Fraction Decomposition

Let $A(x)$ and $B(x)$ be two polynomials with real coefficients, and such that *the degree of $A(x)$ is smaller than the degree of $B(x)$*. Then it follows from a theorem in algebra that the quotient $A(x)/B(x)$ can be written as the following sum

$$\frac{A(x)}{B(x)} = F_1(x) + F_2(x) + \cdots + F_p(x)$$

where each $F_i(x)$ has one of the forms

$$\frac{A_k}{(ax + b)^k} \quad \text{or} \quad \frac{A_m x + B_m}{(ax^2 + bx + c)^m}$$

where A_k, A_m, and B_m are real constants, k and m are suitable positive integers, and $ax^2 + bx + c$ is irreducible over $\mathbb{R}$. The sum $F_1(x) + F_2(x) + \cdots + F_p(x)$ is called the *partial fraction decomposition* of $A(x)/B(x)$ and each term $F_i(x)$ is called a *partial fraction*.

If the degree of $A(x)$ is *greater than or equal* to the degree of $B(x)$, then dividing $A(x)$ by $B(x)$ and using (8.1) we obtain

$$\frac{A(x)}{B(x)} = Q(x) + \frac{R(x)}{B(x)}$$

where $Q(x)$ is a polynomial and $R(x)/B(x)$ is a proper rational expression. Since the degree of $R(x)$ is less than the degree of $B(x)$, we can replace $R(x)/B(x)$ by its partial fraction decomposition. Thus we conclude that every rational function $A(x)/B(x)$ can be written as

$$\frac{A(x)}{B(x)} = Q(x) + F_1(x) + F_2(x) + \cdots + F_p(x)$$

where $Q(x)$ is a polynomial (if the degree of $A(x)$ is greater than or equal to the degree of $B(x)$) and $F_1(x), F_2(x), \ldots, F_p(x)$ are the partial fractions described above.

From now on, we assume that $A(x)/B(x)$ is a proper rational function, that is, the degree of $A(x)$ is smaller than the degree of $B(x)$. In order to find the partial fraction decomposition of $A(x)/B(x)$, it is necessary to write the denominator $B(x)$ as a product of factors of the type $(ax + b)^k$ and $(ax^2 + bx + c)^m$ where $ax^2 + bx + c$ is irreducible. Next we proceed according to the following rules.

> ### Rules for Partial Fraction Decomposition
>
> 1. For each factor of the type $(ax + b)^k$ with $k \geq 1$, the partial fraction decomposition of $A(x)/B(x)$ contains a sum of k terms
>
> $$\frac{A_1}{ax + b} + \frac{A_2}{(ax + b)^2} + \cdots + \frac{A_k}{(ax + b)^k}$$
>
> where $A_1, A_2, \ldots, A_k$ are real constants.
> 2. For each factor of the type $(ax^2 + bx + c)^m$ where $m \geq 1$ and $ax^2 + bx + c$ is irreducible over $\mathbb{R}$, the partial fraction decomposition of $A(x)/B(x)$ contains a sum of m terms
>
> $$\frac{A_1 x + B_1}{ax^2 + bx + c} + \frac{A_2 x + B_2}{(ax^2 + bx + c)^2} + \cdots + \frac{A_m x + B_m}{(ax^2 + bx + c)^m}$$
>
> where $A_1, B_1, \ldots, A_m, B_m$ are real constants.

◆ **Example 1.** Decompose $\dfrac{5x - 1}{2x^2 - 5x - 3}$ into partial fractions.

Solution. First, we factor the denominator and obtain

$$2x^2 - 5x - 3 = (2x + 1)(x - 3).$$

The denominator contains two linear factors: $2x + 1$ and $x - 3$. According to the rule 1 above, the partial fraction decomposition of $\dfrac{5x - 1}{2x^2 - 5x - 3}$ contains terms of the form $\dfrac{A}{2x + 1}$ and $\dfrac{B}{x - 3}$. Since $2x + 1$ and $x - 3$ are the only factors of the denominators, we can write

$$\frac{5x - 1}{2x^2 - 5x - 3} = \frac{A}{2x + 1} + \frac{B}{x - 3}.$$

Next, we have to determine the constants A and B. If we multiply both sides of the last expression by $2x^2 - 5x - 3 = (2x + 1)(x - 3)$, we obtain

$$5x - 1 = A(x - 3) + B(2x + 1).$$

Since the last expression is an identity, it must be satisfied for all values of x. If we set $x = 3$, then the first term on the right-hand side equals zero, and we can obtain B:

$$5(3) - 1 = A(3 - 3) + B[2(3) + 1]$$
$$14 = 7B$$
$$B = 2.$$

If we set $x = -1/2$, then the second term on the right-hand side equals zero, and we obtain the value of A:

$$5\left(-\frac{1}{2}\right) - 1 = A\left(-\frac{1}{2} - 3\right) + B\left[2\left(-\frac{1}{2}\right) + 1\right]$$

$$-\frac{7}{2} = A\left(-\frac{7}{2}\right)$$

$$A = 1.$$

Thus the given expression is decomposed into partial fractions as follows:

$$\frac{5x - 1}{2x^2 - 5x - 3} = \frac{1}{2x + 1} + \frac{2}{x - 3}.$$

◇ **Practice Exercise 1.** Decompose the expression $\dfrac{5x - 5}{3x^2 - 7x + 2}$ into partial fractions.

Answer. $\dfrac{5x - 5}{3x^2 - 7x + 2} = \dfrac{1}{x - 2} + \dfrac{2}{3x - 1}.$

◆ **Example 2.** Decompose $\dfrac{6x^3 - 19x^2 + 6x + 5}{2x^2 - 5x - 3}$ into partial fractions.

Solution. Since the degree of the numerator is greater than the degree of the denominator, we have to write the given expression as a sum of a polynomial and a proper rational expression. After performing a long division (whose computation we leave to the reader as an exercise), we obtain that

$$\frac{6x^3 - 19x^2 + 6x + 5}{2x^2 - 5x - 3} = 3x - 2 + \frac{5x - 1}{2x^2 - 5x - 3}.$$

Next, we proceed to decompose the proper rational expression $\dfrac{5x - 1}{2x^2 - 5x - 3}$ into partial fractions. But this is exactly the expression considered in Example 1. Thus the given expression can be written as

$$\frac{6x^3 - 19x^2 + 6x + 5}{2x^2 - 5x - 3} = 3x - 2 + \frac{1}{2x + 1} + \frac{2}{x - 3}.$$

◇ **Practice Exercise 2.** Decompose the rational expression $\dfrac{6x^3 - 11x + 2x - 3}{3x^2 - 7x + 2}$ into partial fractions.

Answer. $\dfrac{6x^3 - 11x + 2x - 3}{3x^2 - 7x + 2} = 2x + 1 + \dfrac{1}{x - 2} + \dfrac{2}{3x - 1}.$

◆ **Example 3.** Decompose $\dfrac{5x^2 + 2x + 2}{x^3 - 1}$ into partial fractions.

Solution.   Factoring the denominator, we have

$$x^3 - 1 = (x - 1)(x^2 + x + 1),$$

where $x^2 + x + 1$ is a quadratic polynomial that is irreducible over the reals. According to steps 1 and 2 above, the partial fraction decomposition of the given rational expression contains only two terms: $\dfrac{A}{x - 1}$ and $\dfrac{Bx + C}{x^2 + x + 1}$.

Writing

$$\frac{5x^2 + 2x + 2}{x^3 - 1} = \frac{A}{x - 1} + \frac{Bx + C}{x^2 + x + 1},$$

we now have to determine the constants A, B, and C. Multiplying both sides by $x^3 - 1$, we obtain

$$5x^2 + 2x + 2 = A(x^2 + x + 1) + (Bx + C)(x - 1).$$

Setting $x = 1$ and solving for A yields

$$5 \cdot 1^2 + 2 \cdot 1 + 2 = A(1^2 + 1 + 1) + (B \cdot 1 + C)(1 - 1),$$
$$9 = 3A,$$
$$A = 3.$$

Substituting 3 for A in the above expression, setting $x = 0$, and solving for C gives us

$$5 \cdot 0^2 + 2 \cdot 0 + 2 = 3(0^2 + 0 + 1) + (B \cdot 0 + C)(0 - 1),$$
$$2 = 3 + C(-1),$$
$$C = 1.$$

Finally, substituting 3 for A and 1 for C, we rewrite the expression for $5x^2 + 2x + 2$ as follows:

$$5x^2 + 2x + 2 = 3(x^2 + x + 1) + (Bx + 1)(x - 1).$$

Now, we may set $x = -1$ and solve for B:

$$5 = 3 + (-B + 1)(-2),$$
$$B = 2.$$

Hence,

$$\frac{5x^2 + 2x + 2}{x^3 - 1} = \frac{3}{x - 1} + \frac{2x + 1}{x^2 + x + 1}.$$

◇ **Practice Exercise 3.**   Decompose the rational expression $\dfrac{5x^2 + 1}{x^3 + 1}$ into partial fractions.

Answer. $\dfrac{5x^2 + 1}{x^3 + 1} = \dfrac{2}{x + 1} + \dfrac{3x - 1}{x^2 - x + 1}.$

◆ **Example 4.**   Decompose $\dfrac{4x^2 - 3x + 5}{(x - 1)^2(x + 2)}$ into partial fractions.

Solution. The denominator contains two linear factors: $x - 1$ and $x + 2$. Since the factor $x - 1$ appears with the exponent 2, then according to step 1 the partial fraction decomposition will contain terms of the form $\dfrac{A}{x - 1} + \dfrac{B}{(x - 1)^2}$. Moreover, the contribution from the factor $x + 2$ will be of the form $\dfrac{C}{x + 2}$. Next, write

$$\frac{4x^2 - 3x + 5}{(x - 1)^2(x + 2)} \doteq \frac{A}{x - 1} + \frac{B}{(x - 1)^2} + \frac{C}{x + 2}.$$

In order to determine the constants, multiply both sides by $(x - 1)^2(x + 2)$ and get

$$4x^2 - 3x + 5 = A(x - 1)(x + 2) + B(x + 2) + C(x - 1)^2.$$

If we set $x = 1$, then

$$6 = 3B,$$
$$B = 2.$$

If we set $x = -2$, then

$$27 = 9C,$$
$$C = 3.$$

Finally, setting $x = 0$, we obtain

$$5 = -2A + 4 + 3,$$
$$A = 1.$$

Therefore,

$$\frac{4x^2 - 3x + 5}{(x - 1)^2(x + 2)} = \frac{1}{x - 1} + \frac{2}{(x - 1)^2} + \frac{3}{x + 2}.$$

◇ **Practice Exercise 4.** Decompose $\dfrac{5x^2 + 15x + 7}{(x - 1)(x + 2)^2}$ into partial fractions.

Answer. $\dfrac{5x^2 + 15x + 7}{(x - 1)(x + 2)^2} = \dfrac{3}{x - 1} + \dfrac{2}{x + 2} + \dfrac{1}{(x + 2)^2}.$

◆ **Example 5.** Decompose $\dfrac{x^3 - 2x^2 + 2x - 4}{(x^2 + x + 1)^2}$ into partial fractions.

Solution. The quadratic polynomial $x^2 + x + 1$ is irreducible over $\mathbb{R}$ (why?). Thus, using rule 2 above, we can write that

$$\frac{x^3 - 2x^2 + 2x - 4}{(x^2 + x + 1)^2} = \frac{Ax + B}{x^2 + x + 1} + \frac{Cx + D}{(x^2 + x + 1)^2}.$$

Multiplying both sides by $(x^2 + x + 1)^2$, we obtain

$$x^3 - 2x^2 + 2x - 4 = (Ax + B)(x^2 + x + 1) + (Cx + D).$$

Now, we use a different technique to determine the coefficients A, B, C, and D. Multiplying and rearranging the terms on the right-hand side, we get

$$x^3 - 2x^2 + 2x - 4 = Ax^3 + (A + B)x^2 + (A + B + C)x + (B + D).$$

According to the theorem about identically equal polynomials (Section 7.1), these two polynomials are identically equal if and only if the coefficients of the terms of the same degree are equal. Thus we obtain the system of linear equations

$$\begin{aligned} A &= 1 \\ A + B &= -2 \\ A + B + C &= 2 \\ B + D &= -4 \end{aligned}$$

in the unknowns A, B, C, and D. This system can be easily solved as follows. We already have $A = 1$. Substituting 1 for A into the second equation and solving for B, we get $B = -3$. Substituting 1 for A and -3 for B into the third equation and solving for C, we obtain $C = 4$. Finally, substituting -3 for B into the last equation gives us $D = -1$. Thus we have the partial fraction decomposition

$$\frac{x^3 - 2x^2 + 2x - 4}{(x^2 + x + 1)^2} = \frac{x - 3}{x^2 + x + 1} + \frac{4x - 1}{(x^2 + x + 1)^2}.$$

◇ **Practice Exercise 5.** Decompose the expression $\dfrac{2x^2 - x + 5}{(x^2 + x + 2)^2}$ into partial fractions.

Answer. $\dfrac{2x^2 - x + 5}{(x^2 + x + 2)^2} = \dfrac{2}{x^2 + x + 2} - \dfrac{3x - 1}{(x^2 + x + 2)^2}.$

 EXERCISES 8.5

In Exercises 1–10, find the constants A, B, C, and D for which the left-hand side is equal to the right-hand side.

1. $\dfrac{4x - 14}{(x - 2)(x - 4)} = \dfrac{A}{x - 2} + \dfrac{B}{x - 4}$

2. $\dfrac{7x + 15}{(x + 3)(x - 4)} = \dfrac{A}{x + 3} + \dfrac{B}{x - 4}$

3. $\dfrac{x + 3}{(x - 4)(3x + 2)} = \dfrac{A}{x - 4} + \dfrac{B}{3x + 2}$

4. $\dfrac{2x + 7}{(x + 6)(2x - 3)} = \dfrac{A}{x + 6} + \dfrac{B}{2x - 3}$

5. $\dfrac{5x^2 + 9x + 3}{x(x + 1)^2} = \dfrac{A}{x} + \dfrac{B}{x + 1} + \dfrac{C}{(x + 1)^2}$

6. $\dfrac{7x + 11}{(x - 2)(x + 3)^2} = \dfrac{A}{x - 2} + \dfrac{B}{x + 3} + \dfrac{C}{(x + 3)^2}$

7. $\dfrac{2x + 3}{(x + 1)(x^2 + x + 1)} = \dfrac{A}{x + 1} + \dfrac{Bx + C}{x^2 + x + 1}$

8. $\dfrac{x^2 - 9x + 1}{(x - 4)(x^2 + 3)} = \dfrac{A}{x - 4} + \dfrac{Bx + C}{x^2 + 3}$

9. $\dfrac{2x^2 - x + 6}{(x^2 - x + 2)^2} = \dfrac{Ax + B}{x^2 - x + 2} + \dfrac{Cx + D}{(x^2 - x + 2)^2}$

10. $\dfrac{x^3 + x}{(x^2 - x + 1)^2} = \dfrac{Ax + B}{x^2 - x + 1} + \dfrac{Cx + D}{(x^2 - x + 1)^2}$

In Exercises 11–20, decompose the given rational expressions into partial fractions.

11. $\dfrac{x - 8}{x^2 - x - 12}$ **12.** $\dfrac{-x + 9}{x^2 + 5x - 6}$ **13.** $\dfrac{-7x + 5}{2x^2 + 5x - 12}$ **14.** $\dfrac{16x + 2}{6x^2 - x - 1}$

15. $\dfrac{9x^2 + 9x - 12}{2x^3 - x^2 - 6x}$ **16.** $\dfrac{3x^2 + 12x - 2}{3x^3 + 5x^2 - 2x}$ **17.** $\dfrac{4x^2 + x - 2}{x^3 - x^2}$ **18.** $\dfrac{3x^2 - 12x + 11}{x^3 - 5x^2 + 8x - 4}$

19. $\dfrac{2x^3 + 5x^2 + 8x + 1}{(x^2 + 2x + 2)^2}$ **20.** $\dfrac{x^3 + 4x^2 + 2x}{(x^2 + 1)^2}$

CHAPTER SUMMARY

A polynomial with real coefficients and degree n,

$$A(x) = a_n x^n + a_{n-1} x^{n-1} + \cdots + a_1 x + a_0, \quad a_n \neq 0,$$

can be viewed either as an algebraic expression or as a function defined for all real values of x.

Two polynomials $A(x)$ and $B(x)$ are *identically equal* if $A(x) = B(x)$ for all values of x. For two polynomials to be identically equal, they must have the same degree, and coefficients of terms of the same degree must be equal.

Polynomials can be added, subtracted, multiplied, and divided. The division algorithm for polynomials states that when we divide a polynomial $A(x)$ (the *dividend*) of degree n by a polynomial $B(x)$ (the *divisor*) of degree m, with $n \geq m$, we obtain a polynomial $Q(x)$ (the *quotient*) of degree $n - m$ and a polynomial $R(x)$ (the *remainder*) of degree less than m. In that case we have $A(x) = B(x) \cdot Q(x) + R(x)$. When $R(x) = 0$, then $A(x)$ is said to be *divisible* by $B(x)$ and $B(x)$ is said to be a *factor* of $A(x)$.

Of special interest is the case when the divisor is a first degree polynomial of the form $x - a$. The Remainder Theorem states that if we divide $A(x)$ by $x - a$, then the remainder is equal to the function value of $A(x)$ at $x = a$, that is, $R = A(a)$. The Factor Theorem tells us that a linear polynomial $x - a$ is a factor of $A(x)$ exactly when $A(a) = 0$. For these reasons, it is useful to know methods that simplify the computation of function values of a polynomial. One such method, easy to use with a calculator, is *Horner's method*. Another is the method of *synthetic division*, where quotient and remainder are obtained without performing a long division.

A number r is said to be a *zero* or *root* of a polynomial $P(x)$ if $P(r) = 0$. The Fundamental Theorem of Algebra states that every polynomial of degree greater than zero has at least one root (real or complex). Also, every polynomial of degree $n \geq 1$ has exactly n zeros, counting multiplicities.

Polynomials with integral or rational coefficients may have *rational zeros*. The *test for rational zeros*, in Section 8.4, tells us how to determine such zeros.

Every polynomial with real coefficients can always be written as a product of *linear* and *quadratic polynomials* with real coefficients. This is the Factorization Theorem. As an application, the *partial fraction decomposition* of rational functions is obtained: every rational expression can be written as a sum of a polynomial and simpler rational expressions. The method of partial fraction decomposition is used in integral calculus.

REVIEW EXERCISES

Use long division to find the quotient and remainder in the following divisions.

1. $(6x^3 - 4x^2 + 5x - 2) \div (2x^2 - 1)$

2. $(9x^3 - 6x^2 - 4x - 2) \div (3x^2 + x)$

3. $(2x^6 - 3x^5 + x^4 - 5x^3 - 4x^2 + 2x - 3) \div (x^2 - 3x + 1)$

4. $(3x^6 - 4x^5 + 2x^4 - 2x^3 + 3x^2 + 4x - 1) \div (x^2 + 2x + 2)$

Find the quotient and the remainder by synthetic division.

5. $(2x^3 - 3) \div (x - 4)$

6. $(4x^3 - 2x) \div (x + 5)$

7. $(2x^4 - 3x^2 + 1) \div (x + 3)$

8. $(3x^5 - 4x + 1) \div (x - 2)$

c 9. $(0.3x^3 - 0.2x^2 + 0.1x + 1) \div (x + 0.4)$

c 10. $(2.1x^3 + 3.2x^2 - 1.5x - 1.3) \div (x - 1.2)$

In Exercises 11–16, show that the linear polynomial is a factor of the given polynomial. Factor the polynomial.

11. $2x^3 - 13x^2 + 16x - 5; \ x - 5$

12. $3x^3 + 25x^2 + 36x - 36; \ x + 6$

13. $2x^3 + 3x^2 - 8x + 3; \ 2x - 1$

14. $3x^3 - 2x^2 - 12x + 8; \ 3x - 2$

c 15. $0.5x^3 + 0.8x^2 - 0.1x + 0.6; \ x + 2$

c 16. $x^3 + 2.6x^2 - 5.2x + 1.6; \ x - 0.4$

In Exercises 17–20, verify that the given number is a zero of the given polynomial and find the other zeros. Express the polynomial as a product of linear factors.

17. $5x^2 + 13x - 6, \ \dfrac{2}{5}$

18. $x^2 - 4x + 1, \ 2 - \sqrt{3}$

19. $x^4 - 2x^3 + 4x^2 + 2x - 5, \ 1 - 2i$

20. $x^4 - 4x^3 + x^2 + 16x - 20, \ 2 + i$

21. Find a third degree polynomial with real coefficients and having zeros -1 and $3 - 2i$.

22. Find a third degree polynomial with real coefficients and having zeros $1 + i\sqrt{3}$ and 0.

23. Show that 1 is a double root of $x^4 - 6x^3 + 12x^2 - 10x + 3$ and find the other roots.

24. Show that -1 is a double root of $x^4 - 2x^2 + 1$ and find the other roots.

In Exercises 25–28, find a polynomial $P(x)$ of lowest degree with the indicated zeros and satisfying the given conditions.

25. $-1, -2, -3, P(1) = 12$

26. 2 (multiplicity 1), 3 (multiplicity 2), leading coefficient 4

27. $2 + i, 2 - i, 3$ (multiplicity 2), leading coefficient 1

28. $1 + i\sqrt{2}, 1 - i\sqrt{2}, 0$ (multiplicity 3), leading coefficient 2

Find all rational roots of the following polynomials.

29. $x^3 - 5x^2 + 3x + 9$

30. $x^3 - 9x^2 + 26x - 24$

31. $2x^3 - 7x^2 + 4x + 4$

32. $3x^3 - 17x^2 + 28x - 12$

Show that the equations in Exercises 33–36 have no rational root.

33. $2x^2 - 3 = 0$

34. $4x^2 - 3 = 0$

35. $x^2 - x - 3 = 0$

36. $3x^2 - x - 1 = 0$

Decompose the rational expressions given in Exercises 37–40 into partial fractions.

37. $\dfrac{2x^2 - x}{(x - 1)(x^2 - x + 1)}$ **38.** $\dfrac{4x^2 + 10x + 14}{(x + 3)(x^2 + x + 4)}$ **39.** $\dfrac{3x^2 + 7x + 1}{x(x + 1)^2}$ **40.** $\dfrac{x^3 - x^2 + 4x - 3}{(x^2 + 2)^2}$

41. Find all fifth roots of 2.

42. Find all fourth roots of 3.

43. Verify that i is a zero of the polynomial $x^2 - 3ix - 2$, but the complex conjugate $-i$ of i is not a zero. Why doesn't this contradict the second theorem of Section 8.3? Find the other root.

44. Verify that $1 + i$ is a zero of the polynomial $x^2 - (1 + 3i)x + 2(i - 1)$, but that $1 - i$ is not a zero. Find the other zero.

● **45.** Prove that every polynomial of odd degree with real coefficients has at least one real root.

● **46.** Prove that if n is even, then $x + a$ is a factor of $x^n - a^n$.

● **47.** Prove that if n is odd, then $x + a$ is a factor of $x^n + a^n$.

● **48.** Prove that $x - a$ is a factor of $x^n - a^n$ for all positive integers n.

● **49.** Prove that if $r_1, r_2, \ldots, r_n$ are the zeros of the polynomial

$$P(x) = x^n + a_{n-1}x^{n-1} + \cdots + a_1 x + a_0,$$

then $r_1 + r_2 + \cdots + r_n = -a_{n-1}$.

● **50.** Under the assumptions of Exercise 49, prove that $r_1 r_2 \cdots r_n = (-1)^n a_0$.

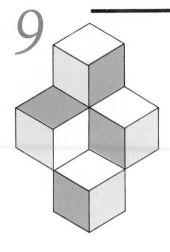

9 Mathematical Induction, Sequences, Series, and Probability

In this chapter, we state the principle of mathematical induction and discuss the method of proof by mathematical induction. Next, we derive the Binomial Theorem and some of its applications. We also discuss the notions of sequences and series and study arithmetic and geometric series. Finally, permutations, combinations, set partitioning, and elementary probability are discussed in the last two sections of the chapter.

9.1 THE PRINCIPLE OF MATHEMATICAL INDUCTION

Suppose that we wanted to prove the validity of the formula

$$(9.1) \quad 1 + 2 + 3 + \cdots + n = \frac{n(n + 1)}{2}$$

for every natural number n.

As a first attempt, we could check this formula for several values of n. For example, if $n = 1$, 2, and 3, we have

$$1 = \frac{1(1 + 1)}{2},$$

$$1 + 2 = 3 = \frac{2(2 + 1)}{2},$$

$$1 + 2 + 3 = 6 = \frac{3(3 + 1)}{2}.$$

Thus formula (9.1) is true for $n = 1$, 2, and 3. As an exercise, you could also check the validity of the formula for $n = 4$, 5, and 6.

At this point, we might ask ourselves whether this method of checking, sometimes called *experimental induction*, constitutes a mathematical proof. In other words, from the fact that (9.1) is true whenever we choose a particular *n*, can we conclude that (9.1) is true for *all n*? The answer is negative for the following reasons. First, this method of checking *never ends*. Second, even if we had checked the validity of formula (9.1) for all values of *n* from 1 to 1000, there is no guarantee that the formula would be true for $n = 1001$.

To see that experimental induction is an unreliable and misleading method of proof, we consider a different example. Let

$$p(n) = n^2 - n + 11$$

be a quadratic function defined over the set of natural numbers. Is $p(n)$ a prime number for every natural number *n*? Let us try experimental induction. If $n = 1$, then $p(1) = 11$ is a prime number. If $n = 2$ and 3, then $p(2) = 13$ and $p(3) = 17$, which are also prime numbers. Continuing the checking, we can verify that $p(n)$ is a prime number for $n = 1, 2, 3, \ldots, 10$. These checks might lead us to believe that $p(n)$ is a prime number for all *n*. However,

$$p(11) = 11^2 - 11 + 11 = 11^2$$

is not a prime number! Although $p(n)$ is a prime for the first 10 choices of the natural number *n*, it fails to be a prime when $n = 11$.

This shows that experimental induction cannot be used to prove statements or theorems. Thus, if we want to prove formula (9.1), a more rigorous method, such as the method of *mathematical induction*, must be used.

The Axiom of Induction

Let *S* be a set of natural numbers satisfying the following properties:
i) 1 is an element of *S*;
ii) if the natural number *k* is an element of *S*, the successor $k + 1$ of *k* is also an element of *S*.
  Then *S* is the set of all natural numbers.

This is a basic axiom of the natural numbers system. Every mathematical proof based upon the axiom of induction is said to be a proof by mathematical induction.

◆ Example 1. Prove the formula

$$1 + 2 + 3 + \cdots + n = \frac{n(n + 1)}{2}$$

by mathematical induction.

Solution. Let *S* be the set of all natural numbers for which the given formula is true. As we already mentioned, the formula is true for $n = 1$. Thus 1 is an element of *S*, which is part i) of the Axiom of Induction. Next, to show that *S* satisfies part ii) of the Axiom, we must show that whenever we assume that a certain number k belongs to S—that is, whenever we assume that $1 + 2 + \cdots + k = k(k + 1)/2$ is true—we can deduce from that fact that $k + 1$ must also belong to S, i.e., that $1 + 2 + \cdots + (k + 1) = (k + 1)[(k + 1) + 1]/2$ must also be true. To do this, write

$$1 + 2 + \cdots + (k + 1) = 1 + 2 + \cdots + k + (k + 1).$$

Since formula (9.1) is true for k substituted for n, we have

$$
\begin{aligned}
1 + 2 + &\cdots + (k + 1) \\
&= \frac{k(k + 1)}{2} + (k + 1) \\
&= \frac{k(k + 1) + 2(k + 1)}{2} \qquad \text{[Writing both terms with the same denominator]} \\
&= \frac{(k + 1)(k + 2)}{2} \qquad \text{[Factoring]} \\
&= \frac{(k + 1)[(k + 1) + 1]}{2}.
\end{aligned}
$$

and this is formula (9.1) with $n = k + 1$. Thus $k + 1$ is an element of S. We conclude that the set S satisfies properties i) and ii) of the Axiom of Induction. Therefore S coincides with the set $\mathbb{N}$ of all natural numbers and, consequently, the statement (9.1) is true for every natural number.

◇ **Practice Exercise 1.** Using induction, prove the formula

$$1 + 3 + 5 + \cdots + (2n - 1) = n^2.$$

(*Hint*: Check the formula for $n = 1$. Assume it to be true for k and prove it for $k + 1$.)

In what follows, the notation $P(n)$ will express a statement about natural numbers, such as, "every even number is of the form $2n$," "the sum of the first n natural numbers is $n(n + 1)/2$," and so on. The following is an equivalent version of the axiom of induction.

The Principle of Mathematical Induction

Let $P(n)$ express a statement about natural numbers, such that:
i) $P(1)$ is a true statement;
ii) for each natural number k, the truth of $P(k)$ implies the truth of $P(k + 1)$.

Then, $P(n)$ is a true statement for all natural numbers.

◆ **Example 2.** Prove by induction that $5^n - 2^n$ is divisible by 3 for any natural number n.

Solution. Let $P(n)$ be the statement "$5^n - 2^n$ is divisible by 3." First we claim that $P(1)$ is true. Indeed, if $n = 1$, then $5 - 2 = 3$ is divisible by 3. Next let us assume that $P(k)$ is true and let us prove that $P(k + 1)$ is true. Write

$$
\begin{aligned}
5^{k+1} - 2^{k+1} &= 5^k \cdot 5 - 2^k \cdot 2 \\
&= 5^k \cdot (3 + 2) - 2^k \cdot 2 \\
&= 5^k \cdot 3 + 5^k \cdot 2 - 2^k \cdot 2 \\
&= 5^k \cdot 3 + (5^k - 2^k) \cdot 2
\end{aligned}
$$

We now examine each term on the right-hand side. The first term $5^k \cdot 3$ is clearly divisible by 3. The second one, $(5^k - 2^k) \cdot 2$, is also divisible by 3, since we are assuming that $P(k)$ is a true statement. Both terms being divisible by 3, their sum $5^{k+1} - 2^{k+1}$ is also divisible by 3. Thus, we have proved that $P(k + 1)$ is true. From the principle of induction, it follows that $5^n - 2^n$ is divisible by 3 for all natural numbers.

◇ **Practice Exercise 2.** Using induction, prove that 2 is a factor of $5^n - 3^n$ for all n.

On many occasions, it is useful to have the following version of the principle of induction.

The Modified Principle of Induction

Let $P(n)$ express a statement about a natural number n, and let m be a fixed natural number such that:
i) $P(m)$ is a true statement;
ii) for each natural number $k \geq m$, the truth of $P(k)$ implies the truth of $P(k + 1)$.
    Then, $P(n)$ is a true statement for all natural numbers $n \geq m$.

Notice that when $m = 1$, the modified principle of induction coincides with the principle of induction.

◆ **Example 3.** Prove by induction that $2^n < n!$ for all $n \geq 4$.

Solution. Recall that $n! = 1 \cdot 2 \cdot \ldots \cdot n$. Let $P(n)$ be the statement "$2^n < n!$" and let $m = 4$. Since $2^4 = 16 < 4! = 24$, it follows that $P(4)$ is true. Suppose now that $P(k)$ is true for $k \geq 4$, and let us prove that $P(k + 1)$ is true. We have

$$2^{k+1} = 2^k \cdot 2 < k! \cdot 2,$$

because we have assumed that $2^k < k!$. Since $k \geq 4$, we see that $k + 1 > 2$. Replacing this quantity above, we get

$$2^{k+1} < k! \cdot 2 < k! \cdot (k + 1);$$

hence

$$2^{k+1} < (k+1)!,$$

that is, $P(k+1)$ is true. By the modified principle of induction it follows that $2^n < n!$ for all $n \geq 4$.

You should check that the inequality of Example 3 is false for $n = 1, 2$ and 3.

◇ **Practice Exercise 3.**   Using induction, prove that $2^{n+1} < n!$ for all $n \geq 5$.

The Summation Notation

A sum such as $1 + 2 + 3 + \cdots + n$ is often represented in shorthand notation as follows

$$1 + 2 + 3 + \cdots + n = \sum_{j=1}^{n} j.$$

The symbol Σ is the Greek capital letter *sigma*, and it stands for *sum*. The letter j under the sum is called the *index of summation*, and the numbers 1 and n are called the *limits of summation*. The symbol $\sum_{j=1}^{n} j$ is read as "the sum of j from 1 to n." For example,

$$\sum_{j=1}^{6} j = 1 + 2 + 3 + 4 + 5 + 6.$$

Other examples are

$$\sum_{j=1}^{4} (2j - 1) = 1 + 3 + 5 + 7$$

and

$$\sum_{j=1}^{5} j^2 = 1^2 + 2^2 + 3^2 + 4^2 + 5^2.$$

More generally, if $a_1, a_2, \ldots, a_n$ are n numbers, their sum is represented by

$$a_1 + a_2 + \cdots + a_n = \sum_{j=1}^{n} a_j.$$

◆ **Example 4.**   Evaluate each of the following sums:

a) $\displaystyle\sum_{j=1}^{5} (2j + 3)$ b) $\displaystyle\sum_{j=3}^{6} \frac{2j + 1}{j - 1}$

Solution. **a)** In the expression $2j + 3$, first replace j by 1, then 2, and so on, until finally j is replaced by 5; compute the numbers and add.

$$\sum_{j=1}^{5} (2j + 3) = (2 \cdot 1 + 3) + (2 \cdot 2 + 3) + (2 \cdot 3 + 3)$$
$$+ (2 \cdot 4 + 4) + (2 \cdot 5 + 3)$$
$$= 5 + 7 + 9 + 11 + 13$$
$$= 45.$$

b) We have:

$$\sum_{j=3}^{6} \frac{2j + 1}{j - 1} = \frac{2 \cdot 3 + 1}{3 - 1} + \frac{2 \cdot 4 + 1}{4 - 1} + \frac{2 \cdot 5 + 1}{5 - 1} + \frac{2 \cdot 6 + 1}{6 - 1}$$
$$= \frac{7}{2} + \frac{9}{3} + \frac{11}{4} + \frac{13}{5}$$
$$= \frac{711}{60}.$$

◇ **Practice Exercise 4.** Find the following sums:

a) $(2j - 1)$ **b)** $\displaystyle\sum_{j=1}^{4} \frac{j - 1}{2j}$

Answer.   **a)** 25   **b)** $\dfrac{23}{24}$.

Properties of Sums

Let n be a natural number, let a_j and b_j be real numbers for $j = 1, 2, \ldots, n$, and let c be any real number. The following are useful and important properties of sums:

1. $\displaystyle\sum_{j=1}^{n} c = n \cdot c$

2. $\displaystyle\sum_{j=1}^{n} (a_j + b_j) = \sum_{j=1}^{n} a_j + \sum_{j=1}^{n} b_j$

3. $\displaystyle\sum_{j=1}^{n} ca_j = c \sum_{j=1}^{n} a_j.$

Formula (1) says that if we add n terms equal to the constant c, then the sum equals the product $n \cdot c$.

Each one of these three formulas can be proved by induction. As an illustration, we prove formula (1). First, it is understood that

$$\sum_{j=1}^{1} a_j = a_1.$$

Thus, if $n = 1$, we have

$$\sum_{j=1}^{1} c = c = 1 \cdot c,$$

and formula (1) holds for $n = 1$. Assuming it to be true for k, let us prove it for $k + 1$. We have

$$\sum_{j=1}^{k+1} c = \left(\sum_{j=1}^{k} c \right) + c$$
$$= k \cdot c + c \qquad \text{[by our induction assumption]}$$
$$= (k + 1) \cdot c. \qquad \text{[factoring } c\text{]}$$

Thus, formula (1) holds for $k + 1$. By the principle of induction, formula (1) is true for *every* n.

The proofs of formulas (2) and (3) are left as exercises.

◆ **Example 5.**   Find the sum $\displaystyle\sum_{j=1}^{5} (2j + 3)$.

Solution.   This is a repeat of the sum in Example 4a that we now compute using formulas (1), (2), and (3). We have

$$\sum_{j=1}^{5} (2j + 3) = \sum_{j=1}^{5} 2j + \sum_{j=1}^{5} 3$$
$$= 2 \left(\sum_{j=1}^{5} j \right) + 5 \cdot 3$$
$$= 2(1 + 2 + 3 + 4 + 5) + 5 \cdot 3$$
$$= 2 \cdot 15 + 5 \cdot 3$$
$$= 45.$$

◇ **Practice Exercise 5.**   Find the sum $\displaystyle\sum_{j=1}^{5} (2j + 1)$.

Answer.   35.

EXERCISES 9.1

In Exercises 1–10, use mathematical induction to prove each formula for all natural numbers.

1. $2 + 4 + 6 + \cdots + 2n = n(n + 1)$

2. $1 + 3 + 5 + \cdots + (2n - 1) = n^2$

3. $1 \cdot 2 + 2 \cdot 3 + 3 \cdot 4 + \cdots + n(n + 1) = \dfrac{n(n + 1)(n + 2)}{3}$

4. $\dfrac{1}{1 \cdot 2} + \dfrac{1}{2 \cdot 3} + \dfrac{1}{3 \cdot 4} + \cdots + \dfrac{1}{n(n + 1)} = \dfrac{n}{n + 1}$

5. $1^2 + 2^2 + 3^2 + \cdots + j^2 = \dfrac{j(j + 1)(2j + 1)}{6}$

6. $1^3 + 2^3 + 3^3 + \cdots + k^3 = \left[\dfrac{k(k + 1)}{2} \right]^2$

7. $1^3 + 3^3 + 5^3 + \cdots + (2j - 1)^3 = j^2(2j^2 - 1)$

8. $1^2 + 3^2 + 5^2 + \cdots + (2j - 1)^2 = \dfrac{j(4j^2 - 1)}{3}$

9. $\dfrac{1}{2} + 1 + \dfrac{3}{2} + 2 + \cdots + \dfrac{k}{2} = \dfrac{k(k + 1)}{4}$

10. $1 + 8 + 16 + \cdots + 8(k - 1) = (2k - 1)^2$

Use mathematical induction to prove each of the statements in Exercises 11–20.

11. 4 is a factor of $7^k - 3^k$ for all natural numbers k.

12. $9^k - 5^k$ is divisible by 4 for all $k \in \mathbb{N}$.

13. $4^{2n} - 1$ is divisible by 5 for all $n \in \mathbb{N}$.

14. 7 is a factor of $8^n - 1$ for all natural numbers k.

15. $m^3 - m + 3$ is a multiple of 3 for all $m \in \mathbb{N}$.

16. $m^2 + m$ is an even number for all $m \in \mathbb{N}$.

17. $3^{2j} - 1$ is divisible by 8 for all natural numbers j.

18. $j^3 + 2j$ is divisible by 3 for all natural numbers j.

19. $x^n - a^n$ is divisible by $x - a$ for all natural numbers n.
      [*Hint:* $x^{n+1} - a^{n+1} = (x^n - a^n)x + a^n(x - a)$.]

20. $x + a$ is a factor of $x^{2n-1} + a^{2n-1}$ for all natural numbers n.

Find the sums given in Exercises 21–30.

21. $\displaystyle\sum_{j=1}^{6} (j + 1)$

22. $\displaystyle\sum_{j=1}^{5} 3j$

23. $\displaystyle\sum_{j=1}^{5} (2j - 1)$

24. $\displaystyle\sum_{j=1}^{6} (3j - 2)$

25. $\displaystyle\sum_{j=1}^{4} \dfrac{2j + 1}{2}$

26. $\displaystyle\sum_{j=1}^{5} \dfrac{3j - 3}{4}$

27. $\displaystyle\sum_{j=1}^{6} j^2$

28. $\displaystyle\sum_{j=1}^{5} (2j - 1)^2$

29. $\displaystyle\sum_{j=3}^{5} (j^2 - 2j)$

30. $\displaystyle\sum_{j=2}^{4} (2^j + 1)$

In Exercises 31–40, prove each inequality by mathematical induction.

31. $2^n > n$ for all natural numbers n.

32. $3^n \geq 1 + 2n$ for all natural numbers n.

33. $2^{k+3} < (k+3)!$ for all $k \in \mathbb{N}$. [*Hint:* $n! = 1 \cdot 2 \cdot \ldots \cdot (n-1) \cdot n$]

34. $k^2 \geq 2k - 1$ for all $k \in \mathbb{N}$.

35. $j^2 > 2j$ for all $j \geq 3$.

36. $2^j > j^2$ for all $j \geq 5$.

37. $2^{m-1} < m!$ for all $m > 2$.

38. If $0 < a < 1$, then $0 < a^n < 1$ for all natural numbers n.

39. If $a > 1$, then $a^n > 1$ for all natural numbers n.

40. If $0 < b < a$, then $b^n < a^n$ for all natural numbers n.

41. Prove formula (2). [*Hint:* Use mathematical induction to prove that

$$(a_1 + b_1) + (a_2 + b_2) + \cdots + (a_n + b_n)$$
$$= (a_1 + a_2 + \cdots + a_n) + (b_1 + b_2 + \cdots + b_n).]$$

42. Prove formula (3).

43. Let a and b be real numbers. Use mathematical induction to prove that $(ab)^n = a^n b^n$ for every natural number n.

44. Let a be a real number. Prove that $a^n \cdot a^m = a^{n+m}$ for every pair of natural numbers n and m. [*Hint:* First use induction to show that $a^n \cdot a = a^{n+1}$.]

45. Let a be a real number. Prove that $(a^n)^m = a^{nm}$ for every pair of natural numbers n and m.

● **46.** If z is a complex number, let $\bar{z}$ denote its complex conjugate. Prove that $\overline{z_1 + z_2 + \cdots + z_n} = \bar{z}_1 + \bar{z}_2 + \cdots + \bar{z}_n$.

● **47.** Prove that $\overline{z_1 \cdot z_2 \cdots z_n} = \bar{z}_1 \cdot \bar{z}_2 \cdots \bar{z}_n$.

● **48.** Let a and r be real numbers with $r \neq 1$. Prove that

$$a + ar + ar^2 + \cdots + ar^{n-1} = \frac{a(1 - r^n)}{1 - r}$$

where n is any natural number.

9.2   THE BINOMIAL THEOREM

Consider the binomial expression

$$(a + b)^n$$

where n is a natural number. Using the principle of mathematical induction, we shall prove a formula that gives the expansion of $(a + b)^n$ without having to compute the corresponding products of the binomial $a + b$. This expansion is a sum of products of powers of a and b multiplied by certain coefficients which are called *binomial coefficients*. These coefficients have interesting properties and play an important role in finite probability.

Before explaining how the expansion formula for $(a + b)^n$ can be obtained, we first discuss some properties of the binomial coefficients.

We assume that you already know the notation

$$(9.2) \quad n! = 1 \cdot 2 \cdot \ldots \cdot n.$$

We recall also the special definitions $1! = 1$ and $0! = 1$.

The Binomial Coefficient

If k and n are natural numbers, with $k \leq n$, the *binomial coefficient* $\binom{n}{k}$ is defined by

$$(9.3) \quad \binom{n}{k} = \frac{n!}{k!(n-k)!}.$$

Since $n! = n(n-1)\cdots(n-k+1)(n-k)!$, it follows that

$$\binom{n}{k} = \frac{\cdot\, n!}{k!(n-k)!} = \frac{n(n-1)\cdots(n-k+1)(n-k)!}{k!(n-k)!}.$$

Thus, we can also write the binomial coefficient $\binom{n}{k}$ as follows:

$$(9.4) \quad \binom{n}{k} = \frac{n(n-1)\cdots(n-k+1)}{k!}.$$

◆ **Example 1.** Evaluate $\binom{5}{2}$, $\binom{6}{0}$, and $\binom{7}{1}$.

Solution. According to formula (9.3), we have:

$$\binom{5}{2} = \frac{5!}{2!(5-2)!} = \frac{5!}{2!3!} = \frac{5 \cdot 4 \cdot 3!}{2!3!} = 10$$

$$\binom{6}{0} = \frac{6!}{0!(6-0)!} = \frac{6!}{6!} = 1$$

$$\binom{7}{1} = \frac{7!}{1!(7-1)!} = \frac{7!}{1!6!} = \frac{7 \cdot 6!}{1!6!} = 7.$$

◇ **Practice Exercise 1.** Find the numbers $\binom{7}{2}$, $\binom{6}{3}$, and $\binom{5}{5}$.

Answer. 21, 20, and 1.

Properties of Coefficients

The following are properties of the binomial coefficients.

$$1. \quad \binom{n}{k} = \binom{n}{n-k}$$

This can be seen as follows:

$$\binom{n}{n-k} = \frac{n!}{(n-k)![n-(n-k)]!}$$

$$= \frac{n!}{(n-k)!k!} = \binom{n}{k}.$$

---

2. $\binom{n}{0} = \binom{n}{n} = 1$ for all n.

The first equality is obtained from property 1 by letting $k = 0$. To show the second equality, we write

$$\binom{n}{n} = \frac{n!}{n!(n-n)!} = \frac{n!}{n!0!} = 1.$$

---

3. $\binom{n}{k-1} + \binom{n}{k} = \binom{n+1}{k}.$

This is proved by writing

$$\binom{n}{k-1} + \binom{n}{k} = \frac{n!}{(k-1)!(n-k+1)!} + \frac{n!}{k!(n-k)!}.$$

Since the least common denominator for the last two fractions is $k!(n-k+1)!$, we have

$$\binom{n}{k-1} + \binom{n}{k} = \frac{n!k + n!(n-k+1)}{k!(n-k+1)!}$$

$$= \frac{n![k+n-k+1]}{k!(n-k+1)!} \qquad \text{[Factoring } n!\text{]}$$

$$= \frac{(n+1)!}{k!(n+1-k)!}$$

$$= \binom{n+1}{k}.$$

Powers of a Binomial

We already know from Section 1.5 that

$$(a+b)^2 = a^2 + 2ab + b^2.$$

This formula tells us that the square of the binomial $a + b$ has three terms:

$$a^2 = a^2 b^0, \quad 2ab, \quad \text{and} \quad b^2 = a^0 b^2.$$

As we move from left to right, the powers of a decrease by 1, while the powers of b increase by 1. Moreover, the coefficients of the expansion are, respectively,

$$\binom{2}{0} = 1, \quad \binom{2}{1} = 2, \quad \text{and} \quad \binom{2}{2} = 1.$$

Similarly, if we compute the cube of $a + b$, we get

$$(a + b)^3 = a^3 + 3a^2 b + 3ab^2 + b^3.$$

This expansion has four terms:

$$a^3 = a^3 b^0, \quad 3a^2 b, \quad 3ab^2, \quad \text{and} \quad a^0 b^3 = b^3.$$

From one term to the next, the powers of a decrease by 1 while the powers of b increase by 1. From left to right the coefficients are

$$\binom{3}{0} = 1, \quad \binom{3}{1} = 3, \quad \binom{3}{2} = 3, \quad \text{and} \quad \binom{3}{3} = 1.$$

In view of these two observations, we conjecture that the following formula is true in general.

The Binomial Theorem

If a and b are real numbers and n is a natural number, then

$$(9.5) \quad (a + b)^n = \binom{n}{0} a^n + \binom{n}{1} a^{n-1} b + \binom{n}{2} a^{n-2} b^2$$
$$+ \cdots + \binom{n}{n-1} ab^{n-1} + \binom{n}{n} b^n$$

or, in summation notation,

$$(9.6) \quad (a + b)^n = \sum_{k=0}^{n} \binom{n}{k} a^{n-k} b^k.$$

The proof is done by mathematical induction. First, we prove that (9.6) is true for $n = 1$:

$$(a + b)^1 = \sum_{k=0}^{1} \binom{1}{k} a^{1-k} b^k$$
$$= \binom{1}{0} a^1 b^0 + \binom{1}{1} a^0 b^1$$
$$= a + b.$$

Next, assuming that (9.6) is true for some N, let us prove that it is true for $N + 1$. We have

$$(a + b)^{N+1} = (a + b)(a + b)^N$$

$$= (a + b) \sum_{k=0}^{N} \binom{N}{k} a^{n-k} b^k$$

$$= a \sum_{k=0}^{N} \binom{N}{k} a^{N-k} b^k + b \sum_{k=0}^{N} \binom{N}{k} a^{N-k} b^k$$

$$= \left[\binom{N}{0} a^{N+1} + \binom{N}{1} a^N b + \binom{N}{2} a^{N-1} b^2 + \cdots + \binom{N}{N} ab^N \right]$$

$$+ \left[\binom{N}{0} a^N b + \binom{N}{1} a^{N-1} b^2 + \cdots + \binom{N}{N-1} ab^N + \binom{N}{N} b^{N+1} \right]$$

$$= \binom{N}{0} a^{N+1} + \left[\binom{N}{0} + \binom{N}{1} \right] a^N b + \left[\binom{N}{1} + \binom{N}{2} \right] a^{N-1} b^2$$

$$+ \cdots + \left[\binom{N}{N-1} + \binom{N}{N} \right] ab^N + \binom{N}{N} b^{N+1}$$

Using property 3 and noticing that

$$\binom{N}{0} = \binom{N+1}{0} \quad \text{and} \quad \binom{N}{N} = \binom{N+1}{N+1},$$

we can rewrite the right-hand side of the last expression and get

$$(a + b)^{N+1} = \binom{N+1}{0} a^{N+1} + \binom{N+1}{1} a^N b + \binom{N+1}{2} a^{N-1} b^2$$

$$+ \cdots + \binom{N+1}{N+1}$$

$$= \sum_{k=0}^{N+1} \binom{N+1}{k} a^{N+1-k} b^k,$$

which means that (9.6) is true for $N + 1$. By the principle of induction, the formula is true for every natural number n.

Formulas (9.5) and (9.6) are the *binomial expansion* of $(a + b)^n$.

◆ **Example 2.**   Use the binomial formula to expand $(a + b)^5$.

Solution.   According to (9.6) with $n = 5$, we have

$$(a + b)^5 = \sum_{k=0}^{5} \binom{5}{k} a^{5-k} b^k$$

$$= \binom{5}{0} a^5 + \binom{5}{1} a^4 b + \binom{5}{2} a^3 b^2 + \binom{5}{3} a^2 b^3 + \binom{5}{4} ab^4 + \binom{5}{5} b^5.$$

We now compute the binomial coefficients:

$$\binom{5}{0} = 1,$$

$$\binom{5}{1} = \frac{5!}{1!4!} = 5,$$

$$\binom{5}{2} = \frac{5!}{2!3!} = 10.$$

By property 2, we have

$$\binom{5}{3} = \binom{5}{2}, \quad \binom{5}{4} = \binom{5}{1} \quad \text{and} \quad \binom{5}{5} = \binom{5}{0}.$$

Thus,

$$(a + b)^5 = a^5 + 5a^4b + 10a^3b^2 + 10a^2b^3 + 5ab^4 + b^5.$$

◇ **Practice Exercise 2.**   Expand $(x + y)^4$ by the Binomial Theorem.

Answer. $(x + y)^4 = x^4 + 4x^3y + 6x^2y^2 + 4xy^3 + y^4.$

◆ **Example 3.**   Evaluate $(x - 2y)^4$.

Solution.   The power $(x - 2y)^4$ is of the form $(a + b)^4$ with $a = x$ and $b = -2y$. By the Binomial Theorem, we have

$$(x - 2y)^4 = [x + (-2y)]^4$$

$$= x^4 + \binom{4}{1}x^3(-2y) + \binom{4}{2}x^2(-2y)^2$$

$$+ \binom{4}{3}x(-2y)^3 + \binom{4}{4}(-2y)^4$$

$$= x^4 - \binom{4}{1}2x^3y + \binom{4}{2}4x^2y^2 - \binom{4}{3}8xy^3 + \binom{4}{4}16y^4$$

$$= x^4 - 8x^3y + 24x^2y^2 - 32xy^3 + 16y^4$$

after computing the binomial coefficients.

◇ **Practice Exercise 3.**   Find the binomial expansion of $(2a - b)^5$.

Answer. $(2a - b)^5 = 32a^5 - 80a^4b + 80a^3b^2 - 40a^2b^3 + 10ab^4 - b^5.$

The $(k + 1)$st Term in the Binomial Expansion

The $(k + 1)$st term in the expansion of $(a + b)^n$ is

$$\binom{n}{k}a^{n-k}b^k \quad \text{for} \quad 0 \le k \le n.$$

Note that *the exponent of b in the $(k + 1)$st term is k.*

◆ **Example 4.**   Find the sixth term in the expansion of $(3u - 2v)^8$.

Solution.   Let $a = 3u$ and $b = -2v$. Since the exponent of b in the sixth term is 5, we see that

$$\binom{8}{5}(3u)^{8-5}(-2v)^5 = \frac{8!}{5!3!}(3u)^3(-2v)^5$$

$$= \frac{8!}{5!3!}27u^3(-32v^5)$$

$$= -56 \cdot 27 \cdot 32u^3v^5$$

$$= -48384u^3v^5.$$

◇ **Practice Exercise 4.**   Find the fifth term of the expansion of $(2a + 3b)^7$.

Answer. $\binom{7}{4}(2a)^3(3b)^4 = 22{,}680a^3b^4.$

Pascal's Triangle

The following array of numbers

$$
\begin{array}{ccccccccccccc}
 & & & & & & 1 & & & & & & \\
 & & & & & 1 & & 1 & & & & & \\
 & & & & 1 & & 2 & & 1 & & & & \\
 & & & 1 & & 3 & & 3 & & 1 & & & \\
 & & 1 & & 4 & & 6 & & 4 & & 1 & & \\
 & 1 & & 5 & & 10 & & 10 & & 5 & & 1 & \\
1 & & 6 & & 15 & & 20 & & 15 & & 6 & & 1 \\
\end{array}
$$

. . . .

obtained by writing the binomial coefficients in the expansion of $(a + b)^n$ where $n = 0, 1, 2, \ldots$, is called *Pascal's triangle*. Note that the numbers 1, 2, and 1 in the third row from the top are the binomial coefficients in the expansion of $(a + b)^2$. Similarly, the numbers 1, 5, 10, 10, 5, and 1 in the sixth row are the binomial coefficients of $(a + b)^5$. Also observe that in the fourth row from the top, 6 is the sum of 3 and 3, the two numbers above it. In the fifth row, 10 is the sum of 6 and 4, the two numbers above it, and so on. Can you write the next two lines of Pascal's triangle?

The array of numbers above was named after Blaise Pascal (1623–1662), a French scientist and religious philosopher, who wrote a treatise on the triangular array. However, Pascal's triangle was known to mathematicians even earlier. In 1303, the Chinese mathematician Chu-Shi-kie wrote of these triangular arrays.

⬡ EXERCISES 9.2

Evaluate the binomial coefficients.

1. $\binom{9}{6}$ 2. $\binom{8}{3}$ 3. $\binom{12}{9}$ 4. $\binom{11}{7}$

In Exercises 5–14, expand using the Binomial Theorem.

5. $(a + b)^4$ **6.** $(a + b)^7$ **7.** $(x + y)^6$ **8.** $(x + y)^8$ **9.** $(2a - b)^5$

10. $(3a + 2b)^5$ **11.** $(x - 2)^5$ **12.** $(2x - 5)^4$ **13.** $\left(a - \dfrac{1}{a}\right)^6$ **14.** $\left(3m - \dfrac{2}{m}\right)^5$

15. Find the fifth term in the expansion of $(2x - a)^7$.

16. Find the seventh term in the expansion of $(3x + 4a)^9$.

17. Find the eighth term in the expansion of $\left(\dfrac{u}{2} - 2\right)^{10}$.

18. Find the tenth term in the expansion of $\left(2a + \dfrac{b}{2}\right)^{12}$.

19. Find the term independent of y in the expansion of $\left(2y - \dfrac{1}{2y}\right)^8$.

20. Find the term independent of x in the expansion of $\left(\dfrac{3x}{4} - \dfrac{4}{3x}\right)^{10}$.

21. Find the middle term in the expansion of $(a^{1/4} + b^{1/4})^8$.

22. Find the middle term in the expansion of $(u^{1/2} - \sqrt{2})^{12}$.

23. Find the two middle terms in the expansion of $(xy - a)^7$.

24. Find the two middle terms in the expansion of $(3x - 2)^5$.

25. Find the term involving b^6 in the expansion $(a - 3b^2)^5$.

26. Find the term involving x^3 in the expansion of $\left(x^2 - \dfrac{1}{x}\right)^6$.

c ● **27.** Use the first five terms in the binomial expansion of $(1 + 0.01)^8$ to approximate $(1.01)^8$. Use your calculator to compute $(1.01)^8$ and compare the answers.

c ● **28.** Approximate $(0.99)^{10}$ using the first five terms in the binomial expansion of $(1 - 0.01)^{10}$. Compute $(0.99)^{10}$ by using your calculator and compare your results.

● **29.** Show that $\dbinom{n}{0} = \dbinom{n + 1}{0}$.

● **30.** Show that $\dbinom{n}{n} = \dbinom{n + 1}{n + 1}$.

9.3 SEQUENCES AND SERIES

As we discussed in Chapter 4, a function is a rule that assigns to each element x of a set X a unique element y of a set Y. A *sequence* is a particular kind of function that we now define.

Sequences

Let Y be an arbitrary set. A *sequence of elements of Y* is a function $f: \mathbb{N} \to Y$ whose domain is the set $\mathbb{N}$ of all natural numbers.

The element $f(1)$ of Y is called the *first term* of the sequence, $f(2)$ is the *second term,* and so on. The element $f(n)$ is said to be the *nth term* of the sequence.

It is customary to denote $f(n)$ (the image of n under f) by a letter with a subscript. Thus we may write $f(n) = a_n$, where the subscript n is the argument of the function f. By doing this, we can use any one of the following notations to indicate the sequence f:

$$a_1, a_2, \ldots, a_n, \ldots,$$
$$\{a_n\}_{n=1}^{\infty}, \quad \text{or, simply,} \quad \{a_n\}.$$

We also say that a_n is the *general term* of the sequence $\{a_n\}$.

The set Y where f takes its values is arbitrary. However, we shall consider only *sequences of real numbers*. These correspond to the case when $Y = \mathbb{R}$.

◆ **Example 1.**   Write the first five terms of each sequence:
a) $f(n) = n^2$ **b)** $\{(-1)^n 2^n\}$.

Solution.   As $n = 1, 2, 3, 4,$ and 5, we get
a) 1, 4, 9, 16, 25   **b)** $-2, 4, -8, 16, -32$.

◇ **Practice Exercise 1.**   Write the first four terms of each sequence.

a) $\left\{\dfrac{n}{n+1}\right\}$ **b)** $f(n) = \dfrac{n}{n^2 + 1}$

Answer.   **a)** $\dfrac{1}{2}, \dfrac{2}{3}, \dfrac{3}{4}, \dfrac{4}{5}$ **b)** $\dfrac{1}{2}, \dfrac{2}{5}, \dfrac{3}{10}, \dfrac{4}{17}$.

In Example 1 and Practice Exercise 1, the sequences were given by an explicit formula. This may not always be the case, as the next example shows.

◆ **Example 2.**   Write the first five terms of the sequence defined by
$$a_1 = 3, \quad a_k = a_{k-1} + 5, \quad \text{for all } k \geq 2.$$

Solution.   The expression $a_k = a_{k-1} + 5$, $k \geq 2$, is said to be a *recurrence relation*. It allows us to determine the kth term of the sequence when the $(k-1)$st term is known. Since $a_1 = 3$, then
$$a_2 = a_1 + 5 = 3 + 5 = 8.$$
Also,
$$a_3 = a_2 + 5 = 8 + 5 = 13,$$
$$a_4 = a_3 + 5 = 13 + 5 = 18,$$
$$a_5 = a_4 + 5 = 18 + 5 = 23.$$

Thus the five terms of the sequence are 3, 8, 13, 18, and 23.

In this example it is possible to find an explicit expression for a_n, the nth term of the sequence. To see this, we observe that if $n = 2, 3,$ and 4, then

$$a_2 = a_1 + 5$$
$$a_3 = a_2 + 5 = a_1 + 5 + 5 = a_1 + 2 \cdot 5$$
$$a_4 = a_3 + 5 = a_1 + 2 \cdot 5 + 5 = a_1 + 3 \cdot 5.$$

These relations seem to indicate that

$$a_n = a_1 + (n-1) \cdot 5, \quad \text{for all } n \geq 2.$$

Or, since $a_1 = 3$,

$$a_n = 3 + (n-1) \cdot 5, \quad \text{for all } n \geq 2.$$

This result can be proved by induction, and the proof is left as an exercise.

◇ **Practice Exercise 2.** Write the first five terms of the sequence defined by

$$a_1 = 1, \quad a_k = \frac{1}{2} a_{k-1}, \quad k \geq 2.$$

Write the nth term of this sequence.

Answer. $a_1 = 1, a_2 = \dfrac{1}{2}, a_3 = \dfrac{1}{4}, a_4 = \dfrac{1}{8}, a_5 = \dfrac{1}{16}; a_n = \dfrac{1}{2^n};$ for all $n \geq 2$.

Newton's Method for Approximating Square Roots

An important example of a sequence defined by a recurrence relation is provided by *Newton's method for approximating square roots* of a nonnegative real number.

It is proved in calculus that the recurrence relation

$$(9.7) \quad x_n = \frac{1}{2}\left(x_{n-1} + \frac{a}{x_{n-1}}\right), \quad n \geq 2,$$

defines a sequence of approximations for the square root of a nonnegative number a. We start the sequence of approximations by choosing for x_1 any reasonable approximation of the square root of a. The number x_1 is our *initial guess*. Once x_1 is selected, formula (9.7) gives us x_2, x_3, and so on. These numbers will be better approximations for the square root of a.

The method, named in honor of Isaac Newton (1642–1727), an English mathematician, is very effective and quickly yields decimal approximations to any desired accuracy.

c ◆ **Example 3.** Taking as initial guess $x_1 = 1.5$ and using Newton's method, find four decimal approximations of $\sqrt{2}$.

Solution. If $x_1 = 1.5$, then using (9.7) we get

$$x_2 = \frac{1}{2}\left(1.5 + \frac{2}{1.5}\right) = 1.4166667.$$

Next,

$$x_3 = \frac{1}{2}\left(1.4166667 + \frac{2}{1.4166667}\right) = 1.4142157$$

and, finally,

$$x_4 = \frac{1}{2}\left(1.4142157 + \frac{2}{1.4142157}\right) = 1.4142136.$$

This result already coincides with the approximation for $\sqrt{2}$ given by our calculator. A better approximation is 1.414213562. By comparing these approximations we can see how effective Newton's method is: with only three computations we have obtained an approximation for $\sqrt{2}$ accurate to six decimal places.

Using Your Calculator

If your calculator has a memory, you can generate a sequence of approximations for the square root of 2 (or any other positive number) in a very simple way. First, store the initial guess in the memory and, by recalling it whenever necessary, perform the computations indicated in formula (9.7) to obtain the second approximation of $\sqrt{2}$. Next, store the obtained result and repeat the same operations to obtain the third approximation, and so on. The following keystrokes were implemented in a TI-30-II (a similar procedure could be adapted to other calculators). After entering the initial guess 1.5, storing it, and press the following keys

$$\boxed{\text{RCL}}\boxed{+}\,2\,\boxed{\div}\boxed{\text{RCL}}\boxed{=}\boxed{\div}\,2\,\boxed{=}\boxed{\text{STO}}$$

you obtain the display 1.4166667 which is now stored in the memory. Repeating the same keystrokes, you'll obtain the third element of the approximating sequence 1.4142157, and so on.

◇ **Practice Exercise 3.**   Use Newton's method and find the first four approximations for $\sqrt{3}$. Take as your first guess $x_1 = 1.6$.

Answer. $x_1 = 1.6$, $x_2 = 1.7375$, $x_3 = 1.7320594$, $x_4 = 1.7320508$.

HISTORICAL NOTE

Sir Isaac Newton (1642–1727),

English mathematician, physicist, and natural philosopher, was one of the greatest figures in the entire history of science. His scientific work contained major contributions to mathematics and experimental and theoretical physics, and it profoundly influenced 18th century thought.

Newton's most important discoveries, including the development of differential calculus, the formulation of the laws of gravitation, and the finding that white light is made up of the colors of the spectrum, were made in the 2-year period of 1665 and 1666.

Newton also formulated the laws of motion, invented the reflecting telescope, and made important contributions to optics and astronomy.

Series

With every sequence $a_1, a_2, \ldots, a_n, \ldots$ we can associate another sequence whose general term is defined by

$$s_n = a_1 + a_2 + \cdots + a_n,$$

or

$$s_n = \sum_{j=1}^{n} a_j.$$

In this setting, the number s_n is called the *nth partial sum of the series*

$$\sum_{j=1}^{\infty} a_j$$

and the sequence $\{s_n\}$ is called the *sequence of partial sums of the series*. The elements of the original sequence $\{a_j\}$ are called the *terms of the series*.

Conversely, given a sequence $\{s_n\}$, we may set

$$a_1 = s_1, \quad a_n = s_n - s_{n-1} \quad \text{for } n \geq 2$$

and obtain a sequence of terms whose series will have the given $\{s_n\}$ as their partial sums.

◆ **Example 4.**   Find the first four partial sums of the series whose terms are $1, 2, \ldots, n, \ldots$ (the sequence of all natural numbers). Find a general expression for the *n*th partial sum.

Solution.   We have

$$s_1 = 1$$
$$s_2 = 1 + 2 = 3$$
$$s_3 = 1 + 2 + 3 = 6$$
$$s_4 = 1 + 2 + 3 + 4 = 10.$$

The *n*th partial sum is

$$s_n = 1 + 2 + \cdots + n$$

which, from formula (9.1), is equal to

$$s_n = \frac{n(n + 1)}{2}.$$

◇ **Practice Exercise 4.**   Find the first four partial sums of the series whose terms are $2, 4, 6, \ldots, 2n, \ldots$ (the sequence of all even numbers). Also find s_n, the *n*th partial sum.

Answer. $s_1 = 2$, $s_2 = 6$, $s_3 = 12$, $s_4 = 20$; $s_n = n(n + 1)$.

◆ **Example 5.**   Find a general expression for the nth partial sum of the series whose terms are $\left\{ \dfrac{1}{n(n+1)} \right\}$.

Solution.   We have

$$s_n = \frac{1}{1 \cdot 2} + \frac{1}{2 \cdot 3} + \cdots + \frac{1}{(n-1)n} + \frac{1}{n(n+1)}.$$

Since

$$\frac{1}{n(n+1)} = \frac{1}{n} - \frac{1}{n+1},$$

we can rewrite s_n as follows:

$$s_n = \left(\frac{1}{1} - \frac{1}{2} \right) + \left(\frac{1}{2} - \frac{1}{3} \right) + \cdots + \left(\frac{1}{n-1} - \frac{1}{n} \right) + \left(\frac{1}{n} - \frac{1}{n+1} \right).$$

Simplifying, we obtain

$$s_n = 1 - \frac{1}{n+1} = \frac{n}{n+1}.$$

◇ **Practice Exercise 5.**   Find s_n the nth partial sum of the series whose terms are $\{n(n+1)\}$. (*Hint:* See Exercise 3 of Section 9.1.)

Answer. $s_n = \dfrac{n(n+1)(n+2)}{3}$.

Explicit Expression for the Partial Sums of a Series

Whenever possible, it is very convenient to have an explicit expression for the nth partial sum of a series. For example, we saw in Example 4 that

$$s_n = \frac{n(n+1)}{2}$$

is the nth partial sum of the series whose terms are $\{n\}$. Thus if we wish to find the partial sum

$$1 + 2 + 3 + \cdots + 25,$$

we use the formula above with $n = 25$, reducing our computations to a minimum:

$$1 + 2 + 3 + \cdots + 25 = s_{25} = \frac{25 \cdot 26}{2} = 325.$$

As another example, suppose that we want to compute the sum

$$\sum_{j=1}^{19} \frac{1}{j(j+1)}.$$

According to Example 5, we have

$$s_n = \sum_{j=1}^{n} \frac{1}{j(j+1)} = \frac{n}{n+1}.$$

Thus

$$s_{19} = \frac{19}{20} = 0.95.$$

Unfortunately, in most cases it is difficult, if not impossible, to find an explicit formula for the nth partial sum of a series. However, in the next two sections, we shall study *arithmetic* and *geometric series*. In both cases, explicit expressions for the general term will be derived.

 EXERCISES 9.3

Write the first five terms of the sequences.

1. $a_n = 5 - 2n$ **2.** $a_n = 3n - 7$ **3.** $a_n = (-1)^n n^2$ **4.** $a_n = \dfrac{(-1)^n n}{n+1}$

5. $a_n = (-3)^n$ **6.** $a_n = 3(2^{-n})$ **7.** $a_n = \left(1 + \dfrac{1}{n}\right)^n$ **8.** $a_n = \dfrac{(-1)^{n+1}}{n^2}$

9. $a_n = \dfrac{(-1)^n 2^n}{2^n - 1}$ **10.** $a_n = 1 + (-1)^n$

The sequences in Exercises 11–20 are defined by a recurrence relation. Find the first five terms of each sequence.

11. $a_1 = -5, \; a_k = 2a_{k-1} + 7$ **12.** $a_1 = 2, \; a_k = a_{k-1} + 3$ **13.** $a_1 = 0, \; a_k = 4 - 3a_{k-1}$

14. $a_1 = 2, \; a_k = \dfrac{a_{k-1}}{4}$ **15.** $a_1 = 3, \; a_k = \dfrac{1}{a_{k-1}^2}$ **16.** $a_1 = 7, \; a_k = \dfrac{k}{a_{k-1}}$

17. $a_1 = 3, \; a_k = (k-1)a_{k-1}$ **18.** $a_1 = 1, \; a_k = 2a_{k-1}$ **19.** $a_1 = 2, \; a_k = \dfrac{3}{4}a_{k-1}$

20. $a_1 = 3, \; a_k = \dfrac{1}{3}a_{k-1}$

In Exercises 21–28, the nth partial sum s_n of a series is given. Find a_n the general term of the series.

21. $s_n = 2^{n+1} - 2$ **22.** $s_n = 1 - \left(\dfrac{1}{2}\right)^n$ **23.** $s_n = n(n+1)$ **24.** $s_n = n^2$

25. $s_n = \dfrac{n(n+1)}{4}$ **26.** $s_n = (2n-1)^2$ **27.** $s_n = 3^n - 1$ **28.** $s_n = \dfrac{n(5n-1)}{2}$

c ● 29. Use Newton's method to find four decimal approximations for $\sqrt{5}$. Take as initial guess $x_1 = 2$.

c ● 30. Use Newton's method to find four decimal approximations for $\sqrt{6}$. Take as initial guess $x_1 = 3$.

9.4 ARITHMETIC SEQUENCES AND SERIES

Arithmetic Sequences

A sequence $\{a_n\}$ is called an *arithmetic sequence* if there is a number d such that

$$a_n - a_{n-1} = d$$

for every $n \in \mathbb{N}$.

The number d is called the *common difference* of the sequence. Arithmetic sequences are also called *arithmetic progressions*.

The simplest example of an arithmetic sequence is the sequence $1, 2, \ldots, n, \ldots$ of all natural numbers, where $d = 1$. The sequence

$$13, 18, 23, \ldots, 5n - 2, \ldots \quad n \geq 3$$

(see Example 2 of Section 9.3) is an arithmetic sequence with $d = 5$.

If $\{a_n\}$ is an arithmetic sequence whose common difference is d, then

$$a_n = a_{n-1} + d, \quad \text{for all } n \in \mathbb{N}.$$

This is a recurrence relation that yields the successive elements of the sequence, such as

$$a_1$$
$$a_2 = a_1 + d$$
$$a_3 = a_2 + d = a_1 + 2d$$
$$a_4 = a_3 + d = a_1 + 3d.$$

It can be proved, by mathematical induction, that the general term of the arithmetic sequence is given by

---

$$(9.8) \quad a_n = a_1 + (n - 1)d.$$

◆ **Example 1.** If 9 and 12 are the fifth and the sixth terms of an arithmetic progression, find the first and the tenth terms of the progression.

Solution. We have $a_5 = 9$ and $a_6 = 12$, so

$$d = a_6 - a_5 = 12 - 9 = 3.$$

From (9.8) with $n = 5$, $a_5 = 9$, and $d = 3$, we have

$$9 = a_1 + (5 - 1) \cdot 3,$$
$$9 = a_1 + (4) \cdot 3,$$
$$a_1 = -12 + 9 = -3.$$

Next, we find a_{10}:

$$a_{10} = a_1 + (10 - 1)d$$
$$= (-3) + (10 - 1)3$$
$$= -3 + 9 \cdot 3$$
$$= 24$$

◇ **Practice Exercise 1.** Suppose that 43 and 40 are the fourth and fifth terms of an arithmetic sequence. Find the first and the fifteenth terms of the sequence.

Answer. $a_1 = 52$, $a_{15} = 10$.

◆ **Example 2.** If the third and seventh terms of an arithmetic sequence are 6 and -2, respectively, find the twentieth term of the sequence.

Solution. We have $a_3 = 6$ and $a_7 = -2$. According to (9.8), we get the linear system of equations

$$6 = a_1 + (3 - 1)d$$
$$-2 = a_1 + (7 - 1)d$$

whose solution is $a_1 = 10$ and $d = -2$. We can now find the term a_{20}:

$$a_{20} = 10 + (20 - 1) \cdot (-2)$$
$$= 10 + 19 \cdot (-2)$$
$$= -28.$$

◇ **Practice Exercise 2.** Let 23 and 33 be the fifth and tenth terms of an arithmetic sequence. Find the eighteenth term of the sequence.

Answer. $a_{18} = 49$.

A Property of Arithmetic Sequences

The following property of the terms of an arithmetic sequence will be used later. If $n > 1$ is a fixed natural number, and r is such that $1 \le r \le n - 1$, write the first n elements of the sequence as follows:

$$\underbrace{a_1, \ldots, a_r,}_{r \text{ terms}} a_{r+1}, \ldots, a_{n-r}, \underbrace{a_{n-r+1}, \ldots, a_n}_{r \text{ terms}}$$

There are r terms to the left of a_{r+1} and r terms to the right of a_{n-r}. We want to show that

> **(9.9)** $a_{r+1} + a_{n-r} = a_1 + a_n$, for all $1 \le r \le n - 1$.

According to (9.8), we have

$$a_{r+1} = a_1 + rd$$
$$a_{n-r} = a_1 + (n - r - 1)d$$

By adding both expressions and simplifying, we get

$$a_{r+1} + a_{n-r} = (a_1 + rd) + [a_1 + (n - r - 1)d]$$
$$= a_1 + rd + a_1 + (n-1)d - rd$$
$$= a_1 + a_n,$$

which is formula (9.9).

Arithmetic Series

An *arithmetic series* is a series whose sequence of terms is an arithmetic sequence. If $a_1, a_2, \ldots, a_n, \ldots$ is an arithmetic sequence with common difference d, then the nth partial sum of the arithmetic series is given by

$$(9.10) \quad s_n = \frac{(a_1 + a_n)n}{2}.$$

To prove formula (9.10), we write s_n in two different ways:

$$s_n = a_1 + a_2 + \cdots + a_{n-1} + a_n$$

and

$$s_n = a_n + a_{n-1} + \cdots + a_2 + a_1.$$

By adding term by term and using the commutativity and associativity of the sum, we obtain

$$2s_n = (a_1 + a_n) + (a_2 + a_{n-1}) + \cdots + (a_{n+1} + a_2) + (a_n + a_1).$$

By virtue of (9.9), each term between parentheses on the right-hand side is equal to $a_1 + a_n$. Since there are n such terms, we have

$$2s_n = (a_1 + a_n) \cdot n;$$

hence (9.10)

$$s_n = \frac{(a_1 + a_n) \cdot n}{2}.$$

Alternatively, substituting $a_n = a_1 + (n - 1)d$ into the last expression yields the formula

---

$$(9.11) \quad s_n = [2a_1 + (n - 1)d]\frac{n}{2}.$$

◆ **Example 3.**   Find the sum of all the odd numbers between 25 and 87, inclusive.

Solution. The odd numbers from 25 to 87 form an arithmetic progression with $a_1 = 25$ and $d = 2$. We use (9.8) to find the place of 87 in such progression:

$$87 = 25 + (n - 1)2,$$

so

$$n = 32.$$

Next, we use (9.10) to find s_{32}:

$$s_{32} = \frac{(25 + 87) \cdot 32}{2} = 1792.$$

◇ **Practice Exercise 3.** Find the sum of all even numbers from 18 to 42 inclusive.

Answer. 390.

◆ **Example 4.** Find the sum of the first 28 terms of an arithmetic progression if the first term is 105 and the common difference is -3.

Solution. Use (9.11) with $n = 28$, $a_1 = 105$, and $d = -3$ to obtain

$$s_{28} = [2 \cdot 105 + (28 - 1)(-3)]\frac{28}{2}$$
$$= (210 - 81) \cdot 14$$
$$= 1806.$$

◇ **Practice Exercise 4.** The first term of an arithmetic sequence is 25 and the common difference is 4. Find the sum of the first 30 terms of the sequence.

Answer. 2490.

 EXERCISES 9.4

In Exercises 1–6, find the common difference and determine the next two terms of each arithmetic progression.

1. 2, 4, 6, . . .

2. 3, 6, 9, . . .

3. 2, 0.5, -1, . . .

4. $\dfrac{1}{3}, \dfrac{5}{6}, \dfrac{4}{3}, \ldots$

5. log 2, log 4, log 8, . . .

6. $\log 10^{-1}, \log 10^{-2}, \log 10^{-3}, \ldots$

7. If 4 and 7 are the first two terms of an arithmetic sequence, find the fifteenth term.

8. Find the tenth term of an arithmetic progression whose first and second terms are -3 and 5, respectively.

9. If 15 and 6 are the second and fifth term of an arithmetic progression, find the first and the ninth terms.

10. The third and the fifth terms of an arithmetic sequence are 12 and 20, respectively. Find the eleventh term.

In Exercises 11–14, a_1 and d are the first term and the common difference of an arithmetic progression. Find the sum s_n for each given n.

11. $a_1 = -10$, $d = 4$, $n = 12$

12. $a_1 = 42$, $d = -3$, $n = 15$

13. $a_1 = \dfrac{1}{2}$, $d = \dfrac{1}{3}$, $n = 16$

14. $a_1 = \dfrac{2}{3}$, $d = -\dfrac{3}{4}$, $n = 21$

In Exercises 15–20, find the sums.

15. $\displaystyle\sum_{n=1}^{25} (2n + 3)$

16. $\displaystyle\sum_{n=1}^{30} (3n + 5)$

17. $\displaystyle\sum_{n=1}^{30} (4 - 3n)$

18. $\displaystyle\sum_{n=1}^{22} (12 - 4n)$

19. $\displaystyle\sum_{n=1}^{15} \left(\dfrac{n}{2} + 3\right)$

20. $\displaystyle\sum_{n=1}^{21} \left(\dfrac{3n}{4} - 2\right)$

21. Find the sum of all even numbers between 21 and 95.
22. Find the sum of all odd numbers between 36 and 112.
23. How many multiples of 4 are there between 15 and 94? Find their sum.
24. How many natural numbers divisible by 5 are there between 27 and 108? Find their sum.
25. A boy saves 10¢ on July 1, 20¢ on July 2, 30¢ on July 3, and so on. How much does he save during July?
26. A clock strikes once at 1:00, twice at 2:00, and so on. Assuming that the clock does not strike on the half-hour, how many times does the clock strike between 12:30 a.m. and 12:30 p.m.?
● 27. A pile of logs has 20 logs in the bottom layer, 19 in the second layer, and so on, until there is one log in the last layer. How many logs are there in the pile?
● 28. Find three numbers in arithmetic progression such that the sum of the first and the third is 40 and the product of the first and the second is 300.
● 29. Using formula (9.10), show that the sum of the first n odd natural numbers is n^2.
● 30. Using formula (9.10), show that the sum of the first n even natural numbers is $n(n + 1)$.

9.5   GEOMETRIC SEQUENCES AND SERIES

Geometric Sequences

A sequence $\{a_n\}$ is called a *geometric sequence* if there is a number $r \neq 0$ such that

$$(9.12) \qquad \frac{a_{n+1}}{a_n} = r$$

for every $n \in \mathbb{N}$.

The number r is called the *common ratio* of the sequence. Geometric sequences are also called *geometric progressions*.

The relation (9.12) can also be written as

$$a_{n+1} = a_n r, \quad n \in \mathbb{N}.$$

Setting $n = 1, 2, \ldots$ we obtain the successive elements of the sequence:

$$a_1$$
$$a_2 = a_1 r$$
$$a_3 = a_2 r = a_1 r^2$$
$$a_4 = a_3 r = a_1 r^3, \quad \text{and so on.}$$

This indicates that the general term of the sequence is given by

$$(9.13) \quad a_n = a_1 r^{n-1}.$$

This formula can be proved by induction.

As an example, suppose that n_0 denotes the initial number of bacteria in a culture where the number of bacteria is doubling every hour. Then the numbers

$$n_0, 2n_0, 2^2 n_0, 2^3 n_0, \ldots$$

representing the initial number of bacteria, the number of bacteria after 1 hour, 2 hours, etc., form a geometric sequence of ratio 2.

Similarly, the numbers

$$1, \frac{1}{2}, \frac{1}{2^2}, \frac{1}{2^3}, \ldots$$

form a geometric sequence of ratio 1/2.

◆ **Example 1.** If the first term of a geometric progression is 2 and the ratio is 3, find the first four terms of the progression.

Solution. We have $a_1 = 2$ and $r = 3$. Thus,

$$a_1 = 2$$
$$a_2 = 2 \cdot 3 = 6$$
$$a_3 = 2 \cdot 3^2 = 18$$
$$a_4 = 2 \cdot 3^3 = 54.$$

◇ **Practice Exercise 1.** The common ratio of a geometric sequence is 1/4. If the first term is 32, find the next four terms of the sequence.

Answer. $8, 2, \dfrac{1}{2}, \dfrac{1}{8}.$

◆ **Example 2.** If the fourth and the ninth terms of a geometric sequence are $-1/8$ and $1/256$, find the common ratio r.

Solution. By using formula (9.13) with $n = 4$ and $n = 9$, respectively, we have:

$$-\frac{1}{8} = a_1 r^3$$

and

$$\frac{1}{256} = a_1 r^8.$$

Dividing term by term, we get

$$\frac{-1/8}{-1/256} = \frac{a_1 r^3}{a_1 r^8},$$

$$-32 = \frac{1}{r^5},$$

$$r^5 = -\frac{1}{32}.$$

Therefore

$$r = -\frac{1}{2}.$$

◇ **Practice Exercise 2.** The seventh term of a geometric sequence is 405 and the tenth term is 10935. Find the common ratio of the sequence.

Answer. 3.

Geometric Series

A *geometric series* is a series whose sequence of terms is a geometric sequence. If $a_1, a_2, \ldots, a_n, \ldots$ is a geometric sequence, with common ratio $r \neq 1$, then the nth partial sum of the geometric series is given by

$$(9.14) \quad s_n = \frac{a_1(1 - r^n)}{1 - r}.$$

This formula can be proved by induction (Exercise 50 of Section 9.1) or as follows. Let

$$s_n = a_1 + a_1 r + a_1 r^2 + \cdots + a_1 r^{n-1} = \sum_{k=1}^{n} a_1 r^{k-1}.$$

Multiplying both members by r, we obtain

$$rs_n = a_1r + a_1r^2 + \cdots + a_1r^n.$$

Subtracting the second equation from the first one and simplifying, we obtain

$$s_n - rs_n = (a_1 + a_1r + \cdots + a_1r^{n-1}) - (a_1r + \cdots + a_1r^{n-1} + a_1r^n)$$
$$= a_1 - a_1r^n$$

or

$$s_n(1 - r) = a_1(1 - r^n).$$

Hence

$$s_n = \frac{a_1(1 - r^n)}{1 - r}$$

which is formula (9.14).

c ◆ **Example 3.** Find the sum of the first 15 terms of the geometric progression whose first term is 1 and whose ratio is 2.

Solution. Apply formula (9.14) with $n = 15$, $a_1 = 1$, and $r = 2$, to obtain

$$s_{15} = \frac{1 \cdot (1 - 2^{15})}{1 - 2} = 2^{15} - 1 = 32{,}767.$$

c ◇ **Practice Exercise 3.** The first term of a geometric sequence is 3 and the common ratio is 0.2. Find the sum of the first 5 terms of the sequence.

Answer. 3.7488.

Infinite Geometric Series

About 450 B.C. the Greek philosopher Zeno of Elea (495–435 B.C.) published a book of paradoxes as a challenge to the philosophers and mathematicians of his time. One of them can be paraphrased as follows. In order for a runner to go a certain distance, he must first run half of the distance; in order to go half the distance, he must first run a quarter of the distance; in order to go a quarter of the distance, he must first run one eighth of the distance, and so on. Since to go any distance at all the runner has an infinite number of distances to run, it appears impossible ever to begin running. Thus, concluded Zeno, motion itself is impossible!

This is clearly a paradox, since we know that motion is possible and that once started, the runner will eventually reach his destination. The difficulty with Zeno's paradox results, perhaps, from our perception that it is impossible to do infinitely many things in a finite length of time.

From a mathematical viewpoint, we can say that the distance of 1 mile is the "sum" of half a mile plus a quarter of a mile plus an eighth of a mile, and so on *ad infinitum* (Figure 9.1, from right to left). Thus, we may write

$$\frac{1}{2} + \frac{1}{4} + \frac{1}{8} + \cdots = 1$$

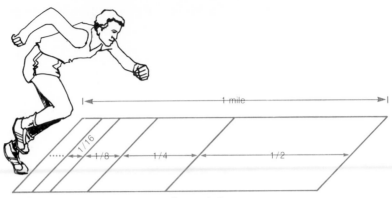

Figure 9.1

where the dots indicate that we are computing a "sum" containing infinitely many terms. (Note that the operation of sum is only defined for a finite number of terms.)

To justify this concept, consider the geometric series whose terms are the elements of the geometric sequence $\{1/2, 1/2^2, 1/2^3, \ldots\}$. According to formula (9.14), with $a_1 = 1/2$ and $r = 1/2$, the nth partial sum of this series is

$$s_n = \frac{\frac{1}{2}\left(1 - \frac{1}{2^n}\right)}{1 - \frac{1}{2}}.$$

Simplifying, we obtain

$$s_n = \frac{\frac{1}{2}\left(1 - \frac{1}{2^n}\right)}{\frac{1}{2}}$$

$$= 1 - \frac{1}{2^n},$$

for every natural number n. Now as n increases without bound, 2^n increases without bound, so that $1/2^n$ decreases to zero. We conclude that s_n approaches 1 as n increases without bound. On the other hand,

$$s_n = \frac{1}{2} + \frac{1}{4} + \frac{1}{8} + \cdots + \frac{1}{2^n}$$

and, as n increases without bound, the number of terms of this sum increases without bound. Thus we can write

$$\frac{1}{2} + \frac{1}{4} + \frac{1}{8} + \cdots + \frac{1}{2^n} + \cdots = 1 \quad \text{or} \quad \sum_{j=1}^{\infty} \frac{1}{2^j} = 1$$

and say that 1 is *the sum of all terms of the geometric sequence* $\{1/2^n\}$.

Sum of an Infinite Geometric Series

If the ratio r of a geometric sequence $\{a_n\}$ is such that $|r| < 1$, then the sequence of partial sums $\{s_n\}$ approaches the number

$$\frac{a_1}{1 - r},$$

as n increases without bound.

To see this, we write (9.14) as follows:

$$s_n = \frac{a_1}{1 - r} - \frac{a_1}{1 - r} r^n.$$

Now, it can be proved that if $|r| < 1$, then r^n *approaches* 0 *as n increases without bound*. Thus the second term of the difference above approaches 0 and, consequently, s_n approaches $a_1/(1 - r)$, as n increases without bound.

Thus we set

$$(9.15) \quad s = \frac{a_1}{1 - r}$$

and call this number *the sum of the infinite geometric series.*

Alternative ways of writing the sum of the infinite series are:

$$(9.16) \quad a_1 + a_1 r + a_1 r^2 + \cdots + a_1 r^{n-1} + \cdots = \frac{a_1}{1 - r}$$

or, using the summation notation,

$$(9.17) \quad \sum_{k=1}^{\infty} a_1 r^{k-1} = \frac{a_1}{1 - r}.$$

Note that if $|r| \geq 1$, the infinite series has no sum because the term $[a_1/(1 - r)]r^n$ in the expression for s_n does not exhibit any regular behavior as n becomes large.

◆ **Example 4.** Find the infinite sum $\sum_{n=1}^{\infty} \left(\frac{3}{4}\right)^{n-1}$.

Solution. This is an infinite series whose first term is $a_1 = 1$ and whose ratio is $r = 3/4$. According to (9.17), we obtain

$$\sum_{n=1}^{\infty} \left(\frac{3}{4}\right)^{n-1} = \frac{1}{1 - \dfrac{3}{4}} = \frac{1}{\dfrac{1}{4}} = 4.$$

◇ **Practice Exercise 4.**   Find the sum of the series $\displaystyle\sum_{n-1}^{\infty}\left(\frac{2}{3}\right)^{n}$.

Answer.   2.

◆ **Example 5.**   A ping-pong ball dropped from a height of 1 meter rebounds one-half the distance after each fall. Find the total distance the ball travels before coming to rest.

Solution.   The ball travels 1 meter when dropped. After the first rebound the ball travels 1/2 m up and 1/2 m down, that is, 1 m; after the second rebound it travels $2 \cdot 1/4$ m $= 1/2$ m; after the third rebound it travels $2 \cdot 1/8 = 1/4$ m, and so on. The total distance traveled is then given by the infinite series

$$1 + 1 + \frac{1}{2} + \frac{1}{4} + \cdots = 1 + \frac{1}{1 - \dfrac{1}{2}} \qquad \begin{array}{l}\text{[using (12.15) with } a_1 = 1 \\ \text{and } r = 1/2]\end{array}$$

$$= 1 + 2 = 3 \text{ m.}$$

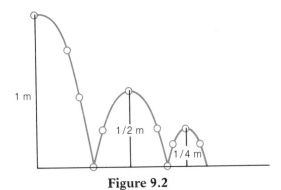

Figure 9.2

◇ **Practice Exercise 5.**   A soccer ball dropped from a height of 1 meter rebounds 3/4 the distance after each fall. What is the total distance traveled by the ball before coming to rest?

Answer.   7 meters.

Decimal Numbers as Fractions

Infinite geometric series can also be used to convert repeating decimals into rational numbers; in fact, a repeating decimal can be viewed as being an expression for a certain kind of geometric series.

◆ **Example 6.**   Write the repeating decimal $0.\overline{36}$ as a fraction.

Solution.   Since

$$0.\overline{36} = 0.363636\ldots$$
$$= 0.36 + 0.0036 + 0.000036 + \cdots$$

the right-hand side is the sum of an infinite geometric series whose first term is 0.36 and whose ratio is 0.01. Using formula (9.15), we get

$$0.\overline{36} = \frac{0.36}{1 - 0.01} = \frac{0.36}{0.99} = \frac{4}{11}.$$

Check this result by dividing 4 by 11.

◇ **Practice Exercise 6.**   Write the repeating decimal $0.\overline{45}$ as a fraction.

Answer. $\dfrac{5}{11}$.

 EXERCISES 9.5

In Exercises 1–6, find the ratio and determine the next two terms of each geometric progression.

1. 2, 4, 8, . . .

2. 3, 9, 27, . . .

3. $1, -\dfrac{1}{2}, \dfrac{1}{4}, \ldots$

4. $8, -2, \dfrac{1}{2}, \ldots$

5. $\dfrac{1}{2}, \dfrac{1}{3}, \dfrac{2}{9}, \ldots$

6. $\dfrac{2}{3}, -\dfrac{1}{2}, \dfrac{3}{8}, \ldots$

7. If the first term of a geometric sequence is 5 and the ratio is 4, find the first four terms of the sequence.

8. The first term of a geometric sequence is -2 and the ratio is $-3/5$. Find the first four terms of the progression.

9. Find the fifth term of a geometric progression whose first two terms are 3/8 and 1/4.

10. Find the seventh term of a geometric progression whose first two terms are 6 and 9.

11. The third and sixth terms of a geometric sequence are 2 and $-1/4$, respectively. Find the first term.

12. The second and the sixth terms of a geometric sequence are 4 and 1/64. Find the ninth term.

c 13. Find the sum of the first 10 terms of a geometric sequence whose first term is 6 and whose ratio is -2.

c 14. Find the sum of the first 4 terms of a geometric sequence whose first term is 100 and whose ratio is 3/4.

In Exercises 15–18, find each sum.

15. $\displaystyle\sum_{k=1}^{12} (-2)^{k-1}$

16. $\displaystyle\sum_{k=1}^{8} \left(\dfrac{1}{3}\right)^{k-1}$

17. $\displaystyle\sum_{k=1}^{10} 3^{k-1}$

18. $\displaystyle\sum_{k=1}^{15} \left(\dfrac{1}{2}\right)^{k-1}$

In Exercises 19–26, find the sum of each geometric series.

19. $1 - \dfrac{1}{2} + \dfrac{1}{4} - \dfrac{1}{8} + \cdots$

20. $\dfrac{2}{3} + \dfrac{4}{9} + \dfrac{8}{27} + \dfrac{16}{81} + \cdots$

21. $3 + \dfrac{3}{4} + \dfrac{3}{16} + \dfrac{3}{64} + \cdots$

22. $2 - \dfrac{2}{5} + \dfrac{2}{25} - \dfrac{2}{125} + \cdots$

23. $\displaystyle\sum_{n=1}^{\infty} \left(\dfrac{5}{6}\right)^{n-1}$

24. $\displaystyle\sum_{n=1}^{\infty} 2\left(-\dfrac{3}{4}\right)^{n-1}$

25. $\displaystyle\sum_{n=1}^{\infty} \left(-\dfrac{4}{5}\right)^{n}$

26. $\displaystyle\sum_{n=1}^{\infty} \left(\dfrac{2}{3}\right)^{n}$

Write each repeating decimal as a fraction.

27. $0.\overline{18}$

28. $0.\overline{15}$

29. $0.6\overline{3}$

30. $0.\overline{12}$

31. $0.3\overline{18}$

32. $0.4\overline{06}$

33. $0.12\overline{42}$

34. $0.25\overline{45}$

35. A tennis ball dropped from a height of 6 ft rebounds 0.8 the distance after each fall. Determine the total distance covered by the ball before coming to rest.

36. The first swing of the bob of a pendulum is 2 ft. In each subsequent swing the bob travels 3/4 of the preceding swing. Find how far the bob will travel before coming to rest.

● **37.** Assume that each New Yorker spends 3/4 of his or her income in New York, and saves or spends the rest elsewhere. Suppose also that during a convention, out-of-town conventioneers spent a total of $200,000 in New York. What is the total amount of money spent in New York as a result of the convention? (This is called the *multiplier effect* in economics.)

● **38.** If a population p_0 grows at the constant rate of $100r$ percent per year, show that the population $p(t)$ after t years is given by $p(t) = p_0(1 + r)^t$. How long will it take for the population of a certain country to double if it is increasing at a rate of 3% per year?

● **39.** A wheel rotating at a rate of 1200 revolutions per minute is slowing down. If in each subsequent minute it rotates 1/3 as many times as in the preceding minute, find how many revolutions the wheel will make before stopping.

● **40.** A company estimates that as a result of advertising 1,000,000 people will buy a new product being marketed. It also estimates that satisfied buyers will induce 50% more people to buy the product, and that the new buyers will induce another 50% more people to buy the product, and so on. How many people are estimated to buy the product?

9.6   PERMUTATIONS, COMBINATIONS, AND SET PARTITIONING

Suppose that four people are members of a school committee. In how many ways can a president and vice-president be chosen? To answer this question, let us label the people as *A*, *B*, *C*, and *D*. If *A* is chosen to be the president, then the vice-president can be chosen in *three* different ways: either *B*, or *C*, or *D*. If *B* is the president, then there are also three possible choices for a vice-president: *A*, or *C*, or *D*, and so on. To help us count the number of all possibilities, we form a *tree diagram* as follows.

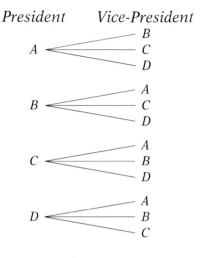

Figure 9.3

Therefore, we have a total of 12 possibilities.

This number is obtained as follows. There are 4 possible choices for a committee president. Once a president is selected, there are 3 possible choices for a vice-president. Altogether the total number of choices is $4 \cdot 3 = 12$.

This example illustrates the Fundamental Principle of Counting, which we now state.

Fundamental Principle of Counting

Suppose that two operations O_1 and O_2 are performed in order, with n_1 possible outcomes for the first operation and n_2 possible outcomes for the second operation. Then, there are $n_1 n_2$ possible combined outcomes of the first and second operations.

Suppose that k operations $O_1, O_2, \ldots, O_k$ are performed in order with n_1 the number of outcomes for O_1, n_2 the number for $O_2, \ldots, n_k$ the number for O_k. Then,

$$n_1 \cdot n_2 \cdot \ldots \cdot n_k$$

is the total number of outcomes of the operations performed in that order.

◆ **Example 1.**   Five horses are competing in a race for first or second place. In how many different ways can the race be decided, assuming that there are no ties for either first or second place?

Solution.   There are 5 possible outcomes for first place. Once the first place is decided, there are 4 possible outcomes for second place. Thus, according to the Fundamental Principle of Counting, the first and second places can be decided in $5 \cdot 4 = 20$ different ways.

◇ *Practice Exercise 1.*   In how many ways can we choose a chairperson and a vice-chairperson for a committee of six people?

Answer.   30.

◆ *Example 2.*   Ten people form a committee. In how many different ways can a chairperson, a vice-chairperson, and a secretary be selected?

Solution.   Here we have three operations O_1, O_2, and O_3. Operation O_1 consists of selecting a chairperson, O_2 selecting a vice-chairperson, and O_3 a secretary. There are 10 different ways to select the chairperson. Once the chairperson is selected there are 9 ways that a vice-chairperson can be selected. Finally, after a chairperson and vice-chairperson are selected there are 8 ways to select a secretary. Thus, the total number of selections for a chairperson, vice-chairperson, and secretary is $10 \cdot 9 \cdot 8 = 720$.

◇ *Practice Exercise 2.*   Eight athletes are competing for a gold, silver, or bronze medal in an Olympic event. How many different outcomes can the competition have?

Answer.   336.

Permutations

In all examples discussed so far, we have considered only *ordered arrangements* of objects. For example, if person *A* is elected president of the school committee and *B* vice-president, then we have an arrangement that we may denote by (*A*, *B*). Such an arrangement is, of course, different from the arrangement (*B*, *A*) where *B* would be the president and *A* the vice-president. Arrangements where the *order* distinguishes one arrangement from another are called *permutations*.

Suppose that we have a collection of *n* distinct objects. How many permutations of *k* objects ($k \le n$) can we obtain from the original collection of *n* objects? The number of these permutations will be denoted by $P(n, k)$. How can we determine this number? Notice that we are performing *k* operations in order on a set of *n* objects. The number of outcomes of the first operation is *n*, of the second operation is $n - 1$, of the third operation is $n - 2$, . . . , of the *k*th operation is $n - (k + 1)$. By the Fundamental Principle of Counting, we obtain

(9.18) $P(n, k) = n(n - 1) \cdots (n - k + 1).$

At this point, we recall that $(n - k)! = 1 \cdot 2 \cdot \ldots \cdot (n - k)$. Multiplying and dividing the last expression by $(n - k)!$, we obtain

$$P(n, k) = n(n - 1) \cdots (n - k + 1)\left[\frac{(n - k)!}{(n - k)!}\right]$$
$$= \frac{n(n - 1) \cdots (n - k + 1)(n - k) \cdots 2 \cdot 1}{(n - k)!}$$
$$= \frac{n!}{(n - k)!}$$

Thus we obtain the following formula.

Permutations of k Objects from a Set of n Objects

$$(9.19) \quad P(n, k) = \frac{n!}{(n - k)!}, \quad 0 \le k \le n.$$

---

c ◆ **Example 3.**   Find the number of permutations of 4 objects from a set of 30 objects.

Solution.   We have

$$P(30, 4) = \frac{30!}{(30 - 4)!}$$

$$= \frac{30!}{26!}$$

$$= \frac{30 \cdot 29 \cdot 28 \cdot 27 \cdot 26!}{26!}$$

$$= 657{,}720.$$

c ◇ **Practice Exercise 3.**   What is the number of permutations of 3 objects from a set of 80 objects?

Answer. $\dfrac{80!}{77!} = 492{,}960.$

Permutations of *n* Objects

If in formulas (9.18) and (9.19) the number k is equal to n, we obtain the number of permutations of n objects from a set of n objects.

$$(9.20) \quad P(n, n) = n(n - 1) \cdots 2 \cdot 1 = n!$$

Permutations of n objects from a set of n objects are called, for simplicity, *permutations of n objects*. Formula (9.20) tells us that the number of permutations of n objects is *n factorial*.

c ◆ **Example 4.**   Find the number of all permutations of 8 objects.

Solution.   We have $P(8, 8) = 8! = 8 \cdot 7 \cdot 6 \cdot 5 \cdot 4 \cdot 3 \cdot 2 \cdot 1 = 40{,}320$. Note that you may obtain this number by multiplying 8 by 7, the result by 6, the result by 5, and so on until you reach 1. However, if you use a scientific calculator you can find this result with a few keystrokes. For example, on a TI-30-II, the keystrokes: 8 INV $x!$ will give you 40,320.

c ◇ **Practice Exercise 4.**   What is the number of permutations of 10 objects?

Answer.   3,628,800.

Combinations

Suppose that after choosing the president and vice-president of the school committee (discussed at the beginning of this section), the four committee members go out for dinner. Upon arriving at the restaurant, they are told that the only table available is a table that seats only two people. They decide to wait for a larger table. While they are waiting, one of the committee members figures out that the number of seating arrangements for four people at a table that seats only two is 6. How was this number determined? First, notice that if A and B form a seating arrangement denoted by AB, this arrangement is the same as BA. In other words, the order of the arrangement is now immaterial. Referring to the tree diagram of Figure 9.3, you can see that the following are all possible seating arrangements.

$$
\begin{array}{lll}
AB & & \\
AC & BC & \\
AD & BD & CD
\end{array}
$$

Figure 9.4

that is a total of 6 different arrangements.

Arrangements where the order of the objects is immaterial are called *combinations*. The problem of finding the number of seating arrangements for the school committee members is the same as the problem of finding the *number of combinations of two objects from a set of four objects*. This number is denoted by $C(4, 2)$.

In general, the number of combinations of k objects from a set of n objects is denoted by $C(n, k)$. How can we find this number? First, consider the seating arrangement BC of Figure 9.4. If we arrange the two people B and C in a given order, we obtain the permutations BC and CB. We see that the operation of taking any combination of two objects from a set of four objects followed by the operation of permutation of two objects gives us the set of *all* permutations of 2 objects from a set of four objects. Thus

$$P(4, 2) = C(4, 2) \cdot P(2, 2),$$

so

$$C(4, 2) = \frac{P(4, 2)}{P(2, 2)} = \frac{4!}{2!2!} = 6.$$

If $C(n, k)$ denotes the number of combinations of k objects from a set of n objects, and we multiply this number by $P(k, k) = k!$, the number of permutations of k objects, we obtain $P(n, k)$, the number of permutations of k objects from a set of n objects. That is,

$$P(n, k) = C(n, k) \cdot P(k, k).$$

Hence, we derive

$$(9.21) \quad C(n, k) = \frac{P(n, k)}{P(k, k)}.$$

If we replace $P(n, k)$ with $n!/(n - k)!$ and $P(k, k)$ with $k!$, formula (9.21) can be written as follows.

Combinations of k Objects from a Set of n Objects

$$(9.22) \quad C(n, k) = \frac{n!}{(n - k)!\,k!}.$$

We have seen these numbers before: in Section 9.2 page 441 we defined the binomial coefficient

$$\binom{n}{k} = \frac{n!}{k!(n - k)!}$$

which are the same fractions (except for the order of the factors in the denominator). Thus we have the important relation

$$C(n, k) = \binom{n}{k}.$$

The number $C(n, k)$ of all combinations of k objects from a set of n objects is also denoted by $\binom{n}{k}$, the symbol used in Section 9.2 for the binomial coefficients.

◆ **Example 5.** In how many ways can 8 books be arranged in a bookshelf if the shelf has space available for only three books?

Solution. We have to form sets of 3 books (the order is immaterial) from our given set of 8 books. Thus, we have to find the number of combinations of 3 objects from a set of 8 objects.

$$C(8, 3) = \frac{8!}{(8 - 3)!3!}$$

$$= \frac{8!}{5!3!} = \frac{8 \cdot 7 \cdot 6 \cdot 5!}{5!3!} = \frac{8 \cdot 7 \cdot 6}{3 \cdot 2} = 56.$$

◇ **Practice Exercise 5.** How many four-person subcommittees can be formed from a committee of ten people?

Answer. $C(10, 4) = \dfrac{10!}{6!4!} = 210.$

Partitioning *n* Objects in *k* Cells

Suppose that twelve congressmen are to be assigned to three committees of three, four, and five people in such a way that each one of them is to be assigned to only one committee. In how many ways can this be done? According to the Fundamental Principle of Counting, let us perform three operations in the following order. In the first operation we form committees of three from the twelve congressmen. There are $C(12, 3)$ possible outcomes for the first operation. Once one committee of three is formed, there will be nine congressmen left. In the second operation we form committees of four from nine congressmen, obtaining $C(9, 4)$ committees. Once a committee of three and a committee of four are set up, there will be 5 congressmen left. Our third operation will be to form committees of five members. Since there are only 5 congressmen left, we can form only one committee. Note that $1 = C(5, 5)$. Now, by the Fundamental Principle of Counting, the total number of committees of three, four, and five congressmen is

$$C(12, 3) \cdot C(9, 4) \cdot C(5, 5) = \frac{12!}{3!9!} \cdot \frac{9!}{4!5!} \cdot 1$$

$$= \frac{12!}{3!4!5!} = 27,720.$$

What we just did was to partition 12 objects into three cells with each cell containing three, four, or five distinct objects. (It is important to note that $3 + 4 + 5 = 12$.) The number of all such partitions is denoted by $\Gamma(12, 3, 4, 5)$, and we have $\Gamma(12, 3, 4, 5) = 12!/3!4!5!$. This discussion justifies the following result, which can be derived from the Fundamental Principle of Counting.

Partition of *n* Objects in *k* Cells

If *n* objects are to be partitioned into *k* cells such that
  i) each object belongs to only one cell,
  ii) the first cell contains n_1 objects, the second cell n_2 objects, . . . , the *k*th cell n_k objects,
  iii) $n_1 + n_2 + \cdots + n_k = n$,
then the number of all partitions is

$$(9.23) \quad \Gamma(n, n_1, \ldots, n_k) = \frac{n!}{n_1!n_2! \ldots n_k!}.$$

---

c ◆ **Example 6.**   Two people are playing a card game that uses a 40-card deck. Four cards are turned face up and each person is dealt three cards. Find how many deals are possible.

Solution.   After 4 cards are turned face up, the remaining 36 cards are to be partitioned into three cells. Two cells are the two players receiving three cards each. The

30 cards that remain are placed into the third cell. Thus we have

$$\Gamma(36, 3, 3, 30) = \frac{36!}{3!3!30!}$$
$$= 38,955,840.$$

c ◇ **Practice Exercise 6.** Four people are playing the same card game of Exercise 5. How many deals are possible if each person receives three cards?

Answer. $\Gamma(36, 3, 3, 3, 3, 34) \simeq 4.6262008 \times 10^{14}$.

Distinguishable Permutations

An *anagram* is a transposition of the letters of a word to form another word. How many anagrams can be obtained from the word BOOK? First, notice that if we interchange the two letters O and keep fixed the letters B and K, a permutation of the letters B, O, O, K has been performed without changing the given word. Such a permutation is called a *nondistinguishable permutation*. However, by interchanging the last two letters, a *distinguishable permutation* is performed, giving us the anagram BOKO. Now, if all the letters were different the number of all anagrams would be 4! = 24, the number of permutations of four objects. Since two letters are the same, it follows that the number n of anagrams that can be obtained from the word BOOK must be less than 24. To find this number, notice that to each distinguishable permutation there correspond 2! undistinguishable permutations that are obtained by interchanging the letter O. Thus if we multiply n by 2!, we obtain the number of permutations of four objects; that is, $2!n = 4!$, or

$$n = \frac{4!}{2!} = 12.$$

The same reasoning applies to other cases. Suppose that we want to find the number of anagrams obtained from the word ANACONDA. There are 3! arrangements of the letter A and 2! arrangements of the letter N (a total of 3!2! arrangements) that do not change any given anagram. Thus, if n denotes the number of all anagrams obtained from the word ANACONDA, then $3!2!n$ is equal to 8!, the number of permutations of 8 different objects. It follows that

$$n = \frac{8!}{3!2!}.$$

These two examples illustrate the following general result.

The Number of Distinguishable Permutations

The number of distinguishable permutations of n objects, of which n_1 are equal, n_2 are equal of another kind, . . . , n_k are equal of a further kind, and such that $n = n_1 + n_2 + \cdots + n_k$, is

(9.24) $\Gamma(n, n_1, \ldots, n_k) = \dfrac{n!}{n_1!n_2! \ldots n_k!}.$

◆ **Example 7.** How many distinguishable permutations can be obtained with the letters of the word INDIANA?

Solution. The given word has 7 letters, of which the letter A is repeated twice, the letter I is repeated twice, and the letter N is repeated twice. According to formula (9.24), the number of distinguishable permutations is

$$\Gamma(7, 2, 2, 2, 1) = \frac{7!}{2!2!2!1!} = 630.$$

◇ **Practice Exercise 7.** Find the number of anagrams of the word ARKANSAS.

Answer. $\Gamma(8, 3, 2, 1, 1, 1) = 3360.$

EXERCISES 12.6

In Exercises 1–18, evaluate each number.

1. $P(6, 2)$
2. $P(8, 2)$
3. $P(8, 3)$
4. $P(6, 3)$
c 5. $P(40, 4)$
c 6. $P(36, 5)$
c 7. $P(80, 5)$
c 8. $P(120, 4)$
9. $C(10, 6)$
10. $C(8, 4)$
11. $C(40, 2)$
12. $C(40, 3)$
c 13. $C(52, 13)$
c 14. $C(48, 10)$
15. $\Gamma(10, 3, 3, 4)$
16. $\Gamma(12, 2, 2, 8)$
c 17. $\Gamma(52, 5, 5, 5, 5, 32)$
c 18. $\Gamma(52, 11, 11, 30)$

19. In how many ways can a president and a vice-president be elected from a committee of ten people?
20. Eight horses are running for first, second, or third place. Find the number of different ways in which the first, second, and third place can be decided, assuming that there are no ties.
21. Ten athletes are competing for a gold, silver, or bronze medal in high jump. How many outcomes can this competition have?
22. Twelve people form a committee. In how many different ways can a chairperson, a vice-chairperson, a secretary, and a treasurer be selected?
23. How many three-digit numbers can be obtained from the digits 1, 2, 3, and 4? How many four-digit numbers? (No digit can be used more than once.)
24. How many ordered arrangements containing three letters can you form by using the vowels a, e, i, o, u? How many containing five letters?

(No repetitions of letters in an arrangement are allowed.)

25. If the letters in the word SMILEY are used to form a three-letter code word, find how many code words can be obtained.
26. The numbers 5, 6, 7, 8, and 9 are being used to form three- and four-digit numbers in which no digit is repeated. How many three-digit numbers can be formed? How many four-digit numbers?
27. How many five-person subcommittees can be formed from a committee of twelve people?
28. A box has space available for four books. In how many ways can 8 books be arranged in the box?
29. How many 7-card hands are possible from a standard 52-card deck?
30. How many 9-card hands are possible from a standard 52-card deck?
31. A baseball league is organized with 8 teams. If each team is to play every other team exactly once, find how many games must be scheduled.
32. How many line segments can you draw joining two points at a time if you are given nine points, no three of which lie on a straight line?
33. Ten members of a committee are to be assigned to three subcommittees of three, three, and four people. In how many ways can this be done if each person is to be assigned to only one subcommittee?
34. In how many ways can fifteen people be divided

into three committees of 4, 5, and 6 people, respectively?

35. Four persons are playing bridge, a card game where each player is dealt 13 cards from a 52-card deck. How many different deals are possible?

36. If four people are playing poker and each is dealt 7 cards from a 52-card deck, find how many different deals are possible.

37. Find the number of distinguishable permutations obtained with the letters of the word MISSISSIPPI.

38. How many anagrams can you form with the letters of the word TENNESSEE?

39. Eight boys and six girls were elected to a student committee. How many 4-student subcommittees can be formed? How many subcommittees consisting of 2 boys and 2 girls can be formed?

40. A student council consists of ten girls and six boys. How many 6-student committees are possible? How many committees consisting of 4 girls and 2 boys can be formed.

9.7 ELEMENTARY PROBABILITY

Probability is a mathematical theory whose applications touch nearly every area of human activity. The theory of probability is used in actuarial science, medical statistics, physics and astronomy, meteorology, game theory, decision-making in the business world, and so on. It deals with situations that involve an uncertain future.

In probability, an *experiment* is any process that generates well-defined *outcomes*. For example, the experiment of flipping a coin has two possible outcomes: the coin lands heads up (H) or tails up (T). The tossing of a die has six possible outcomes: the number appearing on the upper face can be 1, 2, 3, 4, 5, or 6.

In probability theory we seek answers to questions of the following type. If a coin is flipped, what are the chances of obtaining heads? If a die is tossed, what is the likelihood of rolling a 5? What is the probability of obtaining four aces if you are dealt five cards from a deck of 52 playing cards?

Throughout this section we shall discuss a few elementary ideas in probability and restrict our discussion to experiments whose outcomes are *equally likely*. That is, if a coin is flipped, the chance that it lands heads up is the same as that of landing tails up. In the experiment of tossing a die, we assume that the probability of any of the six outcomes is exactly the same. The coin and the die are then said to be *fair*.

A *sample space* of an experiment is the set of all possible outcomes. For example, the flipping of a coin has two outcomes H and T, so that the sample space of this experiment is

$$S = \{H, T\}.$$

Analogously, the sample space for the experiment of tossing a die is

$$S = \{1, 2, 3, 4, 5, 6\}.$$

We shall consider only *finite* sample spaces.

An *event* is any subset of the sample space. For example, $E = \{H\}$, a subset of $S = \{H, T\}$, is the event of obtaining heads after flipping a coin. If you toss a die and observe a 5 appearing on the upper face, this is an event denoted by $E_1 = \{5\}$. The event $E_2 = \{1, 2, 3, 4\}$ could be interpreted as observing a number less than 5.

We now give the following definition.

***

Probability of an Event

Let S be a sample space and let E be an event. If $n(S)$ and $n(E)$ denote the number of elements of S and E, respectively, then the *probability of the event E* is defined by the ratio

$$(9.25) \quad P(E) = \frac{n(E)}{n(S)}.$$

***

Since E is always a subset of S, it follows that $n(E) \leq n(S)$, so $0 \leq P(E) \leq 1$. Thus, the probability of any event is always a number less than or equal to 1.

For instance, the sample space of the experiment of flipping a coin is, as we already know, $S = \{H, T\}$, so $n(S) = 2$. If $E = \{H\}$, then $n(E) = 1$. Thus in the flipping of a coin, the probability that heads will turn up is $P(E) = 1/2$. Suppose that we are tossing a die. What is the probability that a number less than 5 will appear on the upper face of the die? As we said above, the subset $E_2 = \{1, 2, 3, 4\}$ represents the event of observing a number less than 5. Since $S = \{1, 2, 3, 4, 5, 6\}$, it follows that $P(E_2) = n(E_2)/n(S) = 4/6 = 2/3$.

◆ **Example 1.**   Two dice are being tossed. What is the probability that   **a)** the sum of the numbers is 5   **b)** the sum of the numbers is 7?

Solution.   Suppose that one die is white and the other is yellow, and we use ordered pairs to represent the outcome of each toss as follows. The pair (2, 3) indicates that a 2 has appeared on the upper face of the white die and a 3 on the yellow die. Now, if 1 appears on the white die, we may get one of the six pairs

$$(1, 1), (1, 2), (1, 3), (1, 4), (1, 5), \text{ and } (1, 6).$$

Similarly, if a 2 shows up on the white die there are six possibilities for the number on the yellow die:

$$(2, 1), (2, 2), (2, 3), (2, 4), (2, 5), \text{ and } (2, 6).$$

Reasoning in the same way with the numbers 3, 4, 5, and 6 we conclude that the total number of ordered pairs is 36. Thus if S is the sample space of this experiment, then $n(S) = 36$.

a) The event E_1 corresponding to the sum of the numbers being 5 is given by the subset

$$E_1 = \{(1, 4), (2, 3), (3, 2), (4, 1)\}.$$

Since $n(E_1) = 4$, we see that

$$P(E_1) = \frac{4}{36} = \frac{1}{9}.$$

This is the probability that the sum of the numbers is 5.

b) The following set contains all pairs whose sum is 7:

$$E_2 = \{(1, 6), (2, 5), (3, 4), (4, 3), (5, 2), (6, 1)\}.$$

We have

$$P(E_2) = \frac{n(E_2)}{n(S)} = \frac{6}{36} = \frac{1}{6}.$$

That is, the probability that the sum of numbers is 7 is 1/6.

◇ **Practice Exercise 1.** If two coins are flipped, find the probability that **a)** both coins will turn up tails **b)** one coin will turn up heads and the other tails.

Answer. **a)** $\dfrac{1}{4}$ **b)** $\dfrac{1}{2}$.

c ◆ **Example 2.** You are dealt five cards from a deck of 52 playing cards. What is the probability of getting four aces?

Solution. Denote by S the sample space of all 5-card hands from the 52-card deck. Since the order of the cards in a hand is immaterial, it follows that $n(S) = C(52, 5)$. There are four aces in a 52-card deck. If the four aces are dealt, then the fifth card can be any one of the 48 remaining cards. Since the event E is getting all four aces, it follows that $n(E) = 48$. Thus

$$P(E) = \frac{n(E)}{n(S)} = \frac{48}{C(52, 5)}$$

$$= \frac{48 \cdot (5!47!)}{52!} \approx 1.85 \times 10^{-5}.$$

c ◇ **Practice Exercise 2.** An experiment consists of drawing 5 cards from a deck of 52 playing cards. What is the probability of getting three jacks?

Answer. $\dfrac{C(4, 3)C(48, 2)}{C(52, 5)} \simeq 1.736 \times 10^{-3}.$

Mutually Exclusive Events

Suppose that you are tossing two dice. Let E_1 be the event of obtaining 5 as the sum of the numbers, and E_2 of obtaining the sum 7. These two events are *mutually exclusive*: you cannot obtain the sum 5 *and* the sum 7 at the same time. As we saw in Example 1,

$$E_1 = \{(1, 4), (2, 3), (3, 2), (4, 1)\},$$
$$E_2 = \{(1, 6), (2, 5), (3, 4), (4, 3), (5, 2), (6, 1)\}$$

and these two sets have no element in common.

In general, let S be the sample space of an experiment and let E_1 and E_2 be two events associated with the experiment. The events E_1 and E_2 are said to be *mutually exclusive* if the sets E_1 and E_2 have no elements in common. In this case, one may also say that the sets E_1 and E_2 are *disjoint*.

Let E be the *union* of these sets, denoted by $E = E_1 \cup E_2$. Since $n(E) = n(E_1) + n(E_2)$, it follows that

$$P(E) = \frac{n(E)}{n(S)} = \frac{n(E_1) + n(E_2)}{n(S)} = \frac{n(E_1)}{n(S)} + \frac{n(E_2)}{n(S)},$$

so

$$P(E) = P(E_1) + P(E_2).$$

Thus the probability of the union of two mutually exclusive (or disjoint) events is the sum of their respective probabilities. Since this result depends upon an obvious fact about counting—that the number of elements in the union of several disjoint sets is the sum of the number of elements in the respective sets—it generalizes to any finite family of mutually exclusive events as follows.

The Probability of a Union of Mutually Exclusive Events

Let $E_1, E_2, \ldots, E_k$ be mutually exclusive events associated with a certain experiment. If

$$E = E_1 \cup E_2 \cup \cdots \cup E_k,$$

then

(9.26) $P(E) = P(E_1) + P(E_2) + \cdots + P(E_k).$

◆ **Example 3.**   Two dice are being tossed. What is the probability that the sum of the numbers is 5 or 7?

Solution.   As we observed above, the two events

$$E_1 = \{(1, 4), (2, 3), (3, 2), (4, 1)\}$$

and

$$E_2 = \{(1, 6), (2, 5), (3, 4), (4, 3), (5, 2), (6, 1)\}$$

are mutually exclusive. The event of obtaining the sum 5 or 7 is the union $E = E_1 \cup E_2$ of these events. Since $n(E_1) = 4$, $n(E_2) = 6$, and $n(S) = 36$ (see Example 1), we obtain from (9.26)

$$P(E) = P(E_1) + P(E_2)$$
$$= \frac{4}{36} + \frac{6}{36}$$
$$= \frac{1}{9} + \frac{1}{6}$$
$$= \frac{5}{18}$$

which is the probability that the sum of the numbers is 5 or 7.

◇ **Practice Exercise 3.** One card is drawn from a 52-card deck. What is the probability of obtaining an ace or a king?

Answer. $\dfrac{4}{52} + \dfrac{4}{52} = \dfrac{2}{13}$.

 EXERCISES 9.7

A six-sided die is tossed. Find the probability of each of the following events.

1. The number showing is odd.
2. The number showing is prime.
3. The number showing is less than 5.
4. The number showing is greater than or equal to 3.
5. The number showing is less than 7.
6. The number showing is greater than 7.

In Exercises 7–14, assume that two dice are tossed and we observe the sum of the numbers showing on the upper faces. Find the probability that the sum satisfies the condition in each of the following cases.

7. Equal to 6.
8. Equal to 7.
9. Less than or equal to 7.
10. Greater than 6.
11. Equal to a prime number.
12. Equal to a multiple of 4.
13. Equal to 7 and at least 11.
14. At most 5 or equal to 10.

15. If two coins are flipped, what is the probability that both coins turn up heads?
16. If two coins are flipped, what is the probability that one coin turns up heads and the other tails?

Exercises 17–20 refer to the following experiment. Marbles marked with the numbers 1, 2, 3, 4, and 5 are placed in a box. After the contents are mixed well, two marbles are simultaneously selected from the box. Find a sample space S with equally likely outcomes for this experiment. Also, find the probability of each of the following events.

17. One marble is marked with an odd number and the other with an even number.

18. Both marbles are marked with an odd number.
19. The sum of the numbers is at least 9.
20. The sum of the numbers is at most 9.

In Exercises 21–24, consider the following situation. You are taking a three-question true-false quiz, and you are guessing all the answer. Produce a sample space for all your possible answers, and find the probability of each of the following events.

21. Your three guesses are right.
22. Only one of your guesses is right.
23. At least two of your guesses are right.
24. Exactly two of your guesses are right.

25. There are five questions on a true-false test. If a student guesses the answer for each question, find the probability that a) 5 answers are correct b) three answers are correct and two are incorrect.
26. Refer to Exercise 25. Find the probability that a) two answers are correct and three are incorrect b) at least three answers are correct.

In Exercises 27–32, our experiment consists of dealing five cards from a deck of 52 playing cards.

c 27. Find the probability of being dealt five cards of the same suit.
c 28. What is the probability of being dealt four cards of the same suit?
c 29. What is the probability of being dealt 3 aces and 2 jacks?
c 30. Find the probability of being dealt 2 aces and three kings.
c 31. In poker, a *straight flush* is a hand of all five cards of the same suit and in consecutive sequence, as the 6, 7, 8, 9, and 10 of hearts. What is the probability of being dealt a straight flush?

c **32.** Referring to Exercise 31, a straight flush 10, J, Q, K, A of one suit is called a *royal flush*. Find the probability of being dealt a royal flush.

33. One card is drawn from a 52-card deck. What is the probability that the card is   a) a 5 or a jack?   b) a diamond or a spade?

34. Two dice are rolled. Find the probability that a) the sum of the numbers is 6 or 10   b) the sum of the numbers is no less than 11 or no more than 3.

35. A box contains 4 white, 6 red, and 8 marbles. If a marble is drawn at random from the box, find the probability that   a) a red marble is drawn   b) a white or a blue marble is drawn.

36. In Exercise 35, what is the probability of a) drawing a blue marble   b) drawing a white or a red marble?

37. Denise and Pat are members of an eight-person committee. If a two-person subcommittee is to be selected at random, what is the probability that Denise and Pat will be selected?

38. Two convention officers are to be selected from a group of five to form a subcommittee to check delegate credentials. Find the probability that Judy and Lynn (both convention officers) will be selected.

39. Three coffee varieties, Java, Mocha, and Santos, are being tasted by a person who has no ability to distinguish a difference between coffees. If the person is asked to rank the three varieties according to taste preference, what is the probability that the person will rank Santos as best?

40. In Exercise 39, what is the probability that the person will rank Santos as best and Mocha as second best?

CHAPTER SUMMARY

The *axiom of induction* is one of the foundations of the natural number system. It implies the *principle of mathematical induction*, one of the most important methods of proof in mathematics. As an application of mathematical induction, we proved the *Binomial Theorem*, which gives a general formula for a power of a binomial. The *binomial expansion* of $(a + b)^n$ is a sum of products of powers of a and b, multiplied by certain coefficients known as the *binomial coefficients*. These are the numbers of *combinations of k objects from a set of n objects*. The binomial coefficients have some interesting properties, such as the formation of *Pascal's triangle*.

A *real sequence* is a function whose domain is the set of natural numbers. With every sequence we can associate the *partial sums* of a *series*. Of particular importance are *arithmetic* and *geometric* sequences and series. In an arithmetic sequence the difference of two consecutive terms is a constant called the *common difference* of the sequence. In a geometric sequence the quotient of a term by the preceding term is a constant called the *common ratio* of the sequence. If the absolute value of the ratio of a geometric sequence is less than one, then the sum of the corresponding *infinite series* is a finite number that can be easily computed. This important result can be used, among other things, to write repeating decimal numbers as fractions.

Given a set of *n* objects, we can arrange them into subsets of *k* objects in different ways. Ordered arrangements are called *permutations*. If the order is immaterial, then the arrangement is called a *combination*. By using the *Fundamental Principle of Counting*, we can derive formulas that give the number of *permutations or combinations of k objects from a set of n objects*, the number of ways in which *n* objects can be *partitioned in k cells*, and the number of *distinguishable permutations*.

The *theory of probability* deals with *experiments*, which have different *outcomes*. A *sample space* of an experiment is the set of all possible outcomes. We have considered only *finite* sample spaces. An *event* is any subset of the sample space. The *probability of an event* is the quotient of the number of elements of an event by the number of elements of the sample space. The theory of probability is of great importance in mathematics and applied sciences. It has applications in nearly every area of human activity.

REVIEW EXERCISES

Evaluate:

1. $\binom{18}{15}$

2. $\binom{25}{24}$

Expand using the Binomial Theorem:

3. $\left(\dfrac{a}{2} + 2b\right)^5$

4. $\left(3x - \dfrac{y}{3}\right)^6$

5. Find the sixth term in the expansion of $(2u - b)^7$.

6. Find the eighth term in the expansion of $(3a - 4b)^{10}$.

7. Find the middle term in the expansion of $(3x - 4)^6$.

8. Find the two middle terms in the expansion of $\left(2x + \dfrac{a}{2}\right)^7$.

9. Find the term containing a^2 in the expansion of $(a^{1/2} + b^{1/2})^8$.

10. Find the term containing u in the expansion of $(u^{1/3} - 2)^7$.

c 11. Use the Binomial Theorem to evaluate $(1.02)^6$.

c 12. Use the Binomial Theorem to evaluate $(0.99)^5$.

In Exercises 13–18, write the first five terms of the sequences.

13. $a_n = 5 - \dfrac{n}{2}$

14. $a_n = \dfrac{n}{3} - 7$

15. $a_n = \dfrac{(-1)^{n-1}}{n!}$

16. $a_n = (-1)^{n+1} n^3$

17. $a_n = \dfrac{(-1)^n(n-1)}{n+1}$

18. $a_n = \dfrac{1}{(2n+1)!}$

In Exercises 19–22, find the general term a_n of a sequence, given the general partial sum s_n of the series whose terms are the $\{a_k\}$.

19. $s_n = 4^n + 1$

20. $s_n = \dfrac{n(3n+1)}{2}$

21. $s_n = \dfrac{n(n+1)(n+2)}{3}$

22. $s_n = \dfrac{n(n+1)(2n+1)}{6}$

c 23. Use Newton's method to find four decimal approximations for $\sqrt{7}$. Take as initial guess $x_1 = 2.1$.

c 24. Use Newton's method to find four decimal approximations for $\sqrt{8}$. Take $x_1 = 3.1$ as initial guess.

In Exercises 25–28, find the common difference and determine the next two terms of each arithmetic progression.

25. $10, 7, 4, \ldots$

26. $2, 0.5, -1, \ldots$

27. $\ln 3, \ln 1, \ln 3^{-1}$

28. $\log 10, \log 100, \log 1000, \ldots$

29. If 6 and 11 are the third and fifth terms of an arithmetic sequence, find the tenth term.

30. If 3.25 and 2.5 are the fourth and seventh terms of an arithmetic sequence, find the second term.

31. How many natural numbers divisible by 6 are there between 15 and 81? Find their sum.

32. Find the sum of all the numbers divisible by 8 that are between 50 and 122.

In Exercises 33–36, find the ratio and determine the next two terms of the geometric progression.

33. $\dfrac{3}{5}, \dfrac{1}{2}, \dfrac{5}{12}, \ldots$

34. $\dfrac{4}{5}, \dfrac{7}{10}, \dfrac{49}{80}, \ldots$

35. $5, 7.5, 11.25, \ldots$

36. $5, 1, 0.2, \ldots$

37. The third and fifth terms of a geometric progression are 4/9 and 16/81. Find the second term.

38. The second and fourth terms of a geometric progression are 80 and 5. Find the sixth term.

Find the sum of each series:

39. $\displaystyle\sum_{n=1}^{\infty} 3\left(-\frac{1}{3}\right)^{n-1}$

40. $\displaystyle\sum_{n=1}^{\infty} 5\left(\frac{2}{5}\right)^{n-1}$

Write the repeating decimals as fractions:

41. $0.\overline{27}$

42. $0.\overline{54}$

43. $0.0\overline{24}$

44. $0.1\overline{72}$

In Exercises 45–52, prove each proposition by mathematical induction.

45. $4 + 8 + 12 + \cdots + 4n = 2n(n+1)$
for all $n \in \mathbb{N}$.

46. $\displaystyle\sum_{j=1}^{\infty} \frac{1}{(2j-1)(2j+1)} = \frac{n}{2n+1}$ for all $n \in \mathbb{N}$.

47. $a(b_1 + \cdots + b_n) = ab_1 + \cdots + ab_n$
for all $n \geq 2$.

48. $\log(a_1 \cdot a_2 \cdot \ldots \cdot a_n) = \log a_1 + \log a_2 + \cdots + \log a_n$, $n \geq 2$.

49. $(1 + x)^n \geq 1 + nx$ if $x > -1$, for all natural numbers n.

50. $(a + b)^n > a^n + b^n$, a and b positive, $n > 1$.

51. $a + b$ is a factor of $a^{2n+1} + b^{2n+1}$ for every natural number n.

52. $x^{2n} - a^{2n}$ is divisible by $x + a$.

53. Prove that $\dbinom{n}{0} + \dbinom{n}{1} + \dbinom{n}{2} + \cdots + \dbinom{n}{n} = 2^n$
for all n.
(*Hint:* Write $2^n = (1 + 1)^n$ and expand using the Binomial Theorem.)

54. Prove that $\displaystyle\sum_{k=0}^{n} (-1)^k \dbinom{n}{k} = 0$ for all n.

55. A free-falling object travels 16 ft during the first second, 48 ft during the second second, and so on. How many feet will the object fall during the eleventh second? How many feet will it fall in 11 seconds?

56. An auditorium has 15 rows of seats. If there are 20 seats in the first row, 22 in the second row, 24 in the third row, and so on, find the total number of seats.

57. In a tapered ladder each rung, from bottom to top, is 1/8 inch shorter than the previous one. If the bottom rung is 12 inches long and the ladder has 11 rungs, find the length of the top rung. Assuming no waste, how many inches of rung material would be needed to build the ladder?

58. The number of bacteria in a colony doubles after each hour. If the initial number of bacteria is 1800, what is the number of bacteria in the colony after 6 hours? After $7\frac{1}{2}$ hours?

● **59.** A weight at the end of a line swings through an arc 2 meters long in its first swing. On each subsequent swing the weight travels 0.8 the length of the preceding swing. How far will the weight travel before coming to rest?

● **60.** To stimulate the economy of a depressed area, the government distributes $2,000,000 through a subsidy program. Assuming that each institution or individual will spend 80% of the amount received, and that 80% of this will be spent next, and so on, find the total amount of money that will be spent as a result of the program.

61. Evaluate the following numbers.
a) $P(8, 3)$ b) $P(10, 6)$
c) $C(10, 5)$ d) $C(12, 4)$
e) $\Gamma(8, 4, 3, 1)$ f) $\Gamma(10, 6, 2, 2)$

c **62.** Evaluate each of the following numbers.
a) $P(30, 4)$ b) $P(96, 3)$
c) $C(15, 8)$ d) $C(20, 10)$
e) $\Gamma(18, 8, 6, 4)$ f) $\Gamma(20, 9, 5, 3, 3)$

63. In how many ways can four flags of different colors be arranged on a pole?

64. In how many ways can five candidates be listed on a ballot?

65. How many distinct lines can be drawn through eight points, no three of which lie on the same line?

66. Three noncollinear points determine a circle. How many distinct circles can be drawn through eight points, no three of which are collinear?

67. A junior soccer league has 8 teams. How many games must be scheduled if each team has to play every other team twice?

68. The technique of paired comparisons is often used in panel tests of consumer products. In this technique each product is paired with each of the other products and the comparison is then made. If a sample consists of 8 products, find how many paired comparisons can be made.

c **69.** In how many ways can a jury of 12 people be selected from a group of 18 prospective jurors? (Assume that the order of selection is immaterial.)

c **70.** In several states the automobile license plate

numbering system consists of three letters followed by three digits. If no repetitions are permitted, find the total number of distinct license plates.

In Exercises 71–72, two dice are rolled. Find the probability of each event.

71. The sum of the numbers is greater than or equal to 8.

72. The sum of the numbers is equal to 7 or at least 10.

In Exercises 73–74, six marbles numbered 1, 2, 3, 4, 5, and 6 are placed in a box. After the marbles are mixed, two are simultaneously picked up. Find the probability of each event.

73. Both marbles are marked with an even number.

74. The sum of the numbers is at most 8.

75. Jack and Jill belong to a 20-member hiking club. Every year two people are selected at random from the club to serve as secretary and treasurer. What is the probability that Jack is selected as secretary and Jill as treasurer?

76. Find the probability of guessing ten correct answers on a ten-question true-false test.

77. Find the probability of being dealt a five-card hand of three diamonds and two spades from a 52-card deck.

78. Five cards are dealt from a standard 52-card deck. What is the probability of obtaining three jacks?

79. A subcommittee of five people is to be chosen from a student committee that consists of 5 girls and 5 boys. What is the probability that there will be three girls and two boys in the subcommittee?

80. A *full house* consists of three cards of one rank (such as three eights or three queens) and two cards of another rank (such as two fours or two kings). What is the probability of being dealt a full house from a standard 52-card deck in 5-card poker?

Tables _____

TABLE 1 Values of e^x and e^{-x}

Source: Raymond A. Barnett,
College Algebra and Trigonometry,
McGraw-Hill Book Co.

TABLE 2 Common Logarithms

Source: W. Fleming and D. Varberg,
Algebra and Trigonometry,
Prentice-Hall, Inc.

TABLE 3 Natural Logarithms

Source: W. Fleming and D. Varberg,
Algebra and Trigonometry,
Prentice-Hall, Inc.

Tables _____

TABLE 1 Values of e^x and e^{-x}

x	e^x	e^{-x}	x	e^x	e^{-x}	x	e^x	e^{-x}
0.00	1.000	1.000	0.35	1.419	0.705	0.70	2.014	0.497
0.01	1.010	0.990	0.36	1.433	0.698	0.71	2.034	0.492
0.02	1.020	0.980	0.37	1.448	0.691	0.72	2.054	0.487
0.03	1.031	0.970	0.38	1.462	0.684	0.73	2.075	0.482
0.04	1.041	0.961	0.39	1.477	0.677	0.74	2.096	0.477
0.05	1.051	0.951	0.40	1.492	0.670	0.75	2.117	0.472
0.06	1.062	0.942	0.41	1.507	0.664	0.76	2.138	0.468
0.07	1.073	0.932	0.42	1.522	0.657	0.77	2.160	0.463
0.08	1.083	0.923	0.43	1.537	0.651	0.78	2.182	0.458
0.09	1.094	0.914	0.44	1.553	0.644	0.79	2.203	0.454
0.10	1.105	0.905	0.45	1.568	0.638	0.80	2.226	0.449
0.11	1.116	0.896	0.46	1.584	0.631	0.81	2.248	0.445
0.12	1.127	0.887	0.47	1.600	0.625	0.82	2.270	0.440
0.13	1.139	0.878	0.48	1.616	0.619	0.83	2.293	0.436
0.14	1.150	0.869	0.49	1.632	0.613	0.84	2.316	0.432
0.15	1.162	0.861	0.50	1.649	0.607	0.85	2.340	0.427
0.16	1.174	0.852	0.51	1.665	0.600	0.86	2.363	0.423
0.17	1.185	0.844	0.52	1.682	0.595	0.87	2.387	0.419
0.18	1.197	0.835	0.53	1.699	0.589	0.88	2.411	0.415
0.19	1.209	0.827	0.54	1.716	0.583	0.89	2.435	0.411
0.20	1.221	0.819	0.55	1.733	0.577	0.90	2.460	0.407
0.21	1.234	0.811	0.56	1.751	0.571	0.91	2.484	0.403
0.22	1.246	0.803	0.57	1.768	0.566	0.92	2.509	0.399
0.23	1.259	0.795	0.58	1.786	0.560	0.93	2.535	0.395
0.24	1.271	0.787	0.59	1.804	0.554	0.94	2.560	0.391
0.25	1.284	0.779	0.60	1.822	0.549	0.95	2.586	0.387
0.26	1.297	0.771	0.61	1.840	0.543	0.96	2.612	0.383
0.27	1.310	0.763	0.62	1.859	0.538	0.97	2.638	0.379
0.28	1.323	0.756	0.63	1.878	0.533	0.98	2.664	0.375
0.29	1.336	0.748	0.64	1.896	0.527	0.99	2.691	0.372
0.30	1.350	0.741	0.65	1.916	0.522	1.00	2.718	0.368
0.31	1.363	0.733	0.66	1.935	0.517	1.01	2.746	0.364
0.32	1.377	0.726	0.67	1.954	0.512	1.02	2.773	0.361
0.33	1.391	0.719	0.68	1.974	0.507	1.03	2.801	0.357
0.34	1.405	0.712	0.69	1.994	0.502	1.04	2.829	0.353

TABLE 1   (continued)

x	e^x	e^{-x}	x	e^x	e^{-x}	x	e^x	e^{-x}
1.05	2.858	0.350	1.40	4.055	0.247	1.75	5.755	0.174
1.06	2.886	0.346	1.41	4.096	0.244	1.76	5.812	0.172
1.07	2.915	0.343	1.42	4.137	0.242	1.77	5.871	0.170
1.08	2.945	0.340	1.43	4.179	0.239	1.78	5.930	0.169
1.09	2.974	0.336	1.44	4.221	0.237	1.79	5.989	0.167
1.10	3.004	0.333	1.45	4.263	0.235	1.80	6.050	0.165
1.11	3.034	0.330	1.46	4.306	0.232	1.81	6.110	0.164
1.12	3.065	0.326	1.47	4.349	0.230	1.82	6.172	0.162
1.13	3.096	0.323	1.48	4.393	0.228	1.83	6.234	0.160
1.14	3.127	0.320	1.49	4.437	0.225	1.84	6.297	0.159
1.15	3.158	0.317	1.50	4.482	0.223	1.85	6.360	0.157
1.16	3.190	0.313	1.51	4.527	0.221	1.86	6.424	0.156
1.17	3.222	0.310	1.52	4.572	0.219	1.87	6.488	0.154
1.18	3.254	0.307	1.53	4.618	0.217	1.88	6.553	0.153
1.19	3.287	0.304	1.54	4.665	0.214	1.89	6.619	0.151
1.20	3.320	0.301	1.55	4.712	0.212	1.90	6.686	0.150
1.21	3.353	0.298	1.56	4.759	0.210	1.91	6.753	0.148
1.22	3.387	0.295	1.57	4.807	0.208	1.92	6.821	0.147
1.23	3.421	0.292	1.58	4.855	0.206	1.93	6.890	0.145
1.24	3.456	0.289	1.59	4.904	0.204	1.94	6.959	0.144
1.25	3.490	0.287	1.60	4.953	0.202	1.95	7.029	0.142
1.26	3.525	0.284	1.61	5.003	0.200	1.96	7.099	0.141
1.27	3.561	0.281	1.62	5.053	0.198	1.97	7.171	0.139
1.28	3.597	0.278	1.63	5.104	0.196	1.98	7.243	0.138
1.29	3.633	0.275	1.64	5.155	0.194	1.99	7.316	0.137
1.30	3.669	0.273	1.65	5.207	0.192	2.00	7.389	0.135
1.31	3.706	0.270	1.66	5.259	0.190	2.01	7.463	0.134
1.32	3.743	0.267	1.67	5.312	0.188	2.02	7.538	0.133
1.33	3.781	0.264	1.68	5.366	0.186	2.03	7.614	0.131
1.34	3.819	0.262	1.69	5.420	0.185	2.04	7.691	0.130
1.35	3.857	0.259	1.70	5.474	0.183	2.05	7.768	0.129
1.36	3.896	0.257	1.71	5.529	0.181	2.06	7.846	0.127
1.37	3.935	0.254	1.72	5.585	0.179	2.07	7.925	0.126
1.38	3.975	0.252	1.73	5.641	0.177	2.08	8.004	0.125
1.39	4.015	0.249	1.74	5.697	0.176	2.09	8.085	0.124

TABLE 1 (continued)

x	e^x	e^{-x}	x	e^x	e^{-x}	x	e^x	e^{-x}
2.10	8.166	0.122	2.40	11.023	0.091	2.70	14.880	0.067
2.11	8.248	0.121	2.41	11.134	0.090	2.71	15.029	0.067
2.12	8.331	0.120	2.42	11.246	0.089	2.72	15.180	0.066
2.13	8.415	0.119	2.43	11.359	0.088	2.73	15.333	0.065
2.14	8.499	0.118	2.44	11.473	0.087	2.74	15.487	0.065
2.15	8.585	0.116	2.45	11.588	0.086	2.75	15.643	0.064
2.16	8.671	0.115	2.46	11.705	0.085	2.76	15.800	0.063
2.17	8.758	0.114	2.47	11.822	0.085	2.77	15.959	0.063
2.18	8.846	0.113	2.48	11.941	0.084	2.78	16.119	0.062
2.19	8.935	0.112	2.49	12.061	0.083	2.79	16.281	0.061
2.20	9.025	0.111	2.50	12.182	0.082	2.80	16.445	0.061
2.21	9.116	0.110	2.51	12.305	0.081	2.81	16.610	0.060
2.22	9.207	0.109	2.52	12.429	0.080	2.82	16.777	0.060
2.23	9.300	0.108	2.53	12.554	0.080	2.83	16.945	0.059
2.24	9.393	0.106	2.54	12.680	0.079	2.84	17.116	0.058
2.25	9.488	0.105	2.55	12.807	0.078	2.85	17.288	0.058
2.26	9.583	0.104	2.56	12.936	0.077	2.86	17.462	0.057
2.27	9.679	0.103	2.57	13.066	0.077	2.87	17.637	0.057
2.28	9.777	0.102	2.58	13.197	0.076	2.88	17.814	0.056
2.29	9.875	0:101	2.59	13.330	0.075	2.89	17.993	0.056
2.30	9.974	0.100	2.60	13.464	0.074	2.90	18.174	0.055
2.31	10.074	0.099	2.61	13.599	0.074	2.91	18.357	0.054
2.32	10.176	0.098	2.62	13.736	0.073	2.92	18.541	0.054
2.33	10.278	0.097	2.63	13.874	0.072	2.93	18.728	0.053
2.34	10.381	0.096	2.64	14.013	0.071	2.94	18.916	0.053
2.35	10.486	0.095	2.65	14.154	0.071	2.95	19.106	0.052
2.36	10.591	0.094	2.66	14.296	0.070	2.96	19.298	0.052
2.37	10.697	0.093	2.67	14.440	0.069	2.97	19.492	0.051
2.38	10.805	0.093	2.68	14.585	0.069	2.98	19.688	0.051
2.39	10.913	0.092	2.69	14.732	0.068	2.99	19.886	0.050
						3.00	20.086	0.050

Source: Raymond A. Barnett, *College Algebra and Trigonometry*, McGraw-Hill Book Co.

TABLE 2   Common Logarithms

n	.00	.01	.02	.03	.04	.05	.06	.07	.08	.09
1.0	.0000	.0043	.0086	.0128	.0170	.0212	.0253	.0294	.0334	.0374
1.1	.0414	.0453	.0492	.0531	.0569	.0607	.0645	.0682	.0719	.0755
1.2	.0792	.0828	.0864	.0899	.0934	.0969	.1004	.1038	.1072	.1106
1.3	.1139	.1173	.1206	.1239	.1271	.1303	.1335	.1367	.1399	.1430
1.4	.1461	.1492	.1523	.1553	.1584	.1614	.1644	.1673	.1703	.1732
1.5	.1761	.1790	.1818	.1847	.1875	.1903	.1913	.1959	.1987	.2014
1.6	.2041	.2068	.2095	.2122	.2148	.2175	.2201	.2227	.2253	.2279
1.7	.2304	.2330	.2355	.2380	.2405	.2430	.2455	.2480	.2504	.2529
1.8	.2553	.2577	.2601	.2625	.2648	.2672	.2695	.2718	.2742	.2765
1.9	.2788	.2810	.2833	.2856	.2878	.2900	.2923	.2945	.2967	.2989
2.0	.3010	.3032	.3054	.3075	.3096	.3118	.3139	.3160	.3181	.3201
2.1	.3222	.3243	.3263	.3284	.3304	.3324	.3345	.3365	.3385	.3404
2.2	.3424	.3444	.3464	.3483	.3502	.3522	.3541	.3560	.3579	.3598
2.3	.3617	.3636	.3655	.3674	.3692	.3711	.3729	.3747	.3766	.3784
2.4	.3802	.3820	.3838	.3856	.3874	.3892	.3909	.3927	.3945	.3962
2.5	.3979	.3997	.4014	.4031	.4048	.4065	.4082	.4099	.4116	.4133
2.6	.4150	.4166	.4183	.4200	.4216	.4232	.4249	.4265	.4281	.4298
2.7	.4314	.4330	.4346	.4362	.4378	.4393	.4409	.4425	.4440	.4456
2.8	.4472	.4487	.4502	.4518	.4533	.4548	.4564	.4579	.4594	.4609
2.9	.4624	.4639	.4654	.4669	.4683	.4698	.4713	.4728	.4742	.4757
3.0	.4771	.4786	.4800	.4814	.4829	.4843	.4857	.4871	.4886	.4900
3.1	.4914	.4928	.4942	.4955	.4969	.4983	.4997	.5011	.5024	.5038
3.2	.5051	.5065	.5079	.5092	.5105	.5119	.5132	.5145	.5159	.5172
3.3	.5185	.5198	.5211	.5224	.5237	.5250	.5263	.5276	.5289	.5302
3.4	.5315	.5328	.5340	.5353	.5366	.5378	.5391	.5403	.5416	.5428
3.5	.5441	.5453	.5465	.5478	.5490	.5502	.5514	.5527	.5539	.5551
3.6	.5563	.5575	.5587	.5599	.5611	.5623	.5635	.5647	.5658	.5670
3.7	.5682	.5694	.5705	.5717	.5729	.5740	.5752	.5763	.5775	.5786
3.8	.5798	.5809	.5821	.5832	.5843	.5855	.5866	.5877	.5888	.5899
3.9	.5911	.5922	.5933	.5944	.5955	.5966	.5977	.5988	.5999	.6010
4.0	.6021	.6031	.6042	.6053	.6064	.6075	.6085	.6096	.6107	.6117
4.1	.6128	.6138	.6149	.6160	.6170	.6180	.6191	.6201	.6212	.6222
4.2	.6232	.6243	.6253	.6263	.6274	.6284	.6294	.6304	.6314	.6325
4.3	.6335	.6345	.6355	.6365	.6375	.6385	.6395	.6405	.6415	.6425
4.4	.6435	.6444	.6454	.6464	.6474	.6484	.6493	.6503	.6513	.6522
4.5	.6532	.6542	.6551	.6561	.6571	.6580	.6590	.6599	.6609	.6618
4.6	.6628	.6637	.6646	.6656	.6665	.6675	.6684	.6693	.6702	.6712
4.7	.6721	.6730	.6739	.6749	.6758	.6767	.6776	.6785	.6794	.6803
4.8	.6812	.6821	.6830	.6839	.6848	.6857	.6866	.6875	.6884	.6893
4.9	.6902	.6911	.6920	.6928	.6937	.6946	.6955	.6964	.6972	.6981
5.0	.6990	.6998	.7007	.7016	.7024	.7033	.7042	.7050	.7059	.7067
5.1	.7076	.7084	.7093	.7101	.7110	.7118	.7126	.7135	.7143	.7152
5.2	.7160	.7168	.7177	.7185	.7193	.7202	.7210	.7218	.7226	.7235
5.3	.7243	.7251	.7259	.7267	.7275	.7284	.7292	.7300	.7308	.7316
5.4	.7324	.7332	.7340	.7348	.7356	.7364	.7372	.7380	.7388	.7396

TABLE 2 (continued)

n	.00	.01	.02	.03	.04	.05	.06	.07	.08	.09
5.5	.7404	.7412	.7419	.7427	.7435	.7443	.7451	.7459	.7466	.7474
5.6	.7482	.7490	.7497	.7505	.7513	.7520	.7528	.7536	.7543	.7551
5.7	.7559	.7566	.7574	.7582	.7589	.7597	.7604	.7612	.7619	.7627
5.8	.7634	.7642	.7649	.7657	.7664	.7672	.7679	.7686	.7694	.7701
5.9	.7709	.7716	.7723	.7731	.7738	.7745	.7752	.7760	.7767	.7774
6.0	.7782	.7789	.7796	.7803	.7810	.7818	.7825	.7832	.7839	.7846
6.1	.7853	.7860	.7868	.7875	.7882	.7889	.7896	.7903	.7910	.7917
6.2	.7924	.7931	.7938	.7945	.7952	.7959	.7966	.7973	.7980	.7987
6.3	.7993	.8000	.8007	.8014	.8021	.8028	.8035	.8041	.8048	.8055
6.4	.8062	.8069	.8075	.8082	.8089	.8096	.8102	.8109	.8116	.8122
6.5	.8129	.8136	.8142	.8149	.8156	.8162	.8169	.8176	.8182	.8189
6.6	.8195	.8202	.8209	.8215	.8222	.8228	.8235	.8241	.8248	.8254
6.7	.8261	.8267	.8274	.8280	.8287	.8293	.8299	.8306	.8312	.8319
6.8	.8325	.8331	.8338	.8344	.8351	.8357	.8363	.8370	.8376	.8382
6.9	.8388	.8395	.8401	.8407	.8414	.8420	.8426	.8432	.8439	.8445
7.0	.8451	.8457	.8463	.8470	.8476	.8482	.8488	.8494	.8500	.8506
7.1	.8513	.8519	.8525	.8531	.8537	.8543	.8549	.8555	.8561	.8567
7.2	.8573	.8579	.8585	.8591	.8597	.8603	.8609	.8615	.8621	.8627
7.3	.8633	.8639	.8645	.8651	.8657	.8663	.8669	.8675	.8681	.8686
7.4	.8692	.8698	.8704	.8710	.8716	.8722	.8727	.8733	.8739	.8745
7.5	.8751	.8756	.8762	.8768	.8774	.8779	.8785	.8791	.8797	.8802
7.6	.8808	.8814	.8820	.8825	.8831	.8837	.8842	.8848	.8854	.8859
7.7	.8865	.8871	.8876	.8882	.8887	.8893	.8899	.8904	.8910	.8915
7.8	.8921	.8927	.8932	.8938	.8943	.8949	.8954	.8960	.8965	.8971
7.9	.8976	.8982	.8987	.8993	.8998	.9004	.9009	.9015	.9020	.9025
8.0	.9031	.9036	.9042	.9047	.9053	.9058	.9063	.9069	.9074	.9079
8.1	.9085	.9090	.9096	.9101	.9106	.9112	.9117	.9122	.9128	.9133
8.2	.9138	.9143	.9149	.9154	.9159	.9165	.9170	.9175	.9180	.9186
8.3	.9191	.9196	.9201	.9206	.9212	.9217	.9222	.9227	.9232	.9238
8.4	.9243	.9248	.9253	.9258	.9263	.9269	.9274	.9279	.9284	.9289
8.5	.9294	.9299	.9304	.9309	.9315	.9320	.9325	.9330	.9335	.9340
8.6	.9345	.9350	.9355	.9360	.9365	.9370	.9375	.9380	.9385	.9390
8.7	.9395	.9400	.9405	.9410	.9415	.9420	.9425	.9430	.9435	.9440
8.8	.9445	.9450	.9455	.9460	.9465	.9469	.9474	.9479	.9484	.9489
8.9	.9494	.9499	.9504	.9509	.9513	.9518	.9523	.9528	.9533	.9538
9.0	.9542	.9547	.9552	.9557	.9562	.9566	.9571	.9576	.9581	.9586
9.1	.9590	.9595	.9600	.9605	.9609	.9614	.9619	.9624	.9628	.9633
9.2	.9638	.9643	.9647	.9652	.9657	.9661	.9666	.9671	.9675	.9680
9.3	.9685	.9689	.9694	.9699	.9703	.9708	.9713	.9717	.9722	.9727
9.4	.9731	.9736	.9741	.9745	.9750	.9754	.9759	.9763	.9768	.9773
9.5	.9777	.9782	.9786	.9791	.9795	.9800	.9805	.9809	.9814	.9818
9.6	.9823	.9827	.9832	.9836	.9841	.9845	.9850	.9854	.9859	.9863
9.7	.9868	.9872	.9877	.9881	.9886	.9890	.9894	.9899	.9903	.9908
9.8	.9912	.9917	.9921	.9926	.9930	.9934	.9939	.9943	.9948	.9952
9.9	.9956	.9961	.9965	.9969	.9974	.9978	.9983	.9987	.9991	.9996

Source: W. Fleming and D. Varberg, *Algebra and Trigonometry*, Prentice-Hall, Inc.

TABLE 3 Natural Logarithms

n	.00	.01	.02	.03	.04	.05	.06	.07	.08	.09
1.0	0.0000	0.0100	0.0198	0.0296	0.0392	0.0488	0.0583	0.0677	0.0770	0.0862
1.1	0.0953	0.1044	0.1133	0.1222	0.1310	0.1398	0.1484	0.1570	0.1655	0.1740
1.2	0.1823	0.1906	0.1989	0.2070	0.2151	0.2231	0.2311	0.2390	0.2469	0.2546
1.3	0.2624	0.2700	0.2776	0.2852	0.2927	0.3001	0.3075	0.3148	0.3221	0.3293
1.4	0.3365	0.3436	0.3507	0.3577	0.3646	0.3716	0.3784	0.3853	0.3920	0.3988
1.5	0.4055	0.4121	0.4187	0.4253	0.4318	0.4383	0.4447	0.4511	0.4574	0.4637
1.6	0.4700	0.4762	0.4824	0.4886	0.4947	0.5008	0.5068	0.5128	0.5188	0.5247
1.7	0.5306	0.5365	0.5423	0.5481	0.5539	0.5596	0.5653	0.5710	0.5766	0.5822
1.8	0.5878	0.5933	0.5988	0.6043	0.6098	0.6152	0.6206	0.6259	0.6313	0.6366
1.9	0.6419	0.6471	0.6523	0.6575	0.6627	0.6678	0.6729	0.6780	0.6831	0.6881
2.0	0.6931	0.6981	0.7031	0.7080	0.7130	0.7178	0.7227	0.7275	0.7324	0.7372
2.1	0.7419	0.7467	0.7514	0.7561	0.7608	0.7655	0.7701	0.7747	0.7793	0.7839
2.2	0.7885	0.7930	0.7975	0.8020	0.8065	0.8109	0.8154	0.8198	0.8242	0.8286
2.3	0.8329	0.8372	0.8416	0.8459	0.8502	0.8544	0.8587	0.8629	0.8671	0.8713
2.4	0.8755	0.8796	0.8838	0.8879	0.8920	0.8961	0.9002	0.9042	0.9083	0.9123
2.5	0.9163	0.9203	0.9243	0.9282	0.9322	0.9361	0.9400	0.9439	0.9478	0.9517
2.6	0.9555	0.9594	0.9632	0.9670	0.9708	0.9746	0.9783	0.9821	0.9858	0.9895
2.7	0.9933	0.9969	1.0006	1.0043	1.0080	1.0116	1.0152	1.0188	1.0225	1.0260
2.8	1.0296	1.0332	1.0367	1.0403	1.0438	1.0473	1.0508	1.0543	1.0578	1.0613
2.9	1.0647	1.0682	1.0716	1.0750	1.0784	1.0818	1.0852	1.0886	1.0919	1.0953
3.0	1.0986	1.1019	1.1053	1.1086	1.1119	1.1151	1.1184	1.1217	1.1249	1.1282
3.1	1.1314	1.1346	1.1378	1.1410	1.1442	1.1474	1.1506	1.1537	1.1569	1.1600
3.2	1.1632	1.1663	1.1694	1.1725	1.1756	1.1787	1.1817	1.1848	1.1878	1.1909
3.3	1.1939	1.1970	1.2000	1.2030	1.2060	1.2090	1.2119	1.2149	1.2179	1.2208
3.4	1.2238	1.2267	1.2296	1.2326	1.2355	1.2384	1.2413	1.2442	1.2470	1.2499
3.5	1.2528	1.2556	1.2585	1.2613	1.2641	1.2669	1.2698	1.2726	1.2754	1.2782
3.6	1.2809	1.2837	1.2865	1.2892	1.2920	1.2947	1.2975	1.3002	1.3029	1.3056
3.7	1.3083	1.3110	1.3137	1.3164	1.3191	1.3218	1.3244	1.3271	1.3297	1.3324
3.8	1.3350	1.3376	1.3403	1.3429	1.3455	1.3481	1.3507	1.3533	1.3558	1.3584
3.9	1.3610	1.3635	1.3661	1.3686	1.3712	1.3737	1.3762	1.3788	1.3813	1.3838
4.0	1.3863	1.3888	1.3913	1.3938	1.3962	1.3987	1.4012	1.4036	1.4061	1.4085
4.1	1.4110	1.4134	1.4159	1.4183	1.4207	1.4231	1.4255	1.4279	1.4303	1.4327
4.2	1.4351	1.4375	1.4398	1.4422	1.4446	1.4469	1.4493	1.4516	1.4540	1.4563
4.3	1.4586	1.4609	1.4633	1.4656	1.4679	1.4702	1.4725	1.4748	1.4770	1.4793
4.4	1.4816	1.4839	1.4861	1.4884	1.4907	1.4929	1.4952	1.4974	1.4996	1.5019
4.5	1.5041	1.5063	1.5085	1.5107	1.5129	1.5151	1.5173	1.5195	1.5217	1.5239
4.6	1.5261	1.5282	1.5304	1.5326	1.5347	1.5369	1.5390	1.5412	1.5433	1.5454
4.7	1.5476	1.5497	1.5518	1.5539	1.5560	1.5581	1.5602	1.5623	1.5644	1.5665
4.8	1.5686	1.5707	1.5728	1.5748	1.5769	1.5790	1.5810	1.5831	1.5831	1.5872
4.9	1.5892	1.5913	1.5933	1.5953	1.5974	1.5994	1.6014	1.6034	1.6054	1.6074
5.0	1.6094	1.6114	1.6134	1.6154	1.6174	1.6194	1.6214	1.6233	1.6253	1.6273
5.1	1.6292	1.6312	1.6332	1.6351	1.6371	1.6390	1.6409	1.6429	1.6448	1.6467
5.2	1.6487	1.6506	1.6525	1.6544	1.6563	1.6582	1.6601	1.6620	1.6639	1.6658
5.3	1.6677	1.6696	1.6715	1.6734	1.6753	1.6771	1.6790	1.6808	1.6827	1.6845
5.4	1.6864	1.6882	1.6901	1.6919	1.6938	1.6956	1.6974	1.6993	1.7011	1.7029

$$\ln(N \cdot 10^m) = \ln N + m \ln 10, \quad \ln 10 = 2.3026$$

TABLE 3 (continued)

n	.00	.01	.02	.03	.04	.05	.06	.07	.08	.09
5.5	1.7047	1.7066	1.7084	1.7102	1.7120	1.7138	1.7156	1.7174	1.7192	1.7210
5.6	1.7228	1.7246	1.7263	1.7281	1.7299	1.7317	1.7334	1.7352	1.7370	1.7387
5.7	1.7405	1.7422	1.7440	1.7457	1.7475	1.7492	1.7509	1.7527	1.7544	1.7561
5.8	1.7579	1.7596	1.7613	1.7630	1.7647	1.7664	1.7682	1.7699	1.7716	1.7733
5.9	1.7750	1.7766	1.7783	1.7800	1.7817	1.7834	1.7851	1.7867	1.7884	1.7901
6.0	1.7918	1.7934	1.7951	1.7967	1.7984	1.8001	1.8017	1.8034	1.8050	1.8066
6.1	1.8083	1.8099	1.8116	1.8132	1.8148	1.8165	1.8181	1.8197	1.8213	1.8229
6.2	1.8245	1.8262	1.8278	1.8294	1.8310	1.8326	1.8342	1.8358	1.8374	1.8390
6.3	1.8406	1.8421	1.8437	1.8453	1.8469	1.8485	1.8500	1.8516	1.8532	1.8547
6.4	1.8563	1.8579	1.8594	1.8610	1.8625	1.8641	1.8656	1.8672	1.8687	1.8703
6.5	1.8718	1.8733	1.8749	1.8764	1.8779	1.8795	1.8810	1.8825	1.8840	1.8856
6.6	1.8871	1.8886	1.8901	1.8916	1.8931	1.8946	1.8961	1.8976	1.8991	1.9006
6.7	1.9021	1.9036	1.9051	1.9066	1.9081	1.9095	1.9110	1.9125	1.9140	1.9155
6.8	1.9169	1.9184	1.9199	1.9213	1.9228	1.9242	1.9257	1.9272	1.9286	1.9301
6.9	1.9315	1.9330	1.9344	1.9359	1.9373	1.9387	1.9402	1.9416	1.9430	1.9445
7.0	1.9459	1.9473	1.9488	1.9502	1.9516	1.9530	1.9544	1.9559	1.9573	1.9587
7.1	1.9601	1.9615	1.9629	1.9643	1.9657	1.9671	1.9685	1.9699	1.9713	1.9727
7.2	1.9741	1.9755	1.9769	1.9782	1.9796	1.9810	1.9824	1.9838	1.9851	1.9865
7.3	1.9879	1.9892	1.9906	1.9920	1.9933	1.9947	1.9961	1.9974	1.9988	2.0001
7.4	2.0015	2.0028	2.0042	2.0055	2.0069	2.0082	2.0096	2.0109	2.0122	2.0136
7.5	2.0149	2.0162	2.0176	2.0189	2.0202	2.0215	2.0229	2.0242	2.0255	2.0268
7.6	2.0282	2.0295	2.0308	2.0321	2.0334	2.0347	2.0360	2.0373	2.0386	2.0399
7.7	2.0412	2.0425	2.0438	2.0451	2.0464	2.0477	0.0490	2.0503	2.0516	2.0528
7.8	2.0541	2.0554	2.0567	2.0580	2.0592	2.0605	2.0618	2.0631	2.0643	2.0656
7.9	2.0669	2.0681	2.0694	2.0707	2.0719	2.0732	2.0744	2.0757	2.0769	2.0782
8.0	2.0794	2.0807	2.0819	2.0832	2.0844	2.0857	2.0869	2.0882	2.0894	2.0906
8.1	2.0919	2.0931	2.0943	2.0956	2.0968	2.0980	2.0992	2.1005	2.1017	2.1029
8.2	2.1041	2.1054	2.1066	2.1078	2.1090	2.1102	2.1114	2.1126	2.1138	2.1150
8.3	2.1163	2.1175	2.1187	2.1199	2.1211	2.1223	2.1235	2.1247	2.1258	2.1270
8.4	2.1282	2.1294	2.1306	2.1318	2.1330	2.1342	2.1353	2.1365	2.1377	2.1389
8.5	2.1401	2.1412	2.1424	2.1436	2.1448	2.1459	2.1471	2.1483	2.1494	2.1506
8.6	2.1518	2.1529	2.1541	2.1552	2.1564	2.1576	2.1587	2.1599	2.1610	2.1622
8.7	2.1633	2.1645	2.1656	2.1668	2.1679	2.1691	2.1702	2.1713	2.1725	2.1736
8.8	2.1748	2.1759	2.1770	2.1782	2.1793	2.1804	2.1815	2.1827	2.1838	2.1849
8.9	2.1861	2.1872	2.1883	2.1894	2.1905	2.1917	2.1928	2.1939	2.1950	2.1961
9.0	2.1972	2.1983	2.1994	2.2006	2.2017	2.2028	2.2039	2.2050	2.2061	2.2072
9.1	2.2083	2.2094	2.2105	2.2116	2.2127	2.2138	2.2148	2.2159	2.2170	2.2181
9.2	2.2192	2.2203	2.2214	2.2225	2.2235	2.2246	2.2257	2.2268	2.2279	2.2289
9.3	2.2300	2.2311	2.2322	2.2332	2.2343	2.2354	2.2364	2.2375	2.2386	2.2396
9.4	2.2407	2.2418	2.2428	2.2439	2.2450	2.2460	2.2471	2.2481	2.2492	2.2502
9.5	2.2513	2.2523	2.2534	2.2544	2.2555	2.2565	2.2576	2.2586	2.2597	2.2607
9.6	2.2618	2.2628	2.2638	2.2649	2.2659	2.2670	2.2680	2.2690	2.2701	2.2711
9.7	2.2721	2.2732	2.2742	2.2752	2.2762	2.2773	2.2783	2.2793	2.2803	2.2814
9.8	2.2824	2.2834	2.2844	2.2854	2.2865	2.2875	2.2885	2.2895	2.2905	2.2915
9.9	2.2925	2.2935	2.2946	2.2956	2.2966	2.2976	2.2986	2.2996	2.3006	2.3016

Source: W. Fleming and D. Varberg, *Algebra and Trigonometry*, Prentice-Hall, Inc.

Answers to Odd-Numbered Exercises

CHAPTER 1

EXERCISES 1.1 **1.** $2 \cdot 2 \cdot 2 \cdot 2 \cdot 3 \cdot 5$ **3.** $2 \cdot 2 \cdot 2 \cdot 2 \cdot 3 \cdot 3 \cdot 3$ **5.** $2 \cdot 3 \cdot 5 \cdot 7 \cdot 11$ **7.** $11/15$ **9.** $1/10$
11. $0.\overline{428571}$ **13.** 0.525 **15.** $0.2\overline{7}$ **17.** 0.125 **19.** $3/20$ **21.** $41/333$ **23.** $23/55$ **25.** $73/66$
27. $0.5\overline{9}$ **29.** $0.\overline{324415}$

EXERCISES 1.2 **1.** Associativity of the sum. **3.** Commutativity of the product.
5. Distributivity of the product over the sum. **7.** Distributive property and commutativity of the product.
9. $11/40$ **11.** $-1/18$ **13.** $4/135$ **15.** $5/3$ **17.** $21/2$ **19.** $-5/14$ **21.** -2 **23.** $10/27$
25. $-3/56$ **27.** $7/260$ **29.** No; $\sqrt{2} + (-\sqrt{2}) = 0$, which is a rational number.
35. a) rational, b) irrational, c) rational, d) irrational. **37.** $22/7 > \pi$

EXERCISES 1.3 **1.** 14641 **3.** $1/125$ **5.** $16/9$ **7.** -512 **9.** $-1/343$ **11.** $-3/64$ **13.** $13/36$
15. $11/6$ **17.** $1/(4x^2y^6)$ **19.** $(1+x^4)/x^2$ **21.** $(3x^2z^5)/y^4$ **23.** $12/x^6$ **25.** $c^8/(256a^4b^8)$ **27.** $1/(4x^4y^6)$
29. $3a^9b^9$ **31.** $1/(3a^3b^4c^4)$ **33.** x^4/a^8y^8 **35.** b^n/a^n **37.** 5.15×10^9 **39.** 1.8×10^{-9}
41. 5.142×10^{-10} **43.** 5.55 to 5.65 ft **45.** 15.755 to 15.765 km **47.** 21.3145 to 21.3155 kg
49. 3 **51.** 5 **53.** 4 **55.** 2 **57.** 3.4×10^{-19} **59.** 3.5×10^{11} **61.** 3.2×10^5 **63.** 3.5×10^{-12}
65. 5.983×10^{21} metric tons **67.** 1.983×10^{20} N **69.** 1.862×10^5 miles per second

EXERCISES 1.4 **1.** $x^2 - 2x - 3$ **3.** $(1/6)x^4 + 5x^2 - 7/10$ **5.** $x^7 + 5x^5 + 4x^4 + 3x^3 + 2x^2 + x + 2$
7. $t^5 - t^4 + 6t^3 + 3t^2 + 3t + 1$ **9.** $3x^4 + 2x^3 - 4x^2 - x + 4$ **11.** $5x^3 + 4x^2 - 7x^2y + 9xy^2 - 4y^3 + 7xy - x + 5$
13. $8x^3 - 16x^2 + 10x - 2$ **15.** $3x^4 - 8x^3 + 14x^2 - 8x + 3$ **17.** $(1/6)x^4 - (1/12)x^3 - (2/3)x^2 + (25/12)x - 2$
19. $x^7 + x^6 - x^4 - x^3 + x^3 + x + 1$ **21.** $x^2 + 2xy + y^2 - 1$ **23.** $x^4 - y^4 - x^3y - xy^3$
25. $x^3 - 6x^2 + 11x - 6$ **27.** $8x^3 + 12x^2 + 6x + 1$ **29.** $9x^4 - 13x^2 + 1$ **31.** $3x^3 + 4x - 5$
33. $3x^5 - 4x^3 - x^2 + 2x$ **35.** $4x^6 - 5x^4 + 3x^2$ **37.** $5x^2 - 3xy$ **39.** $x - 2xz^2 + 3x^2yz^2$

EXERCISES 1.5 **1.** $4x^2 + 8x + 3$ **3.** $4x^2 + 4ax - 3a^2$ **5.** $4a^2x^2 - 20abx + 25b^2$ **7.** $16y^2 - 24by + 9b^2$
9. $x^4 - 25$ **11.** $x^2 - 2$ **13.** $a - b$ **15.** $x^2 - 2xy + y^2 + 2ax - 2ay + a^2$ **17.** $(x+2)(x+4)$
19. $2(x+3)(x+5)$ **21.** $5(x+2)(x-2)$ **23.** $3a(x-1)(x+3)$ **25.** $(8a + 3b)(8a - 3b)$
27. $(4x^2 + 9)(2x+3)(2x-3)$ **29.** $(x-3)(x^2 + 3x + 9)$ **31.** $8(x+2)(x^2 - 2x + 4)$
33. $2a(2x+1)(4x^2 - 2x + 1)$ **35.** $(2x-1)(x+3)$ **37.** $(2x-1)^2$ **39.** $(2x-1)(3x-1)$
41. $(x+3)(3x+2)$ **43.** $(c-3d)(2a+b)$ **45.** $(2x-3y)(3a-b)$ **47.** $(x^2 + 2a)(y - 3b)$
49. $(a+b-2)(a^2 + 2ab + b^2 + 2a + 2b + 4)$ **51.** $(x+2)(x^2 - 2x + 4)(x-2)(x^2 + 2x + 4)$
53. $(x + 2y + 2z)(x + 2y - 2z)$ **55.** $(2x+1)(4x^2 - 8x + 7)$ **57.** $4ab$ **59.** $(u+4)(u-4)/2$
61. $(x + 1/3)(x - 1/2)$ **63.** $(x+2)(x^2 - 2x + 4)/2$ **65.** 641.8

EXERCISES 1.6 **1.** $5ax^2$ **3.** $2x^5 + 3x^3 - x$ **5.** $(x-3)/(x-4)$ **7.** $1/(x+8)$ **9.** $(u-3)/(u+3)$
11. $2x/(2x-1)$ **13.** $(4x^2 + 7x + 4)/(4x^2 + 11x + 6)$ **15.** $(4x+5)/(x+2)$ **17.** $(x^2 - 6x + 8)/(x-3)$
19. $(13x + 12)/(x-4)(x+4)$ **21.** $8/(x+4)$ **23.** $(x^2 + 5x - 1)/(x+3)^2$ **25.** $(2x^2 - 4)/(2x+1)(x+1)(x-1)$
27. $(5x+1)/4(x-1)(3x-1)$ **29.** $(4x-9)/(x+1)(x+2)(x-3)$ **31.** $1/(x-1)$ **33.** $y/2(y+2)$ **35.** $1/4b^2$
37. $(x+2)(x+1)/(x-1)(2x+1)$ **39.** $(2x+1)/(x-1)$ **41.** $x^2/2$ **43.** $x(x-1)$ **45.** $x/2(2x-1)$
47. $5(2x-1)(x+3)$ **49.** $(3x+1)(x+1)/(2x-1)(x+4)$ **51.** $x^2/2(x+1)(x-1)$ **53.** $(5-x)/(3x-7)$
55. $(2x^2 + 5x - 3)(4x^2 - 6x - 2)$ **57.** $-(x+3)/(x+2)$ **59.** $(4+x)/(x+3)$

EXERCISES 1.7 **1.** 9 **3.** -2 **5.** 4/5 **7.** -1 **9.** 29 **11.** $10 - 2\sqrt{21}$ **13.** $2a^2$ **15.** $-3a\sqrt[3]{a^2}$
17. $4x^2$ **19.** $3\sqrt{2}\,a^2b^2$ **21.** $4a^2x^3$ **23.** $2x^2y^3$ **25.** $2abc^2$ **27.** $4ab^2/c^3$ **29.** $2ab\sqrt[5]{ab^2}/c$
31. $\sqrt{2ab}/2a^2b$ **33.** $\sqrt[3]{12am^2n^2}/3n$ **35.** $2abx\sqrt[6]{2ab^2x^2}$ **37.** $m + n$ **39.** $5a\sqrt[4]{25a}$ **41.** $13\sqrt{2}$
43. $-10\sqrt{3}$ **45.** $10\sqrt{5} - 11\sqrt{2}$ **47.** $8\sqrt[3]{2}$ **49.** $3\sqrt{x}$ **51.** $10a\sqrt{2a}$ **53.** $(\sqrt{2} + \sqrt{6})/2$
55. $(a\sqrt{c} + \sqrt{bc})/c$ **57.** $-3 - 3\sqrt{2}$ **59.** $(\sqrt{15} + \sqrt{6})/3$ **61.** $(\sqrt{ac} - c)/c$ **63.** $(x^2 + x^3)/x^2$
65. $-2\sqrt{6} + \sqrt{30} - 4\sqrt{5} + 8$ **67.** $a(\sqrt{a} - 2)/(a - 4)$ **69.** $(x - \sqrt{xy})/(x - y)$ **71.** $(\sqrt{x - 1} + x - 1)/(2 - x)$
73. $(a - \sqrt{b})/(a^2 - b)$ **75.** $\sqrt{x^2 + 1} - x$ **77.** $\sqrt{8} + 2\sqrt{15}$

EXERCISES 1.8 **1.** $\sqrt{8^3}$ **3.** $1/\sqrt[4]{64}$ **5.** $\sqrt[5]{(a^2b^3)^3}$ **7.** $\sqrt{x^2 + y^2}$ **9.** $5^{2/3}$ **11.** $x^{3/4}$ **13.** $a^{5/2}$
15. $a^{17/4}$ **17.** 4 **19.** 1/9 **21.** -27 **23.** $15u^2$ **25.** $25x^4$ **27.** $m^{1/3}n^{1/2}$ **29.** $25a^{1/2}/b$
31. $27/x^6y^3$ **33.** $1/6a^4b$ **35.** $1/16a^{4/3}b^2$ **37.** $z^{3/2}/6xy^2$ **39.** a^4b^3 **41.** $4a^3/5x^2$
43. $27a^{1/2}/8x^6$ **45.** $a^{1/2}/b^{3/4}$ **47.** $\sqrt[6]{72}$ **49.** $\sqrt[4]{2a^3}$ **51.** 1 **53.** $\sqrt[4]{2a^3}$ **55.** $\sqrt{2xy}$
57. $\sqrt[6]{mv}$ **59.** $\sqrt[8]{8a^3}$

EXERCISES 1.9 **1.** $-1 + 5i$ **3.** $8 - 11i$ **5.** $2 - 13i$ **7.** $2 - 5i$ **9.** $7 - 9i$ **11.** $24 + 7i$
13. $-31 - 12i$ **15.** $(15/2) + (5/2)i$ **17.** 34 **19.** 145/144 **21.** -21 **23.** $-9 + 19i$ **25.** $30 - 15i$
27. $(1/2) - (1/2)i$ **29.** $-(3/4)i$ **31.** $-(21/29) + (20/29)i$ **33.** $(1/10) - (7/10)i$ **35.** $-(4/15) - (7/15)i$

REVIEW EXERCISES **1.** 0.4375 **3.** $0.\overline{81}$ **5.** 0.234375 **7.** 259/50 **9.** 47/333 **11.** 29/22
13. $1.\overline{4}$ **15.** $0.\overline{68}$ **17.** 15/2 **19.** 1/4 **21.** $-35/19$ **23.** No; for example $\sqrt{2} - \sqrt{2} = 0$.
25. a) rational, b) irrational, c) irrational, d) rational **27.** $12x^2/a^4y^3$ **29.** $3a^2c^4/2b^5$ **31.** $3m/2y^5$
33. $m^6p^3q^{15}/8$ **35.** b^2x^6/a^4 **37.** $1/5a^3x^2$ **39.** $x^{1/2}y$ **41.** $(2^{2/3}a^{7/2})/(3^{3/2}x^2)$ **43.** $b^{5/2}y^{3/4}$
45. 2.03×10^4 **47.** 5.73×10^{-22} **49.** 2.961×10^{12} **51.** 9.468×10^{12} km **53.** 1.318×10^{25} lbs
55. $u^2 - 3u - 10$ **57.** $x^4 + x^2 - 6$ **59.** $9a^4 - b^4$ **61.** $x^2 - a$ **63.** $x - 4$ **65.** $(v - 5)^2$
67. $(3x - 2)(x + 4)$ **69.** $(2a + 5b)(x - 4y)$ **71.** $(2u - 3v)(b + a)$ **73.** $2x(2x + 3a + 3)(2x - 3a - 3)$
75. $(7x - 27)/(x - 6)(x - 3)$ **77.** $(6x + 2)/(x + 2)(x - 2)$ **79.** $(x + 7)/(x - 1)^2(x + 1)$ **81.** $y^2/(x^3 - y^3)$
83. $5/6(x - 1)$ **85.** $-5(x + 5)/x$ **87.** $(x + 1)/(x - 1)$ **89.** $4\sqrt{3}xy^3$ **91.** $5ab\sqrt[4]{ab^3}$ **93.** $ab^5\sqrt{6ab}$
95. $2\sqrt[5]{ab^3c^4}/c^2$ **97.** $2abc\sqrt[6]{b^2c^3}$ **99.** $9a^2x^4y^2$ **101.** $(4\sqrt{3} + \sqrt{15})/9$ **103.** $(a + \sqrt{ab})/a$
105. $2 + \sqrt{6} + \sqrt{10} + \sqrt{15}$ **107.** $(2x + 2\sqrt{y} + \sqrt{xy} + y)/(x - y)$ **109.** $-\sqrt{x^2 - 1} - \sqrt{x^2 + x} - \sqrt{x^2 - x} - x$
111. 1 **113.** $8 - 13i$ **115.** $-31 - 12i$ **117.** 13 **119.** $34i$ **121.** $72 + 96i$ **123.** $(4/41) + (5/41)i$
125. $(8/89) + (5/89)i$

CHAPTER 2

EXERCISES 2.1 **1.** 10/3 **3.** $-9/8$ **5.** $3\sqrt{2}/2$ **7.** -4 **9.** 1 **11.** 3 **13.** $3\sqrt{3}/2$
15. $8\sqrt{3} + 8\sqrt{2}$ **17.** 1 **19.** 1 **21.** 18/11 **23.** 45/31 **25.** No solution **27.** 7/4 **29.** 3/5
31. 4.742 **33.** 9.869 **35.** 6.29×10^{-14} **37.** $C = (5/9)(F - 32)$ **39.** a) $T = PV/C$ b) $V = CT/P$
41. a) $y = P/2 - x$ b) 100 cm **43.** $r = P/tP_0 - 1/t$ **45.** $r = S/2\pi h$ **47.** $m_1 = r^2F/GM_2$ **49.** $R = V/I$

EXERCISES 2.2 **1.** 111, 112, 113, 114 **3.** 24 and 25 **5.** 7.5 ft, 15 ft **7.** 31 cm, 18 cm **9.** 78
11. \$125 **13.** \$80 **15.** 1320 people **17.** 8 dimes, 12 quarters, 15 nickels **19.** \$1,183.60
21. \$22,000 at 7.1%, \$20,000 at 14.5% **23.** 450 grams of chemical A, 270 grams of chemical B
25. 375 ft³ of cement, 250 ft³ of sand, 250 ft³ of stone **27.** 40 gallons of acid A, 120 gallons of acid B **29.** 12 miles
31. a) 750 miles, b) 3 hours **33.** $R = 36$ mi/h **35.** 2.5 hours, 12:00 noon **37.** 12/5 hours
39. 12/7 hours

EXERCISES 2.3 **1.** $-2, \sqrt{3}, 2$ **3.** $-\sqrt{2}, -2/3, 1/2$ **5.** $1/4, \sqrt{3}, \sqrt{5}, 4$ **7.** $-5/2, -\sqrt{2}, 1, \sqrt{6}$
9. $-2.45, -\sqrt{6}, \sqrt{2}, 17/12$ **11.** $\{x \in \mathbb{R}: -2 \le x \le 3\}$ **13.** $\{x \in \mathbb{R}: 2 \le x < 8\}$

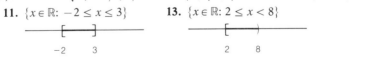

15. $\{x \in \mathbb{R}: x < 5\}$

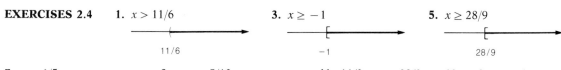

17. $(-2, 7)$

19. $(-3, 3)$

21. $[-2, +\infty)$

23. $5, -5$ **25.** $7/2, -7/2$ **27.** -2 **29.** $4, -2$ **31.** $-1/2, 7/2$

33. $2, 2/3$ **35.** 15 **37.** 5 **39.** 14 **41.** $\pi - 25/8$ **43.** $\sqrt{3} - \sqrt{2}$ **45.** $3 - a$ **47.** $10 - 2x$
49. $d(A, B) = 2, d(A, C) = 5, d(B, C) = 3$ **51.** $d(A, B) = 5/2, d(A, C) = 6, d(B, C) = 7/2$
53. $d(A, B) = 0.76, d(A, C) = 0.42, d(B, C) = 1.18$

EXERCISES 2.4   **1.** $x > 11/6$ **3.** $x \geq -1$ **5.** $x \geq 28/9$

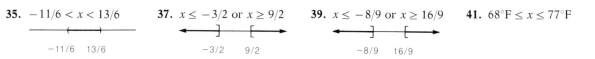

7. $x \geq 6/5$ **9.** $x < -5/12$ **11.** $11/2 < x < 23/2$ **13.** $-9 < x \leq 6$

15. $x > -2/5$ **17.** $x > 1/5$ **19.** $x > -3$ **21.** $x < -4$ **23.** $x > 1/4$ **25.** $x < 3$ or $x > 5$

27. $-10/3 < x < 8/3$ **29.** $x < 1/6$ or $x > 5/6$ **31.** $-1 < x < 7$ **33.** $-2 < x < 7$

35. $-11/6 < x < 13/6$ **37.** $x \leq -3/2$ or $x \geq 9/2$ **39.** $x \leq -8/9$ or $x \geq 16/9$ **41.** $68°F \leq x \leq 77°F$

43. 10 ohms $\leq R \leq 20$ ohms   **45.** $4/3 \leq V \leq 4$ **47.** $10.5 \leq MA \leq 21$ **49.** 50 mi/h $\leq r \leq 60$ mi/h

EXERCISES 2.5   **1.** ± 2 **3.** ± 4 **5.** $\pm(3/2)\sqrt{2}$ **7.** $-1 \pm \sqrt{2}$ **9.** $(2 \pm \sqrt{11})/5$ **11.** $\pm i\sqrt{15}$
13. $-3 \pm -i\sqrt{10}$ **15.** $(2 \pm 2i\sqrt{5})/3$ **17.** $2, 4$ **19.** $-5, -3$ **21.** $5, 1$ **23.** $0, 3$ **25.** $-(3/4), 2$
27. $4, -2$ **29.** $-5, -3$ **31.** $-8, -2$ **33.** $2/3, 1/3$ **35.** $(-3 \pm \sqrt{13})/4$ **37.** $4, 2$ **39.** $(3 \pm i\sqrt{7})/4$
41. $-(1/3) \pm (i\sqrt{2})/6$ **43.** $-1 \pm \sqrt{7}/2$ **45.** $1, -2$ **47.** $1, -3/2$ **49.** $(-1 \pm \sqrt{7})/2$
51. $(1 \pm i\sqrt{19})/5$ **53.** $(1 \pm i\sqrt{7})/2$ **55.** $(1 \pm i\sqrt{35})/3$ **57.** $0.25 \pm 0.33i$ **59.** $1.22 \pm 1.55i$
61. $\pm 3; 2x^2 - 18 = 2(x - 3)(x + 3)$ **63.** $-2, 1/2; 2x^2 + 3x - 2 = 2(x - \frac{1}{2})(x + 2) = (2x - 1)(x + 2)$
65. $1 \pm \sqrt{2}; x^2 - 2x - 1 = (x - 1 + \sqrt{2})(x - 1 - \sqrt{2})$

EXERCISES 2.6   **1.** $3, 4$ **3.** $2/3, 1/3$ **5.** 1 **7.** 11 **9.** $12, 9$ **11.** 3 **13.** 7 **15.** 2
17. $-3, 1$ **19.** $-2, 1$ **21.** 4 **23.** $(1 \pm i\sqrt{7})/2$ **25.** $3/2, -2$ **27.** $\pm\sqrt{2}, \pm 2$ **29.** $-2, -(1/2)$
31. $\pm\sqrt{3} \pm \sqrt{17}$ **33.** $\pm\sqrt{-3 \pm \sqrt{13}}/2$ **35.** $\pm\sqrt{2}, \pm(i\sqrt{2})/2$ **37.** 4 **39.** $4, 169$ **41.** $1/16, 1/81$
43. $8, 64$ **45.** $\sqrt[3]{-3}, \sqrt[3]{-5}$ **47.** $2, 6$ **49.** $1, 2$

EXERCISES 2.7   **1.** $16, 17$ **3.** $24, 26$ **5.** $9, 18$ **7.** $5, 1/5$ **9.** $6, 10$ **11.** 6 in, 9 in
13. $8, 12$ **15.** $5, 18$ **17.** $14, 14$ **19.** 1 m   **21.** 2 ft   **23.** 5% **25.** 300 **27.** 2 mi/h
29. Mary takes 2 days, Peter takes 3 days   **31.** 50 and 60 mi/h   **33.** a) 10 s, 50 s,   b) 60 s   **35.** 80 in
37. $\$1.50$ **39.** 1.732×10^5 m

EXERCISES 2.8   **1.** $x < -2$ or $x > 5$ **3.** $x \leq -1/2$ or $x \geq 1$ **5.** $x < 2$ or $x > 2$
7. $-\sqrt{6}/2 < x < \sqrt{6}$ **9.** $\mathbb{R}$ **11.** $x \leq -3$ or $x \geq 0$ **13.** $x < -5$ or $x > 2$ **15.** $x < 2$ or $x > 5$

17. $x \le 2 - \sqrt{3}$ or $x \ge 2 + \sqrt{3}$ **19.** $(-1 - \sqrt{15})/2 < x < (-1 + \sqrt{15})/2$ **21.** $-3 < x < 5$

23. $\mathbb{R}$ **25.** $-3 < x \le 1$ **27.** $0 \le x \le 2$ or $x > 4$ **29.** $x < -3$ or $x > 5$ **31.** $x < -5$ or $x \ge 3/2$

33. $-1 < x < 0$ or $x > 2$ **35.** $-\sqrt{3} < x < 0$ or $0 < x < \sqrt{3}$ **37.** $x < -3$ or $1 \le x \le 2$

39. $x < -3$ or $-1 < x < 1/2$ or $x > 4$ **41.** $-2 < x < 1/2$

43. a) $5\,s < t < 55\,s$ b) $7.79\,s < t < 51.21\,s$ c) $0 \le t \le 6\,s$ or $t > 54\,s$ **45.** $0 < p < 150$ **47.** $t > 10$

REVIEW EXERCISES   **1.** 1   **3.** $-16/5$ **5.** $4/13$ **7.** 5   **9.** 18   **11.** 1   **13.** 3, 4

15. $2/3, -(1/2)$ **17.** $1/3$ **19.** 7   **21.** $-8 \pm \sqrt{70}$ **23.** $1 \pm \sqrt{3}/3$ **25.** $3/4$ **27.** $(-5 \pm \sqrt{385})/2$

29. $1/2, -2$ **31.** 16, 81   **33.** $(29 \pm \sqrt{31})/9$ **35.** 5, $29/7$ **37.** $b_1 = 2A/h - b_2$ **39.** $t = (L - L_0)/\alpha L_0$

41. $t = (v_0 \pm \sqrt{v_0 + 2as})/2a$ **43.** $r = (-h \pm \sqrt{h^2 - 2s/\pi})/2$ **45.** $x < -65$ **47.** $x \ge -(41/23)$

49. $x > -3$ **51.** $5 < x \le 15/2$ **53.** $1/2 < x < 5/3$ or $x > 3$ **55.** $-2 < x < 1/3$ or $3 < x < 7$

57. 35, 36, 37, 38   **59.** 1964 voters   **61.** 216   **63.** 2 gallons   **65.** \$400   **67.** 5.5 m, 8 m   **69.** 8 m

71. 1152   **73.** 5, 12, 13   **75.** 30 feet   **77.** 5.222 cm   **79.** 162.76 foot-candles

CHAPTER 3

EXERCISES 3.1   **1.**
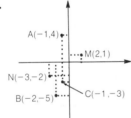
   3. $(-3/2, 3/2)$ **5.** $(-3/8, 3/8)$ **7.** $\left(\dfrac{2 + a}{2}, 0\right)$

9. $4\sqrt{2}$ **11.** $4\sqrt{2}$ **13.** $7/3$ **15.** $\sqrt{26}$ **17.** $2\sqrt{1 + b^2}$ **19.** 5   **21.** $\simeq 31.333$

23. $d(A, B) = 5, d(A, C) = \sqrt{125}, d(B, C) = 10$ **25.** 25   **27.** $d(A, B) = d(B, C) = \sqrt{41}$

29. $d(A, B) = d(A, D) = d(C, D) = d(B, C) = 5\sqrt{2}$ **31.** 2   **33.** $(4, 2)$ **35.** $d(A, P) = d(B, P) = \sqrt{26}$

37. On the line segment   **39.** Not on the line segment

EXERCISES 3.2   **1.** 1   **3.** $-1/12$ **5.** The three points lie on the same line; common slope is $3/2$

7. Common slope is $-1/2$; the points are collinear   **9.** $x = -2$ **11.** $y = 3x - 2$ **13.** $2x + 5y = 0$

15. $y = -2$ **17.**
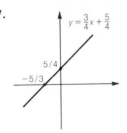
   19.
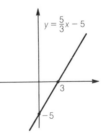
   21.
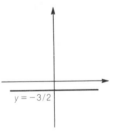

23.
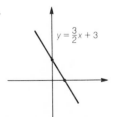
   25.
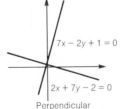
   27.
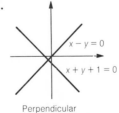
   29.
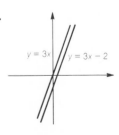

31.
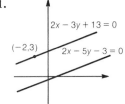
$2x - 3y + 13 = 0$
$(-2,3)$
$2x - 5y - 3 = 0$

33.
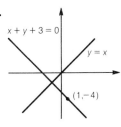
$x + y + 3 = 0$
$y = x$
$(1,-4)$

35.
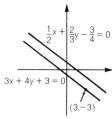
$\frac{1}{2}x + \frac{2}{3}y - \frac{3}{4} = 0$
$3x + 4y + 3 = 0$
$(3,-3)$

37. $-3/4$ **39.** 5

41. $5x - 19y + 85 = 0$ **43.** $7/3$ **45.** $v = \frac{5}{3}u + \frac{5}{3}$, $v = 5$ **47.** 2   **49.** $w = 5s$ or $s = \frac{1}{5}w$, 0.4 in

51. $p = -200t + 1000$, \$400   **53.** 96 ft/s, 6 seconds

EXERCISES 3.3   **1.**
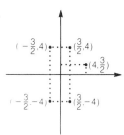
$(-2,5)$ $(2,5)$
$(-5,2)$
$(-2,-5)$ $(2,-5)$

3.
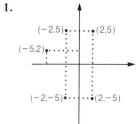
$(0,4)$
$(-4,0)$ $(4,0)$

5.
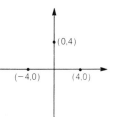
$(-2,6)$ $(2,6)$
$(6,2)$
$(-2,-6)$ $(2,-6)$

7.
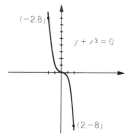
$\left(-\frac{3}{2},4\right)$ $\left(\frac{3}{2},4\right)$
$\left(4,\frac{3}{2}\right)$
$\left(-\frac{3}{2},-4\right)$ $\left(\frac{3}{2},-4\right)$

9.
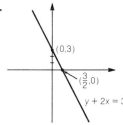
$(0,3)$
$\left(\frac{3}{2},0\right)$
$y + 2x = 3$
No symmetry

11.
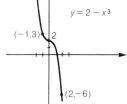
$2y - 3x^2$
Symmetry with respect to the y − axis.

13.
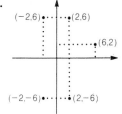
$x - y^2 - 4 = 0$
$(8,2)$
$(5,1)$
4
$(8,-2)$
Symmetry with respect to the x − axis.

15.
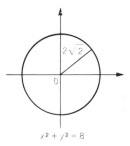
$(-2,8)$
$y + x^3 = 0$
$(2,-8)$
Symmetry with respect to the origin.

17.
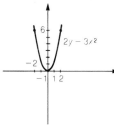
$y = 2 - x^3$
$(-1,3)$
2
$(2,-6)$
Symmetry with respect to the point (0,2).

19.
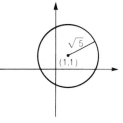
$\sqrt{5}$
$(1,1)$
$(x - 1)^2 + (y - 1)^2 = 5$

21.
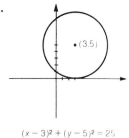
$2\sqrt{2}$
0
$x^2 + y^2 = 8$

23.
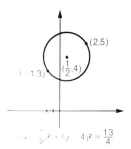
$(3,5)$
$(x - 3)^2 + (y - 5)^2 = 25$

25.

$(2,5)$
$\left(\frac{1}{2},4\right)$
$(-1,3)$
$(x + 1)^2 + (y - 4)^2 = \frac{13}{4}$

27.

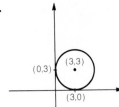

$(x - 3)^2 + (y - 3)^2 = 9$

29.

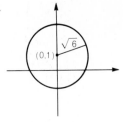

$C(0,1), r = \sqrt{6}$

31.

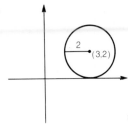

$C(0,0), r = 3$

33.

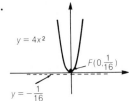

$C(3,2), r = 2$

35.

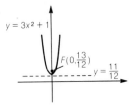

$C(0,2), r = 3$

37.

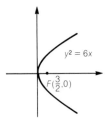

$C(-3,-2), r = 0$

39.

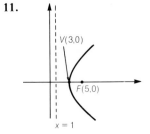

$C(-\frac{3}{2},-2), r = \frac{\sqrt{33}}{2}$

EXERCISES 3.4

(*Note:* P.A. = principal axis, D. = directrix)

1.

$V(0,0), F(0,1/16)$
P.A.: $x = 0$
D.: $y = -1/16$

3.

$V(0,1), F(0,13/12)$
P.A.: $x = 0$
D.: $y = 11/12$

5.

$V(1,-1), F(1,-3/4)$
P.A.: $x = 1$
D.: $y = -5/4$

7.

$y^2 = 6x$

$F(\frac{3}{2},0)$

9.

$(y - 2)^2 = 4(x - 1)$

11.

$y^2 = 8(x - 3)$

13.

$(x + 1)^2 = -2(y + 1)$

15. Parabola opening upwards, $V(0, 3/4)$, $F(0, 7/4)$, D.: $y = -1/4$, P.A.: $x = 0$

17. Parabola opening to the right, $V(-1, 1)$, $F(-3/4, 1)$, D.: $x = -5/4$, P.A.: $y = 1$ **19.** $c < 4$

21. $(1, 3)$ and $(-2, -3)$ **23.**   **25.**   **27.**

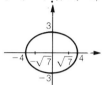

$C(0,0), V(\pm 4, 0), F(\pm\sqrt{7}, 0)$
Major axis: x − axis
Minor axis: y − axis

$C(0,0), V(\pm 3.0), F(\pm\sqrt{5}, 0)$
Major axis: x − axis
Minor axis: y − axis

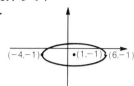

$C(1,-1, V(-4,-1), V_2(6,-1)$
$F_1(1+\sqrt{21},-1), F_2(1-\sqrt{21},-1)$
Major axis: $y = -1$
Minor axis: $x = 1$

29. $\dfrac{x^2}{16} + \dfrac{y^2}{12} = 1$ **31.** $\dfrac{(x-1)^2}{25} + \dfrac{(y-2)^2}{16} = 1$ **33.** $\dfrac{(x-5)^2}{25} + \dfrac{y^2}{16} = 1$ **35.** $\dfrac{x^2}{225} + \dfrac{y^2}{200} = 1$

37. $\dfrac{x^2}{9} + \dfrac{y^2}{5} = 1$ **39.** $C(1, -1)$, major axis: $x = 1$, minor axis: $y = -1$

41. $C(2, -1)$, major axis: $y = -1$, minor axis: $x = 2$ **43.** 9.135×10^7 miles, 9.445×10^7 miles

45. $C(0, 0)$, $V(\pm 8, 0)$, $F(\pm 10, 0)$

47. $C(2, -3)$, $a = 9$, $b = 12$, $c = 15$, $V_1(-7, -3)$, $V_2(11, -3)$, $F_1(-13, -3)$, $F_2(17, -3)$ **49.** $\dfrac{x^2}{9} - \dfrac{y^2}{12} = 1$

51. $\dfrac{(x-2)^2}{16} - \dfrac{(y-1)^2}{9} = 1$ **53.** $x^2 - \dfrac{y^2}{4} = 1$ **55.** $(x-1)^2 - \dfrac{(y-1)^2}{24} = 1$ **57.** $C(0, 1)$, $V_1(0, 0)$, $V_2(0, 2)$

59. $C(2, -1)$, $V_1(-1, -1)$, $V_2(5, -1)$

REVIEW EXERCISES   **1.** $(5/2, 3/2)$ **3.** $(13/24, -3/20)$ **5.** $(6, 1)$ **7.** $(2 \pm \sqrt{11}, 0)$

9. $d(A, B) = \sqrt{125}$, $d(A, C) = 10$, $d(B, C) = 5$, area $= 7.5$ **11.** $d(A, B) = d(C, D) = 4\sqrt{2}$, $d(A, D) = d(B, C) = 6\sqrt{2}$

13. $7/3$ **15.**   **17.**   **19.** $12/7$ **21.** $\simeq 0.041$

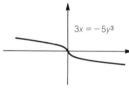

Symmetry with respect
to the origin

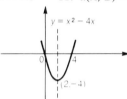

Symmetry with respect
to the line $x = 2$.

23. $x + 2y + 20 = 0$ **25.** $10x - 6y + 9 = 0$ **27.** $10x + y - 6 = 0$ **29.** $3.95x - y + 16.50 = 0$

31. $3x + 2y - 15 = 0$ **33.** 8 **35.** $\simeq 1.61$, $\simeq 0.54$ **37.** $x + y - 8 = 0$ **39.** $y = 20t + 15$, 3.5

41. $V = 1200t + 20,000$ **43.** $x^2 + y^2 = 10$ **45.** $(x + 1)^2 + (y - 1)^2 = 25$

47. $(x + \sqrt{10})^2 + (y + \sqrt{10})^2 = 10$ **49.** $C(0, 0)$, $r = 4$ **51.** $C(0, 3/2)$, $r = \sqrt{13}/2$

53. $C(3/2, -5/2)$, $r = 3$ **55.** $x^2 = 12y$ **57.** $y^2 = 8x$ **59.** $(x - 2)^2 = \dfrac{4}{3}(y - 1)$

61. $(1, 1)$ and $(1/2, 1/4)$ **63.** $\dfrac{(x+1)^2}{5} + \dfrac{y^2}{9} = 1$ **65.** $\dfrac{(x-1)^2}{25} + \dfrac{(y-3)^2}{16} = 1$

67. $9x^2 + 16y^2 = 25$ **69.** $\dfrac{y^2}{9} - \dfrac{x^2}{16} = 1$ **71.** $\dfrac{x^2}{4} - \dfrac{y^2}{21} = 1$ **73.** $\dfrac{x^2}{8} - \dfrac{y^2}{36} = 1$ **75.** $C(2, 3)$, $r = 3$

77. Ellipse, $C(-1, -1)$, $a = 5/3$, $b = 5\sqrt{2}$ **79.** Parabola opening to the right, $V(-3, 3)$

81. Hyperbola, $C(1, 1)$, $a = \sqrt{17}$, $b = \sqrt{17}/2$ **83.** $\pm 4\sqrt{10}$ **85.** $(-1, 0)$ and $(3, 0)$

87. $(x + 2)^2 + (y - 3)^2 = 10$ **89.** 1.285×10^8 miles, 1.549×10^8 miles

CHAPTER 4

EXERCISES 4.1   **1.** Domain: all real numbers. Range: all real numbers
3. Domain: all real numbers. Range: all real numbers   **5.** Domain: all real numbers. Range: all numbers ≤ 4
7. Domain: all real numbers. Range: all numbers ≥ -1 **9.** Domain: all numbers ≥ 0. Range: all numbers ≥ -3
11. Domain: all numbers ≥ 0. Range: all numbers ≥ 1 **13.** Domain: all numbers $\geq 1/2$. Range: all numbers ≥ 0
15. $-10, -4, -2, \simeq -0.33$ **17.** $-5, 0, 3, 1$ **19.** $\simeq 1.59, -1, \simeq 1.13$, not defined, $5/2$
21. not defined, $1, 2\sqrt{2}, \simeq 1.80, 1/2$ **23.**

23.

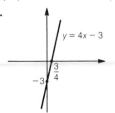

Domain: all real numbers

25.

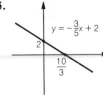

Domain: all real numbers

27.

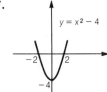

Domain: all real numbers

29.

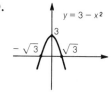

Domain: all real numbers

31.

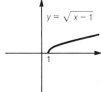

Domain: $[1, +\infty)$

33.

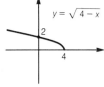

Domain: $(-\infty, 4]$

35.

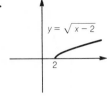

Domain: $[2, +\infty)$

37.

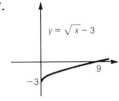

Domain: $[0, +\infty)$

39.

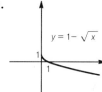

Domain: $[0, +\infty)$

EXERCISES 4.2   **1.** All real numbers except 2   **3.** $(-\infty, 4/3]$ **5.** $x \neq 1$
7. All real numbers except 1 and 2   **9.** $[-1, 1]$ **11.** $9, 6, 8, \simeq 9.67$ **13.** $-1, -1, -3, -(3 + \sqrt{2})$
15. $0, 4, 2/3$, not defined   **17.** $4, -2, 0, \simeq 0.539$ **19.** $5 - 3a, 5 + 3a, 3a - 5, 5 - 3(a + b), 10 - 3(a + b)$

21. $a^2 - 4, a^2 - 4, 4 - a^2, a^2 + b^2 - 8, (a + b)^2 - 4$ **23.** $\dfrac{a - 1}{a + 1}, \dfrac{a + 1}{a - 1}, \dfrac{-a + 1}{a + 1}, \dfrac{2(ab - 1)}{(a + 1)(b + 1)}, \dfrac{a + b - 1}{a + b + 1}$

25. $\dfrac{a}{1 - 5a}, a - 5, \dfrac{a^2 - 10a + 1}{(a - 5)(1 - 5a)}, \dfrac{a}{a^2 - 5a + 1}$ **27.** $\dfrac{a}{2 - a}, 2a - 1, \dfrac{-a^2 + 4a - 1}{(2a - 1)(2 - a)}, \dfrac{a}{2a^2 - a + 2}$

29. $\dfrac{a^2 - 1}{a^2}, \dfrac{1}{1 - a^2}, -\dfrac{(a^2 - 1)^2}{a^2}, \dfrac{-(2a^2 + 1)}{a^2}$

31. $x^2 + 3x + 2, -x^2 + 3x + 8, 3x^3 + 5x^2 - 9x - 15, \dfrac{3x + 5}{x^2 - 3}$ all domains are $(-\infty, \infty)$, except f/g, which is

$(-\infty, -3) \cup (-3, 3) \cup (3, \infty)$

33. $\dfrac{17x^2 + 10x - 1}{(2x - 1)(5x + 3)}, \dfrac{13x^2 + 8x + 1}{(2x - 1)(5x + 3)}, \dfrac{3x^2 + 3x}{(2x - 1)(5x + 3)}, \dfrac{3x(5x + 3)}{(2x - 1)(x + 1)}$ all domains are

$\left(-\infty, -\dfrac{3}{5}\right) \cup \left(-\dfrac{3}{5}, \dfrac{1}{2}\right) \cup \left(\dfrac{1}{2}, +\infty\right)$

35. $2, -\dfrac{6}{x}$, domain: all $x \neq 0$, $1 - \dfrac{9}{x^2}$, domain: all $x \neq 0$, $\dfrac{x-3}{x+3}$, domain: all $x \neq -3$ **37.** 0   **39.** 3

41. $2(a+h)$ **43.** $2h + 4a - 3$ **45.** $A(x) = x(2x+4)$ **47.** $A(p) = \dfrac{p^2}{16}$ **49.** $V(x) = 2x^3$ **51.** $L(h) = 8\pi\sqrt{h}$

53. $p(t) = 1000 + 60t$ **55.** $f = \dfrac{5}{6}d_2$ **57.** $A = 0.4x + 4.75$ **59.** $L(h) = \dfrac{2}{3}\pi h^2$, 0.13, 2.09, 3.27, 8.48

EXERCISES 4.3   **1.**

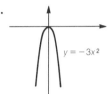

$y = \dfrac{x^2}{2}$

3.

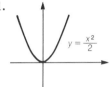

$y = -3x^2$

5.

$y = 3\sqrt{x}$

7.

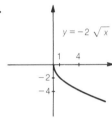

$y = -2\sqrt{x}$

9.

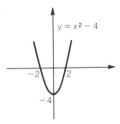

$y = x^2 - 4$

11.

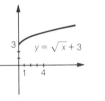

$y = \sqrt{x} + 3$

13.

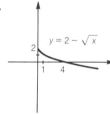

$y = 2 - \sqrt{x}$

15.

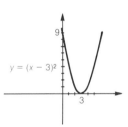

$y = (x - 3)^2$

17.

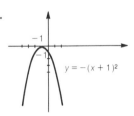

$y = -(x + 1)^2$

19.

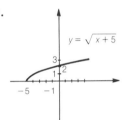

$y = \sqrt{x + 5}$

21.

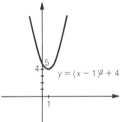

$y = (x - 1)^2 + 4$

23.

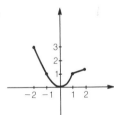

25.

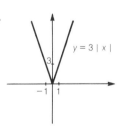

$y = 3|x|$

27.

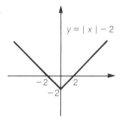

$y = |x| - 2$

29.

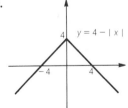

$y = 4 - |x|$

31.

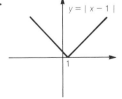

$y = |x - 1|$

33.

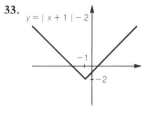

$y = |x + 1| - 2$

35.

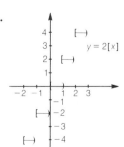

$y = 2[x]$

37.

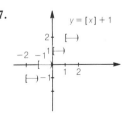

$y = [x] + 1$

39.

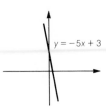

$y = [x - 1]$

41. even   **43.** odd   **45.** even   **47.** even   **49.** even   **51.** odd   **53.** even

55.

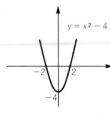

$y = -5x + 3$

Decreasing

57.

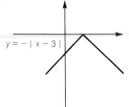

$y = x^2 - 4$

Decreasing on $(-\infty, 0)$
Increasing on $(0, +\infty)$

59.

$y = -|x - 3|$

Increasing on $(-\infty, 3)$
Decreasing on $(3, +\infty)$

61.

$F(x) = \begin{cases} 0 \text{ if } x \geq 0 \\ 2x \text{ if } x < 0 \end{cases}$

Increasing in $(-\infty, 0)$
Neither increasing nor
decreasin on $(0, +\infty)$

63.

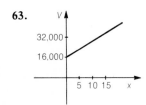

32,000

16,000

5 10 15 x

$V(x) = 16,000(1 + 0.08x)$

65.

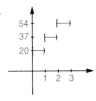

54
37
20
1 2 3

67.

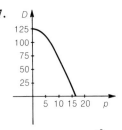

D
125
100
75
50
25
5 10 15 20 p

$D(p) = 125 - \dfrac{p^2}{2}$

EXERCISES 4.4   **1.** $u = 3v$ **3.** $y = 1/3x^2$ **5.** 3   **7.** $\sqrt{3}\sqrt{2}$ **9.** 28/3 inches

11. 250 foot-candles   **13.** 72 ft   **15.** $V = \dfrac{4}{3}\pi r^3$, 47.7 in³   **17.** a) $K = \dfrac{1}{2}mv^2$,   b) 1.16×10^3

19. a) $T = 2\pi\sqrt{\dfrac{l}{g}}$,   b) 1.41 s

REVIEW EXERCISES   **1.** $x \neq 5/4$ **3.** $\left(-\infty, \dfrac{2}{3}\right]$ **5.** $x \neq 2$ and $x \neq -\dfrac{5}{3}$ **7.** $x \neq 2$ and $x \neq -1$

9. $x \geq 3$ or $x \leq -3$ **11.** $x < -2$ or $x \geq 1$ **13.** $(-\infty, 0]$ **15.** $-\dfrac{9}{2}, 1, 14, 14.6$ **17.** $-47, 3, -47, -17$

19. $5, 4\sqrt{3}, \sqrt{5}/2$, not defined   **21.** 3/5, 0, $-3/5$, 0.445   **23.** 0, $\sqrt{11/13}, \sqrt{5}$, 1.14   **25.** -7

27. $10x + 5h - 1$ **29.** $\dfrac{-2}{x(x+h)}$ **31.** $\dfrac{2}{(x+1)(x+h+1)}$ **33.** $\dfrac{2a+b}{b}, \dfrac{2(1-a^3b)}{(1-a^2)(1-ab)}, \dfrac{a+b+a^2b}{a+b-a^2b}$

37. $A(h) = \dfrac{1}{6}h^2$ **39.** $L(h) = \dfrac{12}{25}\pi h^2$ **41.**

8
$y = 2(4 - x^2)$
-2 2

43.

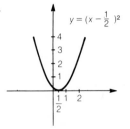

$y = (x - \tfrac{1}{2})^2$
4
3
2
1
$\dfrac{1}{2}$ 1   2

45.

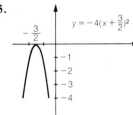

$y = -4(x + \frac{3}{2})^2$

47.

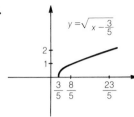

$y = \sqrt{x - \frac{3}{5}}$

49.

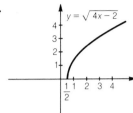

$y = \sqrt{4x - 2}$

51.

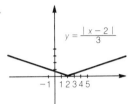

$y = \frac{|x - 2|}{3}$

53.

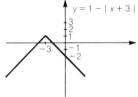

$y = 1 - |x + 3|$

55. odd　　**57.** odd　　**59.** even　　**61.** even

63.

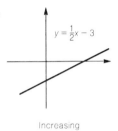

$y = \frac{1}{2}x - 3$

Increasing

65.

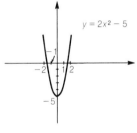

$y = 2x^2 - 5$

Decreasing on $(-\infty, 0)$
Increasing on $(0, +\infty)$

67.

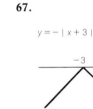

$y = -|x + 3|$

Increasing on $(-\infty, -3)$
Decreasing on $(-3, +\infty)$

69.

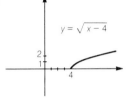

$y = \sqrt{x - 4}$

Increasing on $(4, +\infty)$

71. $P = kd$　　**73.** $w = khr^2$　　**75.** $3\sqrt{3}$　　**77.** by a factor of 4　　**79.** 13.5 gallons

CHAPTER 5

EXERCISES 5.1

1. a)

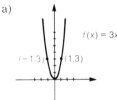

$f(x) = 3x^2$

$(-1,3)$　$(1,3)$

b)

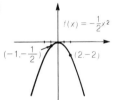

$f(x) = -\frac{1}{2}x^2$

$(-1,-\frac{1}{2})$　$(2,-2)$

c)

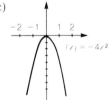

-2　-1　1　2

$f(x) = -4x^2$

d)

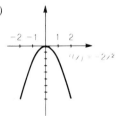

-2　-1　1　2

$f(x) = -2x^2$

3. a)

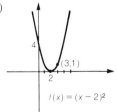

4

$(3,1)$

2

$f(x) = (x - 2)^2$

b)

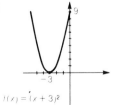

9

-3

$f(x) = (x + 3)^2$

c)

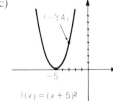

$(-3,4)$

-5

$f(x) = (x + 5)^2$

d)

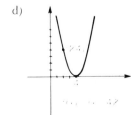

$(2,3)$

5. -16, no x-intercepts, opens upward   **7.** 81, two x-intercepts, opens downward
9. 0, tangent to x-axis, opens upward   **11.** -8, no x-intercepts, opens downward

13.

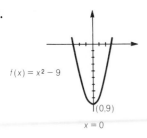

$f(x) = x^2 - 9$

$(0,9)$

$x = 0$

15.

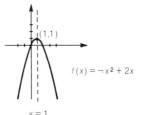

$f(x) = x^2 - 6x + 8$

$(3,-1)$

$x = 3$

17.

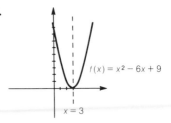

$f(x) = x^2 - 6x + 9$

$x = 3$

19.

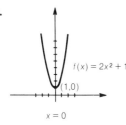

$f(x) = 2x^2 + 1$

$(1,0)$

$x = 0$

21.

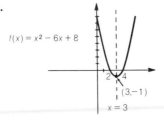

$(1,1)$

$f(x) = -x^2 + 2x$

$x = 1$

23.

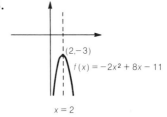

$(2,-3)$

$f(x) = -2x^2 + 8x - 11$

$x = 2$

25. $c = 16$ **27.** $-4 \le b \le 4$ **29.** 18 and 18   **31.** 99,225 m^2 **33.** 200, \$80   **35.** 121 ft, 5.5 s   **37.** 300

EXERCISES 5.2   **1.**

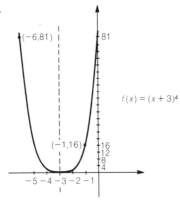

$(-6,81)$ 81

$f(x) = (x + 3)^4$

$(-1,16)$ 16 / 12 / 8 / 4

$-5 -4 -3 -2 -1$

3.

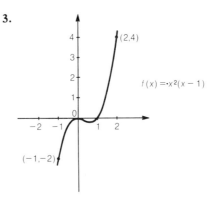

$(2,4)$

$f(x) = x^2(x - 1)$

$(-1,-2)$

5.

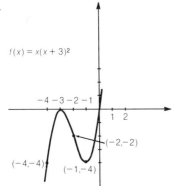

$f(x) = x(x + 3)^2$

$-4 -3 -2 -1$ 1 2

$(-2,-2)$

$(-4,-4)$ $(-1,-4)$

7.

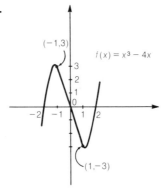

$(-1,3)$

$f(x) = x^3 - 4x$

-2 -1 1   2

$(1,-3)$

9.

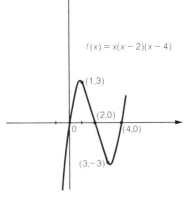

$f(x) = x(x - 2)(x - 4)$

$(1,3)$

$(2,0)$

$(4,0)$

$(3,-3)$

11.

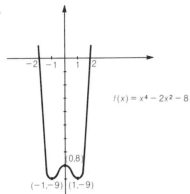

$f(x) = x^4 - 2x^2 - 8$

$(0,8)$

$(-1,-9)$ $(1,-9)$

13.

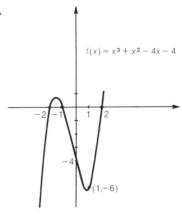

$f(x) = x^3 + x^2 - 4x - 4$

-4

$(1,-6)$

15.

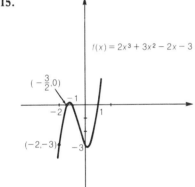

$f(x) = 2x^3 + 3x^2 - 2x - 3$

$\left(-\frac{3}{2},0\right)$

-1

-2 1

$(-2,-3)$ -3

17.

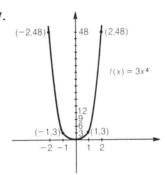

$(-2,48)$ 48 $(2,48)$

$f(x) = 3x^4$

12
9
6
$(-1,3)$ 3 $(1,3)$

-2 -1 1 2

Symmetric with respect
to the line $x = 0$

19.

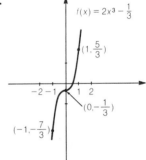

$f(x) = 2x^3 - \frac{1}{3}$

$\left(1,\frac{5}{3}\right)$

-2 -1 1 2

$\left(0,-\frac{1}{3}\right)$

$\left(-1,-\frac{7}{3}\right)$

Symmetric with respect
to the point $(0,-1/3)$

21.

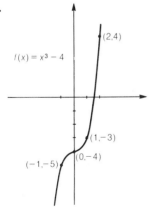

$(2,4)$

$f(x) = x^3 - 4$

$(1,-3)$

$(0,-4)$

$(-1,-5)$

Symmetric with respect to
the point $(0,-4)$

23.

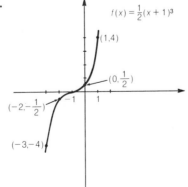

$f(x) = \frac{1}{2}(x + 1)^3$

$(1,4)$

$\left(0,\frac{1}{2}\right)$

-1 1

$\left(-2,-\frac{1}{2}\right)$

$(-3,-4)$

Symmetric with respect to
the point $(-1,0)$

25.

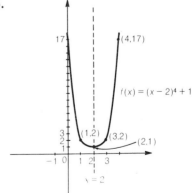

17 $(4,17)$

$f(x) = (x - 2)^4 + 1$

$\frac{3}{2}$ $(1,2)$ $(3,2)$ $(2,1)$
1

-1 0 1 2 3

$x = 2$

Symmetry with respect to
the line $x = 2$

27.

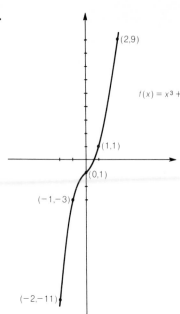

$f(x) = x^3 + x - 1$

(2,9)

(1,1)

(0,1)

(−1,−3)

(−2,−11)

29.

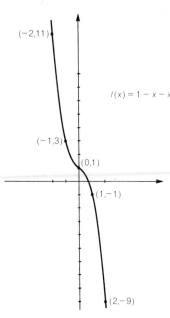

(−2,11)

$f(x) = 1 - x - x^3$

(−1,3)

(0,1)

(1,−1)

(2,−9)

EXERCISES 5.3

(In what follows, VA stands for *vertical asymptote* and HA for *horizontal asymptote*.)

1. VA: $x = -\dfrac{1}{2}$, HA: $y = 0$ **3.** VA: $x = -\dfrac{2}{3}$, HA: $y = \dfrac{2}{3}$ **5.** VA: $x = 2 \pm \sqrt{3}$, HA: $y = 2$ ·

7. No vertical asymptote, HA: $y = 0$ **9.** VA: $x = \dfrac{3}{5}$, no horizontal asymptote

11.

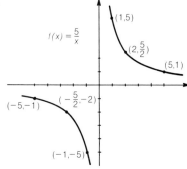

$f(x) = \dfrac{5}{x}$

(1,5)

$\left(2, \dfrac{5}{2}\right)$

(5,1)

$\left(-\dfrac{5}{2}, -2\right)$

(−5,−1)

(−1,−5)

13.

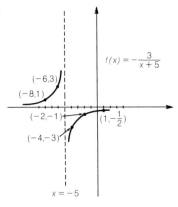

$f(x) = -\dfrac{3}{x+5}$

(−6,3)

(−8,1)

(−2,−1)

(−4,−3)

$\left(1, -\dfrac{1}{2}\right)$

$x = -5$

15.

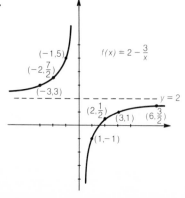

(−1,5)

$\left(-2, \dfrac{7}{2}\right)$

$f(x) = 2 - \dfrac{3}{x}$

(−3,3)

$y = 2$

$\left(2, \dfrac{1}{2}\right)$

(3,1) $\left(6, \dfrac{3}{2}\right)$

(1,−1)

17.

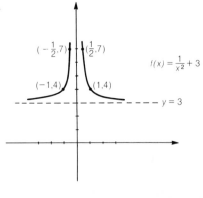

$\left(-\dfrac{1}{2}, 7\right)$ $\left(\dfrac{1}{2}, 7\right)$

$f(x) = \dfrac{1}{x^2} + 3$

(−1,4)   (1,4)

$y = 3$

19.

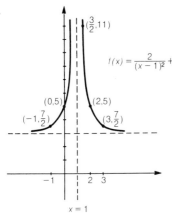

$f(x) = \dfrac{2}{(x-1)^2} + 3$

$\left(\dfrac{3}{2}, 11\right)$

$(0,5)$ $(2,5)$

$\left(-1, \dfrac{7}{2}\right)$ $\left(3, \dfrac{7}{2}\right)$

-1 2 3

$x = 1$

21.

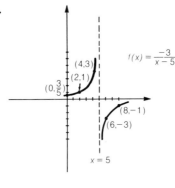

$f(x) = \dfrac{-3}{x-5}$

$(4,3)$

$(2,1)$

$\left(0, \dfrac{3}{5}\right)$

$(8,-1)$

$(6,-3)$

$x = 5$

23.

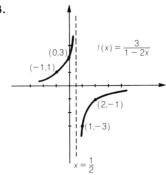

$f(x) = \dfrac{3}{1-2x}$

$(0,3)$

$(-1,1)$

$(2,-1)$

$(1,-3)$

$x = \dfrac{1}{2}$

25.

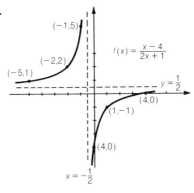

$f(x) = \dfrac{x-4}{2x+1}$

$(-1,5)$

$(-2,2)$

$(-5,1)$

$y = \dfrac{1}{2}$

$(4,0)$

$(1,-1)$

$(4,0)$

$x = -\dfrac{1}{2}$

27.

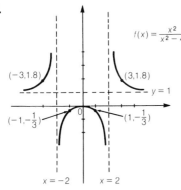

$f(x) = \dfrac{x^2}{x^2-4}$

$(-3,1.8)$ $(3,1.8)$

$y = 1$

$\left(-1, -\dfrac{1}{3}\right)$ 0 $\left(1, -\dfrac{1}{3}\right)$

$x = -2$ $x = 2$

29.

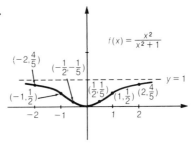

$f(x) = \dfrac{x^2}{x^2+1}$

$\left(-2, \dfrac{4}{5}\right)$

$\left(-\dfrac{1}{2}, -\dfrac{1}{5}\right)$ $y = 1$

$\left(\dfrac{1}{2}, \dfrac{1}{5}\right)$ $\left(2, \dfrac{4}{5}\right)$

$\left(-1, \dfrac{1}{2}\right)$ $\left(1, \dfrac{1}{2}\right)$

-2 -1 1 2

31.

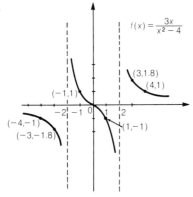

$f(x) = \dfrac{3x}{x^2-4}$

$(3,1.8)$

$(4,1)$

$(-1,1)$

-2 -1 0 1 2

$(1,-1)$

$(-4,-1)$

$(-3,-1.8)$

33.

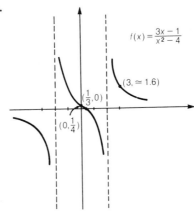

$f(x) = \dfrac{3x-1}{x^2-4}$

$(3, \approx 1.6)$

$\left(\dfrac{1}{3}, 0\right)$

$\left(0, \dfrac{1}{4}\right)$

35.

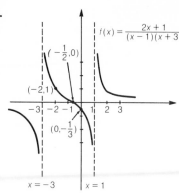

37.

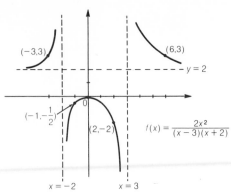

39.

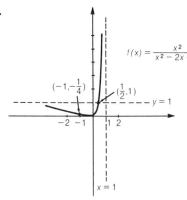

41.

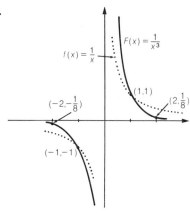

43. $y = \dfrac{3}{5}$ **45.** $y = 0$ **47.** $y = 0$ **49.** a) $y = \dfrac{1}{3}$,   b) $y = 0$

EXERCISES 5.4   **1.** $6x - 5, 6x + 5, 4x + 9$ **3.** $3\sqrt{x + 5} + 1, \sqrt{3x + 6}, 9x + 4$

5. $-\dfrac{(5x^2 + 8)}{x^2 + 2}, \dfrac{1}{4x^2 - 20x + 27}, 4x - 15$ **7.** $x, x, x^9 + 3x^6 + 3x^3 + 2$ **9.** 2   **11.** -0.197

13. 2.6   **15.** $F = f \circ h$ **17.** $H = g \circ h$ **19.** $U = g \circ g$ **21.** $f^{-1}(x) = \dfrac{2x - 1}{x}, x \neq 0$

23. $f^{-1}(x) = \sqrt{x + 4}, x \geq -4$ **25.** $f^{-1}(x) = \dfrac{x^3 + 4}{2}, x \in \mathbb{R}$ **27.** $(f \circ f)(x) = x$

29.

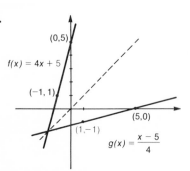

31.

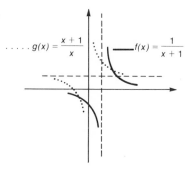

33.

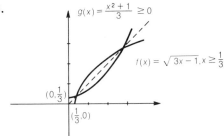

$$g(x) = \frac{x^2 + 1}{3} \geq 0$$

$$f(x) = \sqrt{3x - 1}, x \geq \frac{1}{3}$$

$(0, \frac{1}{3})$

$(\frac{1}{3}, 0)$

35. $D(c) = -0.018c^2 + 0.24c + 299.2$

REVIEW EXERCISES

1.

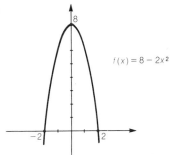

8

$f(x) = 8 - 2x^2$

-2 2

x-intercepts: -2 and 2
y-intercept: 8
Axis of symmetry: $x = 0$
$V(0.8)$

3.

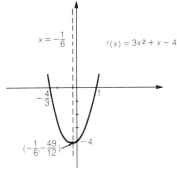

$x = -\frac{1}{6}$

$f(x) = 3x^2 + x - 4$

$-\frac{4}{3}$ 1

$(-\frac{1}{6}, -\frac{49}{12})$ -4

x-intercepts: $-4/3$ and 1
y-intercept: -4
Axis of symmetry: $x = -1/6$

$V(-\frac{1}{6}, -\frac{49}{12})$

5.

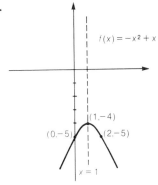

$f(x) = -x^2 + x - 5$

$(1, -4)$

$(0, -5)$ $(2, -5)$

$x = 1$

7.

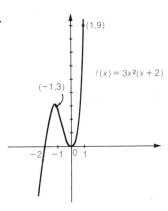

$(1, 9)$

$f(x) = 3x^2(x + 2)$

$(-1, 3)$

-2 -1 0   1

9.

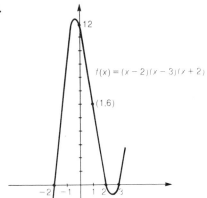

12

$f(x) = (x - 2)(x - 3)(x + 2)$

$(1, 6)$

-2 -1 1   2   3

11.

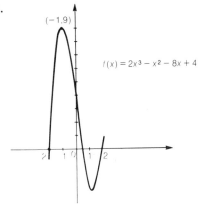

$(-1, 9)$

$f(x) = 2x^3 - x^2 - 8x + 4$

-2 -1 0   1   2

13.

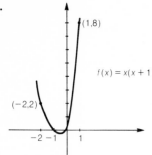

$f(x) = x(x + 1)^3$

$(1,8)$

$(-2,2)$

15.

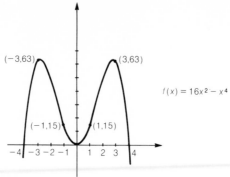

$f(x) = 16x^2 - x^4$

$(-3,63)$ $(3,63)$

$(-1,15)$ $(1,15)$

17.

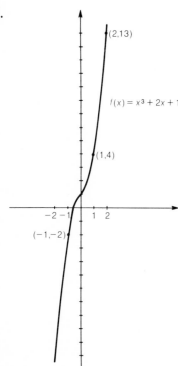

$(2,13)$

$f(x) = x^3 + 2x + 1$

$(1,4)$

$(-1,-2)$

19.

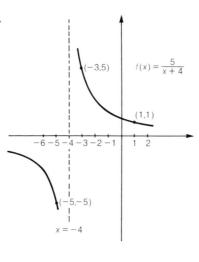

$(-3,5)$ $f(x) = \dfrac{5}{x + 4}$

$(1,1)$

$(-5,-5)$

$x = -4$

21.

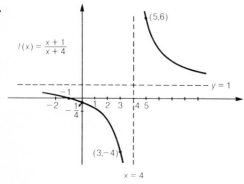

$(5,6)$

$f(x) = \dfrac{x + 1}{x + 4}$

$y = 1$

$-\dfrac{1}{4}$

$(3,-4)$

$x = 4$

23.

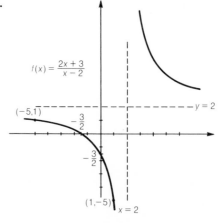

$f(x) = \dfrac{2x + 3}{x - 2}$

$y = 2$

$(-5,1)$

$-\dfrac{3}{2}$

$-\dfrac{3}{2}$

$(1,-5)$ $x = 2$

25.

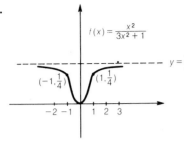

27.

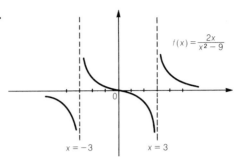

29.

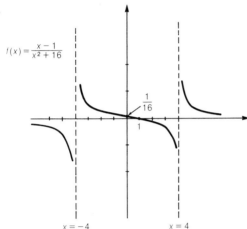

31.

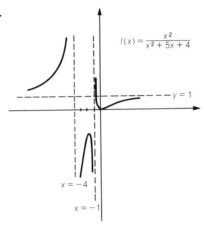

33. $y = \dfrac{3}{2}$ **35.** $y = \dfrac{4}{3}$ **37.** $y = 0$ **39.** $\dfrac{1}{3\sqrt{2x-3}+1}$, $\sqrt{\dfrac{-(9x+1)}{3x+1}}$, $\dfrac{3x+1}{3x+4}$ **41.** x, x, $\dfrac{2x-2}{10-2x}$

43.

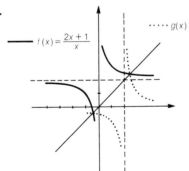

45.

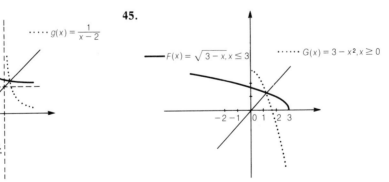

47. $f^{-1}(x) = \dfrac{2+5x}{3x}$, $x \neq 0$ **49.** $f^{-1}(x) = \dfrac{x^2+4}{3}$, $x \geq 0$ **51.** $f^{-1}(x) = -\sqrt{9-x}$, $0 \leq x \leq 9$ **55.** $y = -1$

57. $P(t) = -60\sqrt{t} - 20t + 55$

CHAPTER 6

EXERCISES 6.1   **1.**   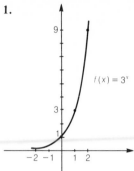   **3.**   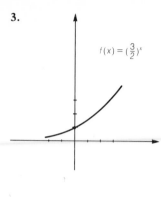   **5.**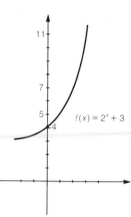

7.   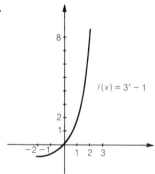   **9.**   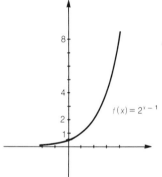   **11.**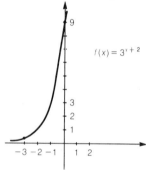

13.   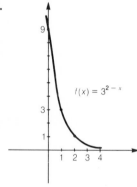   **15.**   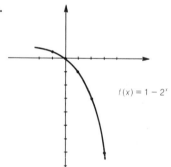   **17.** 16, 1/16, 2   **19.** 15, 5/3, 45

21. 19, 13/4, 4   **23.** 7.389   **25.** 1.649   **27.** 0.607   **29.** 1.051   **31.**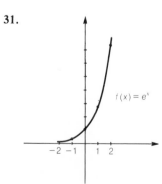

33.

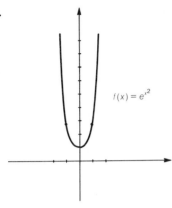

$f(x) = e^{x^2}$

35.

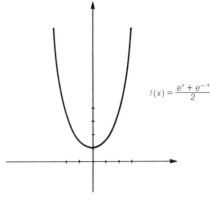

$f(x) = \dfrac{e^x + e^{-x}}{2}$

37. 4.729   **39.** 0.045

41. 1.633   **43.** 36.462   **45.** 7.767   **47.** $-0.358, 0.01, 0.726$ **49.** 2.44141, 2.59374, 2.71692, 2.71825

EXERCISES 6.2   **1.** a) \$10,771.36,   b) \$11,602.22 and a) \$10,778.84,   b) \$11,618.34   **3.** \$6,336.75
5. 180 million   **7.** 21,224   **9.** $p(t) = 15.000 \cdot 2^t$, 120,000
11. a) 605 millimeters of mercury,   b) 429 millimeters of mercury
13. $Q(t) = 100(1/2)^{t/1620}$,   a) 76 mg   b) 25 mg   **15.** 0.67 lumens   **17.** 23,000 years   **19.** 30 s, 45 s

EXERCISES 6.3   **1.** 4   **3.** -3 **5.** 4   **7.** -2 **9.** -3 **11.** $\log_2 32 = 5$ **13.** $\log_4(1/16) = -2$
15. $\log_{1/3}(1/18) = 4$ **17.** $\log_q c = b$ **19.** $2^7 = 128$ **21.** $2^{-10} = 1/1024$ **23.** $10^2 = 100$ **25.** $10^r = q$
27. 1   **29.** 1.26   **31.** 1.3   **33.** -0.3 **35.** 0.52   **37.** 0.9   **39.** 1/2   **41.** 216   **43.** 81/16
45. $\sqrt[5]{4}$ **47.** 1   **49.** 2   **51.** $\log 4$ **53.** $\log_5 36$ **55.** $\log(\sqrt[3]{y}/x^2)$ **57.** $\log[(x + 1)x^2/(x - 2)]$
59. $\log[x^2(x + 2)^3/(x + 1)^4]$ **61.** $2 \log x - 3 \log(x - 2)$ **63.** $2 \log x + 4 \log(x - 2)$ **65.** $(3/2)\log(x - 2) + 2 \log x$
67. $2 \log x + 3 \log(x + 1) - 5 \log(x - 2)$ **69.** $(1/3)\log(x + 1) - \log x - (2/3)\log(x - 2)$

EXERCISES 6.4   **1.** 0.5705   **3.** 3.5119   **5.** 4.8531   **7.** -1.6478 **9.** -0.2907 **11.** 3460
13. 84,198   **15.** 5.45×10^3 **17.** 20.8   **19.** 2.11×10^{-3} **21.** a) 5.0,   b) 6.5
23. a) 1.6×10^{-8},   b) 10^{-5} **25.** 120 db; yes   **27.** $R = 6.5$ **29.** 1.91

EXERCISES 6.5   **1.** $\log 2/\log 7 \simeq 0.356$ **3.** $\log 8/\log 6 \simeq 1.161$ **5.** $-(\log 8/\log 3) \simeq -1.893$
7. $(\log 21 - 1)/5 \simeq 0.064$ **9.** $(\log 3 + \log 2)/(3 \log 3 - 2 \log 3) \simeq 0.938$ **11.** 20   **13.** 3   **15.** $(-3 + \sqrt{41})/2$
17. 1 and 10^{-2} **19.** 10^{10} **21.** 0   **23.** $\ln 3/2 \simeq 0.5493$ **25.** 0.875   **27.** 1.26   **29.** 1.888   **31.** 1.185
33. 3.333   **35.** 2.2   **37.** 1.75   **39.** -1.333 **41.** 1.0686   **43.** 2.4332   **45.** 0.1884   **47.** -2.7210
49. 3.4   **51.** a) 2.5,   b) 4   **53.** 83 min   **55.** -1.1552×10^{-12} **57.** a) 3351.60,   b) 6.25%
59. a) 5.78 mi,   b) 9.1551 mi

REVIEW EXERCISES   **1.** a) 8,   b) 7,   c) 5,   d) 2   **3.** a) 1/64, 1/16, 16,   b) $0.3536, 0.9931 \times 10^{-4}, 0.1768$
5. 10, 2, 80   **7.** a) 16, 3,   b) 100   **9.** a)

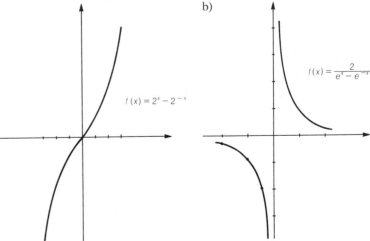

$f(x) = 2^x - 2^{-x}$

b)

$f(x) = \dfrac{2}{e^x - e^{-x}}$

11. 14   **13.** 3/2   **15.** $-7/4$ **17.** 8   **19.** 16/15   **21.** 3/2   **23.** $-\log 48/\log(9/64) \simeq 1.9734$ **25.** 0
27. $\ln(28/5)$ **29.** $\ln(32768/25)$ **31.** $\log(a^7 \sqrt[4]{c^3}/\sqrt{b})$ **33.** $\log[x^4(x+1)^2/(x+2)^3]$
35. $\ln x + 2\ln(x-1) - 3\ln(2x+1)$ **37.** $(1/2)\ln x + (3/2)\ln(x-1) - 4\ln(2x+1)$
39. a) -1.4930,   b) -2.5899,   c) -2.2038,   d) 6.6548   **41.** a) 2.3,   b) 4.30   **43.** 80 db   **45.** 120 db
47. $I_1 = 3.16 \times 10^{-2} I_2$ **49.** 4.29   **51.** 8.69×10^{10} gram   **53.** \$6,860.15   **55.** \$237, \$105.35
57. 143.7°F

CHAPTER 7

EXERCISES 7.1   **1.** (5, 4)   **3.** (3, 2)   **5.** (6, −3)   **7.** (1/4, −1/2)   **9.** (5, 4)   **11.** (2/3, 1/2)
13. (10, 6)   **15.** (6, 4)   **17.** (3, 2)   **19.** (1, −1/2)   **21.** (1/2, 1)   **23.** 5 kg of 20% alloy, 25 kg of 50% alloy
25. boat: 12 mi/h, current: 4 mi/h   **27.** 18 nickels, 12 dimes   **29.** 40 cm, 30 cm

EXERCISES 7.2   **1.** (3, 4, 3)   **3.** (1, 2, 3)   **5.** (2, 1, 3)   **7.** (4, 1, −2)   **9.** (1/2, 1, 1/3)
11. (−1, 2, 1)   **13.** (1, 1, 0, 1)   **15.** (1/3, 1/2)   **17.** (1, 1/2)   **19.** (1, 0, −1)   **21.** (5, −3, −2)
23. (−3z, −1 − 2z, z)   **25.** (2 − z, z, z)   **27.** (2z/3, 5z/3, z)   **29.** (−2, −3)   **31.** No solution
33. (0, 0, 0)   **35.** six 11-cent, ten 18-cent, and sixteen 40-cent stamps
37. 40 kg of alloy A, 40 kg of alloy B, 20 kg of alloy C
39. 30 lb of \$2.60 coffee, 30 lb of \$2.80 coffee, 40 lb of \$3.20 coffee

EXERCISES 7.3   **1.** $\begin{bmatrix} 6 & 1 \\ -2 & 1 \end{bmatrix}$ **3.** $\begin{bmatrix} -1 \\ -2 \\ 4 \end{bmatrix}$ **5.** $\begin{bmatrix} 2 & 4 & -6 \\ 3 & 0 & 6 \end{bmatrix}$ **7.** $\begin{bmatrix} -1 & -5 \\ 3 & -3 \\ 2 & -3 \\ 2 & -1 \end{bmatrix}$

9. $\begin{bmatrix} 3 & -1 \\ 5 & -5 \end{bmatrix}$ $\begin{bmatrix} 1 & 3 \\ 1 & 1 \end{bmatrix}$ $\begin{bmatrix} 6 & 3 \\ 9 & -6 \end{bmatrix}$ $\begin{bmatrix} 1 & 8 \\ 12 & 5 \end{bmatrix}$

11. $\begin{bmatrix} 0 & 0 & 3 \\ 0 & 0 & 2 \end{bmatrix}$ $\begin{bmatrix} 4 & -2 & 7 \\ -2 & 6 & -6 \end{bmatrix}$ $\begin{bmatrix} 6 & -3 & 15 \\ -3 & 9 & -6 \end{bmatrix}$ $\begin{bmatrix} 10 & -5 & 16 \\ -5 & 15 & -16 \end{bmatrix}$

13. $\begin{bmatrix} -1 \\ 1 \\ 5 \end{bmatrix}$ $\begin{bmatrix} 3 \\ -5 \\ -3 \end{bmatrix}$ $\begin{bmatrix} 3 \\ -6 \\ 3 \end{bmatrix}$ $\begin{bmatrix} 8 \\ -13 \\ -10 \end{bmatrix}$ **15.** $\begin{bmatrix} 4 \\ 2 \\ 2 \\ 4 \end{bmatrix}$ $\begin{bmatrix} -2 \\ 2 \\ -2 \\ 0 \end{bmatrix}$ $\begin{bmatrix} 3 \\ 6 \\ 0 \\ 6 \end{bmatrix}$ $\begin{bmatrix} -7 \\ 4 \\ -6 \\ -2 \end{bmatrix}$ **17.** $\begin{bmatrix} 1 & 1 \\ 6 & 0 \end{bmatrix}$ $\begin{bmatrix} 2 & -2 \\ -2 & -1 \end{bmatrix}$

19. $\begin{bmatrix} 3 & -9 & 6 \\ -2 & 6 & -4 \\ 4 & -12 & 8 \end{bmatrix}$ [17]   **21.** $\begin{bmatrix} 5 & 0 \\ 3 & -4 \end{bmatrix}$ $\begin{bmatrix} -6 & 0 & -2 \\ -5 & 2 & -2 \\ 12 & -6 & 5 \end{bmatrix}$ **23.** $\begin{bmatrix} 0 & 0 & 0 \\ 2 & 2 & 2 \\ 0 & 0 & 0 \end{bmatrix}$ $\begin{bmatrix} 0 & 2 & 2 \\ 0 & 0 & 0 \\ 0 & 2 & 2 \end{bmatrix}$

25. [14] $\begin{bmatrix} 8 & 0 & -4 & 12 \\ 2 & 0 & -1 & 3 \\ 0 & 0 & 0 & 0 \\ 4 & 0 & -2 & 6 \end{bmatrix}$ **27.** $\begin{bmatrix} 1 & -1 \\ 0 & 1 \end{bmatrix}$ **29.** $\begin{bmatrix} 1/3 & 1/3 \\ -1/3 & 2/3 \end{bmatrix}$ **31.** $\begin{bmatrix} 1/2 & -1/2 \\ 0 & 1/2 \end{bmatrix}$

33. $\begin{bmatrix} 1 & 0 & -1 \\ 0 & 1 & 0 \\ 0 & 0 & 1 \end{bmatrix}$ **35.** $\begin{bmatrix} 0 & -2 & 1 \\ 0 & 1 & -2 \\ 0 & 0 & 1 \end{bmatrix}$ **37.** No inverse   **39.** $\begin{bmatrix} 1 & 0 & 0 \\ 0 & 1/2 & 0 \\ 0 & 0 & 1/3 \end{bmatrix}$ **41.** (−1/2, −3)
43. (3, −2)   **45.** (1, 2, −1)   **47.** (3, 4, 5)   **49.** (−28, −63, 116)

EXERCISES 7.4   **1.** [−2], −2, −2; [−1], −1, 1

3. $\begin{bmatrix} -1 & 2 \\ -2 & 1 \end{bmatrix}$, 3, 3; $\begin{bmatrix} 1 & 0 \\ 3 & 2 \end{bmatrix}$, 2, −2; $\begin{bmatrix} 3 & -1 \\ 0 & -2 \end{bmatrix}$, −6, −6

5. $\begin{bmatrix} 1 & 0 & 3 \\ 0 & 0 & -1 \\ 0 & 1 & 2 \end{bmatrix}, -1, -1;$ $\begin{bmatrix} 2 & 1 & 3 \\ -2 & 0 & -1 \\ -3 & 0 & 2 \end{bmatrix}, 7, 7;$ $\begin{bmatrix} 1 & 0 & 2 \\ 2 & 1 & 3 \\ -2 & 0 & -1 \end{bmatrix}, 3, -3$ **7.** -1 **9.** -7

11. -2 **13.** -9 **15.** -2 **17.** 24 **19.** 0 **21.** -3 **23.** $\begin{vmatrix} x & y & 1 \\ 2 & 3 & 1 \\ -1 & -2 & 1 \end{vmatrix} = 0$ **25.** -8 **27.** -2

35. $(5, 6)$ **37.** $(9/105, 39/105)$ **39.** $(2, 1, -2)$

EXERCISES 7.5 **1.** $(9, 3), (0, 0)$ **3.** $(-2, -4), (1, -1)$ **5.** $(4, 4), (-1, 1/4)$ **7.** $(1, 3), (-1, -3)$
9. $(2, 1/4), (1/2, 1)$ **11.** $((1 + \sqrt{5})/2, (1 - \sqrt{5})/2), ((1 - \sqrt{5})/2, (1 + \sqrt{5})/2)$ **13.** $(2, 2), (-2, 2)$
15. $(2, 3), (-2, 3)$ **17.** $(2, \sqrt{3}), (2, -\sqrt{3})$ **19.** $(\sqrt{3}/3, \sqrt{6}/6), (\sqrt{3}/3, -\sqrt{6}/6), (-\sqrt{3}/3, \sqrt{6}/6), (-\sqrt{3}/3, -\sqrt{6}/6)$
21. $(\sqrt{21}/7, \sqrt{7}/7), (\sqrt{21}/7, -\sqrt{7}/7), (-\sqrt{21}/7, \sqrt{7}/7), (-\sqrt{21}/7, -\sqrt{7}/7)$ **23.** $(1/2, \sqrt{3}/2), (1/2, -\sqrt{3}/2)$
25. 12 and 20 **27.** 8 m, 10 m **29.** 20 m, 15 m

EXERCISES 7.6 **1.** $(1, -3)$ no, $(-2, 3)$ yes **3.** $(1/3, 1/2)$ yes, $(-2, 3)$ no **5.**

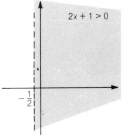

7.

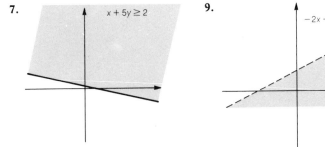

9.

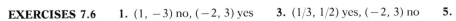

11.

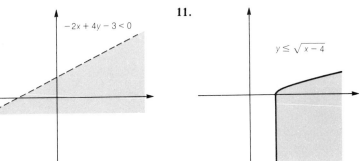

13.

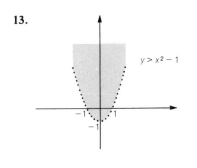

15.

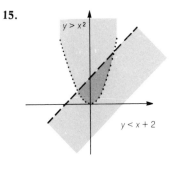

17.

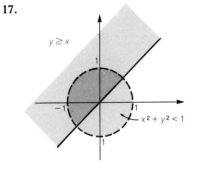

19.

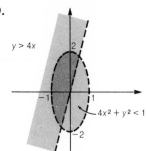

$y > 4x$

$4x^2 + y^2 < 1$

21.

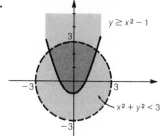

$y \geq x^2 - 1$

$x^2 + y^2 < 3$

23.

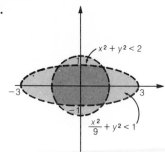

$x^2 + y^2 < 2$

$\frac{x^2}{9} + y^2 < 1$

25.

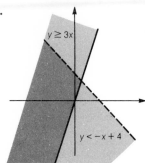

$y \geq 3x$

$y < -x + 4$

27.

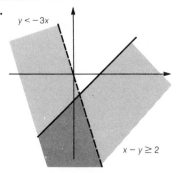

$y < -3x$

$x - y \geq 2$

29.

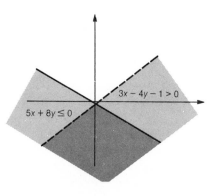

$3x - 4y - 1 > 0$

$5x + 8y \leq 0$

31.

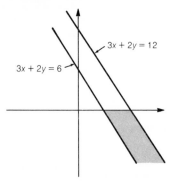

$3x + 2y = 12$

$3x + 2y = 6$

33.

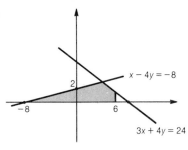

$x - 4y = -8$

$3x + 4y = 24$

35. Maximum 6 at (1, 3), minimum -5 at (0, 0)   **37.** Maximum 13 at (3, 0), minimum -5 at (3, 9)
39. Minimum 23 at (3, 2), no maximum   **41.** 400 barrels of oil, 2100 barrels of gasoline
43. 3 Birdie Customs, 4 Bogey De Luxe, $1410   **45.** 40 grams of A, 30 grams of B
47. Barley: 240 acres, Corn: 120 acres, $13,800   **49.** Center A: 3 days, Center B: 6 days.

REVIEW EXERCISES   **1.** $\begin{bmatrix} 5 & 0 \\ 7 & -4 \end{bmatrix}$ $\begin{bmatrix} -1 & 7 \\ -7 & -9 \end{bmatrix}$ $\begin{bmatrix} 5 & -10 \\ 15 & 10 \end{bmatrix}$

3. $\begin{bmatrix} 10 & 2 \\ -2 & -10 \end{bmatrix}$ $\begin{bmatrix} 5 & -6 \\ 6 & -5 \end{bmatrix}$ $\begin{bmatrix} 0 & 10 \\ -10 & 0 \end{bmatrix}$ **5.** $[1]$ $\begin{bmatrix} 0 \\ 2 \\ 3 \\ 4 \end{bmatrix}$ **7.** $\begin{bmatrix} 7 & -2 \\ 15 & 2 \end{bmatrix}$ $\begin{bmatrix} 11 & -3 & 12 \\ -2 & 2 & -4 \\ -7 & -1 & -4 \end{bmatrix}$

9. $\begin{bmatrix} 2/7 & -1 \\ -3/7 & -2 \end{bmatrix}$ **11.** $\begin{bmatrix} 1 & -1 & 0 \\ 0 & 1/2 & -1/2 \\ 0 & 0 & 1/3 \end{bmatrix}$ **13.** $\begin{bmatrix} 1 & 0 & 0 & 0 \\ -2 & 1 & 0 & 0 \\ 3 & -3 & 1 & 0 \\ -8 & 8 & 4 & 1 \end{bmatrix}$ **15.** 30   **17.** -10 **19.** -16

21. $\begin{bmatrix} 1/a & 0 \\ 0 & 1/b \end{bmatrix}$ **23.** $(-19/5, -18/5)$ **25.** $(4, -2, 2)$ **27.** $(3, -2, 4)$ **29.** 3 **31.** 4 **33.** $(5, -3)$
35. $(1, 2)$ **37.** \$2983.87 at 8%, \$2483.87 at 7.5% **39.** 16 dimes, 64 nickels
41. 6.8% on first \$15,000, 7.4% over \$15,000 **43.** 12 units of P, 8 units of Q **45.** $(2, -3, 1)$
47. 60 lb peanuts, 20 lb almonds, 20 lb cashews **49.** 25 one-dollar, 20 five-dollar, and 30 ten-dollar bills
51. 72 workers at \$6.00, 36 workers at \$8.00, 12 workers at \$10.00 **59.** x_1, x_2, and x_3 **61.** $(0, 3)$
63. $(-1/3, 1/6), (2, 6)$ **65.**

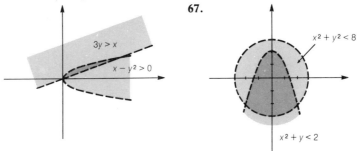

67.

69.

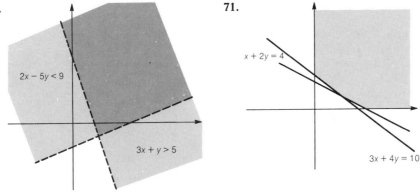

71.

73. Maximum 15 at $(1, 3)$, minimum 0 at $(0, 0)$ **75.** Maximum -4 at $(0, 0)$, minimum -22 at $(0, -6)$
77. 18 and 25 **79.** 10 Standard, 5 De Luxe, \$1000

CHAPTER 8

EXERCISES 8.1 **1.** a) 2, b) 15, c) 2/15, d) 0, e) 49 **3.** 6 **5.** $A = 3, B = -4, C = 5$
7. $A = 3, B = 2, C = 5$ **9.** $A = 1, B = 0, C = 0, D = 1$ **11.** $3x - 2, 14$ **13.** $(1/3)x^2 - (2/9)x + 28/27, 85/27$
15. $x^2 + 3x - 1, 4$ **17.** $2x^3 + (1/2)x + 1, (-5/2)x^2 + 2x + 3$ **19.** $(1/2)x^3 + 1, -x^2 + 3$
21. $x - 3 + 2/(x - 1)$ **23.** $x^2 - 2x + 2 - 10/(2x + 4)$ **25.** $(1/2)x^3 + (1/4)x^2 + (21/8)x + (21/16) + 69/16(2x - 1)$
27. $5x^2 + (5/2) + (5x^3 + (15/2)x^2 - x + 5/2)/(2x^4 - x^2 - 1)$ **29.** $3x^2 + x + 5 + (8x + 8)/(x^2 - x - 1)$

EXERCISES 8.2 **1.** 16 **3.** 1/8 **5.** 22 **7.** -4 **9.** -0.3175 **11.** $2x - 2, -8$
13. $-x^2 - (3/2)x + (7/4), 1/8$ **15.** $4x^2 + 16x + 61, 248$ **17.** $x^4 - 2x^3 + 3x^2 - 6x + 10, 18$
19. $x^3 - x^2 + x - 1, 0$ **21.** $0.2x^2 - 0.13x + 0.275, -0.3175$ **23.** $(x - 3)(x - 2)(x - 1)$
25. $(x + 1)(x - 2)(2x + 1)$ **27.** 29 **29.** -1.158968 **31.** 0.83056 **33.** -20 **35.** -6 **37.** 4
39. $3x + 3a - 2, 3a^2 - 2a + 1$

EXERCISES 8.3 **1.** no **3.** yes **5.** yes **7.** no **9.** yes **11.** $(x - 2)(2x^2 + 3)$
13. $(x + 3)(2x^3 + 5x - 1)$ **15.** $2 + i$ **17.** $2 - 3i$ **19.** $(-3 \pm \sqrt{3}i)/2$ **21.** $-2i, 1, -1$

23. $(-1 \pm \sqrt{3}\,i)/2$ **25.** 5, multiplicity 1; $-2/3$, multiplicity 1
27. $-3/2$, multiplicity 2; 1, multiplicity 3; 4/3, multiplicity 1   **29.** $-1 \pm i$, multiplicity 1; 0, multiplicity 2
31. $1 - i\sqrt{3}$ and $1 + i\sqrt{3}$, both with multiplicity 2   **33.** $x^3 - 3x^2 - 4x + 12$ **35.** $x^3 - 5x^2 + 2x + 8$
37. $x^4 - 4x^3 - 2x^2 + 12x + 9$ **39.** $-x^3 + 5x^2 - 11x + 15$

EXERCISES 8.4   **1.** $-2, -1, 2$ **3.** $-2, 2, 3$ **5.** $-1, 0, 1$ **7.** No rational roots   **9.** $-1, -1/2, 1$
11. $-1/2, 0, 1/2$ **13.** No rational roots   **15.** $x(x + 1)(x - 1)$ **17.** $(1/2)x(x - 2)(x + 2)$
19. $(x + 1)(x + 2)(x - 2)$ **21.** $x(x + 3)(x - 2)$ **23.** $(x + 2i)(x - 2i)(x + 2)(x - 2)$ **25.** $(x - 4)(x + i)(x - i)$
27. $2(x - (3/2))(x + \sqrt{2}\,i)(x - \sqrt{2}\,i)$ **29.** $(3/4)(x + 2)(x - 2)(x - 1 - i)(x - 1 + i)$
31. $(1/4)(x - 1)(x - 2 - \sqrt{2})(x - 2 + \sqrt{2})$

EXERCISES 8.5   **1.** $A = -2, B = 6$ **3.** $A = 1/2, B = -1/2$ **5.** $A = 3, B = 2, C = 1$
7. $A = 1, B = -1, C = 2$ **9.** $A = 0, B = 2, C = 1, D = 2$ **11.** $[11/7(x + 3)] - [4/7(x - 4)]$
13. $-1/(2x - 3) - 3/(x + 4)$ **15.** $(2/x) - 1/(2x + 3) + 3/(x - 2)$ **17.** $(1/x) + (2/x^2) + 2/(x - 1)$
19. $(2x + 1)/(x^2 + 2x + 2) + (2x - 1)/(x^2 + 2x + 2)^2$

REVIEW EXERCISES   **1.** $3x - 2 + (8x - 4)/(2x^2 - 1)$
3. $2x^4 + 3x^3 + 8x^2 + 16x + 36 + (94x - 39)/(x^2 - 3x + 1)$ **5.** $2x^2 + 8x + 32, 125$
7. $2x^3 + 18x^2 - 57x + 171, -512$ **9.** $0.3x^2 - 0.32x + 0.228, 0.9088$ **11.** $(x - 5)(2x - 1)(x - 1)$
13. $2(2x - 1)(x + 3)(x - 1)$ **15.** $0.1(x + 2)(5x + 3)(x - 1)$ **17.** $5(x - (2/5))(x + 3)$
19. $(x - 1 + 2i)(x - 1 - 2i)(x - 1)(x + 1)$ **21.** $x^3 - 5x^2 + 7x + 13$ **23.** $(x - 1)^3(x - 3)$
25. $(1/2)x^3 + 3x^2 + (11/2)x + 3$ **27.** $x^4 - 10x^3 + 38x^2 - 66x + 45$ **29.** $-1, 3$ (multiplicity 2)
31. $-1/2, 2$ (multiplicity 2)   **37.** $1/(x - 1) + (x + 1)/(x^2 - x + 1)$ **39.** $1/x + 2/(x + 1) + 3/(x + 1)^2$
41. $2^{1/5}, 2^{1/5}(\cos(2\pi/5) + i\sin(2\pi/5)), 2^{1/5}(\cos(4\pi/5) + i\sin(4\pi/5)), 2^{1/5}(\cos(6\pi/5) + i\sin(6\pi/5)),$
$2^{1/5}(\cos(8\pi/5) + i\sin(8\pi/5))$

CHAPTER 9

EXERCISES 9.1   **21.** 27   **23.** 25   **25.** 12   **27.** 91   **29.** 26

EXERCISES 9.2   **1.** 84   **3.** 220   **5.** $a^4 + 4a^3 + 6a^2b^2 + 4ab^3 + b^4$
7. $x^6 + 6x^5y^2 + 20x^3y^3 + 15x^2y^4 + 6xy^5 + y^6$ **9.** $32a^5 - 80a^4b + 80a^3b - 40a^2b + 10ab^4 - b^5$
11. $x^5 - 10x^4 + 40x^3 - 80x^2 + 80x - 32$ **13.** $a^6 - 6a^4 + 15a^2 - 20 + 15/a^2 - 6/a^4 + 1/a^6$ **15.** $280x^3a^4$
17. $-1920u^3$ **19.** 70   **21.** $70ab$ **23.** $-35x^4y^4a^3 + 35x^3y^3a^4$ **25.** $-270a^2b^6$ **27.** 1.0828567

EXERCISES 9.3   **1.** $3, 1, -1, -3, -5$ **3.** $-1, 4, -9, 16, -25$ **5.** $-3, 9, -27, 81, -243$
7. $2, 9/4, 64/27, 625/256, 7776/3125$ **9.** $-2, 4/3, -8/7, 16/15, -32/31$ **11.** $-5, -3, 1, 9, 25$
13. $0, 4, -8, 28, -80$ **15.** $1/9, 81, 1/6561, 43{,}046{,}721$ **17.** $3, 3, 6, 18, 72$ **19.** $2, 3/2, 9/8, 27/32, 81/128$
21. 2^n **23.** $2n$ **25.** $n/2$ **27.** $2(3^{n-1})$ **29.** $2, 2.25, 2.2361, 2.236068$

EXERCISES 9.4   **1.** $d = 2; 8, 10$ **3.** $d = -1.5; -2.5, -4$ **5.** $d = \log 2; \log 16, \log 32$ **7.** 46
9. $18, -6$ **11.** 204   **13.** 48   **15.** 725   **17.** -1275 **19.** 105   **21.** 2146   **23.** 1080   **25.** \$49.60
27. 210

EXERCISES 9.5   **1.** $r = 2; 16, 32$ **3.** $r = -1/2; -1/8, 1/16$ **5.** $r = 2/3; 4/27, 8/81$ **7.** $5, 20, 80, 320$
9. $2/27$ **11.** 8   **13.** $-2{,}046$ **15.** $-1{,}365$ **17.** 29,524   **19.** $2/3$ **21.** 4   **23.** 6   **25.** $-4/9$
27. $2/11$ **29.** $7/11$ **31.** $7/22$ **33.** $41/330$ **35.** 54 feet   **37.** \$800,000   **39.** 1,800 revolutions

EXERCISES 9.6   **1.** 30   **3.** 336   **5.** 2,193,360   **7.** 2,884,801,920   **9.** 210   **11.** 780
13. 6.3501356×10^{11} **15.** 4,200   **17.** $1.47822628 \times 10^{24}$ **19.** 90   **21.** 720   **23.** 24, 24   **25.** 120
27. 792   **29.** 133,784,560   **31.** 28   **33.** 4,200   **35.** 5.3644738×10^{28} **37.** 34,650   **39.** 1001, 420

EXERCISES 9.7   **1.** $1/2$ **3.** $2/3$ **5.** 1   **7.** $5/36$ **9.** $7/12$ **11.** $5/12$ **13.** 0   **15.** $1/4$ **17.** $3/5$
19. $1/10$ **21.** $1/8$ **23.** $1/2$ **25.** a) $1/32$,   b) $5/16$ **27.** 1.98×10^{-3} **29.** 9.23×10^{-6}
31. 1.385×10^{-5} **33.** $2/13, 1/2$ **35.** $1/3, 2/3$ **37.** $1/28$ **39.** $1/3$

REVIEW EXERCISES **1.** 816 **3.** $(1/32)a^5 + (5/8)a^4b + 5a^3b^2 + 20a^2b^3 + 40ab^3 + 40ab^4 + 32b^5$
5. $-84u^2b^5$ **7.** $-34560x^3$ **9.** $70a^2b^2$ **11.** 1.1261624 **13.** 9/2, 4, 7/2, 3, 5/2
15. 1, $-1/2$, 1/6, $-1/24$, 1/120 **17.** 0, 1/3, $-1/2$, 3/5, $-2/3$ **19.** $3 \cdot 4^{n-1}$ **21.** $n(n + 1)$
23. 2.1, 2.716, 2.6466768, 2.6457514 **25.** $d = -3$; 1, -2 **27.** $d = \ln 3$; $\ln(1/9)$, $\ln(1/27)$ **29.** 47/2
31. 11,528 **33.** $r = 5/6$; 25/72, 125/432 **35.** $r = 1.5$; 16.875, 25.3125 **37.** 2/3 **39.** 9/4 **41.** 3/11
43. 4/165 **55.** 336 ft, 1936 ft **57.** 43/4 in, 1001/8 in **59.** 10 meters
61. a) 336, b) 151,200, c) 252, d) 495, e) 280, f) 1,260 **63.** 24 **65.** 28 **67.** 56 **69.** 18,564
71. 5/12 **73.** 1/5 **75.** 1/190 **77.** 0.0086 **79.** 0.3968

Index